Physical Processes in Hot Cosmic Plasmas

NATO ASI Series

Advanced Science Institutes Series

A Series presenting the results of activities sponsored by the NATO Science Committee, which aims at the dissemination of advanced scientific and technological knowledge, with a view to strengthening links between scientific communities.

The Series is published by an international board of publishers in conjunction with the NATO Scientific Affairs Division

A	Life Sciences	Plenum Publishing Corporation
B	Physics	London and New York
C	Mathematical and Physical Sciences	Kluwer Academic Publishers Dordrecht, Boston and London
D	Behavioural and Social Sciences	
E	Applied Sciences	
F	Computer and Systems Sciences	Springer-Verlag
G	Ecological Sciences	Berlin, Heidelberg, New York, London,
H	Cell Biology	Paris and Tokyo

Series C: Mathematical and Physical Sciences - Vol. 305

Physical Processes in Hot Cosmic Plasmas

edited by

Wolfgang Brinkmann

Max-Planck Institut für Physik und Astrophysik,
Institut für extraterrestrische Physik,
Garching, FRG

Andrew C. Fabian

Institute of Astronomy,
Cambridge, UK

and

Franco Giovannelli

Istituto di Astrofisica Spaziale,
Frascati, Italy

Kluwer Academic Publishers

Dordrecht / Boston / London

Published in cooperation with NATO Scientific Affairs Division

Proceedings of the NATO Advanced Research Workshop on
Physical Processes in Hot Cosmic Plasmas
Vulcano, Sicily, Italy
May 29 – June 2, 1989

Library of Congress Cataloging in Publication Data
NATO Advanced Research Workshop on Physical Processes in Hot Cosmic
    Plasmas (1989 : Isola Vulcano, Italy)
    Physical processes in hot cosmic plasmas / edited by Wolfgang
Brinkmann, Andrew C. Fabian, and Franco Giovannelli.
       p.   cm. -- (NATO ASI series. Series C, Mathematical and
physical sciences ; vol. 305)
    "Proceedings of the NATO Advanced Research Workshop on Physical
Processes in Hot Cosmic Plasmas, Vulcano (Sicily) May 29-June 2,1989."
    ISBN-13:978-94-010-6732-4   e-ISBN-13:978-94-009-0545-0
    DOI: 10.1007/978-94-009-0545-0

    1. Plasma astrophysics--Congresses. 2. Hot temperature plasmas-
-Congresses. 3. Space plasmas--Congresses.   I. Brinkmann, W.
(Wolfgang), 1941-   . II. Fabian, A. C., 1948-
III. Giovannelli, Franco. IV. Title. V. Series: NATO ASI series.
Series C, Mathematical and physical sciences ; no. 305.
QB462.7.N38   1989
523.01--dc20   90-4070

ISBN-13:978-94-010-6732-4

Published by Kluwer Academic Publishers,
P.O. Box 17, 3300 AA Dordrecht, The Netherlands.

Kluwer Academic Publishers incorporates the publishing programmes of
D. Reidel, Martinus Nijhoff, Dr W. Junk and MTP Press.

Sold and distributed in the U.S.A. and Canada
by Kluwer Academic Publishers,
101 Philip Drive, Norwell, MA 02061, U.S.A.

In all other countries, sold and distributed
by Kluwer Academic Publishers Group,
P.O. Box 322, 3300 AH Dordrecht, The Netherlands.

Printed on acid-free paper

TABLE OF CONTENTS

PREFACE

Gas at temperatures exceeding one million degrees is common in the Universe. Indeed it is likely that most of the gas in the Universe exists in intergalactic space in this form. Such highly-ionized gas, or plasma, is not restricted to the rarefied densities of intergalactic space, but is also found in clusters of galaxies, in galaxies themselves, in the expanding remnants of exploded stars and at higher densities in stars and the collapsed remains of stars up to the highest densities known, which occur in neutron stars. The abundant lower-Z elements, at least, in such gas are completely ionized and the gas acts as a highly conducting plasma. It is therefore subject to many cooperative phenomena, which are often complicated and ill-understood. Many of these processes are, however, well-studied (if not so well-understood) in laboratory plasmas and in the near environment of the Earth. Astronomers therefore have much to learn from plasma physicists working on laboratory and space plasmas and the parameter range studied by the plasma physicists might in turn be broadened by contact with astronomers. With that in mind, a NATO Advanced Research Workshop on *Physical Processes in Hot Cosmic Plasmas* was organized and took place in the Eolian Hotel, Vulcano, Italy on May 29 to June 2 1989. This book contains the Proceedings of that Workshop.

The first Figure in the contribution by Roland Svensson indicates the breadth of the plasma conditions that were considered. They range from densities of 10^{-7} cm^{-3} and less in the case of intergalactic gas to 10^{39} cm^{-3} and greater in the case of neutron stars. The magnetic fields range from $< 10^{-6}$ to 10^{13} Gauss and temperatures from $\sim 10^6$ K and above. Transport properties (diffusion, conduction etc.) were considered in some detail from the theoretical point of view and from laboratory results. Owing to the interdisciplinary nature of the Workshop, we demanded that speakers (and writers) present their work in a pedagogic manner so that those unfamiliar with their perspective might follow the arguments more easily. We are pleased that most of the invited contributions contained in this book are written in this manner and hope that it is thereby of use to a much broader community. Nevertheless, it is clear that more progress can be made before theoretical plasma physicists and observational astronomers understand each other completely, or even appreciate each others' problems, but the whole area is ripe for advance.

We are grateful to the NATO Science Committee for their generous financial aid which made it possible to invite scientists from all over the world. We further thank the Max-Planck Institut für extraterrestrische Physik and the IAS Frascati for additional funding. W. Brinkmann further wants to thank RIKEN, Tokyo, for the experienced support.

Garching, November 1989

Wolfgang Brinkmann

Andrew C. Fabian

Franco Giovannelli

List of Participants

Dr. A. Achterberg	Sterrekundig Instituut, Utrecht, Holland
Prof. G. Auriemma	Universita di Roma, Rome, Italy
Dr. M. Badiali	Istituto di Astrofisica Spaziale, Frascati, Italy
Dr. R. Bandiera	Osservatorio Astrofisico di Arcetri, Firenze, Italy
Dr. X. Barcons	Universidad de Cantabria, Santander, Spain
Dr. M. Baring	MPI für Astrophysik, Garching, W.-Germany
Dr. J. Beall	Naval Research Laboratory, Washington DC, U.S.A.
Prof. D. Biskamp	MPI für Plasmaphysik, Garching, W.-Germany
Dr. W. Brinkmann	MPI für extraterrestrische Physik, Garching, W.-Germany
Prof. C. Canizares	M.I.T., Cambridge Massachusetts, U.S.A.
Dr. D. Cioffi	NASA/GSFC Maryland, U.S.A.
Dr. L. Demeio	Virginia Polytechnic Inst., Blacksburg, Virginia, U.S.A.
Dr. C. Done	Institute of Astronomy, Cambridge, England
Dr. C. Dum	MPI für extraterrestrische Physik, Garching, W.-Germany
Prof. A.C. Fabian	Institute of Astronomy, Cambridge, England
Dr. R. Fusco Femiano	Istituto di Astrofisica Spaziale, Frascati, Italy
Dr. G. Fußmann	MPI für Plasmaphysik, Garching, W.-Germany
Dr. P. Galeotti	Istituto di Cosmogeofisica, Torino, Italy
Dr. G. Ghisellini	Institute of Astronomy, Cambridge, England
Dr. F. Giovannelli	Istituto di Astrofisica Spaziale, Frascati, Italy
Prof. M. Haines	Imperial College, London, England
Prof. N. Itoh	Sofia University, Tokyo, Japan
Dr. J.J. Keady	Los Alamos Nat. Lab., Los Alamos, U.S.A.
Prof. R.M. Kulsrud	Princeton University, Princeton, U.S.A.
Dr. B. Lembege	CRPE / CNET, Paris, France
Prof. D. Lortz	MPI für Plasmaphysik, Garching, W.-Germany
Dr. S. Massaglia	Universita' di Torino, Italy
Prof. R. Mewe	SRON, Utrecht, Holland
Prof. P. Mulser	Technische Hochschule, Darmstadt, W.-Germany
Prof. H. Oegelman	MPI für extraterrestrische Physik, Garching, W.-Germany
Dr. A. Piepenbrink	MPI für extraterrestrische Physik, Garching, W.-Germany
Prof. A.A. Ruzmaikin	IZMIRAN, Troitsk, USSR
Dr. R. Svensson	Nordita, Copenhagen, Denmark
Dr. A. Ulla	University Tromso, Norway
Dr. A. Zdziarski	Space Telescope Science Institute, Baltimore, U.S.A.

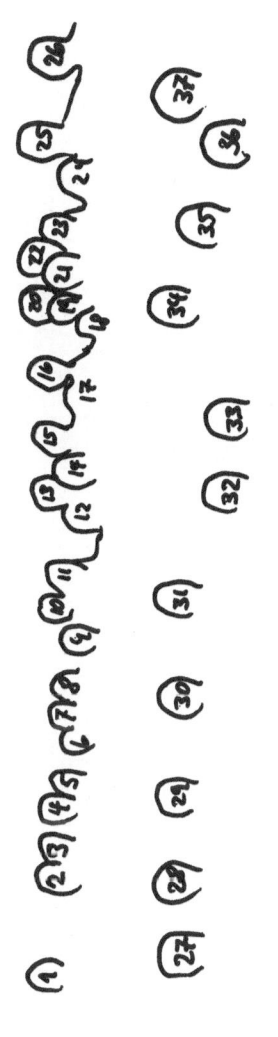

1. R. Kulsrud (USA)
2. B. Lembege (F)
3. S. Massaglia (I)
4. C. Dum (G)
5. R. Bandiera (I)
6. A. Ulla (N)
7. D. Lortz (G)
8. D. Cioffi (USA)
9. C. Canizares (USA)
10. R. Mewe (NL)
11. A.C. Fabian (UK)
12. X. Barcons (Sp)

13. G. Fußmann (G)
14. A. Ruzmaikin (USSR)
15. W. Brinkmann (G)
16. J.J. Keady (USA)
17. L. Demeio (USA)
18. C. Done (UK)
19. H. Ögelman (G)
20. G. Ghisellini (UK)
21. N. Itoh (Japan)
22. A. Achterberg (NL)
23. L. Barbanera (I)
24. Mrs. Zdziarski (USA)
25. A. Zdziarski (USA)

26. M. Baring (G)
27. M. Haines (UK)
28. P. Mulser (G)
29. D. Biskamp (G)
30. J. Beall (USA)
31. F. Giovannelli (I)
32. R. Morani (I)
33. M. Badiali (I)
34. R. Svensson (Dk)
35. A. Piepenbrink (G)
36. Mrs. Baring (G)
37. R. Fusco Femiano (I)

SUPERNOVA REMNANTS AS PROBES OF THE INTERSTELLAR MEDIUM

DENIS F. CIOFFI
NAS/NRC Resident Research Associate
Laboratory for High Energy Astrophysics
Code 665
NASA/Goddard Space Flight Center
Greenbelt, Maryland 20771
U.S.A.

ABSTRACT. Astronomers can use supernova remnants to learn about the environment in which they exist. Simple thermodynamics is first used to describe the evolution of supernova remnants from early to late times. A dynamical formalism is also presented. I then describe a method for calculating the cooling, and hence the luminosity, of the hot shocked gas. Possible complications to this simple picture are considered. The utilization of these theoretical models employing supernova remnants in nearby galaxies is discussed.

1. Introduction

In seconds, a stellar explosion called a supernova (SN) can release about 100 times more energy (10^{53} ergs) than a solar-type star will discharge during its entire 10^{10} yr lifetime of continuous emission.[1] Only $\sim 10^{-4}$ of the energy is emitted as light, but throughout the first days of its existence a SN will shine with a luminosity comparable to the total luminosity from all the other stars in any normal galaxy. When massive stars explode, neutrinos, which drive the explosion itself, quickly carry away almost all ($\sim 99\%$) of the energy, but do not affect the surrounding medium. In this paper we concern ourselves with the remaining 1% of the energy which interacts vigorously with the ambient material, forming a supernova remnant (SNR; recent reviews include Trimble [1983] and Raymond [1984]).

I shall present an overview of the theoretical aspects of continuous, though somewhat idealized, SNR evolution. Throughout I shall try to emphasize the

---

[1] General references to supernovae include Trimble (1982); Bethe and Brown (1985); Woosley and Weaver (1986). The recent SN in the Large Magellanic Cloud caused frenzied activity, e.g., McCray and Wei (1988); Woosley (1988); Woosley and Weaver (1989); Bahcall, Kirshner, and Woosley (1990).

1

W. Brinkmann et al. (eds.), Physical Processes in Hot Cosmic Plasmas, 1–16.
© 1990 *Kluwer Academic Publishers.*

physics and omit most of the computational details. I shall try to assume little prior astronomical knowledge, with the idea that this article may be useful as a primer to the workings of the mechanism which most strongly affects the development of the interstellar medium (ISM).

Indeed, SNRs may dominate the large-scale structure of the ISM of spiral and irregular galaxies, i.e., those galaxies with active star formation (e.g., McKee and Ostriker 1977). When the explosion's $\simeq 10^{51}$ ergs of (initially) kinetic energy is distributed over a cubic parsec ~ 100 yr later, the resulting energy density is approximately 10^7 times larger than the typical interstellar starlight and cosmic ray densities of ~ 1 ev cm^{-3} (e.g., Spitzer 1978). When compared with the average interstellar value of ~ 1 nucleon cm^{-3}, the mass density increases by a factor of ~ 30.

To first order, a single physical process – that is, the heating and subsequent cooling of shocked, cosmic gas – drives SNR evolution, and the application of simple physical laws will help us understand the essential dynamical properties, which in turn determine the kinematics and the luminosity of an evolving remnant. If we can understand the evolution of these expanding balls of hot gas as they interact with the surrounding ISM, we can use SNRs to probe the structure that previous generations of remnants have done so much to create (e.g., Cioffi and Shull 1990).

We proceed with the premise that this basic physical approach allows us to describe SNR evolution with enough accuracy to acquire the desired statistical information. In these calculations, the relevant astrophysical parameter which shows the most variation is the interstellar mass density, ρ_o. It will set the overall X-ray luminosity at a given age (size) of a remnant. With a theoretical knowledge of a remnant's X-ray evolution, we hope to invert the problem and infer the large-scale density structure in nearby galaxies. In addition, by reproducing the observed SNR size distribution, we can find extragalactic SN rates.

I shall first depict the various discrete stages of SNR evolution with a thermodynamic analogy. A slightly more complicated description follows, illustrating the exact dynamics that Cioffi, McKee, and Bertschinger (1988; hereafter CMB) used to study continuous SNR kinematics. This precision demanded taking account of the cooling of the interior gas, and led to an integrated X-ray luminosity calculation, which I shall also describe; details can be found in Cioffi and McKee (1990; hereafter CM).

I shall mention some physics that I have neglected which could have large effects on the luminosity estimates. Further astrophysical complications (§3.2) mean that these solutions may not describe a particular remnant with great accuracy, but they should represent the observations of the majority of SNRs in a galactic collection. Finally, after a short description of how SNRs fit into the "ecology" of a galactic ISM, I shall describe how we intend to use these theoretical models with remnants seen in nearby galaxies such as the Large Magellanic Cloud.

2. Simple Physical Descriptions of SNR Evolution

2.1 ASSUMPTIONS

Because we wish to model the evolution over the entire lifetime of the SNR, with great changes in its physical conditions, these calculations proceed with the simplest assumptions. We shall assume a spherical expansion into a medium of constant density ("uniform"), with no interstellar clouds ("homogeneous"; see §3.2). We assume no thermal conduction, and ignore the presence of magnetic fields. Since the Alfvén velocity is proportional to the strength of the magnetic field, we can re-state this last condition more precisely by saying that when the shock velocity is much greater than the Alfvén velocity we may ignore the dynamical effects of any magnetic fields. Similarly, the magnetic pressure is proportional to the square of the field, but the high pressure behind the high-velocity shock justifies our neglect of both the thermal and magnetic components of the pressure of the external ISM until the end of SNR evolution, when the remnant has slowed to about the sound speed of the ambient ISM.

Two aspects of the cooling introduce the most uncertainty into the calculations. We neglect cooling by dust, and we assume ionization equilibrium; see §3. Both of these err in the same direction: we underestimate the cooling rate of the gas. The added cooling of these processes will have only a minor influence on the dynamics.

2.2 THERMODYNAMIC DESCRIPTION

The explosion in the interior of the pre-supernova star generates a shock wave which then travels through the outer atmosphere of the star at speeds $\sim 10,000\,\mathrm{km\,s^{-1}}$. This strong shock heats and accelerates the stellar material. When the shock blasts through the surface of the star, the ejected matter expands freely and a rarefaction wave propagates inward, quickly cooling the ejecta and resulting in a radial expansion with a total energy $E \simeq 10^{51}$ ergs. The ejecta first interact with the circumstellar medium that borders the star (Figure 1, A).

The completely kinetic energy of the ejecta controls the initial evolution of the remnant, and so we often refer to this first, free-expansion stage of SNR evolution as the "ejecta" stage. The mass of the ejected material, M_{ej}, is typically several times the mass of the sun, $M_\odot = 2 \times 10^{33}$ g, and the expansion velocity at the edge of the remnant is proportional to $\sqrt{E/M_{ej}} \sim 10,000\,\mathrm{km\,s^{-1}}$. This velocity far exceeds the sound speed of the ISM ($10-100\,\mathrm{km\,s^{-1}}$), and a strong shock leads the ejecta. The pressure behind this main shock will drive a reverse shock back into the ejecta (§2.5), reheating it to X-ray emitting temperatures ($\gtrsim 10^6$ K).

We can understand the gross properties of SNR evolution by making an analogy to a simple system with ideal gas behavior (Cioffi 1985). Here we shall regard the SNR as a "black box," where, for the purpose of this analogy, we do not worry about the effects inside the radius of the forward shock, R_s. We track conservation of energy with the first law of thermodynamics, which states that the change in internal energy of a gas, dU, equals the difference between the heat added, dQ, and the PdV work done by the gas,

$$dU = dQ - PdV. \tag{2.1}$$

STANDARD SNR EVOLUTION

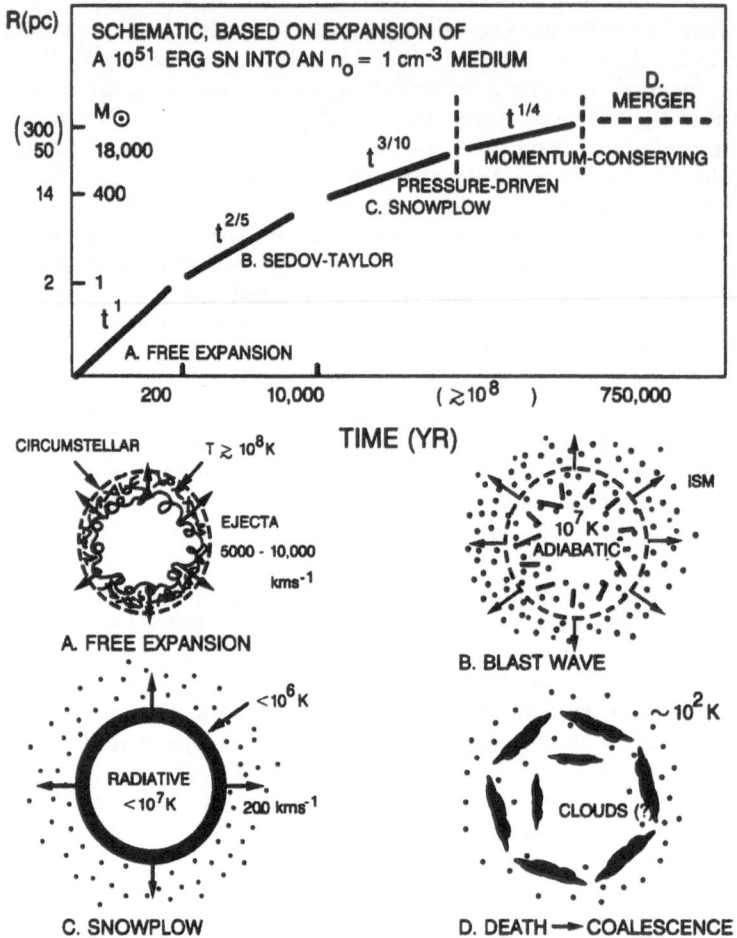

Figure 1. Schematic representing the explosion of a 10^{51} erg supernova into a homogeneous medium of hydrogen density $n_o = 1.0\,\mathrm{cm}^{-3}$. The vertical axis shows the radius of the remnant in parsecs ($1\,\mathrm{pc} = 3.1 \times 10^{18}\,\mathrm{cm}$), and the amount of swept material, in solar masses ($1\,\mathrm{M_\odot} = 2.0 \times 10^{33}\,\mathrm{g}$). The horizontal axis shows the time. The axes are not to scale. The drawings at the bottom portray the evolutionary stage; the labels A through D correspond to the lines on the graph. The remnant will normally merge with the ISM long before the momentum-conserving snowplow stage can occur.

In the freely-expanding ejecta stage just described, the SNR radiates no energy and does no work on the surrounding material, so the internal energy does not change. Until encounter with the reverse shock, the thermodynamic temperature of the adiabatically expanding ejecta keeps dropping with the continued conversion of the random thermal motions to directed kinetic energy. With no loss of energy, if one could convert this kinetic energy back into random thermal motions, one would re-establish the initial temperature of the ejected material, and thus, in this restricted sense, the ejecta stage is "isothermal."

Because of the relatively low mass density of the ISM ($\rho_o \sim 10^{-24}\,\mathrm{g\,cm^{-3}}$; hydrogen number density $n_o \sim 1\,\mathrm{cm^{-3}}$), the ejecta do not immediately notice the surrounding medium, and the radius of the forward shock, with its constant velocity, is directly proportional to time:

$$R_s = v_{ej}\, t, \tag{2.2}$$

where

$$v_{ej} = \left[\frac{5}{3}\frac{2E}{M_{ej}}\right]^{1/2}. \tag{2.3}$$

The ejecta slowly interact with the ambient material, and in the limit that swept matter governs the actions of the remnant, the classic Sedov-Taylor (ST; Sedov 1959; Taylor 1950) self-similar blast-wave solution obtains.[2]

Although now the SNR will effect an *internal* redistribution of the interstellar material, $PdV = 0$ because the remnant does no work on the external ISM, allowing it to pass unobstructed through the shock and be incorporated into the SNR volume (Figure 1, B). Thus, in the above limit of the neglect of the ejecta mass, the average internal mass density equals the constant density of the ISM. The average internal pressure, $\bar{P}$, is therefore directly proportional to the average internal temperature, $\bar{T}$. The high shock speed maintains a high postshock temperature, which insures the continued adiabaticity of the postshock gas, so again $dQ = 0$ and the total internal energy U of this gas remains constant (at a fixed fraction of the total explosion energy E [e.g., Ostriker and McKee 1988]). Because of the self-similar nature of this stage, $\bar{T}$, which measures the average energy per particle, must be proportional to the immediate postshock temperature, T_s, which measures the energy given to the new particles entering the interior through the strong shock; $T_s \propto v_s^2$. Finally, we know that the pressure is proportional to the internal energy per unit volume, $\bar{P} \propto U/R_s^3$. Hence we have

$$\bar{P} \propto \bar{T} \propto T_s \propto v_s^2 \propto \frac{U}{R_s^3} \rightarrow v_s^2 R_s^3 = \text{constant}. \tag{2.4}$$

---

[2] As noted by Jones, Smith, and Straka (1981), the ST solution also describes the previously-mentioned shock wave which travels through the outer atmosphere of the pre-supernova star.

If one assumes that the radius grows as a power law with time, $R_s \propto t^\eta$, the above relationship implies that $\eta = 2/5$ in the ST solution. (One can also find this exponent from straightforward dimensional arguments.)

The SNR proceeds in this fashion until, shortly behind the shock, a certain fluid element suddenly cools (CM), forming a dense shell (Figure 1, C). The gas in the interior of the remnant pushes against the shell and loses internal energy because of this PdV work. The SNR now "snowplows" the ISM, with the swept material accumulating at the shell. Since the postshock material does not flow into the hot interior, the previous arguments about the pressure dependence no longer hold, but the following dynamical argument shows that again $\bar{P} \propto v_s^2$: in the frame of the shell, one feels interior gas pushing with pressure $\bar{P}$ and incoming fluid with an identical ram pressure proportional to $\rho_0 v_s^2$. When viewed from an outside frame, we know that the internal pressure actually drives the shell into a stationary external fluid, causing that ram pressure, and so, with a constant density ambient medium, $\bar{P} \propto v_s^2$.

To zeroth order, the hot interior still does not cool: $dQ \approx 0$. Unlike the free-expansion stage, this stage is thermodynamically "reversible." For a reversible adiabatic process we may write $\bar{P}V^\gamma =$ constant, where $V \propto R_s^3$ is the volume of the SNR and $\gamma = 5/3$ is the ratio of specific heats for an ideal monatomic gas. For an "isolated" *pressure-driven snowplow* (PDS) stage, with $R_s \propto t^\eta$, we would then have (McKee and Ostriker 1977)

$$v_s^2 R_s^5 = \text{constant} \rightarrow \eta = \frac{2}{7}. \tag{2.5}$$

If the interior were to cool completely, the conserved momentum of the shell would continue to expand the remnant. CMB found, however, that though the interior does cool, this *momentum-conserving snowplow*, with $\eta = 1/4$, does not occur under typical interstellar conditions because of the slowing of the expansion speed to the sound speed of the medium, which ends the integral existence of the SNR (Figure 1, D).

2.3 NEWTON'S SECOND LAW: DYNAMICAL DESCRIPTIONS

To simplify the evolutionary descriptions, and to maintain continuity between stages, we say that the SNR has entered a new stage of evolution when it deviates from the conditions that have defined the current stage, and we do not separately classify the transitional periods. For example, the initially-constant shock speed slows before the dominance of the swept material gives the complete, internal, ST solution (CMB), and we formally end the ejecta stage when the ratio of the swept-up mass, $M_s = 4/3\pi R_s^3 \rho_0$, to the ejected mass, M_{ej}, is $\simeq 2/5$ (CM). Also, the interior has begun to cool before the shell exists, and we formally begin the pressure-driven snowplow stage a factor of e (the base of the natural logarithm) sooner than the time at which the sudden catastrophic cooling creates the shell.

CMB carefully treated this transition from the ST to the PDS stage. One can examine the dynamics of SNR expansion with Newton's Second Law: the time derivative of the momentum of the interior fluid is equal to the pressure force on the "wall" of the SNR. A first approximation for the momentum uses the product of the total swept mass, M_s, and the postshock fluid velocity; for the force one then uses the product of the surface area of the SNR, $4\pi R_s^2$, and the average interior pressure $\bar{P}$. The results of this approximation, which again works only with "isolated," discrete stages, are identical to those derived from the above thermodynamic reasoning.

With a more sophisticated formulation, however, CMB did match the expected smooth fluid kinematics seen, for example, in hydrodynamical simulations. The first two improvements involve accounting for the pressure on the distributed mass in the interior, not just at the edge, and accounting for the distributed velocity of the interior mass. The moments K_P and K_{01} accomplish this task, and are of order unity (their definitions are given in the appendix). They are used in the exact blast wave equation of motion from Ostriker and McKee (1988):

$$\frac{d}{dt}(M_s\,K_{01}\nu_1\,v_s) = 4\pi R_s^2 K_P \bar{P}, \qquad (2.6)$$

where the velocity of the postshock fluid is given by the product of ν_1 and the shock velocity, v_s. Behind the adiabatic shock of the ST stage, $\nu_1 = 3/4$. As the shock becomes fully radiative, $\nu_1 \to 1$, and the fluid velocity becomes equal to the shock velocity.

When we previously set $dQ \approx 0$ to obtain eq. (2.5), we had $\bar{P} \propto R_s^{-5}$. Considering the cooling of the interior gas showed that $\bar{P} \propto R_s^{-5}t^{-4/5}$, and also motivated an approximation for the important transition from $\nu_1 = 3/4 \to 1$ (CM). These further improvements enabled the successful replication of smooth kinematics via eq. (2.6). With the knowledge gained from this solution CMB also defined *continuous* "offset" power laws (i.e., constants of integration are included), and found that an exponent of 3/10 fit better than the heretofore standard 2/7 in the PDS stage: the remnant is overpressured with respect to the conditions that yield the 2/7 law. Figure 1 is a schematic that shows a typical SNR evolving through the stages which I have described.

2.4 COOLING OF SHOCKED INTERSTELLAR GAS

Shocks generate entropy. We can identify every interior fluid element by the time t_s at which it encountered the entropy-generating forward shock of the SNR (i.e., a Lagrangian description). Kahn (1976) showed how one can exploit a numerical coincidence to write a form for the entropy of each fluid element which depends *only* on the current time and on the time t_s at which it was shocked.

We can define the astrophysical "cooling function" $\Lambda(T)$ such that the volumetric luminosity of a cooling gas, as a function of its temperature T, is given by $n^2\Lambda(T)(\,\mathrm{ergs\,s^{-1}\,cm^{-3}})$, where n is the hydrogen number density. Gases of "solar" abundance are composed of 90% hydrogen by number, and 10% helium, with traces of oxygen, nitrogen, neon, carbon, iron, etc. providing much of the cooling above 10^4 K. The cooling function does fluctuate, but a power-law approximation frequently suffices over the temperature range of interest.

We next define the "reduced entropy" (McKee 1982), a quantity

$$s \equiv \frac{T^{3/2}}{n}. \tag{2.7}$$

The entropy is proportional to the natural logarithm of s, and if $\Lambda \propto T^{-1/2}$, then

$$s(t_s) - s(t, t_s) \propto t - t_s, \tag{2.8}$$

where $s(t_s)$ is the original value of the entropy of the parcel of gas that was shocked at time t_s, and $s(t, t_s)$ is the value at any later time t; see CM. Although not necessarily the "best" fit,[3] one can approximate the Raymond, Cox, and Smith (1976) cooling function for $10^5 \, \mathrm{K} \gtrsim T \gtrsim 10^{7.6} \, \mathrm{K}$ with the desired power law: $\Lambda = 1.6 \times 10^{-19} T^{-0.5} \, \mathrm{ergs \, cm^3 \, s^{-1}}$ (e.g., McKee 1982).

With the above equation (2.8) and *with the kinematics now properly portrayed*, as described in §2.3, one can trace the cooling of the gas in the interior of the SNR (Cioffi and McKee 1988; Falle 1988; CM). In particular, one can calculate which fluid element will be the first to cool completely, causing the formation of the shell. This time agrees nicely with that from the hydrodynamical simulation of CMB, and is the key fiducial time in the dynamics of radiative SNRs.

Since we can determine the thermal history of every shocked parcel of gas, we can completely account for the energy radiated by the interior. At any time t we can integrate through all the cooling gas, from that shocked at the very beginning, $t_s = 0$, to the most recently shocked gas at $t_s = t$, and find the total luminosity of the SNR:

$$L(t) \propto \int_0^t n(t_s) \, \Lambda\left(T[t_s]\right) \, R_s^2(t_s) \, v_s(t_s) \, dt_s. \tag{2.9}$$

After shell formation this integral splits into three parts: the interior, the cold shell (zero luminosity), and the postshock gas. We must know the pressure to write the density n in terms of s and calculate the integral, but we can assume an isobaric interior because of the high temperature ($T \sim 10^7 \, \mathrm{K}$) of the interior gas, with its consequently short sound travel times.

2.5 THE REVERSE SHOCK

The freely-expanding, relatively cold ejecta feel the high pressure that builds behind the fast forward shock. This high pressure drives a reverse shock back into the ejected material, across the contact interface which separates the ejecta from the outside material swept by the forward shock (Ardavan 1973; McKee 1974). We can calculate the reverse-shock dynamics simply by assuming that the pressure which propels the reverse shock is a constant fraction of the pressure behind the forward shock,

$$\rho_0 v_{rs}^2 = \varphi \rho_0 v_s^2, \tag{2.10}$$

---

[3] One gets a better fit with an exponent $\simeq -0.6$.

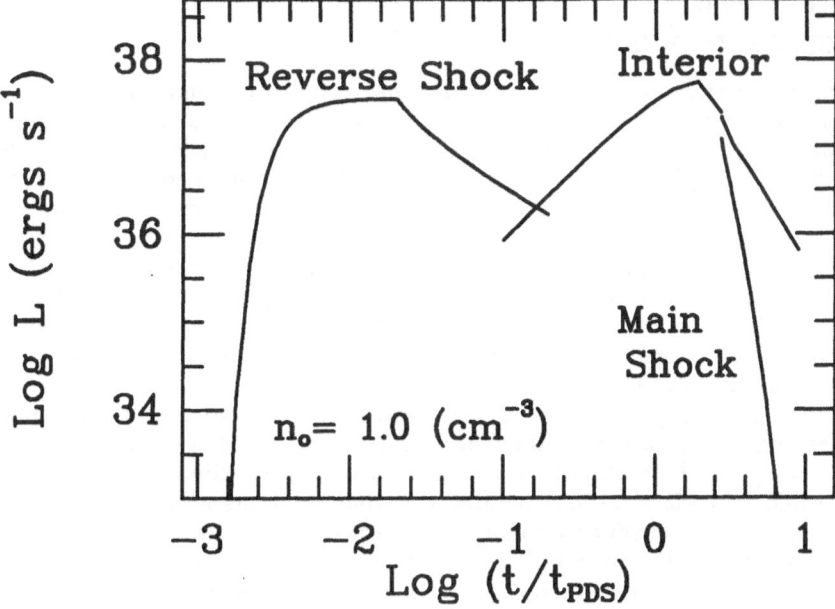

Figure 2. X-ray luminosity above 0.1 keV, from the reverse shock and the interior, from an SNR expanding into a medium with hydrogen number density $n_o = 1\,\mathrm{cm}^{-3}$. Time is normalized to the onset of the PDS stage at $13,300\,\mathrm{yr}$. After the shell has formed, the interior emission is separated from that of the gas behind the forward shock.

where v_{rs} is the velocity of the reverse shock and $\varphi \simeq 0.2$ (Cioffi and McKee 1989; CM).

We calculate the luminosity of the reverse-shocked ejecta in the same spirit as that of the interior, but with negligible cooling we assume that the entropy of the shocked gas is always given by the initial entropy, $s(t_{rs})$, where t_{rs} is the time at which the reverse shock hits a particular fluid element of ejected material. Figure 2 illustrates the importance of the reverse shock with respect to our goal of probing the ISM: the X-ray luminosity of a remnant peaks *twice* because in the early evolution of the SNR the reverse-shocked ejecta outshine the forward-shocked interstellar gas. This duality complicates the problem because a remnant at a given luminosity can be at any one of four places in its evolution. Additional astrophysical information should help resolve this dilemma for a given SNR.

3. Complications

3.1 NEGLECTED PHYSICS

First, the calculations that I have described make no distinction between postshock ion (T_i) and electron (T_e) temperatures. As with the ionization equilibrium that I shall discuss below, the higher density in the reverse-shocked ejecta allows the establishment of approximate temperature equilibrium there (Itoh 1977). In contrast, behind the forward shock early in the ST stage one might find that $T_e \ll T_i$, and this would have a notable effect on the X-ray emission above 3 keV (Itoh 1977). Yoshida and Hanami (1988) have applied a two-temperature model to the earliest times, as the ejecta of SN1987A in the Large Magellanic Cloud interact with circumstellar material presumably from the progenitor star itself. In any case, McKee (1974) has suggested that plasma instabilities in these collisionless shocks tend to equilibrate T_e and T_i.

The assumption of postshock ionization equilibrium is justified if the time needed for the establishment of the equilibrium is much less than dynamical timescales of interest. In general, young SNRs do not meet this criterion (e.g., Shull 1982). Gronenschild and Mewe (1982) pointed out that applying a specific equilibrium calculation to observed young remnants leads to energy and temperature estimates that are too low for the observed luminosity, and to an ambient density estimate that is too high. Canizares (1989) has stressed that remnants as old as the Cygnus Loop ($\simeq 20,000$ yr old) can be "young" in the sense that they are under-ionized with respect to their electron temperatures. (This notion of relative youth does not contradict the dynamical age, since with an ISM of hydrogen density $n_o = 1 \, \text{cm}^{-3}$, a shell will not form until about 35,000 yr after the explosion.)

Hamilton, Sarazin, and Chevalier (1983) calculated a broad range of non-equilibrium models with a Sedov-Taylor solution, and found that they could characterize detailed X-ray spectra with two parameters, the shock temperature T_s and a collisional time scale parameter, $n_o^2 E$. They also found that the ionization time scale, $n_o t$, sets the total X-ray emissivity. Hamilton and Sarazin (1984) later showed that one can predict X-ray spectra by using this ionization time scale and the emission integral $\int n_e^2 dV$, where n_e is the electron density and V is the volume. These prescriptions suggest that our simple models of total, integrated X-ray luminosity can be made to accommodate the additional emissivity due to non-equilibrium ionization.

3.2 ASTROPHYSICAL DISORDER

Calculating the evolution of SNRs using a more realistic representation of the ISM is more difficult, but should not dramatically change the picture that I have painted above. Several additional ingredients come immediately to mind: circumstellar material from a progenitor star, clouds, dust, and magnetic fields. Below, I comment upon these in turn.

Almost all stars have winds. Massive, luminous stars, which eventually explode as supernovae, throw off the most material, with total mass $\lesssim M_{ej}$. The SN ejecta initially interact with this medium rather than with the "pristine" ISM (McKee 1988; Chevalier 1989). Although the shock's adiabatic journey through the wind material allows us to modify the dynamics of the ejecta stage by adjusting M_{ej} to include the mass of the pre-supernova outflow, this situation temporarily thwarts any attempt to probe the true ISM; criteria must be established to disassociate these instances from the SNRs that have evolved beyond the influence of the pre-supernova system.

Most of the mass of the ISM is contained in "clouds" i.e., regions with mass density much higher than the average. Broadly speaking, clouds come in two types. Stars form in the huge ($\sim 10^{20}$ cm), massive ($\sim 10^6 M_\odot$) complexes called Giant Molecular Clouds (e.g., Shu, Adams, and Lizano 1987). Because of the high density within molecular clouds, SNRs will quickly pass through their initial two stages into the radiative stage, and will emit much infrared radiation (Shull 1980).

The smaller ($\sim 10^{19}$ cm) atomic, interstellar clouds may be sheet-like (e.g., Jahoda, McCammon, and Lockman 1986) but the total volume that they occupy in the ISM is uncertain. The SNR shock will engulf these clouds, leaving them to evaporate in the interior (e.g., McKee 1988). This evaporation can cool the interior and modify the early expansion to $R_s \propto t^{3/5}$ (McKee and Ostriker 1977), but in the limit that the clouds fill most of the volume of the ISM, we again approach a homogeneous medium. In the opposite limit of a negligible filling factor, the interaction between the SNR shock and the clouds does not noticeably affect the expansion dynamics, but the shock may "light up" the clouds. These lit clouds could mark the true extent of the SNR forward shock, but the shock might also have passed beyond them (Braun and Strom 1986). Emission from the radiative shocks that are driven into these clouds may provide further information about both the primary SNR shock and the nearby ISM (Cioffi 1986).

Infrared radiation from dust grains in young SNRs can surpass the emission from all other wavelengths (Dwek 1988). The action of the shock on the grains hastens their destruction, however, so this radiation should last only for a brief time. Dwek (1981) estimated that the duration of the ST stage might be shortened slightly (10%), but more work on grain survivability and the effects on SNR dynamics is needed.

Because of the weakness of the interstellar Galactic magnetic field ($\sim 4 \times 10^{-6}$ Gauss), its influence on SNR evolution has long been ignored. This view is being challenged by Cox (1988) and Cox and Slavin (1989), who argue that because magnetic fields can greatly reduce the compression ratio behind radiative shocks, SNRs cannot sweep large amounts of dense, cold material into thin shells.

4. SNRs in Galaxies

4.1 THE ROLE OF SNRs IN GALACTIC ECOLOGY

An analogy between stellar activity within the "violent" ISM (McCray and Snow 1979) and an ecological system (e.g., Cioffi and Shull 1990) helps clarify the influence of SNRs on the large-scale interstellar system. Figure 3 diagrams some of the

A PART OF GALACTIC ECOLOGY

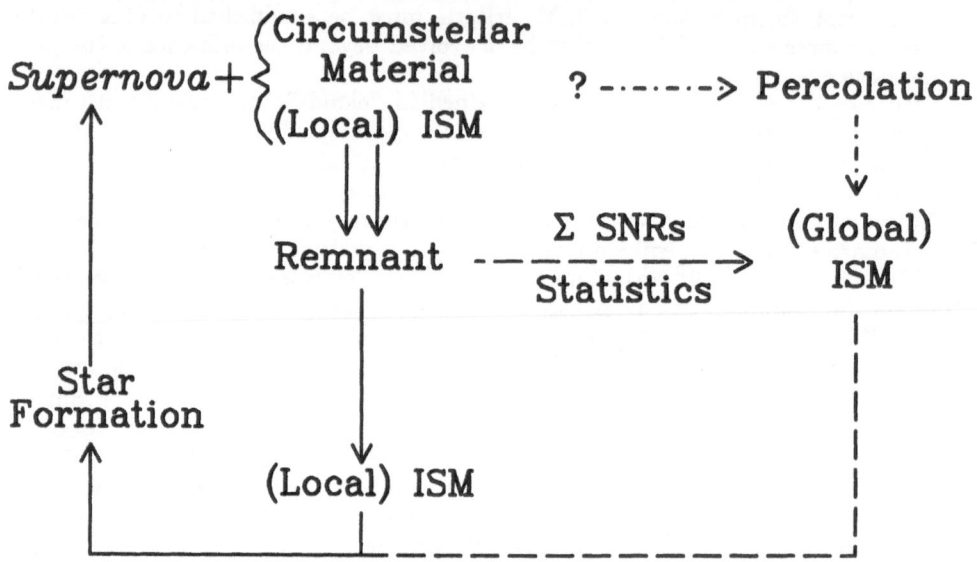

Figure 3. This flow diagram shows the interstellar life and death cycle from supernovae and their resulting remnants. Individual SNRs affect their local ISM, as shown by the left-hand side (solid lines). The right-hand side (dashed lines) shows how the sum of the actions of many SNRs may *create* the large-scale ISM. New stars eventually form from the metal-enriched ISM, and the cycle continues.

interactions, with the loop on the left-hand side referring to a single SNR. The right-hand-side of the diagram shows possible consequences of many SNRs.

The SN explosion first combines with the immediate circumstellar environment. As the remnant continues to expand, it encounters the true local ISM. Some 500,000 to three-quarters of a million years later (CMB; Shull, Fesen, and Saken 1989), this combination of SN ejecta and processed ISM, once an SNR, again becomes the local ISM.

Given enough time and material, the ISM will collect itself and begin anew the process of star formation, which eventually leads to new supernovae. A high rate of SN might continually agitate the ISM, however, disrupting the formation of clouds and preventing star formation. An even higher rate could blow the medium out of the galaxy in a galactic wind, leaving little material behind.

If the SN has exploded within a cluster of stars, as massive SN are wont to do, a previous SNR may have cleared the surrounding volume, and the new SNR shock will lose little velocity in travelling through the rarefied medium to a large radius. If enough rarefied volumes connect, the system is said to have "percolated" (e.g., Cox and Smith 1974; Heiles 1987; Cioffi and Shull 1990). With the pressure that the SNR shocks impart, they might even assist the progress of such a connected system by triggering star formation.

4.2 UTILIZING THESE MODELS

We plan to use this knowledge of SNR evolution to obtain a "global" (i.e., galaxy-wide) description of extragalactic systems. First, with X-ray observations of many SNRs we can *directly* find the large-scale density distribution within a galaxy's ISM. Further, we shall use the fact that SNRs evolve through distinct stages to extract SN rate information from an ensemble of SNRs.

I have indicated how we can calculate the X-ray luminosity of an SNR (§2.4). Although quantities such as the explosion energy and the ejected mass vary among supernovae, and interstellar properties such as the metallicity (heavy element content) and homogeneity of the medium also vary, our current understanding of the ISM within the Milky Way suggests that the interstellar hydrogen number density shows the greatest range, from as low as $n_o = 10^{-3}\,\mathrm{cm}^{-3}$ in rarefied regions to as high as $n_o = 10^6\,\mathrm{cm}^{-3}$ within dense molecular clouds. Buried in the (unwritten) proportionality constant of eq. (2.9) is the density of the ISM, which, given the X-ray luminosity of a resolved SNR, we obtain.

Because the expansion rate slows with time (Figure 1), one might expect to see more SNRs at larger radii. The question is complicated, however, because one must account for the varying luminosity of the SNR and the probability that not all the remnants have evolved into a medium of the same density (Hughes, Helfand, and Kahn 1984; Cioffi and Shull 1990). Over-simplified analyses of this number-radius relationship[4] will not work. The analytic theory outlined in this article should make possible a proper attack using X-ray observed SNRs. Matching the observed distribution of SNR size and luminosity will yield both the SN rate and a consistency check on the spread of the directly-derived densities.

The SNRs in our own Galaxy do not have well-determined distances, and thus do not have well-determined sizes, which makes most attempts at statistical work fruitless (Green 1984). Also, Galactic dust obscuration presumably hides SNRs on the other side of the Galaxy. Extragalactic systems, where one can assume the same distance to all the SNRs, and where a more complete sample should be assembled, are therefore preferred. The distance cannot be too great, however, because one must retain both adequate spatial resolution and a flux sufficient for detection. With current technology, Local Group galaxies offer the best promise for such experiments, and upcoming X-ray satellites will thoroughly inspect these nearby galaxies. Recent optical observations (e.g., of M33 by Long *et al.* 1990) are rapidly increasing the number of observed extragalactic SNRs.

---

[4] With radio observations, the relationship between radio surface brightness and diameter is used (e.g., Caswell and Lerche 1979; Duric and Seaquist 1986).

5. Summary

The basic theory of the evolution of supernova remnants depends on our knowledge of the dynamics and cooling of the hot, shocked, interstellar medium. I have described this theory with simple thermodynamics. Figure 1 summarizes the stages of SNR evolution: 1. Ejecta (Free Expansion); 2. Sedov-Taylor; 3. Pressure-Driven Snowplow; 4. Merger. We can calculate the X-ray luminosity of the interior hot gas, the postshock gas, and the reverse-shocked ejecta. With this theory we can estimate the mass density into which an SNR evolves; looking at many SNRs in an external galaxy gives some idea of the density distribution in that galaxy's ISM. Reproducing the observed number of SNRs at a given radius tells us the SN rate in the galaxy.

I acknowledge my valued office-mate, John Blondin, for many shock-filled conversations and for comments which clarified the text, as did the comments of Mike Shull, Chris McKee, and Rich Pisarski. I thank Dick McCray for the idea behind Figure 1.

Appendix

The moments used in the exact blast-wave equation (2.6):

$$K_P \equiv 2 \int_0^{R_s} \frac{P(r)}{\bar{P}} \frac{r}{R_s} \frac{dr}{R_s} \tag{A.1}$$

$$K_{01} \equiv \int_0^{M_s} \frac{v(m)}{v_1 v_s} \frac{dm}{M_s}. \tag{A.2}$$

Analytic approximations to these equations (cf. Ostriker and McKee 1988) were used when CMB integrated the coupled differential equations that resulted from eq. (2.6).

REFERENCES

Ardavan, H. 1973, *Ap. J*, **184**, 435.
Bahcall, J. N., Kirshner, R. P., and Woosley, S. E. 1990, *Ann. Rev. Astr. Ap.*, in press.
Bethe, H. A., and Brown, G. 1985, *Sci. Am.*, **251**, 60.
Braun, R., and Strom, R. G. 1986, *Astr. Ap.*, **164**, 208.
Canizares, C. R. 1989, in *IAU 115, High Resolution X-ray Spectroscopy of Cosmic Plasmas*, eds. P. Gorenstein and M. Zombeck, p. xxx.
Caswell, J. L., and Lerche, I. 1979, *M.N.R.A.S.*, **187**, 201.
Chevalier, R. A. 1989, preprint.
Cioffi, D. F. 1985, *Ph.D. Thesis*, University of Colorado.
_____ 1986, *B.A.P.S.*, **31**, No. 4, 828.

Cioffi, D. F., and McKee, C. F. 1988, in *IAU 101, Supernova Remnants and the Interstellar Medium*, eds. R. S. Roger and T. L. Landecker, (Cambridge: Cambridge University Press), p. 435.

—————————————— 1989, in *IAU 115, High Resolution X-ray Spectroscopy of Cosmic Plasmas*, eds. P. Gorenstein and M. Zombeck, p. xxx.

—————————————— 1990 (CM), in preparation.

Cioffi, D. F., McKee, C. F., and Bertschinger, E. 1988, *Ap. J.*, **334**, 252 (CMB).

Cioffi, D. F., and Shull, J. M. 1990, *Ap. J.*, submitted.

Cox, D. P. 1988, in *IAU 101, Supernova Remnants and the Interstellar Medium*, eds. R. S. Roger and T. L. Landecker (Cambridge: Cambridge University Press), p. 73.

Cox, D. P., and Smith, B. W. 1974, *Ap. J. (Letters)*, **189**, L105.

Cox, D. P., and Slavin, J. D. 1989, preprint.

Duric, N., and Seaquist, E. R. 1986, *Ap.J.*, **301**, 308.

Dwek, E. 1981, *Ap. J.*, **247**, 614.

——————— 1988, in *IAU 101, Supernova Remnants and the Interstellar Medium*, eds. R. S. Roger and T. L. Landecker (Cambridge: Cambridge University Press), p. 363.

Falle, S. A. E. G. 1988, in *IAU 101, Supernova Remnants and the Interstellar Medium*, eds. R. S. Roger and T. L. Landecker (Cambridge: Cambridge University Press), p. 419.

Green, D. A. 1984, *M.N.R.A.S.*, **209**, 449.

Gronenschild, E. H. B. M., and Mewe, R. 1982, *Astr. Ap. Supp. Ser.*, **48**, 305.

Hamilton, A. J. S., and Sarazin, C. L. 1984, *Ap. J.*, **284**, 601.

Hamilton, A. J. S., Sarazin, C. L., and Chevalier, R. A. 1983, *Ap. J. Supp.*, **51**, 115.

Heiles, C. 1987, *Ap. J.*, **315**, 555.

Hughes, J. P., Helfand, D. J., and Kahn, S. M. 1984, *Ap. J. (Letters)*, **281**, L25.

Itoh, H. 1977, *Pub. Astr. Soc. Japan*, **29**, 813.

Jahoda, K., McCammon, D., and Lockman, F. J. 1986, *Ap. J. (Letters)*, **311**, L57.

Jones, E. M., Smith, B. W., and Straka, W. C. 1981, *Ap. J.*, **249**, 185.

Kahn, F. D. 1976, *Astr. Ap.*, **145**, 50.

Long, K. S., Blair, W. P., Kirshner, R. P., and Winkler, P. F. 1990, *Ap. J. Supp.*, submitted.

McCray, R., and Snow, T. P. Jr. 1979, *Ann. Rev. Astr. Ap.*, **17**, 213.

McCray, R., and Wei, H.-W. 1988, in *Guo Shoujing Summer School of Astrophysics: Origin, Structure, and Evolution of Galaxies*, ed. Fang, L.-J. (World Scientific Press: Singapore), p. 8.

McKee, C. F. 1974, *Ap. J.*, **188**, 335.

—————————— 1982, in *Supernovae: A Survey of Current Research*, eds. M. Rees and R. Stoneham (Dordrecht: Reidel), p. 433.

——————————1988, in *IAU 101, Supernova Remnants and the Interstellar Medium*, eds. R. S. Roger and T. L. Landecker (Cambridge: Cambridge University Press), p. 205.

16

McKee, C. F., and Ostriker, J. P. 1977, *Ap. J.*, **218**, 148.

Ostriker, J. P., and McKee, C. F. 1988, *Rev. Mod. Phys*, **60**, 1.

Raymond, J. C. 1984, *Ann. Rev. Astr. Ap.*, **22**, 75.

Raymond, J. C., Cox, D. P., and Smith, B. W. 1976, *Ap. J.*, **204**, 290.

Sedov, L. I. 1959, *Similarity and Dimensional Methods in Mechanics*, (New York: Academic Press).

Shu, F. H., Adams, F. C., and Lizano, S. 1987, *Ann. Rev. Astr. Ap.*, **25**, 23.

Shull, J. M. 1980, *Ap. J.*, **237**, 769.

_____ 1982, *Ap. J.*, **262**, 308.

Shull, J. M., Fesen, R. A., and Saken, J. M. 1989, *Ap. J.*, **346**, xxx.

Spitzer, L. 1978, *Physical Processes in the Interstellar Medium*, (New York: Wiley-Interscience).

Taylor, G. I. 1950, *Proc. Roy. Soc. London*, **201A**, 159.

Trimble, V. 1982, *Rev. Mod. Phys*, **54**, 1183.

_____ 1983, *Rev. Mod. Phys*, **55**, 511.

Woosley, S. E. 1988, *Ap. J.*, **330**, 218.

Woosley, S. E., and Weaver, T. A. 1986, *Ann. Rev. Astr. Ap.*, **24**, 205.

_____ 1989, *Sci. Am*, **261**, 32.

Yoshida, T., and Hanami, H. 1988, *Prog. Th. Phys.*, **80**, 83.

X-RAY OBSERVATIONS OF SUPERNOVA REMNANTS

Claude R. Canizares
Department of Physics & Center for Space Research
Room 37-501
Massachusetts Institute of Technology
Cambridge, Massachusetts 02139 U.S.A.

ABSTRACT. Supernova remnants (SNRs) are low density, coronal plasmas that are shock heated by the explosion of a supernova. Spectroscopic observations of several bright SNRs have allowed us to deduce some of the physical properties of the emitting plasma. I emphasize the question of equilibration between electron temperature, ion temperature and ionization temperature. Inconsistency between several observations leaves open the question of whether or not the electrons and ions have reached equilibrium. On the other hand, both moderate and high resolution spectra show clear departures from ionization equilibrium. Ionization ages deduced using ratios of various emission lines, including the forbidden and resonance lines of He-like ions, are derived for Puppis A, Cas A and the Cygnus loop and found to be consistent with estimates from independent measurements. Even the 20,000 year old Cygnus Loop SNR is found to be underionized for its temperature. We consider possible departures from Maxwellian electron distributions, although there is neither evidence for them nor a compelling reason to expect them. Future missions, such as *ASTRO-D*, *Spectrum X-Gamma*, and particularly *AXAF*, will greatly enhance our ability to study the plasma properties of SNRs. As one example, *AXAF* will permit the use of satellite lines as diagnostics of electron temperature, ionization state and non-Maxwellian electron distribution functions on many dozen SNRs.

1. INTRODUCTION

A supernova (SN) explosion is a cataclysmic event marking the death of a star (for a review see Trimble 1982, 1983; also see Cioffi's contribution to this volume). It can occur in stars of roughly $1\,M_\odot$ (a so-called Type I SN) or in stars of $\gtrsim 7\,M_\odot$ (Type II SN); a star of intermediate mass expires by gracefully puffing off its outer envelope. The Type I SN explosions are thought to result from the release of nuclear energy (carbon deflagration in a white dwarf that has accreted material from a nearby companion star), whereas Type II events are primarily powered by gravitational collapse of a stellar core to form a neutron star or black hole. In both cases, the kinetic energy of the expelled material is roughly 10^{51} erg, making SNe the most energetic events in the present universe after active galactic

17

W. Brinkmann et al. (eds.), Physical Processes in Hot Cosmic Plasmas, 17–28.
© 1990 *Kluwer Academic Publishers.*

nuclei and quasars.

For thousands of years after the outburst the ejecta expands into the surrounding medium at velocities up to 10^4 km s^{-1}, while shocks slowly convert the kinetic energy into thermal energy. In the simplest picture, a so-called 'reverse shock' ('reverse' in the Lagrangian sense) heats the ejecta themselves, while the forward shock heats the surrounding circumstellar or interstellar material, both of which achieve temperatures of 10^6 to 10^8 K and emit X-rays. These X-ray supernova remnants (SNRs), of which about 50 have been seen in the Milky Way, are of interest to astronomers because they convey information about the SN explosion itself and about the nature and composition of both the stellar ejecta and the surrounding material. X-ray images of SNRs, most of which have been obtained with the grazing incidence telescope on the *Einstein* Observatory, show these regions of high temperature plasma with graphic detail (see Figure 1). Even with only approximate temperature information, the observed X-ray surface brightness can be used to deduce the plasma density in the remnant. Typical values of the electron density are $n_e \sim 1-10$ cm^{-3}. Therefore, SNRs are optically thin and fit the definition of a coronal plasma (see Mewe's contribution to this volume), in which collisional processes dominate.

In this paper, I will review what X-ray observations tell us about SNR plasmas. I will make only passing reference to the astronomical importance of these observations, but one should bear in mind that the most significant question is the determination of the elemental composition of the emitting material. The composition of the ejecta is of the greatest interest, because the SNR is the only place where the innards of a star are exposed to public view. But abundance determinations can only be made after the physical properties of the emitting plasma are reasonably well understood. Both the abundance determination and the physical understanding require high quality spectroscopic observations, which to date have only been performed on a few bright, well-studied SNRs. This paper will concentrate on those observations. To calibrate laboratory and solar spectroscopists, I note that a typical X-ray emission line flux at earth from one of these bright SNRs is $\sim 10^{-2}$ photons cm^{-2} s^{-1}.

For reference, additional reviews of X-ray observations of SNRs can be found in Danziger and Gorenstein 1982, Raymond 1984, Canizares 1984a, Aschenbach 1988, Roger and Landecker 1988. Plasma diagnostics of SNRs are reviewed in more detail in Canizares 1984b, 1989.

2. TEMPERATURE AND EQUILIBRIUM

A given region of shock heated plasma in an SNR can be characterized by its electron temperature T_e, ion temperature T_i, and ionization temperature T_z. The ionization temperature is defined independently for each pair of neighboring ionization states of a given element: it is the temperature at which a plasma in statistical equilibrium would have the observed relative populations of those states. In an equilibrium coronal plasma, all these temperatures are the same. However, the collisional timescales in the low density plasma of an SNR are generally long compared to its age, so it may be that all the temperatures are different. Most of the attempts to understand the plasma properties of SNRs have centered on establishing the degree of disequilibrium.

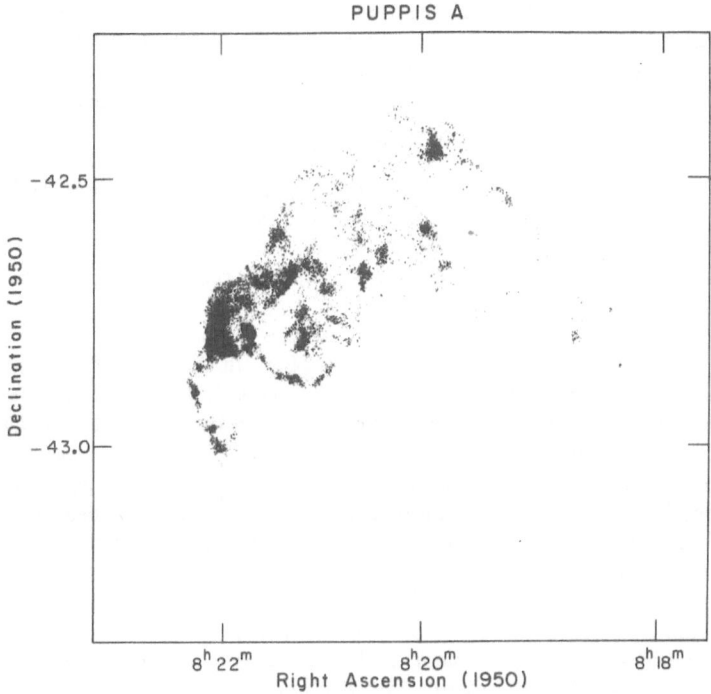

Figure 1. X-ray image of the $\sim$ 3000 year old SNR Puppis A (Petre *et al.* 1982). The remnant is $\sim$ 6000 light years distant and $\sim$ 75 light years across. The brightest region is thought to be a shock-heated interstellar cloud.

2.1 Electron vs. Ion Temperature

The strong, collisionless shocks in an SNR will heat the ions to $T_i = 3Mv_s^2/16k \sim 10^8$ K, where M is the ion mass, v_s the shock velocity (initially $\sim 10^4$ km s^{-1}), and k Boltzmann's constant (see McKee and Holenbach 1980 for a discussion of shocks and T_e vs. T_i equilibration). On the other hand, the timescale for transferring that energy to the electrons by Coulomb collisions is $\sim$ 3 $10^5 T_8^{1.5} n_e^{-1}$ yr, which is much longer than the ages of the observed SNRs (here $T_8 = T/10^8$ K). McKee (1974) suggested that collective processes might effectively couple the electrons and ions, thus equilibrating their temperatures. In the absence of conclusive evidence to support this hypothesis, SNR models (e.g. Hamilton, Sarazin and Chevalier 1983) are generally computed with the alternative assumptions of complete equilibration ('single temperature' models) or of disequilibrium arising from Coulomb coupling only ('two temperature' models) although some intermediate cases have also been considered (Hamilton, Sarazin and Szymkowiak 1986a, 1986b; Itoh, Masai and Nomoto 1988).

The observations are ambiguous. Pravdo and Smith (1979) reported the detection of continuum radiation extending up to energies of 25 keV from two young SNRs, Cas A and Tycho (roughly 300 and 400 years old, respectively). The observations were made with the mechanically collimated, gas filled proportional counters on *HEAO-1*. The existence of such high energy *bremsstrahlung* implies high temperature electrons and was long taken as evidence that effective electron-ion coupling was taking place. However, more recent conventional and gas-scintillation proportional counter observations with the *EXOSAT* (Smith *et al.* 1988, Jansen *et al.* 1988), *Tenma* (Tsunemi *et al.* 1986, Itoh, Masai and Nomoto 1988) and *Ginga* satellites (Tsunemi, private communication), fail to confirm the existence of these hard components at the levels originally reported, leaving the question of electron-ion equilibration open again.

For Tycho's SNR, Itoh *et al.* (1988) can fit the *Tenma* continuum with a three temperature model that allows some electron heating beyond that expected from Coulomb collisions alone, but considerably less than expected for complete electron-ion equilibration (see also Hamilton, Sarazin and Szymkowiak 1986), although they also obtain good fits with no additional electron heating but with partially mixed ejecta. In contrast, Brinkmann *et al.* (1989) find a reasonable fit to the *EXOSAT* data assuming complete electron-ion equilibration together with partial mixing of the ejecta. For Cas A, Jansen *et al.* (1988) argue that models with $T_e = T_i$ give more reasonable results than two temperature models with Coulomb heating only. Their $T_e = 4\ 10^7$ K, well below the Pravdo and Smith (1979) value, is identified with the primary shock in the circumstellar medium and is in agreement with values we find from high resolution spectroscopy (see below).

The ambiguity of these observations is likely to remain unresolved until future missions return data of higher spectral quality (see Section 3.1).

2.2 Non-Equilibrium Ionization

While the degree of electron-ion equilibrium is uncertain, there is no longer any doubt that the ionization state of SNR plasma is far from equilibrium. The degree of ionization of a given element depends on T_e and the 'ionization age' $\tau = n_e t$, where t is time. For a typical case, equilibrium is achieved after $t \sim 2\ 10^4/n_e$ yr, which is longer than the ages of most observed SNRs. This was pointed out by Gorenstein *et al.* (1974) and was incorporated into SNR models by many authors (Itoh 1977, Shull 1983, Hamilton, Sarazin and Chevalier 1983, Mewe, Gronenschild and van den Oord 1985, Hughes and Helfand 1985). These models must follow the time evolution of the SNR as a whole and of each ionization stage.

At the temperatures of interest, most cosmically abundant elements of intermediate atomic number Z are quickly ionized to the He-like stage. Ionization to higher stages proceeds more slowly, so the main effect of non-equilibrium ionization is the preponderance of He-like ions. Higher Z elements like Fe are spread over more stages although they also get stuck in the Ne-like stage. Clearly one must take careful account of such non-equilibrium ionization effects in order to deduce plasma properties and elemental abundances from the emission lines. Furthermore, because the lower ionization stages are efficient radiators, non-equilibrium plasmas also have higher total emissivity — by an order of magnitude in some cases — so it must be accounted for even just to get the energetics right (e.g. Shull 1983).

In low resolution detectors like gas scintillation proportional counters (GSPC), the best evidence for non-equilibrium ionization comes from the centroid energies of the line complexes attributed to Si, S, A, Fe and Ca emission. The centroid energy depends largely on the relative intensities of the He-like and H-like lines, which are not resolved in a GSPC. In fact, the line complexes are not fully resolved from each other. Nevertheless, Tsunemi *et al.* (1986) were able to place consistent constraints on the ionization ages of Tycho and Cas A, as shown in Figure 2. Their results for Cas A are in excellent agreement with independent determinations based on the high resolution spectra described below.

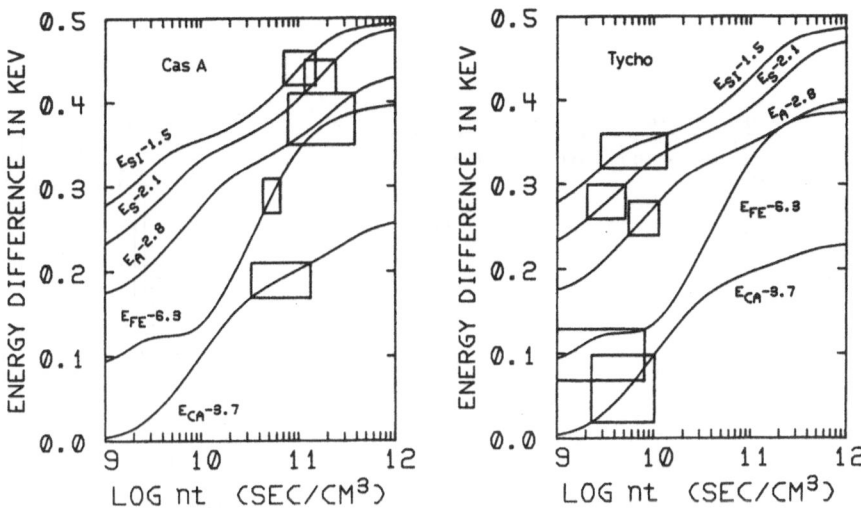

Figure 2. The τ dependence of the mean energies of the $K\alpha$ lines for Si, S, Ca, and Fe (relative to the energies noted). The boxes are the 90% confidence regions for the observed line energies for Cas A (a) and Tycho (b). (From Tsunemi *et al.* 1988 with kind permission).

The first and the least model-dependent measure of non-equilibrium ionization came from high resolution spectra of Puppis A obtained with a Bragg crystal spectrometer on the *Einstein* Observatory (Canizares *et al.* 1983, Winkler, Canizares and Bromley 1983, Canizares 1989a). The primary diagnostic is the ratio of lines from He-like ions. The first excited state of a He-like ion can be either a singlet ($1s2p\ ^1P$) or triplet ($1s2p\ ^3P$ or $1s2s\ ^3S$), whereas the ground state is a singlet ($1s^2\ ^1S$) (see for example Gabriel and Jordan 1969). Decays from the triplet excited state to the singlet ground state are second order transitions and are therefore slower than those from the singlet state, but in plasma at the low densities of SNRs they do eventually occur, giving rise to the forbidden line (from the ^3S state) and the intercombination line (from the ^3P state). Therefore, the relative strengths of these lines compared to that of the resonance line (from the permitted transition between the excited and ground singlet states) depends on the relative populations of the excited states.

For a plasma in ionization equilibrium, the excited states are populated by collisional excitation and by cascades following radiative recombination of the H-like ions. Collisions preferentially populate the singlet state whereas recombination preferentially populates the triplet states (because of their larger statistical weights). This gives rise to line ratios that are somewhat temperature sensitive (e.g. Pradhan 1983). In a plasma that is still ionizing, in other words one that is under-ionized for its electron temperature, the recombination rate is suppressed. This reduces the strengths of the forbidden and intercombination lines relative to the resonance line, giving us the desired diagnostic. The opposite case of an over-ionized, recombining plasma in which the forbidden line is enhanced has been observed in a laboratory plasma (Kallne *et al.* 1984). This case might be of interest in future studies of stellar flares, for example.

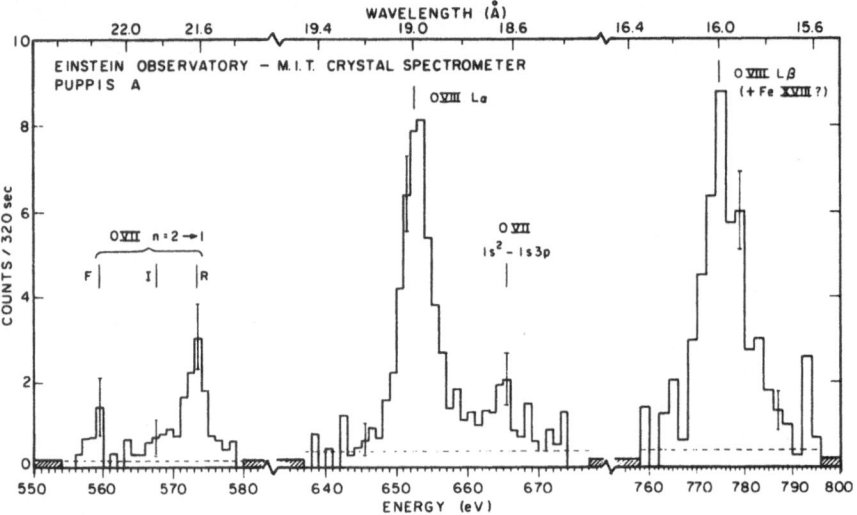

Figure 3. Oxygen line emission from the interior of Puppis A from Winkler *et al.* 1981. The horizontal dashed line indicates the level of the instrumental background.

Figure 3 shows several oxygen lines from the interior of Puppis A, a ∼3000 year old remnant of a Type II SN (Figure 1). The He-like triplet (on the left in the figure) is clearly dominated by the resonance line, as would be expected for an underionized plasma. A comparison of all the line ratios to calculations made with an updated version of the code of Hughes and Helfand (1985; see Fischbach *et al.* 1989 and Canizares 1989a for details) gives constraints on the values of T_e and τ (Figure 4). These results, and those for two other portions of the SNR, give values of τ that are consistent with independently determined values of n_e and t obtained primarily from imaging observations (Fischbach *et al.* 1989). One caveat, however, is that this analysis does not account explicitly for possible small scale inhomogeneities in the plasma, such as those indicated by the *EXOSAT* images of Jansen (1988).

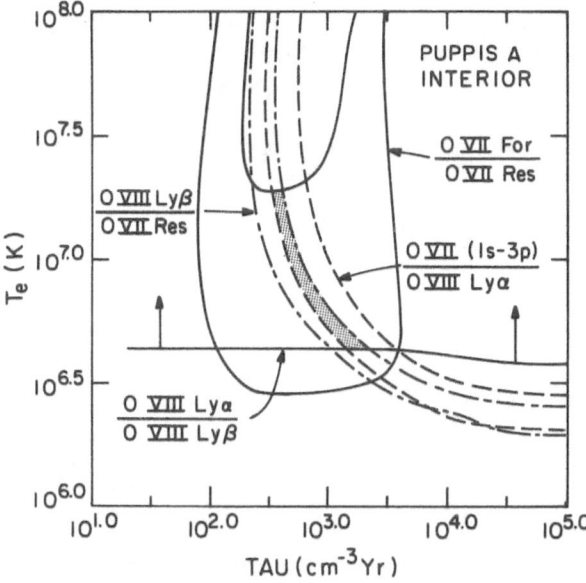

Figure 4. Constraints on T_e and τ for the interior of Puppis A from Fishbach *et al.* 1989. Allowed regions (90% confindence) of the parameter space deduced from various oxygen line ratios are indicated. The shaded region is the intersection of all the allowed regions.

Interstellar absorption extinguishes the oxygen lines in Cas A, but we do resolve the He-like and H-like lines of Si, S and Ne (Markert *et al.* 1989, Canizares 1989a). For each element, the ratio of these lines constrains the values of T_e and τ, as shown in Figure 5, but in themselves they do not require departures from equilibrium. Fixing the temperature at the value measured by Tsunemi *et al.* (1986) and Jensen *et al.* (1988) gives the same value of τ deduced by the former authors, but only for the Si and S lines. This leads us to associate these lines with the higher temperature component, plausibly the shocked circum-stellar medium. This interpretation is also indicated by the spatially resolved temperature measurements of Jansen *et al.* (1988). It is clear from Figure 5 that the Ne lines must come from a separate component. This is presumably the shocked ejecta, which will have lower temperature and higher density, the latter implying a larger τ. It is plausible that the cooler plasma has nearly the same pressure as the hotter component, so the allowed values of T_e and τ must lie near the diagonal line in Figure 5. The intersection of this locus of constant pressure with the contours from the Ne lines fixes the parameters of the shocked ejecta. Note that the higher density ejecta is actually close to ionization equilibrium.

The Cygnus Loop is the most aged remnant that is well studied; it is thought to be 20,000 old. Nevertheless, it too shows a resonance:forbidden line ratio that clearly indicates under ionization of the plasma (Vedder *et al.* 1986). Using the temperature measured by

Ku *et al.* (1984) from the continuum constrains the value of τ to lie between 4,000 to 8,000 yr cm^{-3}, which is reasonable for $n_e \sim 0.2 - 0.5\,\text{cm}^{-3}$.

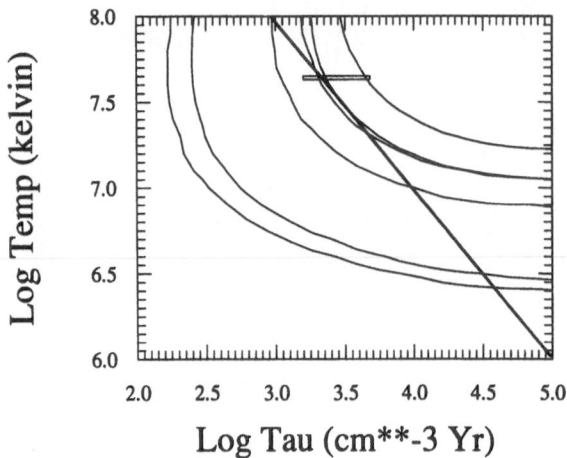

Figure 5. Constraints on T_e and τ for Cas A from Markert *et al.* (1989). Curves indicate the allowed regions (90% confindence) of the parameter space deduced from the line ratios of Si and S (upper right) and Ne (lower left). The box shows the best-fit values from Tsunemi *et al.* (1986), and the diagonal line is a locus of constant pressure.

2.3 Non-Maxwellian Electron Distributions

All the previous analyses have assumed that the electrons do follow a Maxwellian distribution so that T_e is a well defined quantity. The timescale for electron-electron energy equipartition is $\sim 10^3$ times shorter than that for electron-ion equilibration, so the assumption is certainly reasonable. It is possible that in SNRs, as in the solar wind, a given region may acquire suprathermal electrons either through streaming from hotter neighboring regions or through acceleration in some plasma process. Although there is no reason to believe that either mechanism will be very effective, it would be comforting to have observations that confirm the validity of the Maxwellian assumption.

A modest population of suprathermal electrons will not have too drastic an effect on the strengths of most of the stronger emission lines (Owocki and Scudder 1983). But they can seriously alter the ratio of the He-like triplet, as was recently pointed out by Gabriel *et al.* (1988). This is because very energetic electrons will preferentially excite the singlet state, resulting in a stronger resonance-to-forbidden line ratio even if the plasma is in ionization equilibrium. The consistency between the He-like triplet diagnostic and the other line ratios noted above implies that this is not a serious problem in Puppis A. A further argument against suprathermals comes from the broad-band spectrum of Puppis A. If several percent of the electrons in Puppis A had characteristic energies of 15 - 20 keV, which is what would be needed to affect significantly the He-like triplet (Gabriel *et al.* 1988), then they will emit *bremsstrahlung* and give rise to a high energy tail. I believe that such a tail would have been detected by experiments such as Ariel V (Zarnecki *et al.* 1978), and probably *EXOSAT* , *Tenma* and *Ginga* can set even tighter limits.

3. FUTURE PROSPECTS

3.1 Future Missions

Our ability to understand the plasma properties of SNRs has been severely limited by the capabilities of the spectrometers carried on previous missions. This situation will improve dramatically over the next decade. The *ASTRO-D* mission (Tanaka 1989), a Japanese satellite with U.S. collaboration, will carry imaging GSPCs and CCD cameras capable of performing spatially resolved spectroscopy with sufficient resolution and sensitivity to map line complexes across most of the known SNRs. It will be launched in 1993. The Soviet *Spectrum X-Gamma* satellite, scheduled at approximately the same time, will contain imagers and an objective Bragg spectrometer that can map selected lines with very high spectral resolution (Sunyaev 1989). NASAs Advanced X-ray Astrophysics Facility *(AXAF)*, scheduled for launch in 1997, has the most sophisticated and versatile complement of spectrometers. The sensitivity of these instruments for performing high resolution spectroscopy of the kind described here is 1000 times that of *Einstein* . Thus with *AXAF* we will perform detailed plasma diagnostics on many dozens of SNRs in the Galaxy and in our nearest neighbor, the Large Magellanic Cloud (Weisskopf 1987, Canizares 1989b, Weisskopf 1989).

3.2 Satellite Lines

Diagnostics using satellite lines are a concrete example of what a sensitive, high resolution spectrometer can accomplish in the study of SNR plasmas. These lines appear among those in the He-like triplet and are primarily due to transitions in doubly excited Li-like ions in which the additional electron orbital acts as a 'spectator,' shifting slightly the energy of the line. Satellite lines follow excitation by electron impact either through inner shell excitation, $e + Li \rightarrow e + Li^{**}$, or more commonly by dielectronic recombination, $e + He \rightarrow Li^{**}$ (e.g. Gabriel 1972, Bahla, Gabriel and Presnyakov 1975, Gabriel and Phillips 1979, Mewe 1988). The doubly excited Li-like ions can autoionize to the He-like stage (the inverse of dielectronic recombination) or decay radiatively, thereby emitting the satellite line. Radiative decay grows like Z^4 along the Li-like isoelectronic sequence, so the satellite lines are most prominent in higher Z elements, like Fe.

Comparisons of selected satellite lines (e.g. $1s^2 2p - 1s2p^2$) to the He-like resonance line ($1s^2 - 1s2p$; see Section 2.2) provides a sensitive diagnostic of electron temperature. This has been used in studies of the solar corona (Doschek *et al.* 1980) as well as in the laboratory (Bitter *et al.* 1979) and would be an excellent tool to apply to SNRs. The lines from inner shell ionization depend on the presence of Li-like ions, so they can reveal departures from ionization equilibrium. Satellites resulting from dielectronic recombination can also reveal departures from Maxwellian electron distributions (Gabriel and Phillips

1979, Seely, Feldman and Doschek 1987). This is because only those electrons in a very narrow energy band can give rise to dielectronic recombination, so the strength of a given satellite line samples only a narrow slice of the electron energy distribution. In contrast, the resonance line is excited by all electrons above a given threshold. Thus a comparison of the line strengths is proportional to the relative populations of electrons of different energies, which can be compared to that expected for a Maxwellian distribution.

4. CONCLUSION

X-ray astronomers have made a small beginning in their attempts to understand the properties of cosmic plasmas. After many years during which we could only look with envy at the tools of our colleagues who study solar or laboratory plasma, we now anticipate several new and powerful instruments that will allow us to apply many of the techniques developed in those studies of less remote objects. Surely this will increase the degree of overlap between the various disciplines and, I expect, the frequency of meetings such as this one.

I am grateful to Tom Markert and Kathy Fischbach and the members of the group whose unpublished work I have quoted. I also thank Hiroshi Tsunemi for useful discussions.

REFERENCES

Aschenbach, B. 1988, in *Hot Thin Plasmas in Astrophysics,* Pallavicini, B. (ed.), (Dordrecht: Kluwer), p. 185
Bahla, C. P., Gabriel, A., and Presnyakov, G. 1975, *Mon. Not. Roy. Astron. Soc.,* **17**, 359.
Bitter, M.,*et al.* 1979, *Phys. Rev. Lett.,* **43**, 125.
Brinkmann, W., Fink, H., Smith, A., and Haberl, F. 1989, *Astron. Ap.,* , (in press).
Canizares, C. R. 1984a, in Oda, M. and Giacconi, R. (eds.), *X-ray Astronomy '84* (Tokyo:Institute of Space and Astronautical Science, p. 275
Canizares, C. R. 1984b, in Proceedings of, *Course and Workshop on Plasma Astrophysics,* ESA SP-207, p. 159.
Canizares, C. R. 1989a, in Gorenstein, P. and Zombeck, M. (eds), *High Resolution X-ray Spectroscopy of Cosmic Plasmas,* (Cambridge: Cambridge U. Press) in press.
Canizares, C. R. 1989b, in Proceedings of COSPAR Symposium, *Advances in X-ray and Gamma-ray Astronomy* Helsinki 1988, (in press).
Canizares, C. R., Winkler, P. F., Markert, T. H., and Berg, C. 1983 in Danziger, J. and Gorenstein, P. (eds), *Supernova Remnants and Their X-ray Emission* (Dordrecht: Reidel), p. 205.
Danziger, J. and Gorenstein, P. (eds), *Supernova Remnants and Their X-ray Emission,* (Dordrecht: Reidel).
Doschek, G. A., Feldman, U., Kreplin, R. W., and Cohen, L. 1980, *Ap. J.,* **239**, 725.
Fischbach, K. F., Bateman, L. M., Canizares, C. R., Markert, T. H., and Saez, P. J. 1989, in

Gorenstein, P. and Zombeck, M. (eds), *High Resolution X-ray Spectroscopy of Cosmic Plasmas,* (Cambridge: Cambridge U. Press), (in press).

Gabriel, A. 1972, *Mon. Not. Roy. Astron. Soc.,* **160**, 99.

Gabriel, A., Bely-Dubau, F., Faucher, P., and Acton, L. 1988, *Ap. J.,* (in press).

Gabriel, A. and Jordan, C. 1969, *Mon. Not. Roy. Astron. Soc.,* **145**, 241.

Gabriel, A. and Phillips, K. J. H. 1979, *Mon. Not. Roy. Astron. Soc.,* **189**, 319.

Gorenstein, P., Harnden, R., and Tucker, W. 1974, *Ap. J.,* **192**, 661.

Hamilton, A., Sarazin, C., and Chevalier, R. 1983, *Ap. J. Supp.,* **51**, 115.

Hamilton, A., Sarazin, C., and Szymkowiak, A. 1986a, *Ap. J.,* **300**, 698.

———, 1986b , *Ap. J.,* **300**, 713.

Hughes, J. and Helfand, D. 1985, *Ap. J.,* **291**, 544.

Itoh, H., 1977, *Pub. Astron. Soc. Japan,* **29**, 813.

Itoh, H., Massai, K., and Nomoto, K. 1988, *Ap. J.,* **334**, 279.

Jansen, F. 1988, *Ph.D. Thesis,* Leiden State University, The Netherlands

Jansen, F., Smith, A., Bleeker, J. A. M., de Korte, P. A. J., Peacock, A., and White, N. E. 1988, *Ap. J.,* **331**, 949.

Kallne, E. *et al.* 1984, *Phys. Rev. Lett.,* **52**, 2245.

Ku, W. H. M., Kahn, S. M., Pisarski, R., and Long, K. L. 1984, *Ap. J.,* **278**, 615.

Markert T., Blizzard, P., Canizares, C., and Hughes, J. 1988a, in Roger, R. S. and Landecker, T. L. (eds), *Supernova Remnants and the Interstellar Medium,* (Cambridge: Cambridge Univ. Press p. 129.

Markert T., Blizzard, P., Canizares, C., and Hughes, J. 1989, in preparation

McKee, C. 1974, *Ap. J.,* **188**, 335.

McKee, C. and Hollenbach, D. 1980, *Ann. Rev. Astron. Ap.,* **18**, 219.

Mewe, R. 1988, in Brown, R. and Lang, J. (eds.), *Astrohpysical and Laboratory Spectroscopy* (Scottish U. Summer Sch. in Phys.), p. 167.

Mewe, R., Gronenschild, E. H. B. M., and van den Oord, G. H. J. 1985, *Astron. Ap. Supp.,* **62**, 197.

Owocki, S. P. and Scudder, J. D. 1983, *Ap. J.,* **270**, 758.

Petre, R., Canizares, C. R., Kriss, G. A., and Winkler, P. F. 1982, *Ap. J.,* **258**, 22.

Prahdan, A. 1983, *Phys. Rev. A.,* **28**, 2128.

Pravdo, S. and Smith, B. W. 1979, *Ap. J. (Letters),* **234**, L195.

Raymond, J. 1984, *Ann. Rev. Astron. Ap.,* **22**, 75.

Roger, R. S. and Landecker, T. L. (eds.) 1988, *Supernova Remnants and the Interstellar Medium,* (Cambridge: Cambridge Univ. Press)

Seely, J. F., Feldman, U., and Doschek, G. A. 1987, *Ap. J.,* **319**, 541.

Shull, M. 1983, *Ap. J.,* **262**, 308.

Smith, A. *et al.* 1988, *Ap. J.,* **325**, 288.

Sunyaev, R. 1989 in Proceedings of COSPAR Symposium, *Advances in X-ray and Gamma-ray Astronomy,* Helsinki 1988, (in press).

Tanaka, Y. 1989, in Proceedings of COSPAR Symposium , *Advances in X-ray and Gamma-ray Astronomy,* Helsinki 1988, (in press).

Trimble, V. 1982, *Rev. Mod. Phys.,* **54**, 1183.

———, 1983, *Rev. Mod. Phys.,* **55**, 511.

Tsunemi, H., Yamashita, K., Masai, K., Hayakawa, S., and Koyama, K. 1986, *Ap. J.,* **306**, 248.

Tsunemi, H., Manabe, M., Yamashita, K., and Koyama, K. 1988, *Publ. Astron. Soc. Japan*, 40, 449.

Weisskopf, M. 1987, *Astrophysical Lett. Comm.*, 26, 1.

Weisskopf, M. 1989, *Space Sci. Rev.*, (in press).

Winkler, P. F., Canizares, C. R., and Bromley, B. C. 1983 in Danziger, J. and Gorenstein, P. (eds), *Supernova Remnants and Their X-ray Emission*, (Dordrecht: Reidel), p. 205.

Winkler, P. F., Canizares, C. R., Clark, G. W., Markert, T. H., and Petre, R. 1981, *Ap. J.*, 245, 574.

Zarnecki, J. C., Culhane, J. J., Toor, A., Seward, F. D., and Charles, P. A. 1978, *Ap. J. (Letters)*, 219, L17.

SPECTRAL ANALYSIS OF A TOKAMAK PLASMA IN THE VUV-RANGE

G. FUSSMANN , J.V. HOFMANN, G. JANESCHITZ
Max-Planck-Institut für Plasmaphysik
D-8046 Garching, Fed. Rep. of Germany
and
W. PÖFFEL, K.-H. SCHARTNER
I. Physikalisches Institut der Justus Liebig Universität
D-6300 Giessen, Fed. Rep.of Germany

ABSTRACT. The line emission occurring in the important spectral range 100-1100 Å from the ASDEX tokamak plasma is analysed with respect to atomic line identification, radiation losses, and physical relevance. Although plasma densities and temperatures do not differ substantially from those in stellar atmospheres, extreme deviations from corona equilibrium are a common feature in a tokamak. It is shown that in such cases line intensities should be interpreted in terms of fluxes of the particles rather than concentrations.

I. Introduction

Magnetically confined plasmas in nuclear fusion oriented research are objects of spectroscopic investigations for a number reasons:

1. Identification of impurity production and measurement of the corresponding impurity fluxes due to plasma wall interaction (mainly visible spectroscopy).
2. Investigation of dissociation, ionisation and transport in the plasma boundary region (visible to VUV).
3. Investigation of ionisation, recombination and transport of particles in the bulk plasma region (VUV to soft X-ray).
4. Determination of impurity densities and interpretation of observed radiation losses (bolometer measurements).
5. Determination of plasma parameters (particularly temperatures and electron densities; all spectral ranges).

These measurements cover a wide wavelength range, from the visible to the soft X-rays.

In the following we will concentrate mainly on measurements taken at the ASDEX tokamak in the VUV-range which are related to topics 2 to 4 from above. We will first discuss the line features observed in the important spectral range 100 to 1100 Å that is essential for the radiation losses from the plasma boundary region. Emphasis is put on the question of how to interpret the intensities of characteristic lines from light elements and metallic impurities. These questions lead us to a discussion of corona equilibrium and non-equilibrium, a matter of fundamental importance that has been addressed already in a number of papers [1,2]. Whereas usually only small or moderate deviations from equilibrium are assumed, we will show that in the case of VUV-radiation from light elements in a tokamak conditions far from equilibrium are the rule. In these cases the emitted lines represent particle influxes rather than densities.

W. Brinkmann et al. (eds.), Physical Processes in Hot Cosmic Plasmas, 29–37.
© 1990 *Kluwer Academic Publishers.*

II. Line emission in the VUV

The ASDEX tokamak is a toroidal device with circular cross-section of a = 0.4 m minor radius and 1.65 m major radius. For the case of ohmic heating (which we restrict ourselves to in the following discussion) the plasma is characterized by central electron temperatures $T_{eo} \approx 1000$ eV and densities $n_{eo} \approx 6 \times 10^{13}$ cm^{-3}. In the edge region close to the separatrix (r $\approx$ a) these parameters are about $T_{es} \approx 100$ eV , $n_{es} \approx 1 \times 10^{13}$ cm^{-3}. It is this region where the plasma is emitting intensive VUV-radiation. In Fig.1 we show an example of a spectrum taken along a central chord through the plasma covering the range 100 -1100 Å . The spectrum is recorded with a small survey spectrometer (SPRED , 0.35 m Rowland-radius [3,4]), equipped with a 1000 pixel multichannel detector. Spectra are obtained every 20 ms during a discharge of typically 2-3 s duration. The wavelengths assigned to the peaks in Fig.1 are produced by a peak-search program; they fit to the true values within an accuracy of about ± 1 Å. It is, however, not the intention of this paper to describe the details of the evaluation, data processing, calibration [5,6] and comparisons with extended code simulations. Instead we emphasize particular atomic features of the spectrum.

The spectrum shown in Fig.1 is dominated by radiation from carbon and oxygen. Furthermore, from these two elements only three ionisation stages (Li-, Be-, and B-like states) emit bright lines. In Fig. 2 the relevant sections of the Grotrian diagrams with the observable transitions are depicted. The correspondence between the lines from O and C greatly facilitates the intelligibility of the spectrum, and is also essential for consistency checks. As an example, the three lines 150, 173, 184 Å of the Li-like O VI-system have a correspondence to those at 313, 384, 420 Å of the C IV-system, and the line-triplet is in first approximation only shifted and spread to higher wavelengths.

Due to the poor resolution of the spectrometer the doublet structure of O VI is observable only at the highest wavelengths for the 2s-2p transitions (1032, 1038 Å). The associated transitions from C IV at 1551, 1548 Å are out of the range of the instrument. The power losses in these lines can therefore only be estimated by referring again to the correspondence between C IV and O VI. Similarly, the transition ^3P - $^3P^o$ at 1175 Å of the Be-like C III can be estimated. In this case, however, we have additional information from a normal incidence instrument (λ = 500 -1300 Å) from which we learn also that the radiation losses associated with hydrogen L$_\alpha$ (1215 Å) are almost negligible in comparison to those from O and C. These light elements occur in typical abundances of 1 % and 2 %, respectively.

The transitions to the metastable 2s 2p ^3P of the Be-like species are very intense (see Fig. 1 for O IV, 760 Å). A more detailed analysis yields approximately equal populations for the deep lying 2s2p metastable states (6.5 eV, 10.2 eV respectively) and the 2s^2 ground states. In this context it is interesting to note that there are no observable transitions to the metastable quartets of the boron-like system, although the corresponding 2s2p^2 states (8.9 eV, 5.3 eV) are very close to the ground state too.

The spectral analysis is almost complete by taking into account the Na- and Mg-like transitions from Cu and Fe, the most important metallic impurities in ASDEX. These systems resemble very much the Li- and Be-like systems discussed above but show a much larger fine structure splitting. From this point of view the lines 3s-3p from Cu XIX (273.4, 303.6 Å) and Fe XV (335.4, 360.4 Å) correspond to the 2s-2p doublet of O VI at 1032, 1038 Å. From the Mg-like ions Fe XV and Cu XVIII the singlet $^1P^o$ - ^1S transitions at 284.2 Å and 234.2 Å are also clearly detected. On the

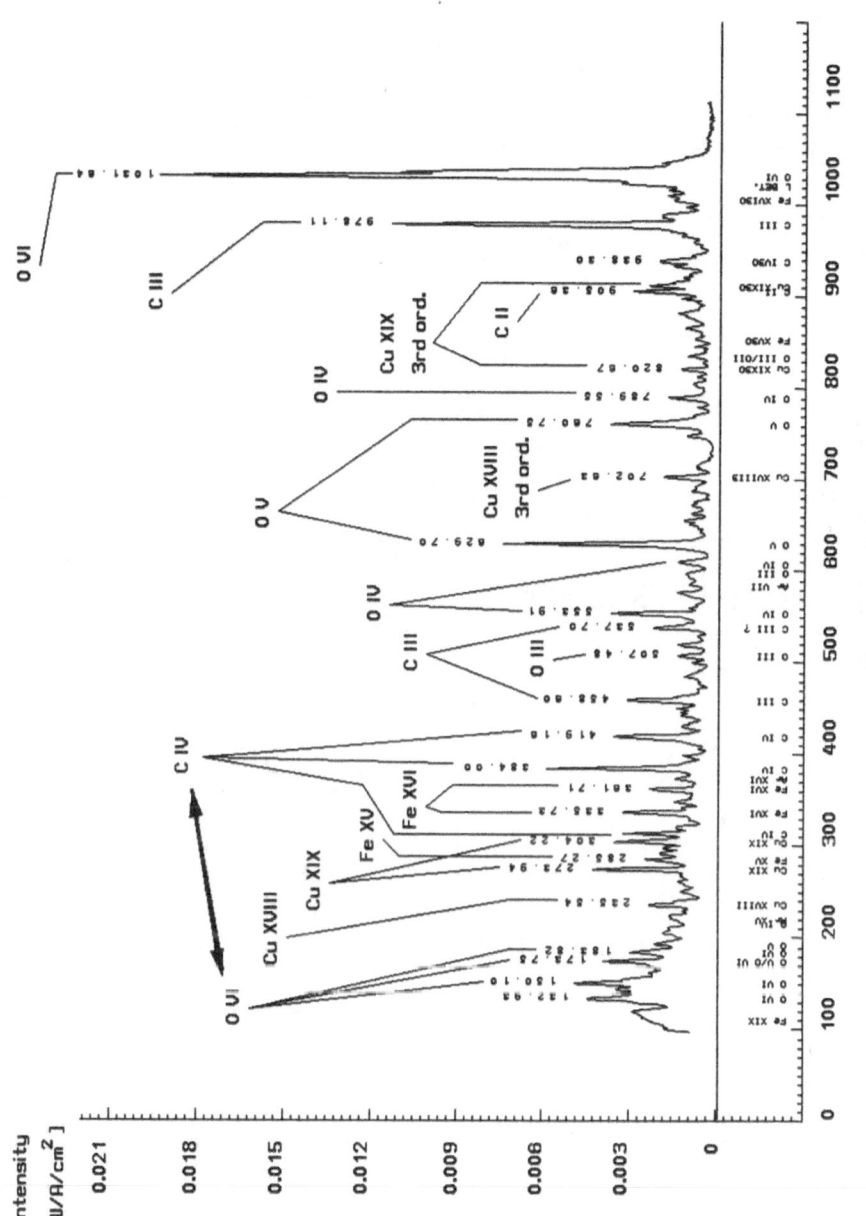

Fig. 1 VUV-spectrum in the range 100-1100 Å for an ohmic discharge in ASDEX (# 25390, 1.442-1.462 s)

32

other hand, lines from the triplet systems (e.g. Fe XV, $^3P^o$ - ^3P around 317-322 Å), do not emerge from the background, most likely because of the increased fine structure splitting that distributes the intensity over a large number of separated lines. The toroidal grating being used shows practically no reflections in second order but third order reflections are relatively strong and produce some of the lines above 400 Å in Fig.1.

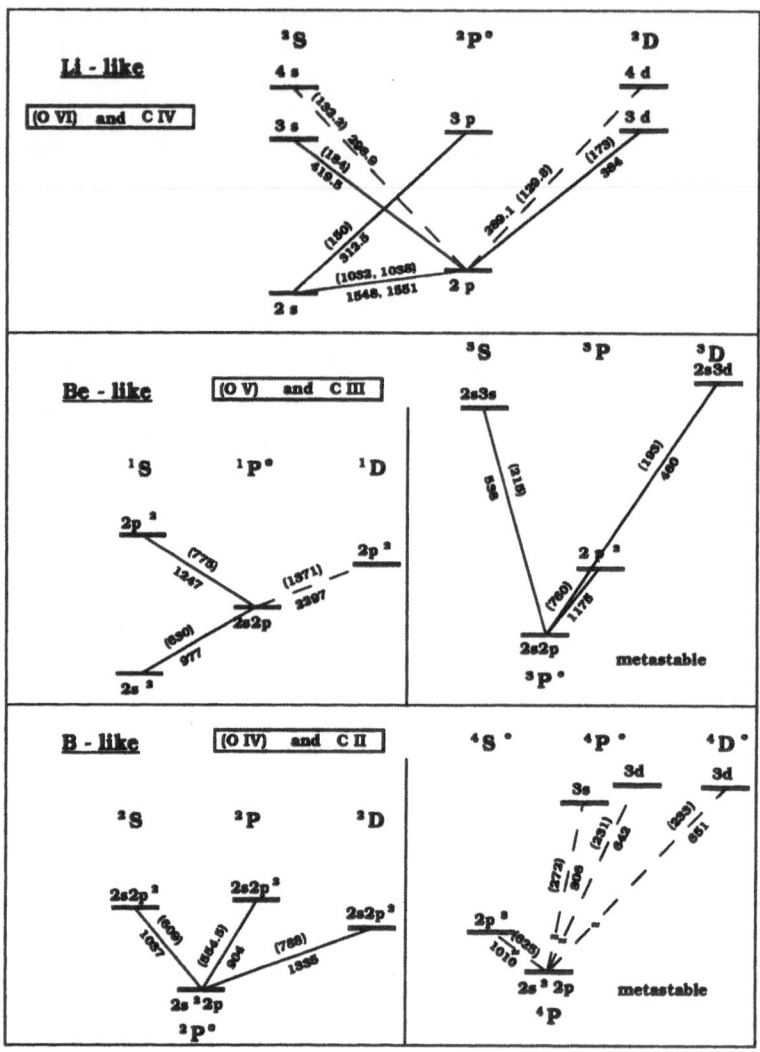

Fig. 2 Grotrian diagrams of Li-, Be-, and B-like spectra. Wavelengths are given for carbon and oxygen (in brackets) ions. Transitions occurring in Fig.1 are marked by solid lines.

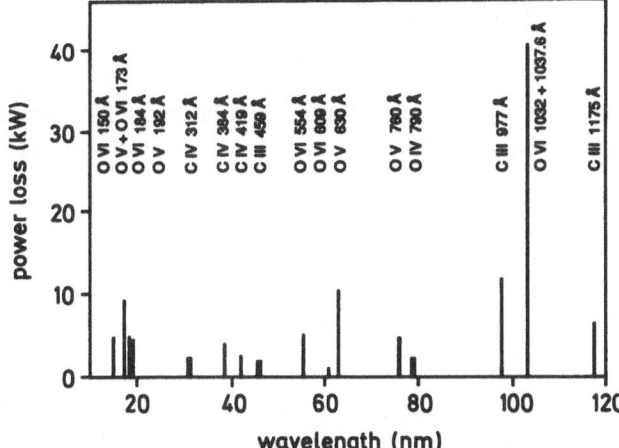

Fig. 3
Radiative power
contributions of
most prominent
VUV lines.

The radiation loss from 16 prominent carbon and oxygen lines in the range 150-1200 Å are visualized in Fig.3 for an ohmically heated plasma with about twice as large concentrations of C and O as the discharge presented in Fig1. In total they represent a power of P_{VUV} =128 kW which is already very close to the bolometrically determined radiation losses of P_{RAD} = 146 kW for the same ohmically heated discharge. Compared to the input power of P_{OH}= 400 kW, radiation is with 37% a substantial, although not dominating part of the power exhaust.

III. Interpretation of line intensities

Similar to most astrophysical objects, the densities in tokamaks are sufficiently low to allow application of the well known *corona excitation model* to calculate line emission from electronic collisional excitation rates $< \sigma v >_{ex}$. Intensities are then obtained by integrating the emissivity along the line of sight:

$$ I_z = B \frac{h \nu}{4 \pi} \int < \sigma v >_{ex.} n_e n_z \, dl $$

Subscript z refers here to the particular ion with ground state density n_z and $B = A_{mn} / \Sigma A_{kn}$ is the branching ratio for the observed transition n -> m. The crucial quantity is the density n_z , which in general is not only a function of temperature, but depends in addition on the influx of impurity neutrals and the transport properties of the plasma perpendicular to magnetic surfaces. For this reason the general problem can only be solved by means of numerical code calculations. There are, however, two limiting cases which lead to considerable simplifications. In the first case of *corona ionisation equilibrium* the fraction $f_z(T_e) = n_z / \Sigma n_z$ is only a function of temperature. The set of coupled differential transport equations for the various ions reduces here to a single one for the total ion density $n_\Sigma = \Sigma n_z$. The second, opposite case, may be termed as the *non-recombination approximation*. In this case each ion or atom that penetrates into the plasma emits a photon with probability

$$P(T_e) = B \frac{< \sigma v >_{ex.}}{< \sigma v >_{Ion.}}$$

before being ionized to the next higher ionisation state. In contrast to the first case where the intensity is proportional to the density n_z, it is now proportional to the local influx Γ_z in state z :

$$I_z = \frac{h v}{4\pi} P \Gamma_z$$

It should be noted that in the latter case I_z is independent of both, plasma transport and - at least approximately - electron density n_e.

We have seen that the two opposite situations , a) *ionisation equilibrium* and b) *non-recombination approximation*, are equivalent to cases where line emission is either *representative of density or flux* . Whether one or the other approximation is an appropriate description of a measurement can be assessed by comparing characteristic residence times τ_{dwell} with relevant atomic times. τ_{dwell} may be defined as the time the particles remain within the temperature region $T_1 < T_0 < T_2$ The high (T_2) and low (T_1) limits specify the positions r_2, r_1 where - under ionisation equilibrium conditions - the corresponding ion density is reduced to 1/e of its maximum value $n_z(T_0)$ at $r = r_0$. τ_{dwell} can thus become small in zones with steep T_e-gradients or high diffusive or convective transport.

There are actually two atomic time scales : $\tau_{rec} = (n_e \alpha_z)^{-1}$ for recombination into state Z-1 , and $\tau_{ion} = (n_e S_z)^{-1}$ for ionisation towards Z+1 , with the rate coefficients $\alpha_z = < \sigma_z v >_{rec}$ and $S_z = < \sigma_z v >_{ion}$. At $r = r_0$ we have $\alpha_z \approx S_z$, however, the T_e-dependence at this position is much steeper in case of S_z. This behaviour is caused by the fact that ionisation is mainly performed by fast electrons from the Maxwellian tail population, whose number increases very rapidly with rising T_e, whereas the number of slow electrons needed for radiative recombination, increases much less dramatically with decreasing T_e. Therefore, the tendency that inward diffusing particles are easily ionized in the hot regions, but on their way out do not recombine . This tendency is further enhanced by the electron density profile which is decreasing towards the plasma boundary. The important atomic time scale is therefore τ_{rec} rather than τ_{ion} .

The described relations are elucidated by Fig.4 where the ionisation rate for O VI ($n_e S_5$) and the recombination rate for O VII ($n_e \alpha_6$) are plotted as functions of radius for an ohmic discharge in ASDEX. We notice that the recombination rate is almost constant in the edge region while the ionisation rate changes by more than four orders of magnitude. The particles need a time as long as $\tau_{rec} \approx 80$ ms to recombine which is much larger than the residence time τ_{dwell} of about 10 ms in this edge region. Recombination is therefore expected to take place in the cooler regions within the divertor chamber or even at the target plates.

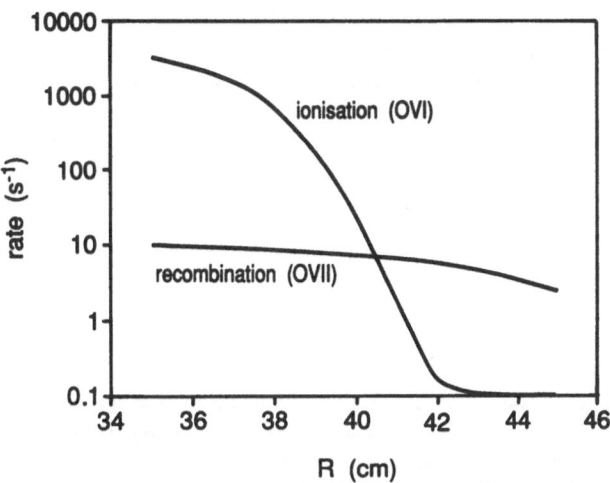

Fig. 4
Recombination and
ionisation rates for O^{+6}
and O^{+5} in the ASDEX
edge region. Notice the
small and nearly constant
recombination rate.

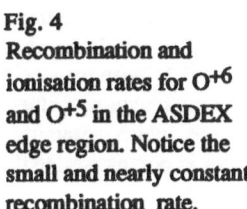

For the purpose of a more general consideration we schematize the relations and divide the plasma cross-section into three zones: the scrape-off layer (I) where the magnetic field lines intersect the target plates and the particles are rapidly lost by streaming along these lines, the edge region (II) around the separatrix with strong T_e- and n_e-gradients, and the hot core region (III) with relatively small gradients. In Table I we compile characteristic data for these three zones .

Table I
Plasma parameters and
relevant times within
three characteristic
regions. The most
abundant ionic states for
O and Fe are also given.

Zone	T_e (eV)	n_e (cm^{-3})	τ_{dwell}(ms)	τ_{rec}(ms)	
				oxygen	iron
I. scrape-off (r=40-45 cm)	1	$0.5 \cdot 10^{13}$	1	70 (O^{+5})	20 (Fe^{+5})
II. edge (r=30-40 cm)	100	$1.0 \cdot 10^{13}$	10	80 (O^{+5})	3 (Fe^{+11})
III. core (r= 0-30 cm)	1000	$5.0 \cdot 10^{13}$	50	60 (O^{+5})	1 (Fe^{+21})

The residence time τ_{dwell} is estimated invoking anomalous transport with a diffusion coefficient of D=0.5 m^2 s $^{-1}$, in rough agreement with experimental observations. The last two columns in Table I contain the recombination times for the most abundant ionic states (according to transport code calculations) of oxygen and iron in these zones. Corona equilibrium, which requires $\tau_{dwell} \gg \tau_{rec}$, is seen to pertain only for Fe in the core region. Conversely, the non- recombination approximation $\tau_{dwell} \ll \tau_{rec}$, holds for O and Fe in the scrape-off region, and in case of O in the edge region too. Oxygen in the core region and iron in the edge zone cannot be well approximated in one or the other way. In view of the statements made above we may generalize these results and in one or the other way. In view of the statements made above we may generalize these results and conclude: VUV-emission lines from the light elements (C I -C VI, O I-O VI) or low ionized metals (Fe I - Fe X) are flux representative. Only radiation from very highly ionized metals can be regarded as density dependent.

We have investigated in a number of experiments plasma discharges with two stationary electron density plateaus differing by a factor of 1.84. During the same plateau phases we observed an increase of the VUV-intensities of C- and O-ions within a range of 1.4 to 2.3. With respect to our conclusions this has to be interpreted by increased impurity influxes at the higher n_e values. This interpretation is in agreement with detailed plasma modelling calculations which, assuming a constant impurity influx, would predict also much too small intensity ratios.in comparison to the measurement.

Finally, some general remarks with respect to the application appear appropriate: For stationary conditions there is of course always a proportionality between influx and density owing to the linear particle balance equation. Hence, knowing the transport coefficients as a function of radius and Z, as well as the n_e- and T_e-profiles, impurity densities can be calculated from measured fluxes and vice versa. For an unchanged background plasma with a fixed proportionality relation all lines are then indicators of the impurity concentration. On the other hand, if the plasma conditions have been changed or are non-stationary (e.g. by applying additional heating) the line emission can change either on account of varying plasma parameters (n_e, T_e) , or particle confinement, or impurity influxes. In such a situation our conclusions from above may be invoked to distinguish between the three possible causes. Lines from low ionisation states give virtually at once an indication of the changes of influxes. As a next step, changes in transport may be identified from the emission of highly ionized atoms in connection with measured n_e-, T_e- profiles.

IV. Summary

For the ASDEX tokamak plasma the VUV range from 100 to 1100 Å has been found most important for radiation losses and with regard to the detection of light impurities (He to Ne) and metals (Ti to Zn). Light elements like carbon and oxygen are observed in the Li-, Be-, and B-like states with pronounced population of the metastable ^{3}P-states of Be-like C^{+2} and O^{+4}. Metals are easily identified in the Na- and Mg-like states where the emission is concentrated in a few lines only. About 90% of the total radiation losses could be recovered by quantitative evaluation of less than 20 lines in the cited spectral range.
In contrast to astrophysical objects which always apply well to corona equilibrium, extreme deviations from such ionisation equilibrium are quite common in a tokamak because of rapid particle transport in regions of large temperature gradients. Of crucial importance in this context is the ineffectivity of radiative recombination in the edge region of the plasma. For this reason, line intensities from low ionisation stages reflect essentially the particle influx. Only the radiation from the highly ionized atoms from the plasma centre is representative of the particle density, as it is the case for stellar plasmas.

References

[1] R.C. Isler, Nucl. Fus., 24, 12,1599-1677 (1984)

[2] R.D. Petrasso, MIT-Report PFC/JA-88-45 ," Tokamak plasmas: A paradigm for coronal equilibrium and disequilibrium", to be published in Proceedings of the Int. Astronomical Union (1988)

[3] B.C. Stratton, R.J. Fonck, K. Ida et al., Rev. Sci. Instr., 57, (8), 2043-2045 (1986)

[4] G. Janeschitz, L.B. Ran, G. Fussmann, K. Krieger, K.-H. Steuer et al. "Impurity Concentrations and Z_{eff} obtained from VUV- Spectroscopy at ASDEX", Report IPP III 147 (1989)

[5] W. Pöffel, K.-H. Schartner and G. Fussmann, G. Janeschitz, Nucl. Instr. and Meth. in Phys. Research, B23,128-130 (1987)

[6] K.-H. Schartner, B. Kraus, W. Pöffel, and K. Reymann, Nucl. Instr. and Meth. in Phys. Research, B 27, 519-526 (1987)

IONIZATION OF HOT PLASMAS

R. Mewe
SRON-Laboratory for Space Research, Utrecht
Beneluxlaan 21, 3527 HS Utrecht, The Netherlands

ABSTRACT

The processes of ionization and recombination in hot ($T \gtrsim 0.1$ MK) astrophysical plasmas are considered, emphasizing the case of optically thin plasmas. A detailed comparison among different computations is made for the coronal model. The emergent radiation spectra are treated as a function of electron temperature. The effects of optical depth, photoionization, high densities, time variations, electromagnetic fields, and non-Maxwellian electron energy distributions are briefly discussed.

1. INTRODUCTION

In a hot optically thin plasma that is sufficiently transparent that the transfer of radiation can be neglected the emergent X-ray spectrum faithfully represents the microscopic emission processes in the plasma and therefore is directly linked to the physical conditions in the plasma. High-resolution spectroscopy in the soft X-ray wavelength region (5–140 Å) with the coming NASA *AXAF* and ESA *XMM* spectrometers will allow to detect and identify discrete spectral features and represents one of the final observational frontiers for high-energy astrophysics. Contained in this band are a multitude of prominent features from nearly all ion stages of many abundant elements, including the K-shell line and continuum transitions of C, N, O, Ne, Mg, and Si, and the L-shell transitions of Si, S, Ar, Ca, Ni, and Fe. Many of these features will be detected either in emission or absorption for a large variety of different astrophysical plasmas including: stellar coronae, isolated hot white dwarfs, cataclysmic variables, X-ray binaries, supernova remnants, the interstellar medium, normal galaxies, clusters of galaxies, and active galactic nuclei. It will be possible to uniquely determine the most important physical parameters for these plasmas including the electron temperature and density distributions, the ion and elemental abundances, mass motions, and the nature of the ambient radiation field. The spectra we expect to obtain will likely provide an enormously important new tool for the study of virtually all known cosmic X-ray sources.

Certain spectral models are needed to infer the physical parameters from the observations. The usual procedure is to convolve theoretical model spectra with the instrumental response and to vary the model parameters optimizing the fit of the model to the observational data. A useful approach is to consider first a simplified plasma model for the X-ray source, neglecting much of the complexity of the temperature and density structure and of the effects of opacity, and to synthesize such models into successively more sophisticated approximations of the source model.

I start with the simplest case: the *coronal model* of a tenuous plasma in a steady state in which electron collisions control the ionization state and emissivity of the gas, assuming a Maxwellian energy distribution for the plasma electrons. Examples of such plasmas are stellar coronae, supernova remnants, and the hot gas in the interstellar medium and in galaxies and clusters, and possibly also the low-density intercloud medium that pervades most of the central

39

W. Brinkmann et al. (eds.), Physical Processes in Hot Cosmic Plasmas, 39–65.
© 1990 *Kluwer Academic Publishers.*

broad-line region in active galactic nuclei.

It is obvious that the physical parameters can be inferred only on the accuracy level of the available rates of ionization, recombination and excitation, so that I will discuss the uncertainty of the currently available rates.

Later on I discuss the effects of relaxing the restrictions by considering photoionization, optical depth effects, high-density effects on the ionization balance, deviations from a Maxwell distribution, and transient plasmas.

2. CORONAL MODEL

The coronal model is a familiar standard model that was first applied to the solar corona (Elwert 1952). From the early discovery in 1948 of X-rays from the solar corona, X-ray spectroscopy has proven to be an invaluable tool in studying hot astrophysical and laboratory plasmas. The assumptions are that the gas is in a state of statistical equilibrium both for the bound atomic states and for the ionization balance. In order to keep the plasma in a stationary state, the radiative power loss has to be compensated by some heating of the plasma. The particles are relaxed to Maxwellian energy distributions with a common temperature, T, a free parameter controlled by external processes. The electron-electron relaxation time $t_{ee} \approx 0.01T^{3/2}n_e^{-1}$ (s) (Spitzer 1962) (T in K, electron density, n_e, in cm^{-3}) is generally short enough to ensure a Maxwellian velocity distribution for the electrons, unless the timescales for energy loss or gain or particle containment are smaller than t_{ee}. If the mechanisms of energy supply to the plasma preferentially heat one kind of particles (e.g., heavy ions in shocks or electrons in microturbulent plasmas) ion and electron temperatures may differ significantly, if the Coulomb collision equilibration time (Spitzer 1962) $t_{ei} \approx 10\ T^{3/2}n_e^{-1}$ (s) is long enough, unless plasma instabilities reduce the equilibration timescale (e.g. in the turbulent shock front in supernova remnants) (Mewe 1984).

In the other extreme case of a high-density plasma in Local Thermodynamic Equilibrium (LTE) every atomic process is as frequent as its inverse process (*principle of detailed balancing*), but in an optically thin plasma where the radiation which originates in the interior can escape the plasma, this cannot longer be the case. A simple description of such a plasma is only possible if we assume that the electron density and the radiation field intensity are so small that in all probability an excited atom will decay by spontaneous radiation and an ionized atom will recombine by radiative or dielectronic recombination. Instead of each collision process being balanced by its inverse collision process as in the LTE model, the balance in the coronal model is between collisional ionization (excitation) and radiative plus dielectronic recombination (spontaneous decay, respectively). In such a tenuous plasma 3-body processes and photoionization and photoexcitation can be neglected.

With approximate formulae for resonance line radiation trapping and line excitation (Mewe 1970) and for radiative transfer (e.g. Athay 1966, Hearn 1966) it can be shown that in many cases astrophysical and laboratory plasmas are effectively optically thin with respect to ionization and emergent X-ray line fluxes. An elucidation of the emergent spectrum in this case requires a knowledge of the cross sections of the fundamental processes (excitation, ionization and recombination) involved.

3. IONIZATION STATE

Much of the temperature sensitivity of the soft X-ray spectrum is associated with the ionization structure. Under the assumptions of the coronal model, all ions can be taken to be in their ground states. The ionization state is controlled by electron impact ionization (including sometimes a contribution from autoionization) and by radiative plus dielectronic recombination. The rate of change of the population density $N_{Z,z}$ (in cm^{-3}) of ion Z^{+z} from element of atomic number Z is given by

$$\frac{dN_{Z,z}}{n_e dt} = N_{Z,z-1}S_{Z,z-1} - N_{Z,z}(S_{Z,z} + \alpha_{Z,z}) + N_{Z,z+1}\alpha_{Z,z+1}, \tag{1}$$

where $S_{Z,z}$ and $\alpha_{Z,z}$ are the total ionization ($z \rightarrow z+1$) and recombination ($z \rightarrow z-1$) rate coefficients (in cm^3 s^{-1}) of ion Z^{+z}, etc.

The ionization structure can be derived by solving for each element Z a set of $Z+1$ coupled rate equations. The population of stage z depends on the four rates which connect it with the neighbouring ionization stages $z-1$ and $z+1$. If ion $z-k$ is most abundant, then in principle all the rates connecting stages between $z-k$ and z would enter. If, for example, all ionization rates would be systematically too high by a factor of two, then the predicted population of ion z would be off by 2^k. Fortunately, if k is more than say 1 or 2, the population is generally to small to matter.

For a steady-state equilibrium the rate of change (left-hand member of Eq. (1)) can be set equal to zero, and the population density ratio $N_{Z,z+1}/N_{Z,z}$ of two adjacent ion stages $Z^{+(z+1)}$ and Z^{+z} can be expressed by

$$N_{Z,z+1}/N_{Z,z} = n_e S_{Z,z}(T)/n_e \alpha_{Z,z+1}(T) = S_{Z,z}/\alpha_{Z,z+1}, \tag{2}$$

which is, to first order, only dependent on T and not on n_e, as long as stepwise ionization in $S_{Z,z}$ and collisional coupling to the continuum in $\alpha_{Z,z+1}$ can be neglected (Mewe 1970, Wilson 1962, Mewe and Schrijver 1978, Mewe and Gronenschild 1981).

3.1. Collisional ionization rates

The inelastic process of ionization (or excitation) is most effective when the relative velocity of the impacting particle is of the order of the velocity of the bound electron in the ion to be ionized (or excited). Thus if electrons and ions have comparable kinetic energies the electrons are much more effective, unless we consider population redistribution between nearby levels where e.g. proton collisions can be efficient. Therefore in the computation of the rate coefficients ionization by impact of ions will be neglected.

We obtain the rate coefficient S by averaging the cross section $Q(v)$ over all velocities v of the electrons in the plasma:

$$S = < vQ(v) > = \int_{E_0}^{\infty} vQ(v)f(E)dE, \tag{3}$$

where $E = \frac{1}{2}mv^2$ is the electron energy, $f(E)$ the electron energy distribution function and E_0 the threshold energy of the collision process. For a Maxwellian distribution with temperature T we have:

$$S = (8kT/\pi m)^{1/2}y^2 \int_1^{\infty} Q(U)\exp(-yU)U\,dU, \tag{4}$$

where $U = E/E_0$ is the initial energy of the impinging electron in terms of the threshold energy, and $y = E_0/kT$ the reduced threshold energy. It turns out that the cross sections for ionization and excitation obey a scaling law by which the cross section can be written in a form

$$Q(U) = \alpha(E_H/E_0)^2 Q_{red}(U)\pi a_0^2, \tag{5}$$

where $Q_{red}(U)$ is a scaled cross section (expressed in $\pi a_0^2 = 8.8\ 10^{-17}$ cm^2), which has approximately the same form for all ions in a given isoelectronic sequence and is nearly independent of atomic parameters. E_H is the ionization energy of hydrogen and α is a numerical factor of the order unity.

In the course of time numerous authors have attempted to represent the functional dependence of the direct electron-impact ionization cross section by a semi-empirical formula (to name a few: Elwert 1952, Drawin 1961, Lotz 1967, 1968–70, see also Mewe 1988), which correctly represents both the behaviour at threshold (U=1), $Q \propto (U-1)$ as well as the quantum-mechanical Born-Bethe dependence at high energies, $Q \propto U^{-1} ln U$.

In astrophysics the rates derived from the semi-empirical formula of Lotz (1967–70) for the cross section for direct impact ionization from the ground state have been widely used. The latter reads

$$Q(U) = \pi a_0^2 \sum_{k=m}^{N} C_m \xi_m (E_H/\chi_m)^2 U_m^{-1} ln(U_m) W(U_m), \tag{6}$$

where the summation is over (sub)shells m of the initial ion Z^{+z} with ionization energy χ_m, $U_m = E/\chi_m$, and where ξ_m is the effective number of equivalent electrons in shell with principal quantum number m. In most cases it is sufficient to set N equal to 1 (outermost shell) or 2 (next inner subshell). The function $W(U) = 1 - b_m \exp[-c_m(U-1)]$ represents the deviation from linear behaviour near threshold (only significant for ions of low charge and for low temperature). From fits to experimental data or extrapolations Lotz (1967–70) found that one can take for ions ionized more than four times $W(U_m)=1$ and for the constant $C_m=2.76$. Then the "Lotz" rate coefficient for direct ionization (DI) (for one shell only and omitting subscript m) follows from Eqs. (4) and (6):

$$S_{DI} = 3.24 \ 10^{-4} \xi \chi_{[eV]}^{-1} T_{[K]}^{-1/2} E_1(\chi/kT) \ (cm^3 s^{-1}), \tag{7}$$

where the first exponential integral $E_1(y) = \int_1^\infty U^{-1} \exp(-yU) dU \simeq e^{-y}/(y+1) \approx e^{-y}/y \ (y \gg 1)$, since ion concentrations typically peak at $y = \chi/kT \sim 5$. Hence direct ionization rates scale with temperature as $S_{DI} \propto T^{1/2} \chi^{-1} \exp(-\chi/kT)$.

Since threshold energies for ionization (and for $\Delta n \neq 0$ excitation) scale as $\chi \propto z_e^2$, where z_e is the effective charge number of the nucleus acting on the electrons in the shell (in particular for hydrogenic ions Z^{+z} : $z_e = z + 1$, for non-hydrogenic ions z_e is defined from $\chi_m = E_H z_e^2/m^2$), the cross section scales as $Q \propto z_e^{-4}$ (see Eqs. (5) and (6)) and the rate coefficient, expressed as a function of *reduced* temperature $\Theta = T/z_e^2$, as $S_{DI} \propto z_e^{-3}$. Calculations based on the Exchange Classical Impact Parameter (ECIP) method (Burgess *et al.* 1977, Summers 1974) predict rates typically about half those given by the Lotz formula, whereas the ionization rates now coming into widest use, those from Distorted Wave calculations (e.g. Younger 1981), appear to lie in between (see discussion by Raymond (1988)). Many ionization cross sections have been measured for the lower ions with 10–20% accuracy (e.g. Gregory *et al.* 1987), while for the higher ions many atomic *rate* coefficients for ionization, recombination, and excitation have been obtained from plasma measurements (e.g. Wang *et al.* 1986–88; for reviews see Kunze 1972, Griem 1988). For cases in which experimental or theoretical data are not available, Burgess and Chidichimo (1983) suggest to use the Lotz formula (Eq. (7)) with $C \simeq 2.3$ and with ξ and χ properly assigned. More recently, Arnaud and Rothenflug (1985) present an extensive compilation of ionization rates for 15 cosmically abundant elements (H,He,C,N,O,Ne,Na,Mg,Al,Si,S,Ar,Ca,Fe, and Ni) based on fits to available experimental cross sections and on Younger's theoretical results.

An additional ionization process is important for ions which have a large number of electrons in the next inner subshell compared with the number in the outermost shell, such as the sodium isoelectronic sequence ($1s^2 2s^2 2p^6 3s$, e.g. Fe XVI). Here an inner electron can be excited to a level above the ionization threshold and subsequently autoionize (AI, probability A_a), or it decays radiatively (probability A_r) to a bound state below the ionization threshold (e.g. Bely and van Regemorter 1970, Mewe 1988). The contribution to the ionization cross section can be written as a sum $\sum_i Q_{iexc} B_i$ over all target bound states lying above the first ionization limit and where the branching ratio $B_i = A_{ai}/[A_{ai} + A_{ri}]$. Generally $A_r \propto Z^4$ along an isoelectronic sequence, whereas $A_a \sim const.$, so that B_i strongly varies along the sequence, while it can change very suddenly along successive ionization stages of one element. Typically $Q \propto E_0^{-2} U^{-1}$, so that the rate coefficient scales as (see Eqs. (4) and (5)) $S_{AI} \propto T^{-1/2} E_0^{-1} \exp(-E_0/kT)$ for $E_0/kT \gg 1$, where E_0 is the

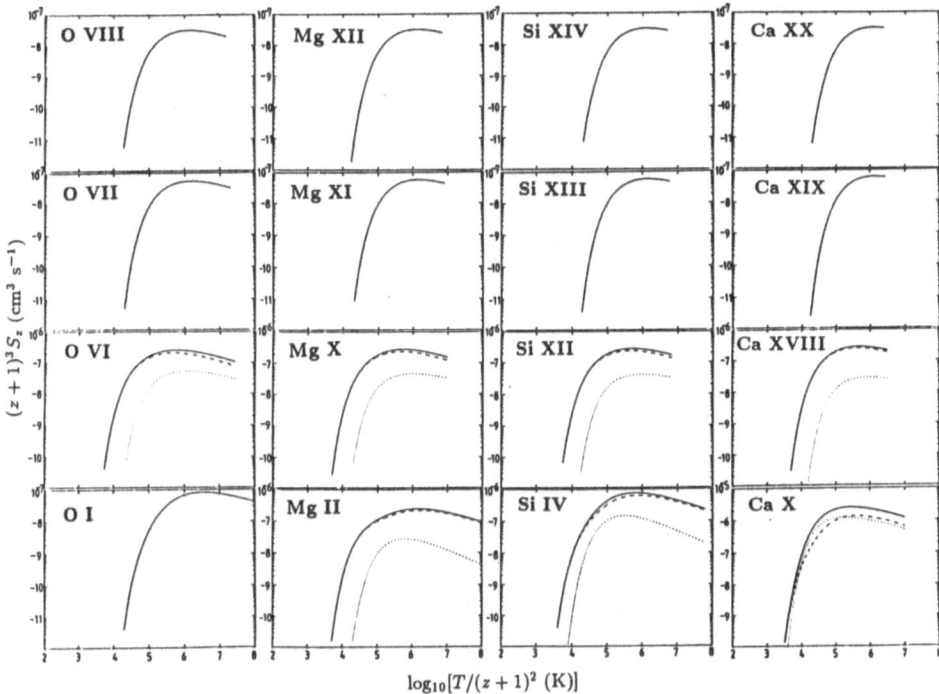

Figure 1. Reduced rate coefficients $(z+1)^3 S_z$ for direct ionization (DI) (dashed line) plus autoion-ization (AI) (dotted) (solid line indicates total rate) of ion Z^{+z} as a function of reduced electron temperature $T/(z+1)^2$ from Arnaud and Rothenflug (1985) (e.g. $z=7$ for ionization of O VIII, etc.).

excitation energy of the autoionizing level. Arnaud and Rothenflug (1985) present a compilation based on semi-empirical fits to the results of calculations by Sampson (1982).

In Figure 1 I have presented, for a number of representative cases covering both low- and high-Z ions, the results of Arnaud and Rothenflug (which I consider as the best ones available for the moment). We present the scaled total rate coefficient $(z+1)^{-3} S_z$ for ionization of ion Z^{+z} in dependence of the reduced temperature $\Theta = T/(z+1)^2$. In cases where the contribution from autoionization becomes also important, we have indicated this in the figure. I guess that in many cases the rates are good to 10–20% and that the overall uncertainty will be at the level of up to ~40% (Raymond 1988).

3.2. Radiative and dielectronic recombination rates

In coronal equilibrium, each ionization is balanced by a recombination process, either radiative or dielectronic. For hot plasmas we can neglect processes of charge transfer, which can be very important for the ionization structure of cooler plasmas, such as photoionized nebulae. The most important charge tranfer process is generally the capture of an electron from neutral hydrogen by a

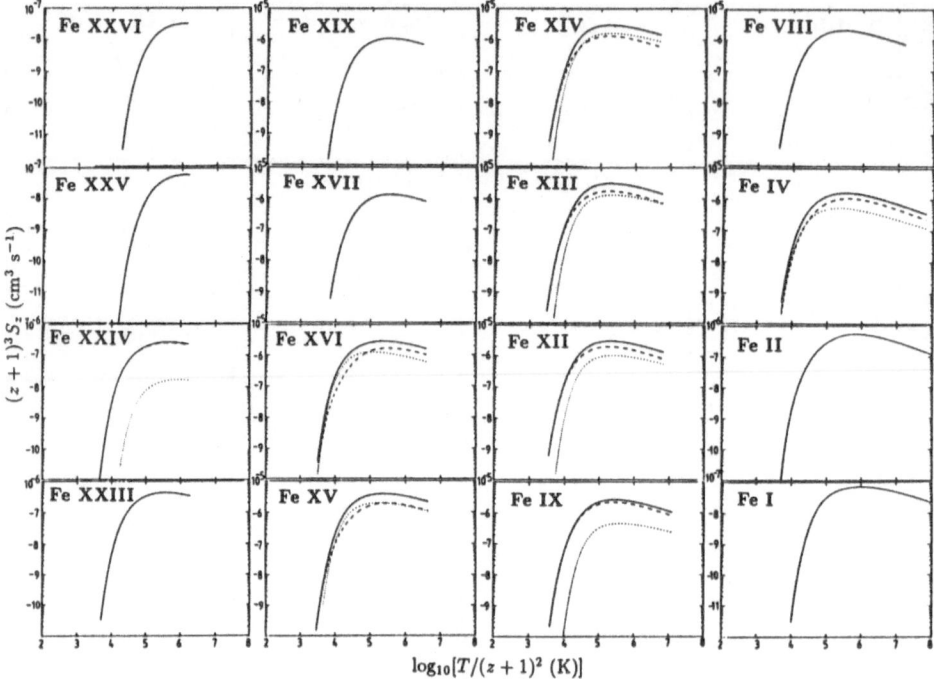

$$\log_{10}[T/(z+1)^2 \ (\mathrm{K})]$$

Figure 1. (continued)

charged ion, resulting in recombination of the ion. For temperatures below ∼0.01 MK this becomes important (e.g. Arnaud and Rothenflug (1985) and review paper by Butler and Dalgarno (1980)). We briefly consider the processes of radiative and dielectronic recombination.

Since radiative recombination occurs when an ion captures a plasma electron, subsequently emitting a photon, this process is the inverse of photoionization. Thus the rate is computed from the photoionization cross section by applying the principle of detailed balancing (Milne relation (e.g. Elwert 1952, Mewe 1988)). For hydrogenic ions it should be accurate to a few percent (Seaton 1959). For the total rate coefficient (inclusive recombination to excited levels) for radiative recombination (RR) of ion $Z^{+(z+1)}$ Elwert (1952) estimated (within ∼ 15%) (see also Mewe 1988), say for $y = \chi_m/kT \gg 1$ (where χ_m is the ionization energy of the recombined ion Z^{+z} in the ground state with principal quantum number m):

$$\alpha_{RR} \approx 3 \ 10^{-11} z_e^2 T_{[\mathrm{K}]}^{-1/2} \ (\mathrm{cm}^3 \ \mathrm{s}^{-1}), \tag{8}$$

where z_e is the effective nuclear charge (e.g. $= z + 1$ for hydrogenic ions and $= m\sqrt{\chi_m/E_H}$ for non-hydrogenic ions). This expression shows that radiative recombination rates in dependence of reduced temperature $\Theta = T/z_e^2$ scale $\propto z_e$.

For recombination towards more complex non-hydrogenic ions to excited levels one can also use the hydrogenic approximation, but for recombination directly to the ground state we can evaluate the rate coefficient from the cross section of photoionization from the ground state (e.g. Reilman and Manson 1979, also Pradhan 1977) using the Milne relation. Arnaud and Rothenflug (1985) present a compilation of fits to results obtained in this way.

The estimated overall uncertainties are probably at the level of 30–40% (see discussion by Raymond 1988).

The recombination of many coronal ions, however, is dominated by dielectronic recombination (for reviews, e.g. Hahn (1985), Hahn and LaGattuta (1988), Bell and Seaton (1985)). Though dielectronic recombination (DR) was already considered quantitatively by Massey and Bates (1942), it was not until 1964, however, when Burgess (1964, 1965) showed its importance in resolving discrepancies in the ionization balance of the solar corona, that DR began to be considered as one of the key processes determining the ionization state in hot thin plasmas. Another aspect is, as was first shown by Gabriel (1972), that this process is responsible for the formation of spectral lines appearing as (mostly) long-wavelength satellites to the resonance lines of highly ionized atoms in hot plasmas (see review by Dubau and Volonté (1980) and also Mewe (1988)). In fact, such lines constitute the only observable effect by which DR manifests itself.

Dielectronic recombination can be considered as a two-step process. It is initiated by the resonant radiationless capture of an energetic plasma electron by an ion $Z^{+(z+1)}$ into a high Rydberg state $n\ell$ of ion Z^{+z}, accompanied by the excitation $p \rightarrow q$ of one of the bound electrons in the core of the recombining ion $Z^{+(z+1)}$, thus forming a doubly excited state in ion Z^{+z}:

$$Z^{+(z+1)}(p) + e^- \rightarrow Z^{+z}(q, n\ell).$$

The initial state p of the recombining ion is usually the ground state in the coronal approximation. The high Rydberg electron ($n\ell$) is just a "spectator" as far as the radiation of the core electron is concerned. Thus, an electron is incident on ion $Z^{+(z+1)}$ with an energy E_0-ε, just ε less than the threshold energy E_0 needed to excite the core electron. As it approaches the ion, it gains kinetic energy in the Coulomb field of the ion, so that in close it has sufficient energy to excite the bound inner core electron, but if it does so, it has not enough energy to escape, and is bound in Rydberg level ($n\ell$) with binding energy ε. Since there is an infinite number of Rydberg states, the DR capture cross section versus electron energy consists of a corresponding series of resonances located just below the threshold for core-electron excitation.

Once the doubly excited state $s = q, n\ell$ has been formed, it has two options. If the energy of the excited core electron is again transferred to the Rydberg electron (the reverse of the capture process), autoionization occurs, the electron leaves, and the event appears as an elastic scattering. If the excited core electron radiates the excess energy away, dielectronic recombination results. The doubly excited state s undergoes a stabilizing radiative transition (satellite line $s \rightarrow k$) directly (or eventually via several cascades) towards a final state $k = r, n\ell$ that lies below the first ionization limit of the recombined ion:

$$Z^{+z}(q, n\ell) \rightarrow Z^{+z}(r, n\ell).$$

The satellite line $s \rightarrow k$ in ion Z^{+z} is slightly shifted to the long-wavelength side of the corresponding parent $q \rightarrow p$ transition in the recombining ion $Z^{+(z+1)}$ due to the presence of the spectator electron.

The dielectronic capture rate coefficient C_c (cm^3 s^{-1}) for the formation of the autoionizing state s is related by detailed balancing to the autoionization rate A_a (s^{-1}), while the branching ratio for the dielectronic recombination channel is $B \equiv A_r/(A_a + A_r)$, where A_r is the radiative decay rate, so that for the dielectronic rate coefficient we can write:

$$\alpha_{DR} = C_c B = \frac{C_c A_r}{A_a + A_r}, \tag{9}$$

with the capture rate given by (e.g. Dubau and Volonté (1980))

$$C_c = 2.07 \; 10^{-16} T_{[K]}^{-3/2}(g_s/g_1) A_a \exp(-E_s/kT) \; (\text{cm}^3 \; \text{s}^{-1}), \tag{10}$$

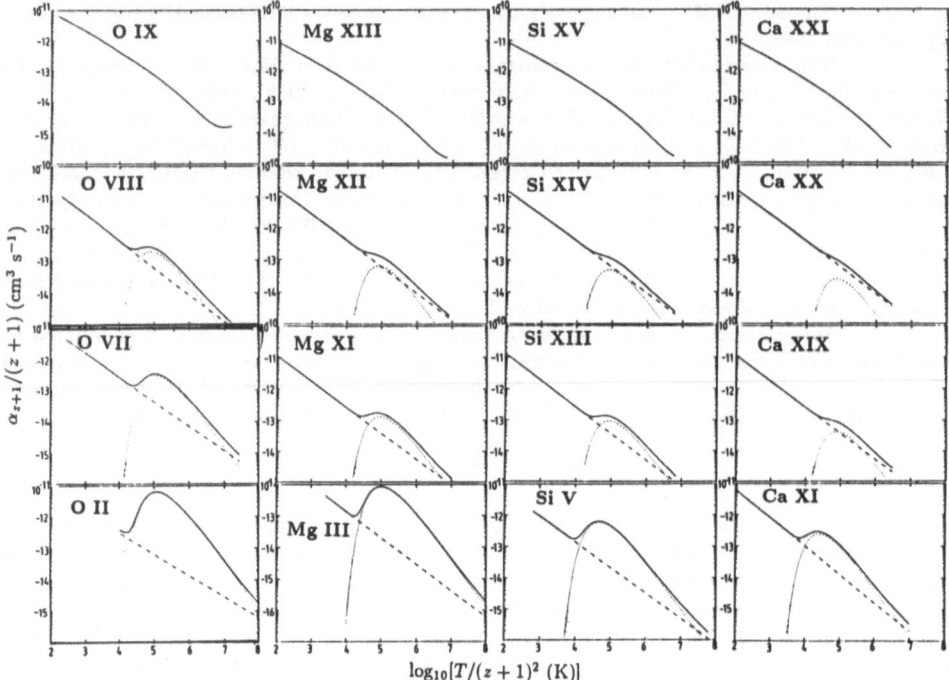

Figure 2. Reduced rate coefficients $\alpha_{z+1}/(z+1)$ for radiative recombination (RR) (dashed line) plus dielectronic recombination (DR) (dotted) (solid line indicates total rate) of ion $Z^{+(z+1)}$ as a function of reduced electron temperature $T/(z+1)^2$ from Arnaud and Rothenflug (1985) (e.g. $z=7$ for recombination of O IX, etc.).

where E_s is the energy difference between the autoionizing state s (with statistical weight $g_s = 2(2\ell+1)$) in ion Z^{+z} and the ground state 1 (stat. weight g_1) in ion $Z^{+(z+1)}$ (E_s is just the kinetic energy E_0-ε of the plasma electron being captured). Then we obtain for the DR rate coefficient (cm^3 s^{-1}) for recombination via level $s = n\ell$:

$$\alpha_{DR}(s) = 2.07 \ 10^{-16}T^{-3/2}\frac{g_s}{g_1}\frac{A_{as}A_{rs}}{A_{as} + A_{rs}}\exp(-E_s/kT). \tag{11}$$

By summing over all possible resonant autoionizing Rydberg states $s = n\ell$ we can obtain the total DR rate.

The behaviour of the DR rate depends sensitively on the branching ratio B, i.e. on the relative magnitudes of A_a and A_r. The A_a can be found by extrapolating the $p \to q$ excitation excitation cross section below threshold. They vary as n^{-3} (e.g. Hahn and LaGattuta 1988). For small n, A_a is likely to be much larger than A_r for the values of ℓ which contribute strongly, and α_{DR} is then $\sim A_r$. As along an isoelectronic sequence the radiative transition probability scales as $A_r \sim Z^4$ (for $\Delta n \neq 0$ transitions), it turns out that for low-Z ions (say $Z \lesssim 10$), A_a still exceeds A_r for large n values up to several hundred (and ℓ values up to ~ 6), so that many resonant

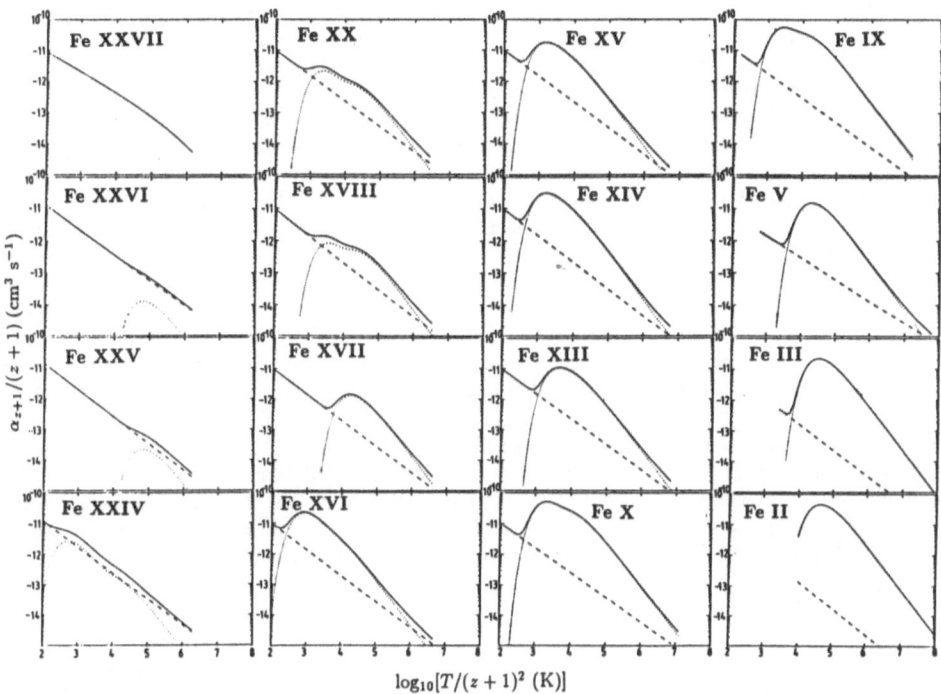

$$\log_{10}[T/(z+1)^2 \text{ (K)}]$$

Figure 2. (continued)

Rydberg states will contribute. However, as Z increases (say $Z \gtrsim 20$), A_a already becomes smaller than A_r for smaller n, and DR from lower states becomes increasingly important. For instance, recombination of Fe XXV ($Z = 26$) occurs primarily through the $n = 2$ shell (Bely-Dubau et al. (1979)). For a given n, ℓ $A_r(n, \ell)$ is nearly constant with n and ℓ (because the Rydberg electron is effectively only a spectator as far as the core-excited electron is concerned), whereas $A_a(n, \ell) \propto n^{-3}$. For low n and l, $A_a \gg A_r$, then $B \sim A_r n^3 \sim n^3$, and $\alpha_{DR}(n, \ell) \sim g_s A_a B \sim (2\ell + 1)A_r$. Furthermore, $A_a(n, \ell)$ decreases rapidly with ℓ (e.g. $\sim \exp(-0.25\ell^2)$), effectively cutting off the contribution from large ℓ (say above $\ell = \ell_c$ between 5 and 10), since for $A_a \ll A_r$, $\alpha_{DR} \sim A_a$. Thus, though (especially for low-Z ions) there is a large number of resonances which *could* contribute to DR (*if* their autoionization rates were larger) $\sim \sum_{\ell=0}^{n-1} 2(2\ell + 1) = 2n^2$, only ℓ's for $\ell \lesssim \ell_c$ (i.e., $A_a \gtrsim A_r$) will contribute $\sim 2(\ell_c + 1)^2$ (e.g. for n=20 and ℓ_c=10 only 200 states out of the possible 800). This is due to the strong decrease of A_a with ℓ.

Burgess (1964, 1965) was the first to compute DR rates and fit the results to a general semi-empirical formula (similar to the expression (11)), later corrected by Burgess and Tworkowski (1976). The expected accuracy was $\sim$30% and the formula is probably valid only near the temperature T_m at which the ion concentration peaks. For $T \gg T_m$ DR rarely matters, but for very low temperatures ($T \ll T_m$), it can be quite important in photoionized plasmas or in rapidly cooling gas. For low temperatures Nussbaumer and Storey (1983) have computed DR rates for abundant elements up to Si. The Burgess approximation is best for $\Delta n = 0$ inner electron transitions and modified versions of the formula have been proposed for $\Delta n \neq 0$ transitions (Merts et al. 1976,

Hahn 1985). Another modification, as was first pointed out by Jacobs *et al.* (1977) is the inclusion of additional autoionizing decay channels of the doubly excited state into *excited* states of the recombining ion (AI not always occurs only via the true inverse radiationless capture, i.e. into the channel associated with the ground state of the recombining ion). For certain ions this can drastically reduce the DR rate. Though the Jacobs rates are good for low-Z ions, the Jacobs correction may seriously underestimate the rates for high-Z ions (e.g. a factor $\sim$2–3 for recombination of Fe XVII (Smith *et al.* 1985)). It appears that the disagreement among various computations allows an overall accuracy of only $\sim$40% (see discussion by Raymond (1988)).

Unfortunately, for dielectronic recombination rates there is less experimental guidance than for ionization rates, because the crossed-beam experimental results suffer from the influence of electric fields experienced by ion beams crossing a magnetic field (e.g. Hahn 1985). This causes field ionization that limits the n to $\sim$ 30, but more important, Stark mixing of different l levels of a given n state, which may cause a dramatic enhancement of the experimental cross section for low-Z ions of up to an order of magnitude compared to the computed zero-field case (see also §3.3).

Arnaud and Rothenflug (1985) have evaluated many theoretical data and updated them in a number of cases. In Figure 2 I present their results for the DR and the total radiative plus dielectronic rate coefficients in a reduced form, i.e. $\alpha_{z+1}/(z + 1)$, in dependence of reduced temperature $\Theta = T/(z + 1)^2$. This scaling is appropriate for RR (see Eq. (8)), but seems also reasonably valid for DR (Summers 1974, Hahn 1980). It is seen that dielectronic recombination becomes increasingly important for lower Z and at higher temperatures, whereas for the higher ionization stages of high-Z ions radiative recombination becomes dominant.

3.3. Ionization balance

On the basis of the estimated accuracies of the rate coefficients we can consider the accuracy of the overall ionization balance that can be calculated by solving Eq. (2) for each pair of subsequent ions. Given the 20–40% uncertainties in both ionization and recombination rates, we expect the predicted ratio N_{z+1}/N_z of ion concentrations of adjacent ionization stages to be off by a factor, say $\sim$1.5–2. Fortunately, the ionization and recombination rates for the H- and He-like ions, which emit the lines that are among the strongest from hot astrophysical plasmas, are known more accurately than most of the other rates. In the case of ions which have a closed outer electron shell (e.g. He- and Ne-like ions) it doesn't matter very much because such ions cover a broad plateau in dependence of temperature (see Figure 3). This is caused by the fact that this ionization stage can easily be reached from the Li-like stage (only one outer electron with low binding energy) and persists long since the next ionization step towards the H-like stage suddenly needs a much ($\sim$4 times) higher ionization energy. This implies that the adjacent Li-like stage is quite critically dependent on temperature. But such ions can still exist when the plasma is not isothermal, but instead has a broad distribution of temperatures (as is often the case, e.g. in stellar coronae or in supernova remnants).

The effect of a systematic error can be readily estimated. Suppose that all ionization rates are overestimated by say a factor of two. Since the rates typically vary as $\exp(-\chi/kT)$, and since ion Z^{+z} concentrations typically peak at $\chi/kT \sim 20(n/z)^{0.8}$ (n is principal quantum number of ground state), i.e. in the range $\sim$2–7, say $\sim$ 5, the ion fraction peaks will be shifted to $\chi/kT \sim 5.7$, so that all ions will be shifted only by about 0.06 in $\log T$ (i.e 15% in T), which does not look too serious.

In the literature many ionization balance calculations have been reported. References up to 1984 are cited in Mewe (1984), but a few of them are mentioned here: Jordan (1969, 1970), Jacobs (1977, 1978–80) (JA), Mewe and Gronenschild (1981), Mewe (1988) (MG), Raymond and Smith (1977), but later updated (see Raymond 1988) (RS), and Arnaud and Rothenflug (1985) (AR). How do these calculations compare? In Figure 3 I compare the AR results for the ion fractions $N_{Z,z}/N_Z$ (where N_Z is the total element abundance) for iron in dependence of the electron

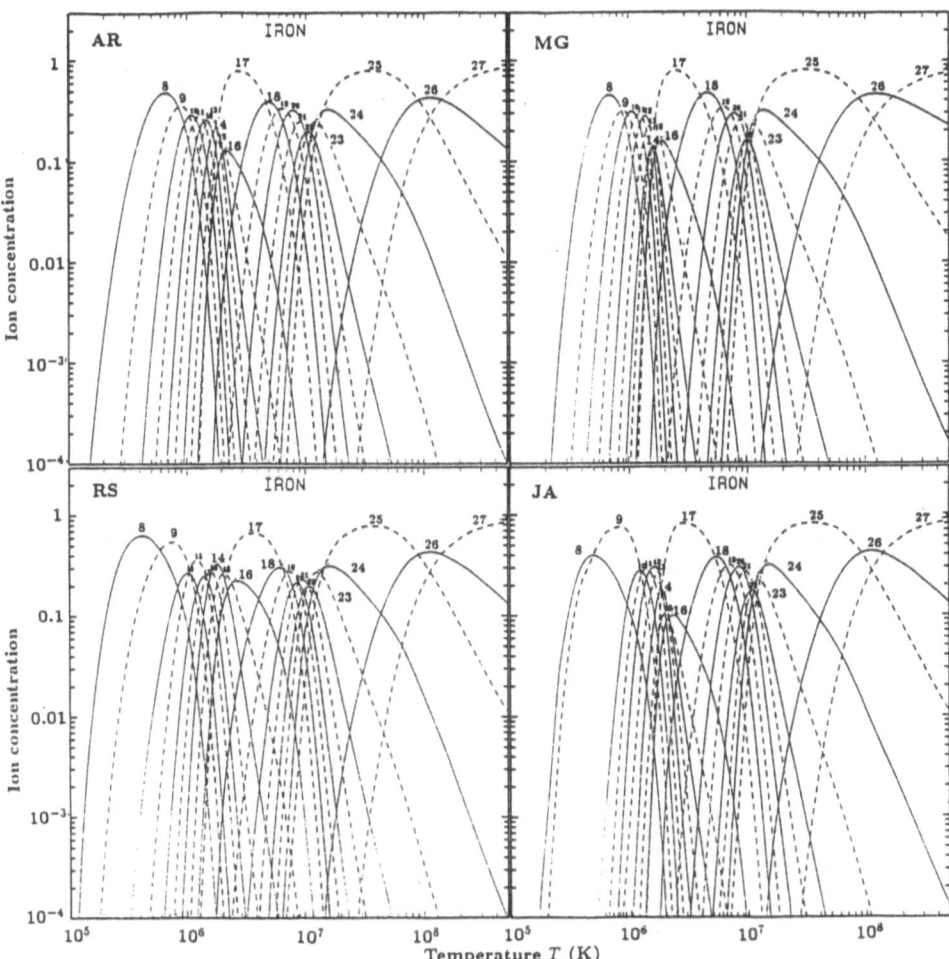

Figure 3a. Ion fractions as a function of temperature for iron as calculated by AR, RS, MG, and JA (see text). Ion stages are designated by numbers, e.g. 8 indicates ion Z^{+7} (Fe VIII), etc.

temperature T with various other calculations (RS, MG, and JA). It is seen that the overall shapes of the curves are quite similar and that the shifts of the ion peaks are generally limited within about $\Delta \log T \approx 0.1$. However, the peak values may differ by about 10–30% and sometimes up to ~50%. The differences mainly result from different dielectronic recombination rates. Comparing the AR and RS results for iron, we notice that the discrepancies are worst for the lower stages of ionization (e.g. Fe VIII and Fe IX), but nearly vanish for the simpler and more thoroughly studied ions of the He- and H-like sequences. Generally, the discrepancy is only about 0.1 in $\log T$, which for many diagnostic purposes is good enough. On the other hand, for a given temperature, the ion abundances may differ by a factor up to about two, especially for the ions around the He-like and

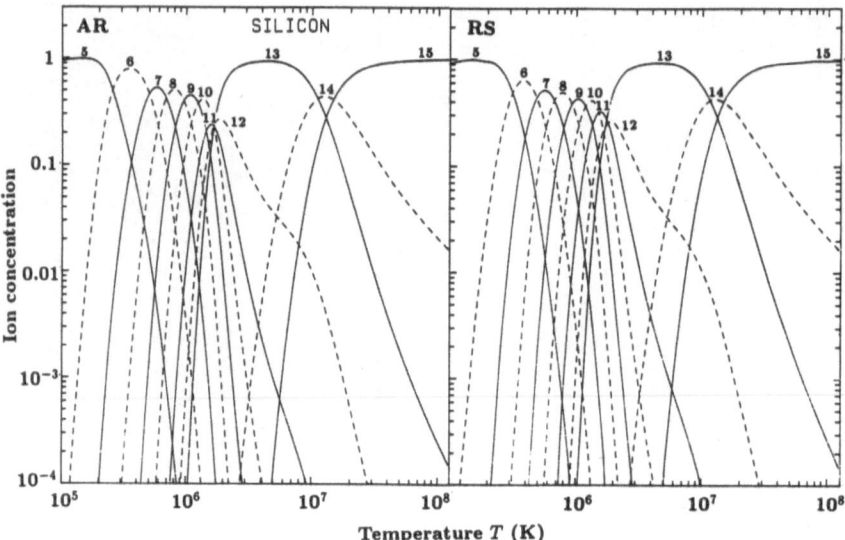

Figure 3b. Ion fractions for silicon.

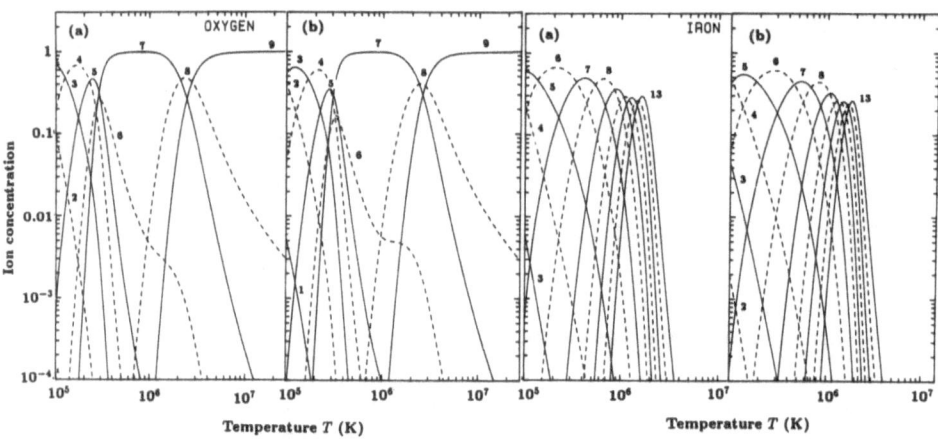

Figure 4. Ion fractions for O and Fe as calculated by AR without (a) and with (b) correction for ℓ level Stark mixing.

Ne-like "plateaus". For diagnostics of tenuous plasmas the AR ionization balance computations can be used with reasonable accuracy.

However, a few words may be said about possible effects of electric fields on DR rates. Experimental breakthroughs occurred about five years ago when DR cross sections could be measured for a few singly ionized atoms (e.g. Mg^+ and Ca^+) in crossed-beam experiments (e.g. review papers cited above and also Müller *et al.* 1987 and Dunn 1986). The measured cross-section values were a factor ~5–10 larger than predicted ones. A possible explanation of this strong discrepancy was given in terms of the effect of an electric field that causes Stark mixing of different ℓ levels of a given n state. In the experiment, a magnetic field B ~200 Gauss was used to focus the electron beam. The ions which cross the electron beam with velocities v ~10^7 cm/s then experience in their rest frame a Lorentz field $E = 10^{-8}vB \approx 20$V/cm, sufficient to produce full Stark mixing. As A_a decreases with ℓ the result of mixing high and low ℓ levels is that A_a is increased for high ℓ and decreased for low ℓ, thus flattening the $A_a(\ell)$ versus ℓ curve, so that more states effectively participate in the recombination process, and the DR cross section is increased. In the actual plasma environment such fields can be generated (see §5.4.). Though the effects of electric fields are yet to be fully incorporated into future DR rate calculations, it is interesting to visualize this effect on the hand of a simple schematic model as given by Müller *et al.* (1987). The typical enhancement factor for the DR recombination of ion $Z^{+(z+1)}$ due to full Stark mixing I approximate by $f = 1 + 10(z+1)^{-1.15}$. (We note that the effects are biggest for the lower ions as here many n states take part in the DR process). In Figure 4 we show the results for the oxygen and iron ionization balance if we correct the DR rates in the AR calculations by the factor f. This may show that this effect can be important for the ions that are formed at lower temperatures ($\lesssim 2$ MK).

4. SPECTRA

Much of the temperature sensitivity of the soft X-ray spectra emitted by optically thin plasmas is associated with the ionization structure. We consider briefly the processes that produce the line and continuum radiation.

4.1. Line radiation

The spectral lines that dominate the soft X-ray spectrum and the cooling of astrophysical plasmas at temperatures up to ~10 MK are mainly excited by electron collisional excitation from the ground state. The formation of a particular spectral line transition $2 \to 1$ emitted by ion Z^{+z} from element of atomic number Z is as follows. The total depopulation rate of the upper level (2) is equal to $N_2 \sum_i A_{2i}$, where N_2 is the density (cm^{-3}) of ions in the upper level 2 and A_{2i} is the probability (s^{-1}) of a spontaneous radiative transition from level 2 towards a lower level i (collisional de-excitation is neglected here). This will be balanced by electron collisional excitations from the ground state (g) at a rate $n_e N_g S_{g2}$, so that the volume emissivity P_{21} (phot cm^{-3} s^{-1}) is given by

$$P_{21} = N_2 A_{21} = n_e N_g B S_{g2} = (N_H/n_e) A_Z n_e^2 \eta_z B S_{g2}, \qquad (12)$$

where S_{g2} is the rate coefficient (cm^3 s^{-1}) for collisional excitation $g \to 2$, $B = A_{21}/\sum_i A_{2i}$ is the radiative branching ratio, N_g the ground state population (cm^{-3}) of ion Z^{+z}, $A_Z = N_Z/N_H$ the abundance of element Z relative to hydrogen (H), η_z the fraction of ions from element Z in ionization stage z and N_H/n_e can be taken equal to 0.85 for a plasma with cosmic abundances. The electron impact excitation cross section is given by Eq. (5) with $\alpha = E_0/(\omega E_H)$ and $Q_{red}(U) = \Omega(U)/U$ (e.g. Mewe 1972). Here Ω is the collision strength for excitation which is a function of $U = E/E_0$, the incident electron energy in terms of the excitation threshold energy E_0 and ω is the statistical weight of the initial (usually ground) level. Integration over the Maxwell distribution (see Eq. (4)) yields for the excitation rate coefficient of a spectral line:

$$S = 8.62 \ 10^{-6} \ \frac{\bar{\Omega}(y)}{\omega T_{[K]}^{1/2}} \exp(-y) \ (\text{cm}^3 \text{ s}^{-1}), \qquad (13)$$

where $y = E_0/kT$ and $\bar{\Omega}(y)$ is the collision strength averaged over the Maxwellian electron energy distribution. A time ago I have introduced (e.g. Mewe 1972, Mewe and Schrijver 1978, Mewe 1988) a parametrized interpolation formula for Ω that represents the correct behaviour near threshold and asymptotically at high energies and that can be integrated analytically over the Maxwellian electron energy distribution. We have used such an expression to fit to many available theoretical and experimental data for excitation cross sections or rates. When no data are available one can make use, for a rough estimate, in the case of optically allowed dipole transitions, of the Gaunt factor formula:

$$\bar{\Omega} = \frac{8\pi}{\sqrt{3}} \frac{E_H}{E_0} \omega f \bar{g} = 197.3 \; \omega f \bar{g} E_{0[\text{eV}]}^{-1}, \tag{14}$$

where f is the absorption oscillator strength and $\bar{g} \equiv \bar{g}(y)$ the average Gaunt factor. With an accuracy of a factor of two or so we may use for $\bar{g}$ typical values of $\sim$0.2 for $\Delta n \neq 0$ transitions and $\sim$1 for $\Delta n = 0$ transitions. More involved expressions for the line formation in an optically thin plasma, including possible cascade effects on the excitation rate and also contributions from recombination ($\propto \eta_{z+1}$), innershell ionization ($\propto \eta_{z-1}$) and unresolved dielectronic recombination satellite lines ($\propto \eta_z$) are given elsewhere (Mewe and Gronenschild 1981, Mewe et al. 1985). For an isothermal plasma the line intensity integrated over the whole source volume V turns out to be proportional to the well-known emission measure $\varepsilon \equiv \int n_e^2 dV$ of the source, but for a more complicated temperature structure in the plasma we should take into account a differential emission measure distribution (e.g. Lemen et al. 1989).

Though in a number of cases the Gaunt factor approximation method gives reasonably accurate results, more accurate values are needed, especially for the weaker lines and optically forbidden transitions. Raymond (1988) gives a brief discussion of the accuracy of several computational methods such as Coulomb-Born (CB) and Distorted Wave (DW) approximations, and the more accurate Close Coupling (CC) method, which properly takes into account the dominant resonances near the threshold of the excitation mentioned in connection with DR (for a theoretical review see Seaton (1975), for a review on crossed-beam experiments see Dolder and Peart (1986), and for a review on plasma measurements of atomic rates see Griem (1988)). CB tends to overestimate the collision strength near threshold by $\sim$20–50%, while DW gives better results, especially for high-Z ions. Few accurate CC results are available for He-like ions (Kingston and Tayal 1983). Recent compilations of excitation collision strengths are reported by Aggarwal et al. (1985) and by Gallagher and Pradhan (1985). For many of the strongest X-ray lines, e.g. from H- and He-like ions and Li- to Ne-like ions, the collision strengths are known with better than 20% accuracy. Near threshold, the strong resonances may spoil the accuracy. However, for applications of rate coefficients in plasma diagnostics, we should not worry too much about this, because in averaging the cross section over the electron energy distribution we smooth out these resonances for a great deal.

4.2. Continuum radiation

Continuum emission in optically thin plasmas is produced by bremsstrahlung, recombination, and sometimes two-photon processes in H- and He-like ions. The total continuum emissivity can be written as the energy emissivity at wavelength λ per unit wavelength interval (Mewe et al. 1986):

$$P_c(\lambda, T) = 2.051 \; 10^{-22} n_e^2 G_c \lambda^{-2} T^{-1/2} \exp[-143.9/(\lambda T)] \; (\text{erg cm}^{-3}\text{s}^{-1}\text{Å}^{-1}), \tag{15}$$

or, alternatively, as the photon number emissivity at photon energy E per unit energy interval:

$$P_c(E, T) = 1.032 \; 10^{-14} n_e^2 G_c E^{-1} T^{-1/2} \exp[-11.6E/T] \; (\text{phot cm}^{-3}\text{s}^{-1}\text{keV}^{-1}), \tag{16}$$

where n_e is the electron density (cm^{-3}), T in MK, λ in Å, and E in keV. The factor G_c is the average total Gaunt factor representing the contributions from free-free (ff), free-bound (fb) and two-photon (2γ) emission. For an easy estimate we can make use of the following approximation (Mewe et al. 1985):

$$G_c \simeq 27.83(T + 0.65)^{-1.33} + 0.15\lambda^{0.34}T^{0.422}, \tag{17}$$

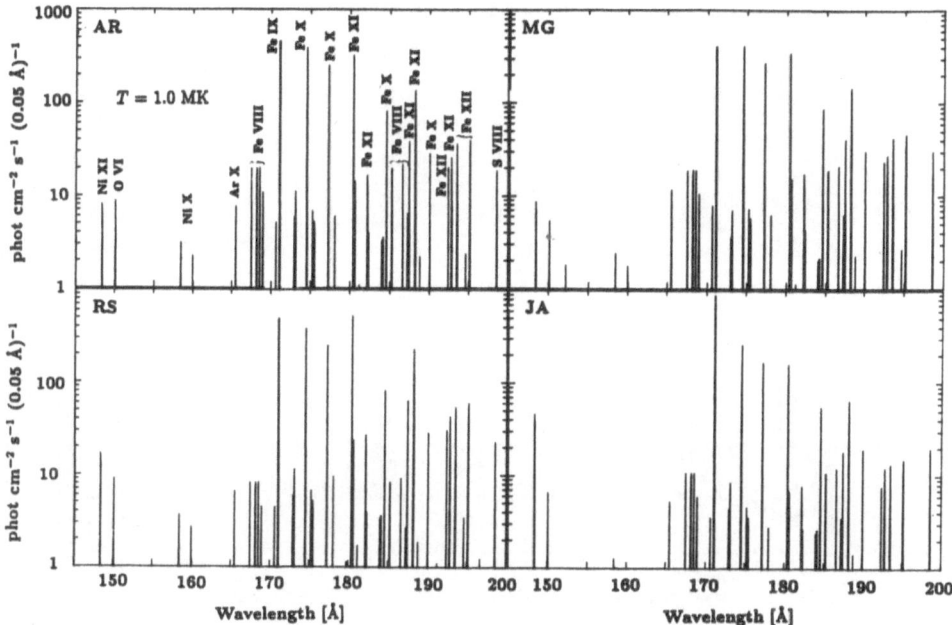

Figure 5a. Comparison of various X-ray spectra calculated for different ionization balances (AR: Arnaud and Rothenflug, RS: Raymond and Smith, MG: Mewe and Gronenschild, JA: Jacobs), for a temperature $T = 1$ MK, reduced emission measure $\varepsilon/d^2 = 10^{50}$ cm^{-3} pc^{-2} (d is source distance), and wavelength region 140–200 Å. The spectra are binned in 0.05 Å intervals. The most prominent lines are labelled with the corresponding ion species. In all calculations excitation rates from Mewe *et al.* (1985) are used.

that gives a good approximation (within ~10–20%) at $T \geq 3$ MK and a reasonable approximation (~30–50%) for 0.2MK $\leq T < 3$ MK for a plasma with cosmic abundances from Allen (1973). However, for wavelengths below the O VIII edge at 16.8 Å, better results can be obtained by properly taking into account the contributions from fb and 2γ emission (Mewe *et al.* 1986). For a further discussion of the continuum emission I may refer to Raymond (1988).

4.3. Calculated spectra and comparison of calculations

Raymond (1988) has compared various spectral calculations and discussed the differences resulting from different ionization balance calculations and from different treatments of the line excitation. He comes to the conclusion that for the strongest X-ray lines from astrophysical plasmas, those of the H- and He-like ions, the agreement generally approaches about 20% (which is important because the He-like lines can be used for density diagnostics), whereas for other cases (e.g. iron lines around 10–12 Å and silicon and sulphur lines around 40–50 Å) discrepancies of a factor of two may exist.

Because the ionization balance strongly determines the overall appearance of the X-ray spectra I made a few calculations with the same (i.e. my) excitation rate data, but using different

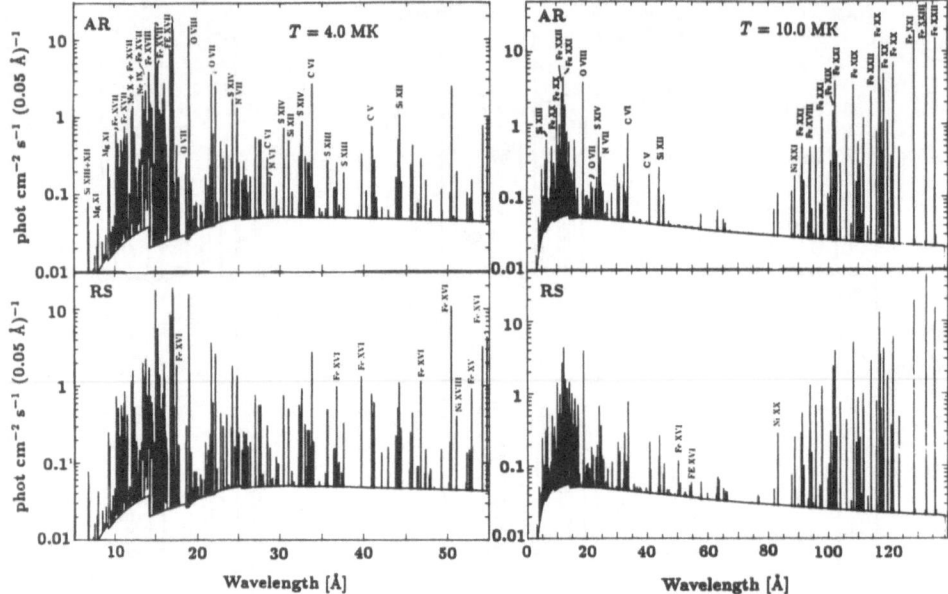

Figure 5b. Comparison of AR and RS calculated spectra for $T = 4$ and 10 MK in the wavelength regions 5–55 Å and 3–140 Å, respectively.

ionization balance calculations. In Figure 5 I compare for a number of temperatures the results in various wavelength regions the results using AR, RS, MG, and JA ionization balance calculations. I used a wavelength resolution of 0.05 Å, which is typical for spectrometers to be flown at the end of this century on the *AXAF* and *XMM* missions. A glance at these pictures confirms the above conclusions. I note that for a worse resolution (e.g. a few Å as for the *EXOSAT* transmission gratings) or for broad-band measurements, the agreement will be somewhat better. Raymond (1988, 1989) has made an attempt to test the existing models by comparing with high-resolution solar X-ray observations. From his comparison of a model calculation with the composite X-ray spectrum of a solar flare he concludes that uncertainties in atomic rates or the breakdown of simplifying model assumptions may lead to errors on the order of ~50 % in the predicted line strengths. A comparison of coronal models with *EXOSAT* spectral X-ray observations with moderate resolution (~3 Å) of a few late-type stars by using differential emission measure distributions (Lemen *et al.* 1989) show a satisfactory agreement, but also indicate that for the interpretation of the future *AXAF* and *XMM* X-ray spectra better model calculations will be needed.

4.4. Temperature diagnostics

Figures 5 show that the soft X-ray spectra sensitively depend on the ionization structure throughout the temperature range 0.1–100 MK so that the mere detection of many lines will provide an accurate "thermometer" for the source emission region. In many sources, however, the X-ray spectrum will not be a unique function of one single temperature, but instead will be determined by a distribution in temperature across the emission region. In those cases, multi-temperature fits to the entire spectrum can yield the differential emission measure (DEM) distribution appropriate to the source. Such a detailed DEM analysis has been performed on the *EXOSAT* transmission grating spectra of late-type stars to constrain the basic properties of stellar magnetic loops (Lemen *et al.* 1989, Schrijver *et al.* 1989). In many cases the DEM analysis will be the primary handle one can use to test models for the physics of the source emitting region. Finally, a sensitive temperature diagnostics is provided by measurements of the ratio of dielectronic recombination satellites to the resonance line, which goes as $T^{-1}\exp[(E_0 - E_s)/kT]$ (see Eqs. (11) and (12), and Mewe (1988)).

4.5. Elemental abundance diagnostics

Fits of theoretical models to the measured spectra, assuming certain abundances for the elements yield the temperature structure of the source plasma. From systematic differences between the fits and measurements in principle we can derive the abundances for those spectral lines for which oscillator strengths and excitation cross sections are sufficiently well known. For instance, to diagnose elemental abundances in a relatively model-independent way, one can measure intensity ratios of pairs of strong (e.g. K- and L-shell) lines of ions from different elements that have comparable ionization potentials (such ions usually coexist in close proximity with respect of temperature). For example, the N/O ratio can be a signature of nuclear processing for stellar and accreting sources (Kahn *et al.* 1984). The Si/Fe and S/Fe ratios may be used to study the evolution of stellar ejecta from young supernova remnants, while the O/Fe ratio can be important in constraining the origin of hot gas in clusters of galaxies and the gravitational settling of heavy elements in the overall cluster potential (Abramopoulos *et al.* 1981).

5. DEVIATIONS FROM THE CORONAL MODEL APPROXIMATION

Apart from the uncertainty level of the basic atomic parameters (typically ~30–50%), other uncertainties can arise from the simplifying assumptions we have made in the coronal model. I will briefly consider several of them.

5.1. Criteria for the optically thin approximation; optical depth effects

In the coronal model we assume that the plasma is optically thin, so that the observed intensities and plasma emissivities are directly proportional. If, once some radiation is absorbed, radiative transfer must be considered.

In the extreme case of very densy plasmas, e.g. those present in the inner regions of accretion flows on compact objects such as hot white dwarfs and neutron stars, the source is optically thick to both continuum and line radiation. The spectrum will resemble, at very high optical depths, blackbody emission.

At intermediate optical depths, the spectral formation is influenced by complicated radiation transfer effects as well as by fundamental atomic processes. Discrete spectral structure is expected which can provide much information about the source (Ross 1979). For X-ray emitting plasmas, Compton scattering plays a significant role as well where transfer through the scattering plasma will broaden and shift line profiles and alter the continuum distribution, depending on temperature and column density (Lightman *et al.* 1981).

For the optically thin approximation we may apply the criterion that the intensity of a given type of radiation should not differ from the value obtained from the optically thin approximation by more than 10%, which can be expressed as (e.g. Cooper 1966):

$$\tau_\lambda(D) \lesssim 0.2, \tag{18}$$

where $\tau_\lambda = \alpha_\lambda D$ is the optical depth at wavelength λ, α_λ the linear absorption coeffcient (cm^{-1}) and D the typical dimension (cm) of the (homogeneous) plasma. We evaluate criterion (18) for the following radiation processes which may contribute to the optical depth (ignoring stimulated emission or stimulated recombinations): (i) photoabsorption of line radiation, (ii) of recombination radiation, (iii) of bremsstrahlung, and (iv) scattering by free electrons (Cooper 1966, Wilson 1962).

(i) For line radiation resonance line absorption is most serious. For a Doppler broadened line profile, the application of criterion (18) to the central wavelength λ (Å) gives (expressing the density $N_{Z,s}$ of the absorbing ion Z^{+s} of element Z in terms of the electron density n_e (cm^{-3}), the ion fraction $\eta_{Z,s}$ and the element abundance $A_Z = N_Z/N_H$, i.e. $N_{Z,s} = 0.85 n_e \eta_{Z,s} A_Z$ for a plasma with cosmic abundances):

$$Dn_e \lesssim 2 \; 10^{13} \frac{\sqrt{T_i/M_i}}{\lambda f \eta_{Z,s} A_Z} \; (\text{cm}^{-2}),$$

where f is the absorption oscillator strength of the line, $T_i(K)$ the ion temperature and M_i the ion mass number. For example, for $\lambda=20$Å, $f=0.5$, $\eta=0.5$, $A=10^{-4}$, $T_i=3 \; 10^6$ K, and $D=10^8$ cm, we have $n_e \lesssim 2 \; 10^{11}$ cm^{-3}, which may emphasize to consider, for instance, very carefully opacity effects in solar and stellar flare and active region conditions. This criterion will be relaxed when additional line broadening from Stark ($\Delta\lambda_S \sim 4 \; 10^{-18}\lambda^2 n_e^{2/3}(z+1)^{-1}$) or Zeeman effect ($\Delta\lambda_Z \sim 10^{-12}\lambda^2 B$) becomes comparable to the Doppler broadening ($\Delta\lambda_D \sim 7 \; 10^{-7}\lambda\sqrt{T_i/M_i}$) at sufficiently large plasma densities or magnetic fields (B in Gauss, line widths $\Delta\lambda$ and wavelengths in Å).

(ii) For photoexcitation we obtain approximately for transitions from the ground state to the absorption edge (where the absorption is maximum):

$$Dn_e \lesssim 5 \; 10^{16} \frac{(z+1)^2}{\eta_{Z,s} A_Z} \; (\text{cm}^{-2}).$$

(iii) Using the free-free absorption coefficient (e.g. Spitzer 1962, Cooper 1966), criterion (18) reduces to

$$Dn_e^2 \lesssim 1.5 \; 10^{46} T^{1/2} \lambda^3 G_c^{-1} \; (\text{cm}^{-5}),$$

where T is the electron temperature (K) and $\bar{G}_c$ the average Gaunt factor (Mewe et al. 1986). The criterion is best evaluated for the region of maximum bremsstrahlung emission, i.e. for $h\nu \sim 2kT$, or $\lambda T \sim 10^8$Å K (here $G_c \sim 1$ for a plasma with cosmic abundances (Mewe et al. 1986)).

(iv) For electron scattering to produce no appreciable optical depth, the Thomson cross section ($\sigma_T = 6.65 \; 10^{-25}$ cm^2) gives

$$Dn_e \lesssim 3 \; 10^{23} \; (\text{cm}^{-2}).$$

With the above criteria (i)–(iv) we can estimate when the intensity at a given wavelength may differ by more than 10% from the optically thin value. Examination of these criteria, indicates that, criterion (i) is often most severe, and criterion (ii) will be so in photoionized plasmas, while the latter two criteria will break down only for very dense plasmas in compact X-ray sources (accretion-powered onto a compact neutron star or black hole) where Compton scattering plays a significant role.

5.1.1. Effects of resonance line scattering

When the plasma becomes optically thick ($\tau \gtrsim 1$) the effect on the intensity of the resonance line is determined by the processes competing with the spontaneous radiative decay to the ground level. In a high-density plasma a resonance photon will be completely destroyed after an absorption, but in a low-density plasma (with $\tau > 1$) the situation can exist that each time a photon is absorbed, only a small fraction b ($\ll 1$) is destroyed. Since on the average a photon is absorbed and re-emitted ("scattered") $\sim \tau$ times before escaping, the plasma can still be considered effectively optically thin as long as $b\tau \ll 1$ (e.g. Hearn 1966). E.g. for Ly α where the loss occurs through electron excitation to the next higher level $b \sim 2 \ 10^{-23} \lambda^3 T^{-1/2} n_e$, but for Ly β the situation is completely different because here the main process competing with Ly β emission is Balmer α emission, hence $b \sim 0.44$, independent of temperature and density.

If the emitting region is not spherical (e.g. a long thin coronal magnetic flux tube), the effects of resonant scattering can drastically enhance the intensity in the direction of the shortest plasma dimension where the re-emitted photons (from absorption in the longitudinal direction) more readily escape (e.g. Acton and Brown 1978, Sylwester et al. 1986, and discussion by Raymond 1988). In principle one can extract from the spectra information about the geometry of the source, although the interpretation may be quite complicated.

5.1.2. Nebular model for photoionized plasmas

Breakdown of criterion (ii) occurs in an important class of cosmic X-ray sources involving pho-toionized nebulae. Examples include accretion-powered sources like X-ray binaries, cataclysmic variables, and active galactic nuclei, where a central X-ray emitting region is surrounded by a cooler, partially ionized medium. Other examples are early-type stars in which X-rays produced in a hot corona must be transferred outward through a stellar wind, and stellar sources located near the nucleus of normal galaxies which are surrounded by the local interstellar medium.

In all these cases photoelectric absorption edges will substantially modify the emergent X-ray spectrum particularly at longer wavelengths. High-resolution spectral measurements of the strength of K-shell absorption edges in combination with emission lines produced by recombination, provide information on the geometry of the medium surrounding the source along the line of sight. The nebular model (e.g. Holt and McCray 1982, McCray 1982, 1984, Kallman and McCray 1982) is the X-ray analogue of a planetary nebula, in which a central continuum source ionizes the surrounding gas. The gas may be optically thick to photoabsorption but not to electron scattering. The ionization and temperature structure of the gas are established by a stationary balance between photoionization (collisional ionization can be neglected) and heating due to the central X-ray source and, on the other hand, (radiative plus dielectronic) recombination and charge exchange and cooling of the gas. When the gas is optically thin, the local radiation field is determined by geometrical dilution of the source spectrum. Then the local state of the gas (at radius R from the central X-ray source) can be parametrized in terms of the scaling parameter $\xi = L/nR^2$ (L is total luminosity of the central source, n is the local gas density; when electron scattering also is important, the ionization parameter is defined as $\xi = L/nR^2\tau$, where τ is the electron scattering optical depth (Ross 1979, Fabian and Ross 1981)).

The model can be applied to a wide variety of astrophysical X-ray sources and ranges from optically thin to optically thick in the photoionization continuum of abundant elements. In the latter case, the transfer of continuum radiation should be taken into account which yields one additional parameter $(Ln)^{1/2}$ which characterizes the continuum optical depth at a given value of ξ.

It may be instructive to contrast the nebular model with the coronal model. In the latter model the mechanism for heating the gas is not specified, but the heat input is coupled directly to the ions and free electrons. The parameters characterizing the coronal model are (electron) temperature, element abundances, and emission measure. At a given temperature only one or two ionization stages of a given element are abundant. In the nebular model the temperature of the gas is not a free parameter, but instead is determined by absorption and emission of radiation in the gas. The elements are primarily ionized by innershell photoionization. As a result a wider range of ionization stages of a given element can simultaneously occur and the elements are more highly ionized ("overionized") at a given temperature than they would be in the coronal model. The emergent X-ray spectrum consists of the central continuum with a low-energy cutoff due to

Figure 6. Model spectra of Fe XVII–XIX from 13–18 Å with 0.052 Å resolution calculated by Liedahl *et al.* (1989). Both spectra are plotted on the same scale, wavelength in Å, intensity in arbitrary units. *Top:* Collisional equilibrium with $T = 5.75$ MK for the electron temperature. The strong lines at 13.5, 14.2, and 15.0 Å (from Fe XIX, XVIII, and XVII, respectively) are collisionally excited $3d$ lines. *Bottom:* Recombination-dominated spectrum at $T=0.11$ MK and ionization structure appropriate to an X-ray photoionized nebula. The prominent lines at 15, 16, and 17 Å (from Fe XIX, XVIII, and XVII) are all recombination-cascade-populated $3s$ lines, which cannot be excited by electron collisions at this low electron temperature. The electron density is in each case 10^{11} cm^{-3}.

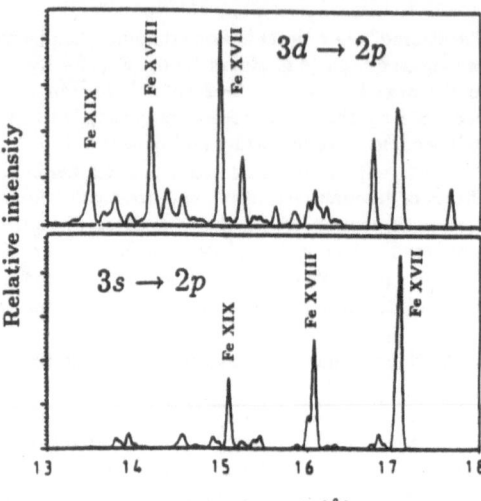

photoabsorption and emission lines due to recombination and fluorescence (in the latter case e.g. K-shell lines from ions with nearly stripped L-shells with reasonable radiative yields such as Li-, Be- and B-like ions (Chen 1986, Chen and Crasemann 1987)) in the nebula, typically near the continuum absorption cutoff.

X-ray photoionized plasmas can differ from collisionally ionized plasmas with similar ion concentrations. Because the photoionized plasma is *overionized* relative to the electron temperature, the excitations of important lines are dominated by recombination, photoexcitation, and cascades as opposed to collisional excitation and dielectronic recombination. This has a drastic effect on the emergent spectrum, as illustrated in Figure 6, where we have plotted the emergent spectra for Fe XVII - XIX ions assuming coronal and photoionized conditions respectively (Liedahl *et al.* 1989). As can be seen, it will be straightforward to distinguish photoionized plasmas from coronal plasmas with such spectra. The relative line intensities detected in the photoionized case can be shown to be sensitive functions of the density and geometry of the emission regions and of the spectral shape of the photoionizing continuum.

5.2. Effects from high density

5.2.1. Transition from coronal to thermal model

The ionization distribution in a low-density coronal plasma is very different from that described by the Saha equation in a very dense plasma in thermodynamic equilibrium, which e.g. for a hydrogenic plasma reads (e.g. Allen 1973, Griem 1964) (Z^{+z} is hydrogenic ion, $Z^{+(z+1)}$ is bare nucleus of charge number $z + 1 = Z$), omitting subscript Z:

$$N_{z+1}/N_z = 2.42 \ 10^{15} T^{3/2} n_e^{-1} \exp[-1.578 \ 10^5 Z^2/T], \qquad (19)$$

where T in K and n_e in cm^{-3}. Whereas in the coronal approximation N_{z+1}/N_z is independent of n_e, this ratio varies as $\sim n_e^{-1}$ in the high-density limit. Clearly, in the latter case Eq. (19) predicts for low-Z ions at lower densities a much higher degree of ionization (i.e. larger N_{z+1}/N_z) compared to the coronal model in which the ions of a given ionization stage can exist at much higher temperatures.

From approximate hydrogenic formulae (Wilson 1962, Griem 1964, McWhirter 1965) it can be estimated whether corona conditions are valid or not. The coronal domain is roughly bounded by the condition (deviations from coronal ionization balance less than about a factor of two):

$$n_e \lesssim 4\ 10^4 (z+1)^2 T^2\ (\text{cm}^{-3}),\tag{20}$$

whereas the extreme high-density limit is reached at densities approximately

$$n_e \gtrsim 1.4\ 10^{15}(z+1)^6 T^{1/2}\ (\text{cm}^{-3}),\tag{21}$$

where T is the electron temperature in K. The latter condition for complete Local Thermal Equilibrium (LTE) may be relaxed by about two orders of magnitude if the resonance transitions are self-absorbed and become optically thick, because this effectively reduces the radiative population rate (e.g. Mewe 1967, 1970).

In the intermediate region between the coronal and thermal domains, the situation is complicated, and the problem is to solve the differential equations describing the population and depopulation of many bound levels (e.g. reviews by Cooper (1966) and Wilson (1962)). Moreover, if the plasma is neither optically thin or thick towards the resonance lines, the rate coefficients depend also on the optical depths and an exact computation would require a solution of the level population rate equations coupled with the equations of radiative transfer.

Because collisional rates between bound levels increase with principal quantum number, and corresponding radiative decay rates decrease, the upper bound states near the ionization limit are strongly collionally coupled to the continuum of free electrons and weakly coupled to the ground state, whereas for the lower levels the reverse is the case. From the work by Wilson (1962) the physical picture emerged that the thermal equilibrium in the continuum extends down to the upper bound levels owing to the high collisional rates between the upper bound levels and the continuum. Since this is imposed on the upper levels by the free electrons, the thermal equilibrium of the bound levels is linked to the continuum and their populations are given by Saha-Boltzmann equations. There is therefore a certain level n_t in the ion, for which upward collisional rates balance the downward radiative decay rates. It is known as the *thermal limit* (TL) because it defines the limit above which the levels are in thermal equilibrium with the continuum and below which the level distributions are approximately coronal. For low n_e, the TL is very close to the ionization limit (coronal approximation valid), for increasing n_e the TL drops down until, at sufficiently high densities it reaches the ground level ($n_t=1$, LTE). For a hydrogenic ion Z^{+z} in an optically thin plasma the thermal limit is given by (Wilson 1962):

$$n_t^7 = 1.4\ 10^{15}(z+1)^6 T^{1/2} n_e^{-1}.\tag{22}$$

For $n_t=1$ Eq. (22) indeed corresponds to Eq. (21).

5.2.2. High-density effects on rate coefficients

Any transition to levels below the TL is equivalent to recombination since the hole that is left is immediately populated by a collisional transition from the continuum; similarly, any excitation from low levels to the thermal levels above the TL is equivalent to ionization, since further excitation and ionization dominates downward radiative decay. If the TL is low enough (say $n_t \lesssim 5$, corresponding to $n_e \gtrsim 10^{12}(z+1)^{7.5}$) effects of density (e.g. stepwise ionization) on the ionization and recombination rate coefficients should be taken into account. In certain cases where metastable levels are important (e.g. low-Z Li-, Be-, and B-like ions) this can occur already at much lower densities (e.g. Vernazza and Raymond 1979).

Bates *et al.* (1962a,b) were the first to make calculations fully taking into account the combined effects of collisions and radiative decay for the case of an optically thin *or* thick hydrogenic plasma. They expressed the net rates of radiative recombination and ionization in terms of binary coefficients which they called "*collisional-radiative*" recombination or ionization coefficients. For the very complicated case in which also trapping of resonance radiation becomes important I made some rough estimates for these rate coefficients with semi-empirical approximation formulae (based on the results of Bates *et al.*) for a highly simplifyed hydrogenic two-level + continuum scheme (Mewe 1970, also 1988). The transition between low- and high-density cases occurs at different densities, roughly at the plasma dimension $D \sim 3 \; 10^{27}(z+1)^{10.5}n_e^{-2}$ cm, which represents the effect of resonance radiation trapping in the case of Doppler broadening. In the high-density limit the ionization rate approaches for this model the excitation rate to the first excited level, the recombination rate approaches the 3-body radiative recombination rate (which is $\propto n_e$), and the ratio of the two coefficients reduces to the Saha equation.

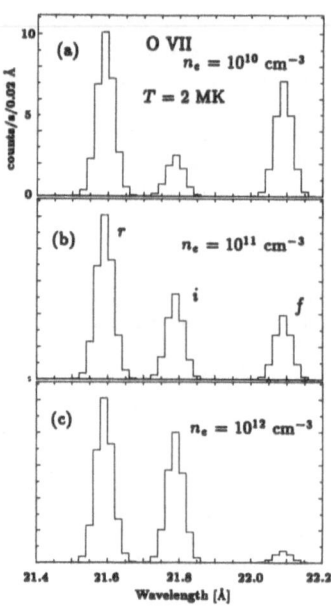

Figure 7. O VII triplet at 22 Å as observed with the model *XMM*reflection grating spectrometer for temperature $T = 2$ MK, electron densities $n_e = 10^{10}$, 10^{11} and 10^{12} cm^{-3} (panels a,b,c), and reduced emission measure $\varepsilon/d^2 = 10^{52}$ cm^{-3} pc^{-2} (d is source distance). The spectra are binned in 0.02 Å bins. Symbols r,i,f indicate the resonance, intercombination, and forbidden line, respectively.

The dielectronic recombination rates are also affected by the density. At increasing density the DR rate will be reduced by the ionization of the highly excited $n\ell$ states. The magnitude of this suppression has a fairly slow n_e dependence and roughly scales as $\sim (n_e/(z+1)^7)^{0.25}$ (e.g. Burgess and Summers 1969, Summers 1974). It will be most important for the lower-Z ions which emit UV emission lines rather than for the X-ray emitting ions (Raymond 1988). The effects set in roughly for densities $n_e \gtrsim (10^3-10^4)(z+1)^7$ cm^{-3}.

5.2.3. Electron density diagnostics

Determination of electron densities in hot cosmic plasmas provides another challenge for X-ray astronomy. In combination with the determination of the emission measure $\varepsilon = <n_e^2 V>$ from spectral fits a value derived for the electron density n_e will provide direct information about the emitting source volume V. In order to test current theoretical models of the X-ray source, it is very important to establish the scale of its size. Electron densities can be measured using density-sensitive spectral lines originating from metastable levels or using innershell excitation satellites to resonance lines (e.g. Mewe 1988). In the first case the He-like 2 → 1 triplet system lines are particularly important (Gabriel and Jordan 1969, Pradhan and Shull 1981, Pradhan 1982, 1985, Mewe *et al.* 1985). The intensity ratio of the forbidden (f) to intercombination (i) lines varies with electron density due to the collisional coupling between the metastable 2^3S upper level of the forbidden line and the 2^3P upper level of the intercombination line. The f/i line intensity ratio in the wavelength region 5–42 Å of He-like ions from C through S can be used to diagnose coronal plasmas in the density range $n_e = 10^8$–10^{15} cm^{-3} and corresponding temperature range $T \sim 1$–20 MK. Figure 7 shows the expected O VII line strengths with 0.05 Å resolution, for a coronal plasma with a temperature of 2 MK and densities of 10^{10}, 10^{11}, and 10^{12} cm^{-3}, respectively.

Deviations from coronal equilibrium can alter these diagnostics through different recombination and ionization effects, but the line intensity ratios remain density-sensitive also for these cases (Pradhan 1985). The f/i ratio does not depend on the model because its density dependence is determined only by the collisional coupling between the upper levels of the two lines, but the ratio $(f+i)/r$ does. E.g., for the O VII triplet it is ~ 1 in the coronal model (where the lines are excited by electron collisions from the ground) and ~ 3 in the photoionized nebular model (where population occurs through recombination, directly or via cascades). Hence this diagnostic tool can be quite generally applied to many astrophysical and laboratory sources, while the singlet/triplet ratio can be used as an indication of the model conditions. Mewe *et al.* (1985) have considered many density-sensitive lines from iron ions ranging from Fe IX–XIV and Fe XVIII–XXIII (and some corresponding nickel lines from Ni XXI–XXIV) covering the wavelength regions 7–13 Å (2ℓ–$3\ell'$, $4\ell'$ transitions), 90–140 Å (2ℓ–$2\ell'$) and 170–275 Å (3ℓ–$3\ell'$). The density dependence is because the upper line level can be excited from various sublevels within the ground state which become collisionally coupled at increasing densities (the same holds for the satellite lines). The Fe and Ni lines can be used as tools for diagnosing plasmas in the density range 10^{10}–10^{15} cm^{-3} and temperature range ~ 0.5–15 MK.

5.3. Transient plasmas

The assumptions of steady-state coronal equilibrium are not always valid for cosmic X-ray sources. The plasma approaches the equilibrium expressed in Eq. (1) over a relaxation time, t_{rel}, comparable to the ionization and recombination times of the relevant ion. In the ionization rate equation (1) are omitted the transport terms relating to the dynamic effects of plasma motions (expansion, diffusion) which are important when steep temperature and density gradients are present. It therefore emphasizes the particular effect of ionization and recombination rather than the fluid aspect. When the physical parameters of the plasma change very quickly due to a variety of processes such as plasma instabilities, shock compression, rapid expansion, rapid heating or cooling of the gas, etc., the assumptions of a steady-state equilibrium break down and the plasma is considered to be in a *transient* state. For example, in cases where the electron density is very low, the time scale t_{rel} for ionization equilibrium to be established at the currently existing electron temperature may greatly exceed the relevant dynamical plasma time scale, t_{pl}, on which *heating* or *cooling* takes place, hence on which the relevant plasma parameters (T, n_e) may vary (e.g. calc. for supernova remnants (SNR): Hamilton and Sarazin 1983, Gronenschild and Mewe 1982, Itoh 1984, Mewe 1984, Raymond 1988); solar flares: Shapiro and Moore 1977). Then the establishment of the ionization balance lags behind the temperature changes and as a result the plasma is *under-* or *over-*ionized compared to the equilibrium state belonging to the instantaneous temperature. This may have dramatic effects on the emergent X-ray spectrum (e.g. soft X-ray *excess* or *deficit* and *enhanced* or *decreased* radiative cooling). Resonance/satellite line ratios (e.g. Mewe 1988) and line/continuum ratios (e.g. Shapiro and Moore 1977) can be used as indicators of deviations from ionization equilibrium.

With approximate formulae for the timescales for ionization (t_{ion}) and recombination (t_{rec}) I made a comparison of $t_{rel} \simeq \min(t_{ion}, t_{rec})$ with the relevant dynamical plasma time scale

t_{pl} for a number of representative cases (Mewe 1984, 1988). Notwithstanding these widely different categories of optically thin astrophysical plasmas including young SNR, hot interstellar media of galaxies, solar or stellar flares, and fusion-oriented laboratory plasmas, span a wide range of dynamical time scales ($t_{pl} \sim 10^{14}$ to 10^{-9} s), they cover only a restricted range in $n_e t_{pl}$ ($\sim 10^9$ to 10^{14} cm^{-3} s). Because the ionization equilibrium time scales $n_e t_{rel}$ range from $\sim 10^{10}$ to 10^{13} cm^{-3} s, the transient situation is believed to apply to many of these cases. Then the time-dependence and the whole plasma *history* must be included in the analysis of the X-ray line spectrum. Current X-ray observations of young SNR clearly confirm that non-equilibrium effects in the ionization balance are very important (e.g. Itoh 1984). In such cases global time-dependent plasma models are applied to find values of temperature and $n_e t$ that can account for the measured line ratios (e.g. review by Bleeker (1989), Hamilton and Sarazin 1983).

5.4. Electric field effects on dielectronic recombination rates

Dielectronic recombination is very sensitive to the presence of electric fields, due to the dependence on very high nl Rydberg states. Few experiments for singly ionized atoms have indicated that electric fields $E \sim 10$–100 V/cm cut off the n by field ionization but fully mix the l for lower n, resulting in a net enhancement of the DR rate by nearly an order of magnitude for lower ions (e.g. Müller *et al.* 1987, cf. §3.3.). In the actual plasma environment the mean thermal drift $v_\perp \sim \sqrt{2kT_i/m_i}$ of ions across a magnetic field B may well generate such Lorentz electric fields. For $E \gtrsim 20$ V/cm this reduces to $B_{[G]}\sqrt{T_{i[MK]}/M_i} \gtrsim 90$, which, e.g., for $M_i \sim 20$ and $T_i \sim 1$ MK implies $B \gtrsim 400$ G. Magnetic fields on the order of this value can exist in solar or stellar coronal loop structures and much higher ($\sim$1000 G) in the lower chromosphere. As the effect is only important for the lowly ionized atoms, it can play a role in cool ($\lesssim 2$ MK) coronal or photoionized plasmas.

5.5. Non-Maxwellian velocity distributions

Departures from a Maxwellian velocity distribution can occur when energy is deposited into (or lost from) the high-energy tail of the distribution at a rate much faster than the Coulomb electron-electron collision rate $t_{ee}^{-1} \sim 100 n_e T^{-3/2}$ s^{-1} (see §2.), resulting in a high-energy excess (or deficit, respectively). Non-Maxwellian distributions are most likely to occur in low-density plasmas and at high velocities where collisional relaxation is slow, or in plasmas confined by strong magnetic fields that inhibit the dissipation process. As the electrons in the high-energy tail are just the ones which excite and ionize ions, the excitation and ionization rates are most strongly affected, whereas recombination rates which rely mostly on the electrons in the bulk of the distribution are not significantly affected.

A deficit of high-energy electrons could arise if the fast electrons lose their energy by exciting or ionizing ions more rapidly than Coulomb collisions shuffle electrons from the bulk of the energy distribution (energy $\sim kT$) to the tail (e.g. Dreicer 1960). The reverse, i.e. the production of a high-energy excess in the tail, may occur in plasmas with a steep temperature gradient where fast electrons can penetrate into cooler regions of the plasma, e.g. in the solar wind (Owocki and Scudder 1983) or in solar flares (Seely *et al.* 1987).

Constraints on the shape of the electron energy distribution may be obtained from measurements of spectral lines from a number of ion charge stages with a range in ionization thresholds or by using a technique of measuring the ratio of two spectral lines that are excited by different portions of the electron energy distribution as suggested by Gabriel and Phillips (1979). The latter method has recently been applied by Seely *et al.* (1987) to measure non-thermal electron energy distributions (e.g. a "bump" in the tail at the time when hard X-ray bursts occur) during the impulsive phase of solar flares by measuring the ratio of the He-like Fe XXV resonance line (excited by all electrons above the threshold energy E_0) and a dielectronic recombination satellite line (produced by resonance excitation within the small autoionization width at energy $E_s \sim 0.7E_0$) (see also Mewe 1988).

6. SUMMARY

High-resolution X-ray spectroscopy has applications to a wide range of optically thin hot astrophysical and laboratory plasmas, but the significance of it as a tool in understanding the physics of these sources will depend much on the reliability of the theoretical models to interpret the spectra. In this paper I have considered the coronal model in describing the ionization of hot plasmas and discussed the various processes of ionization, recombination and excitation, emphasizing the accuracy with which the emergent X-ray spectrum can be predicted. Various effects leading to deviations from this simple model have been discussed. Many of them may well be taken into account if the physical properties and the structure of the emitting source are better understood.

This work has been supported by the Space Research Organization of the Netherlands (SRON).

REFERENCES

Abramopoulos, F., Chanan, G.A., and Ku, W.H.-M.: 1981, *Astrophys. J.* **248**, 429

Acton, L.W., Brown, W.A.: 1978, *Astrophys. J.* **225**, 1065

Aggarwal, K. *et al.*: 1986, *"Report on Recommended Data"* , Atomic Data Workshop, Daresbury Lab.

Allen, C.W.: 1973, *Astrophys. Quantities* , (3rd ed., The Athlone Press, London)

Arnaud, M., Rothenflug, R.: 1985, *Astron. Astrophys. Suppl. Ser.* **60**, 425

Athay, R.G.: 1966, *The Solar Chromosphere and Corona: Quiet Sun* , Reidel, Dordrecht

Bates, D.R., Kingston, A.E., McWhirter, R.W.P.: 1962a, *Proc. Roy. Soc. (London)* **A267**, 297

Bates, D.R., Kingston, A.E., McWhirter, R.W.P.: 1962b, *Proc. Roy. Soc. (London)* **A270**, 155

Bell, R.H., Seaton, M.J.: 1985, *J. Phys. B.: At. Mol. Phys.* **18**, 1589

Bely, O., van Regemorter, H.: 1970, in *Ann. Rev. Astron. Astrophys.* **8**, 329

Bely-Dubau, F., Gabriel, A.H., Volonté, S.: 1979, *MNRAS* **189**, 801

Bleeker, J.A.M.: 1989, *Adv. Space Res.* , in press.

Burgess, A.: 1964, *Astrophys. J.* **139**, 776

Burgess, A.: 1965, *Astrophys. J.* **141**, 1588

Burgess, A. *et al.*: 1977, *MNRAS* **179**, 275

Burgess, A., Chidichimo, M.C.: 1983, *MNRAS* **203**, 1269

Burgess, A., Summers, H.P.: 1969, *Astrophys. J.* **157**, 1007

Burgess, A., Tworkoski, A.S.: 1976, *Astrophys. J.* **205**, L105

Butler, S.E., Dalgarno, A.: 1980, *Astrophys. J.* **241**, 838

Chen, M.H.: 1986, *Atom. Data Nuc. Data Tab.* **34**, 301

Chen, M.H., Crasemann, B.: 1987, *Phys. Rev. A* **35**, 4579

Cooper, J.: 1966, *Rep. Progr. Phys.* **29**, 35

Dolder, K., Peart, B.: 1986, *Adv. At. Mol. Phys.* **22**, 197

Drawin, H.-W.: 1961, *Z. Physik* **164**, 513; **168**, 238

Dreicer, H.: 1960, *Phys. Rev.* **117**, 343

Dubau, J., Volonté, S.: 1980, *Rep. Prog. Phys.* **43**, 199

Dunn, H.: 1986, in *At. Proc. in Electron-Ion and Ion-Ion Collisions* , (ed. F. Brouillard), Plenum Publ. Co., p. 93; in *Electronic and At. Collisions* (ed. D.C. Lorents, W.E. Meyerhof, J.R. Peterson), Elsevier Sc. Publ., p. 23

Elwert, G.: 1952, *Z. Naturf.* **7a**, 432

Fabian, A.C., Ross, R.R.: 1981, *MNRAS* **195**, 29p

64

Gabriel, A.H., Jordan, C.: 1969, *MNRAS* **145**, 241

Gabriel, A.H., Jordan, C.: 1972, *Case Studies in Atomic and Collisional Physics* **2** (eds. E.W. McDaniel, M.R.C. McDowell, N.-H. Publ. Co., Amsterdam), p. 209

Gabriel, A.H., Phillips, K.J.H.: 1979, *MNRAS* **189**, 319

Gallagher, J.H., Pradhan, A.K.: 1985, JILA Data Center Report No. 30, JILA, Univ. of Colorado, Boulder

Gregory *et al.*: 1987, *Phys. Rev. A* **34**, 3657; **35**, 3526

Griem, H.R.: 1964, *Plasma Spectroscopy*, McGraw-Hill, New York; Univ. Microfilms Internatl. 212 00000 7559, Ann Arbor, Mich.

Griem, H.R.: 1988, *J. Quant. Spectr. Rad. Transf.* **40**, 403

Gronenschild, E.H.B.M., Mewe, R.: 1982, *Astron. Astrophys. Suppl. Ser.* **48**, 305

Hahn, Y.: 1985, *Adv. At. Mol. Phys.* **21**, 123

Hahn, Y. *et al.*: 1980, *J. Quant. Spectr. Rad. Transf.* **23**, 65

Hahn, Y., LaGattuta, K.J.: 1988, *Phys. Rep.* **166**, 195

Hamilton, A.J.S., Sarazin, C.L.S., Chevalier, R.A.: 1983, *Astrophys. J. Suppl.* **51**, 115

Hearn, A.G.: 1966, *Proc. Phys. Soc.* **88**, 171

Holt, S., McCray, R.: 1982, *Ann. Rev. Astron. Astrophys.* **20**, 323

Itoh, H.: 1984, *Physica Scripta* **T7**, 19

Jacobs *et al.*: 1977, *Astrophys. J.* **211**, 605; **215**, 690

Jacobs *et al.*: 1978, *J. Quant. Spectr. Rad. Transf.* **19**, 591; 1979, *Astrophys. J.* **230**, 627; 1980, **239**, 1119

Jordan, C.: 1969, *MNRAS* **142**, 501

Jordan, C.: 1970, *MNRAS* **148**, 17

Kahn, S.M., Seward, F.D., Chlebowski, T.: 1984, *Astrophys. J.* **283**, 286

Kallman, T.R., McCray, R.: 1982, *Astrophys. J. Suppl.* **50**, 263

Kingston, A.E., Tayal, S.S.: 1983, *J. Phys. B.* **16**, 3465

Kunze, H.-J.: 1972, *Space Sci. Rev.* **13**, 565

Lemen, J.R., Mewe, R., Schrijver, C.J., Fludra, A.: 1989, *Astrophys. J.* **341**, 474

Liedahl, D.A., Kahn, S.M., Osterheld, A.L., Goldstein, W.H.: 1989, in preparation.

Lightman, A.P., Lamb, D.Q., Rybicki, G.B.: 1981, *Astrophys. J.* **248**, 738

Lotz, W.: 1967, *Astrophys. J. Suppl. Ser.* **14**, 207; *Z. Physik* **206**, 205

Lotz, W.: 1968, *Z. Physik* **216**, 241; :1969, *Z. Physik* **220**, 466; :1979, *Z. Phys.* **232**, 101

Massey, H.S.W., Bates, D.R.: 1942, *Rep. Progr. Phys.* **9**, 62

McCray, R.: 1982, in *Galactic X-ray Sources* (ed. P. Sanford, Wiley & Sons, Chicester), p. 71

McCray, R.: 1984, *Physica Scripta* **T7**, 73

McWhirter, R.W.P.: 1965, in *Plasma diagnostic techniques* (eds. R.H. Huddlestone, S.L. Leonard), Acad. Press, New York, p. 201

Merts, A.L., Cowan, R.D., Magee, N.H.: 1976, Los Alamos Sci. Lab. Rep. LA-6220-MS

Mewe, R.: 1967, *Brit. J. Appl. Phys.* **18**, 107

Mewe, R.: 1970, *Z. Naturf.* **25a**, 1798

Mewe, R.: 1972, *Astron. Astrophys.* **20**, 215

Mewe, R.: 1984, *Physica Scripta* **T7**, 5

Mewe, R.: 1988, in *Astrophysical and Laboratory Spectroscopy* (eds. R. Brown, J. Lang), Scottish Univ. Summer School in Phys. Publ., p. 129

Mewe, R., Gronenschild, E.H.B.M.: 1981, *Astron. Astrophys. Suppl. Ser.* **45**, 11

Mewe, R., Gronenschild, E.H.B.M., van den Oord, G.H.J.: 1985, *Astron. Astrophys. Suppl. Ser.* **62**, 197

Mewe, R., Lemen, J.R., van den Oord, G.H.J.: 1986, *Astron. Astrophys. Suppl. Ser.* **65**, 511

Mewe, R., Schrijver, J.: 1978, *Astron. Astrophys.* **65**, 99

Müller *et al.*: 1987, *Phys. Rev. A* **36**, 599

Nussbaumer, H., Storey, P.J.: 1983, *Astron. Astrophys.* **126**, 75

Owocki, S.P., Scudder, J.D.: 1983, *Astrophys. J.* **270**, 758
Pradhan, A.K.: 1982, *Astrophys. J.* **263**, 477
Pradhan, A.K.: 1985, *Astrophys. J.* **288**, 824
Pradhan, A.K.: 1987, *Physica Scripta* **35**, 840
Pradhan, A.K., Shull, J.M.: 1981, *Astrophys. J.* **249**, 821
Raymond, J.C.: 1988, in *Hot Thin Plasmas in Astrophysics* (ed. R. Pallavicini), Kluwer Acad. Publ., Dordrecht, p. 3
Raymond, J.C.: 1989, in *High Resolution X-ray Spectroscopy of Cosmic Plasmas* (eds. P. Gorenstein, M.V. Zombeck), Proc. IAU Coll. 115, Cambridge, U.S.A., Kluwer Acad. Publ., Dordrecht
Raymond, J.C., Smith, B.W.: 1977, *Astrophys. J. Suppl.* **35**, 419
Reilman, R.F., Manson, S.T.: 1979, *Astrophys. J. Suppl.* **40**, 815
Ross, R.R.: 1979, *Astrophys. J.* **233**, 334
Sampson, D.H.: 1982, *J. Phys. B* **15**, 2087
Schrijver, C.J., Lemen, J.R., Mewe, R.: 1989, *Astrophys. J.* **341**, 484
Seaton, M.J.: 1959, *MNRAS* **119**, 81
Seaton, M.J.: 1975, *Adv. At. Mol. Phys.* **11**, 83
Seely, J.F. *et al.*: 1987, *Astrophys. J.* **319**, 541
Shapiro, P.R., Moore, R.T.: 1977, *Astrophys. J.* **217**, 621
Smith, B.W. *et al.*: 1985, *Astrophys. J.* **298**, 898
Spitzer, L., Jr.: 1962, *Physics of Fully Ionized Gases* (2nd ed.), Intersc. Publ., New York
Summers, H.P.: 1974, *MNRAS* **169**, 663; Appleton Lab. Rep. AL-R-5
Sylwester, B. *et al.*: 1986, *Solar Phys.* **103**, 67
Vernazza, J.E., Raymond, J.C.: 1979, *Astrophys. J.* **228**, L29
Wang, J.-S. *et al.*: 1986, *Phys. Rev. A* **33**, 4293; 1987, **36**, 951; 1988, **38**, 4761
Wilson, R.: 1962, *J. Quant. Spectr. Rad. Transf.* **2**, 477
Younger, S.: 1981, *J. Quant. Spectr. Rad. Transf.* **26**, 329

PARTICLE ACCELERATION NEAR ASTROPHYSICAL SHOCKS

A. Achterberg
Sterrekundig Instituut, Postbus 80.000, NL-3508 TA, Utrecht
&
Center for High Energy Astrophysics, NIKHEF-H, Kruislaan 409,
NL-1098 SL , Amsterdam The Netherlands.

ABSTRACT. The interaction between particles or photons with an astrophysical shock is reviewed. A general equation for these processes is derived from a statistical point of view, and a number of astrophysical applications is considered.

1. Introduction.

About ten years ago, several authors [1-4] independently proposed the mechanism of diffusive shock-acceleration as a possible astrophysical source of energetic particles in our own galaxy (e.g. cosmic rays) or in active galaxies and quasars. In this mechanism, particles gain energy during repeated scattering across a shock front. This energy gain is due to the fact that "scattering centers" advected with the bulk flow in the up- and downstream region near the shock converge relative to each other due to the compression created at the shockfront. The mean energy-gain per shock crossing is of order $\Delta u/v$ with $\Delta u = (u_- - u_+) \cdot n$ the velocity difference across the shock, u_- (u_+) the fluid velocity ahead of (behind) the shock, n is the shock normal and v the particle velocity.

The necessary scattering can be provided by the fluid particles themselves, as is the case with Thomson scattering of photons by electrons in an ionized plasma, and is due to collective effects in the case of gyro-resonant scattering of charged particles by short-wavelength MHD waves (wrinkels in the magnetic field) . The rate at which particles cross the shock is proportional to the scattering rate. The scattering maintains a good coupling between the bulk fluid and the accelerated particles. This results in diffusive propagation which traps particles near the shockfront , and so allows a small fraction of particles to gain energies far in excess of the mean kinetic energy $mu_s^2/2$ per particle in the upstream flow as seen from a shock propagating with velocity u_s.

The simplest version of the mechanism assumes that the scattering mean-free-path $\lambda_{mfp} > d$, with d the thickness of the shock. The shock is then treated as a discontinuity. If this ordering not satisfied, particles or photons no longer see the shock as a discontinuity, but as a continuous flow transition. Particles then gain energy at rate determined by the local compression rate $\nabla \cdot u$ in the flow.

In this paper, I will review some of the features of the shock-acceleration process, with an emphasis on the various astrophysical applications. For more detailed derivations of various equations I refer to the review by Blandford & Eichler [5].

W. Brinkmann et al. (eds.), Physical Processes in Hot Cosmic Plasmas, 67–80.

2. A simple statistical model.

Consider a simple case of a planar, infinitely thin and steady shock. A scattering mechanism operating on both sides of the shock front ensures that particles are scattered repeatedly across the shock. Consider one *cycle*, in which a particle with momentum p crosses the shock from downstream to upstream and back again. One can define the following (suitably averaged) quantities:

$$
\left.
\begin{array}{ll}
\text{momentum change/cycle:} & \Delta(p) \ll p \\
\text{probability of escape/cycle:} & P(p) \ll 1 \\
\text{probability of destruction/cycle:} & A(p) \ll 1
\end{array}
\right\} \tag{1}
$$

Let $dN \equiv n(p)dp$ be the number of particles in the momentum interval $\left(p, p + dp\right)$ *currently involved* in the acceleration process. After completing one cycle, the various quantities change on average according to:

$$
p \longrightarrow \bar{p} = p + \Delta(p) , \tag{2a}
$$

$$
dp \longrightarrow d\bar{p} \approx \left(1 + \frac{\partial \Delta(p)}{\partial p}\right)dp , \tag{2b}
$$

$$
dN \equiv n(p)dp \longrightarrow d\bar{N} \equiv n(\bar{p})d\bar{p} = \left(1 - P(p) - A(p)\right)dN . \tag{2c}
$$

Equation (2c) assumes that no particles are injected into the acceleration process at momentum p. Substituting Eqns. (2a) and (2b) into Eqn. (2c) and making a Taylor expansion to first order yields a differential equation for n(p) which shows how the distribution results from the competition between energy gains and particle losses:

$$
\frac{\partial}{\partial p}\left(\Delta(p)n(p)\right) = - \left(P(p) + A(p)\right)n(p) \equiv -W(p)n(p). \tag{3}
$$

Here I have defined $W(p) \equiv P(p) + A(p)$ which is the total probability for a particle to be removed from the acceleration process during one cycle. Integrating Eqn. (3) gives the differential number density of particles participating in the acceleration:

$$
n(p) = n(p_0) \frac{\Delta(p_0)}{\Delta(p)} \exp\left\{- \int_{p_0}^{p} dp' \left(W(p')/\Delta(p')\right) \right\}. \tag{4}
$$

The integration constant $n(p_0)$ is determined by the injection process. If the escape of particles from the acceleration process is due to advection into the downstream medium, as is usually the case, an observer behind the shock will measure a particle distribution *emerging* into the downstream region proportional to P(p)n(p):

$$
F(p) = x \left(P(p)/\Delta(p)\right) \exp\left\{- \int_{p_0}^{p} dp' \left(W(p')/\Delta(p')\right)\right\}. \tag{5}
$$

Here $F(p)dp$ is the number-density of particles in the momentum interval $(p,p + dp)$ in the downstream region. The constant x must be calculated from a detailed consideration of the boundary conditions.

As a simple example, consider the acceleration of charged particles near a normal shock propagating in a magnetised plasma. I will assume that the scattering process is elastic in the rest-frame of the plasma. When scattering is due to the interaction of the charged particles with low-frequency MHD waves it can be considered elastic provided the wave phase-velocity in the plasma rest frame $|\omega/k| \ll u_s$ with u_s the shock velocity. Scattering keeps the particle distribution nearly isotropic in the rest-frame of the plasma on both sides of the shock. No particles are destroyed in this case so $A(p) = 0$. Escape from the shock is by advection into the downstream region with net flux $\Sigma_+ = u_+ F_+(p) = u_s F_+(p)/r$, with $r \equiv \rho_+/\rho_-$ the compression in the shock. The net flux across the shock back into the upstream region equals $\Sigma_b = (v/4)F_+(p)$, and the chance of escape per cycle equals the ratio $P(p) = \Sigma_+/\Sigma_b$. One then has (e.g. [3]):

$$\Delta(p) = \frac{4(r - 1)}{3r} \frac{u_s}{v} p \quad , \quad A(p) = 0 \ , \quad P(p) = \frac{4u_s}{rv} \quad . \tag{6}$$

Eqn. (5) yields a power-law momentum distribution in the downstream region, with a slope depending only on the compression r of the shock:

$$F_+(p) = F_0 \left(p/p_0\right)^{-\frac{r+2}{r-1}} \quad . \tag{7}$$

The integration constant F_0 follows from an elementary consideration of particle conservation. Let us assume that a seed-distribution of particles is present far upstream of the form $F_-(p) = n_- \delta(p - p_0)$, isotropic in the rest-frame of the plasma. The corresponding flux of particles advected into the shock equals $\Sigma_-(p) = u_s F_-(p)$. The flux into the downstream region equals $\Sigma_+(p) = u_+ F_+(p) = u_s F_+(p)/r$. Since no particles are destroyed, particle conservation in the steady state implies $\int dp \Sigma_-(p) = \int dp \Sigma_+(p)$. All particles are accelerated, so $F_+(p) = 0$ for $p < p_0$. Performing the integrals yields the relation $F_0 = 3r p_0^{-1} n_-/(r - 1)$.

A more involved example is the interaction between photons and a shock in a scattering and absorbing slab of material. Repeated scattering across the shock will lead to a systematic blueshift. But photons can be destroyed by absorption processes and created by emission . Following standard procedure in radiation-transfer theory, I will define a true absorption coefficient α_ν and a pure scattering coefficient σ_ν. The radiation transfer equation can be written in the form:

$$dI_\nu/ds = -(\alpha_\nu + \sigma_\nu)I_\nu + \eta_\nu . \tag{8}$$

Here ds is an infinitesimal pathlength along a ray, η_ν the emissivity and I_ν the intensity. I will assume that strong scattering keeps the photon distribution nearly isotropic. The radiation diffusion coefficient equals $D_\nu = c\lambda_\nu/3$ with $\lambda_\nu = 1/(\alpha_\nu + \sigma_\nu)$ the mean-free-path, while the probability of foton destruction per scattering equals $\varepsilon_\nu = \alpha_\nu/(\alpha_\nu + \sigma_\nu)$. Photons diffuse about one mean-free-path upstream before being overtaken by the shock after a time [6] $\Delta t \approx \lambda_\nu / u_s$. The number of scatterings in that period equals $c\Delta t / \lambda_\nu = c/u_s$. This gives a probability of foton destruction *per cycle* in the upstream region equal to $A_-(\nu) = \varepsilon_\nu^- c/u_s$. Here I have assumed $\varepsilon_\nu \ll u_s/c$ so that a significant fraction of the photons survive their excursion upstream. The probability of photon destruction during its stay in the downstream region can be calculated in a similar fashion, and equals $A_+(\nu) = \varepsilon_\nu^+ c/u_+$.

The net flux away from the shock equals $\Sigma_{\nu+} \approx u_+ I_\nu - D_\nu \mathbf{n} \cdot \nabla I_\nu \approx u_+ I_\nu + (c/3\tau) I_\nu$, where $\tau \equiv (\alpha_\nu + \sigma_\nu)L \gg 1$ is the optical depth of the downstream material taken to be a slab of size L. The flux back across the shock equals as before $(c/4)I_\nu$, so the chance of escape is $P(\nu) = 4\Sigma_{\nu+}/cI_\nu = (4u_s/\tau c) + (4/3\tau)$. Using - and + superscripts to distinguish quantities in the up- and downstream region respectively, one finds per cycle (compare Eqn. (6)) :

$$\frac{\Delta\nu}{\nu} = \frac{4(r-1)}{3r} \frac{u_s}{c} \quad, \quad P(\nu) = \frac{4u_s}{rc}\left(1 + \frac{rc}{3u_s\tau}\right) \quad, \quad A(\nu) = \frac{c}{u_s}\overline{\varepsilon}_\nu \quad . \quad (9)$$

Here I have defined a mean chance of absorption $\overline{\varepsilon}_\nu \equiv \varepsilon_\nu^- + r\varepsilon_\nu^+$. The intensity I_ν at the shock satisfies an equation which is the equivalent of Eqn. (5) with a source term S_ν allowing for the emission of photons adding to the population:

$$I_\nu = \int_O^\nu \frac{d\nu'}{\nu'} S_{\nu'} e^{-\mu(\nu|\nu')} \quad ; \quad \mu(\nu|\nu') \equiv \int_{\nu'}^\nu \frac{d\overline{\nu}}{\overline{\nu}} q(\overline{\nu}), \quad (10a)$$

where:

$$q(\nu) = \frac{W(\nu)}{(\Delta\nu/\nu)} = \frac{3}{r-1}\left(1 + \frac{rc}{3u_s\tau} + \frac{rc^2}{4u_s^2}\overline{\varepsilon}_\nu\right). \quad (10b)$$

The spectral slope $q(\nu)$ defined in Eqn. (10) shows the effect of destruction of photons and finite optical depth on the emerging intensity. If $\overline{\varepsilon}_\nu = 0$ and $\tau \longrightarrow \infty$ one has a power-law intensity $I_\nu \propto \nu^{-3/(r-1)}$, which is the same as solution (5) if one uses the fact that $p = h\nu/c$ and the definition of $I_\nu \propto p F_{ph}(p)$. Even a small amount of absorption $\overline{\varepsilon}_\nu \approx (u_s/c)^2$ can lead to a significant steepening of the emergent intensity. The reason is that about (c/u_s) cycles are needed for one e-folding in frequency, during which a photon undergoes of order $(c/u_s)^2$ scatterings. A more systematic derivation is needed when $(c/u_s)^2\varepsilon_\nu \gtrsim 1$. One has to solve the equations of radiative transfer on either side of the shock in the diffusion approximation [e.g. 7] matching the abberation-corrected normal flux $\mathbf{n} \cdot \left(u I_\nu - D_\nu \nabla I_\nu - u(\partial I_\nu/\partial\ln\nu) \right)$ across the shock ,quite analogous to the correct procedure in the cosmic-ray case [8].

A solution similar to this has been used to explain the occurence of blue-shifted asymmetric line-profiles in the Lyα lines from some distant galaxies as a result of Fermi-acceleration near shocks [9]. These authors assume $P(\nu) = 4/3\tau(\nu)$, and put $A(\nu) = 0$ (pure scattering). This procedure is correct for strong resonance lines when collisional excitation processes can be neglected and $u_s \ll rc/3\tau$.

3. Maximum energy gain for particles due to shock acceleration.

The maximum energy particles can gain in the process of shock acceleration is determined by energy losses. The time it takes a particle to complete one cycle across the shock is given by [6] $\Delta t = t_c \approx (\lambda_- + r\lambda_+)/u_s$. Here λ_- (λ_+) is the scattering mean-free path in the upstream (downstream) medium. The energy-gain per cycle for relativistic particles with Lorentz-factor γ is given by (c.f. Eqn. (4)) $\Delta\gamma = 4(r - 1)(u_s/3c)\gamma$. Introducing a mean energy-loss per unit time - $\langle d\gamma/dt\rangle_{loss}$, a particle subject to shock acceleration experiences a net energy gain according to:

$$\frac{d\gamma}{dt} = \frac{u_s^2}{c\bar{\lambda}}\gamma - \left(\frac{d\gamma}{dt}\right)_{loss} . \tag{11}$$

Here I have defined effective mean-free-path by $\bar{\lambda} \equiv 3r(\lambda_- + r\lambda_+)/4(r - 1)$. Note that $\bar{\lambda} \longrightarrow \infty$ when $r \longrightarrow 1$. For a strong hydrodynamical shock in an ideal gas with specific heat ratio $\Gamma = 5/3$ one has $r = 4$ and $\bar{\lambda} = \lambda_- + 4\lambda_+$. In general, $\bar{\lambda}$ is a function of energy which depends on the details of the scattering process. The maximum Lorentz-factor γ_{max} obtains when the energy gain per cycle equals the mean loss in a cycle time so the two terms on the right-hand- side of Eqn. (11) balance.

In a situation where losses can be neglected the maximum energy achieved by particles depends on the time T (e.g. age of the shock) available for acceleration, and the injection Lorentz factor γ_0. It follows from the implicit equation:

$$T = \int_{\gamma_0}^{\gamma_{max}} d\gamma \left(c\bar{\lambda}(\gamma)/u_s^2\gamma\right). \tag{12}$$

This shows that, although the scattering mechanism is not important for the shape of the distribution of accelerated particles for $\gamma \ll \gamma_{max}$, it is important in that it determines the cycle-time t_c and the maximum energy that can be achieved.

The scattering mechanism most often invoked in the context of diffusive shock acceleration is gyro-resonant pitch-angle scattering on low-frequency MHD waves, such as Alfvén waves [10,11] . Particles with charge Ze gyrating around a magnetic field B with gyro-radius of order $r_g = pc/Ze|B|$ are scattered by waves with frequency ω and wavevector **k** satisfying the resonance condition $\omega - k_\parallel v_\parallel \pm \Omega_c = 0$. Here $k_\parallel \equiv \mathbf{k}\cdot\mathbf{B}/|B|$, $v_\parallel = \mathbf{v}\cdot\mathbf{B}/|B|$ and $\Omega_c \equiv Ze|B|/\gamma mc$ is the cyclotron frequency . A quasi-linear treatment, (presumably) valid for small amplitudes $|\delta B|$ of the wave magnetic field, gives a spatial diffusion coefficient for relativistic particles of order:

$$D(\gamma) \equiv \frac{c^2\tau_{90}}{3} \approx \frac{cr_g}{3}\left(|\delta B(k)|/|B|\right)^{-2}\Big|_{k \approx 1/r_g} . \tag{13}$$

Here τ_{90} is the time it takes to scatter a particle through a 90 degree angle.

An often made (but unsubstantiated) approximation puts $|\delta B(k)|/|B| \approx 1$ at all wavelengths. In that case $\tau_{90} \approx 1/\Omega_c$ so $D(\gamma) \approx cr_g/3 = \gamma\beta mc^2/3Ze|B|$ and the scattering mean-free-path equals the gyro-radius. The rationale behind this estimate is that saturation of the wave-turbulence will occur when the magnetic field perturbations reach an amplitude equal to the mean field. However, recent computer simulations [12] of the wave-particle interactions show that the scaling with $|\delta B(k)|^{-2}$ predicted by Eqn. (13) breaks down for fairly small wave amplitudes . The pitch-angle scattering is strongly affected by non-linear effects, resulting in a scattering mean-free-path of about 5 - 30 gyro-radii. In the estimates below , valid for strong shocks, I parametrize the uncertainty of the details of the scattering process by putting the effective mean-free-path equal to $\bar{\lambda} = r_o/\varepsilon = r_o \gamma\beta/\varepsilon$, where $\varepsilon \equiv 1/\Omega_c \tau_{90} < 1$, and $r_o \equiv mc^2/Ze|B|$ is the gyro-radius of a $\gamma \approx 1$, $\beta = v/c \approx 1$ particle.

In the table below, the most important energy limiting processes invoked in astrophysical context are shown with the associated value of γ_{max} , as can be calculated from Eqns. (11) or (12).

Table 1.

limiting process	$(d\gamma/dt)_{loss}$	γ_{max}
synchrotron losses electrons [13] :　$\tau_s \equiv 6\pi m_e c/\sigma_T B^2$	γ^2/τ_s	$\left(\varepsilon u_s^2 \tau_s/cr_o\right)^{1/2}$
expansion losses ions and electrons: $t_{exp} \approx L/u_s$	γ/t_{exp}	$\varepsilon L u_s/cr_o$
relativistic bremsstrahlung electrons: $t_{bs} \approx 2\pi/n_i \alpha\sigma_T c\langle g\rangle$	γ/t_{bs}	$\varepsilon u_s^2 t_{bs}/cr_o$
proton-proton collisions : 　　$t_{loss} \approx 1/n_p\langle\sigma_{pp}\rangle c$ proton-photon collisions [14,15] : 　　$t_{loss} \approx 1/n_{ph}\langle\sigma_{p-ph}\rangle c$	γ/t_{loss}	$\varepsilon u_s^2 t_{loss}/cr_o$
geometry, e.g. trapping of particles in a magnetic loop of size L [16] : $T < L/u_s$		$\varepsilon L u_s/cr_o$

In this table $\sigma_T \equiv (8\pi/3)(e^2/m_e c^2) \approx 6.65 \times 10^{-25} cm^2$ is the Thomson cross section, L is the size of the source, $\alpha \equiv e^2/\hbar c \approx 1/137$ is the fine-structure constant, n_i the ion density, $\langle g\rangle$ a Gaunt factor of order unity weakly dependent on energy, n_p the non-relativistic proton density and n_{ph} the photon density at $h\nu > m_\pi c^2/\gamma$, with m_π the pion mass.. Proton- proton collisions $p + p \longrightarrow p's + \pi's$ and proton-photon collisions $p + \gamma \longrightarrow p + \pi's$ have a cross section of order $\langle\sigma_{pp}\rangle \approx \langle\sigma_{p-ph}\rangle \approx 10^{-28} cm^2$.

4. Astrophysical sites of shock acceleration.

Shock acceleration is expected to operate at a variety of astrophysical objects. These include blastwaves which result from explosive phenomena such as Solar or stellar flares [17-19] and supernovae [1-4,5 and references therein] , the bowshock of Earth and other planets [20] , the outer edge of shock-bounded bubbles blown into the interstellar medium by stellar winds [21], the so-called super-bubbles which result from sequential supernovae in OB-star associations in the galactic plane [22] and the collimated supersonic outflows associated with Young Stellar Objects [23], galactic X-ray sources such as Sco X-1 , SS433, and Cygnus X-3 [24] , and the nuclei of active galaxies and quasars [25].

Further possible sites of shock acceleration are accretion disks [26] , in which mass slowly spirals towards a compact object such as a white dwarf, neutron star or black hole under the influence of frictional forces and possibly dissipation in shocks. The gravitational binding energy liberated in this fashion is thought to be the primary energy source powering non-thermal emission and the (collimated) outflows from Young Stellar Objects, galactic X-ray sources , active galaxies and quasars. In accretion flows with little or no angular momentum standing accretion shocks can be a site of efficient particle acceleration [27,28]. But since the acceleration process requires a strong coupling between the energetic particles and the bulk flow, there is the danger of particles being "dragged" with the flow to the compact object, none reaching an observer at infinity. This has been discussed recently in some detail in the case of spherical accretion [27].

It is impossible, within the limited confines of this review, to do justice to all the fine points and outstanding problems of the theory of shock-acceleration or particle acceleration in general in all these various objects. In table 2 I have summarized the typical parameters for the shock acceleration process in a number of astrophysical objects. These include the size, age and shock velocity u_s , the mechanism limiting the energy gain and the maximum value of γ that can be reached, according to the expressions given in table 1 of the previous Section.

5. Observational evidence for shock acceleration.

The best *direct* evidence for shock acceleration comes from observations from spacecraft [29,30,31] . In a number of quasi-parallel interplanetary shocks (i.e. $\mathbf{B}$ and $\mathbf{n}$ aligned within 45°) energetic protons and an associated enhancement of MHD wave activity has been recorded more than an hour before the shock passes the spacecraft. The observations can be explained by the theory of diffusive shock acceleration. Accelerated particles "boiling off " the shock form a precursor in the upstream region with a typical dimension $L \approx D/u_s$, the typical scale on which the diffusion of particles away from the shock, and the advection with the flow into the shock balance each other. The angular distribution of the accelerated particles in the precursor region is anisotropic. For super-Alfvénic shocks with $u_s > v_A \equiv |\mathbf{B}|/\sqrt{(4\pi\rho)}$ this anisotropy is sufficient to generate Alfven waves by a gyro-resonant two-stream instability [2,31]. These waves provide the scattering needed to trap particles near the shock.

In all other cases, the evidence is more or less circumstantial. The best studied case is cosmic-ray acceleration by supernova remnants (SNR) in our galaxy e.g. [32—34]. The observed spectrum of galactic cosmic rays, when corrected for the (energy dependent) shielding influence of the heliosphere, is a power-law

Table 2

Object	size L	age t	u_s	B	limiting mechanism		γ_{max}
Solar Flare Type II shock	10^{11} cm	1000 s	1000 km/s	10G	protons: geometry[1] electrons: synchr. loss.		$10^3 \epsilon$ $20\sqrt{\epsilon}$
Earth bowshock	10^{10} cm	----	300 km/s	30μG	protons : geometry		kin. energy $\leq$ 100 keV
Supernova Remnants	10 pc	1000 yr	1000 km/s	1mG	protons: exp. loss. electrons: synchr. loss.		$10^6 \epsilon$ $3 \times 10^3 \sqrt{\epsilon}$
Superbubbles	5 kpc	10^7 yr	10 km/s	0.1 mG	protons: exp. loss. electrons: synchr. loss.		$10^7 \epsilon$ $10^2/\epsilon$
Active Galactic Nuclei 2	$\geq R_g \equiv \dfrac{2GM}{c^2}$ $\approx 10^{13} M_8$ cm	---	$\alpha \sqrt{(R_g/r)} c$	???	protons : foton/pion production		$10^{10} \epsilon \alpha B(G) t_6$

[1] I have labeled with "geometry" all limiting mechanisms where the size of the shock limits the energy gain . This will happen when $D(p)/u_s L \geq 1$, with L the typical size of the shock, and D(p) the diffusion coefficient.

[2] These estimates assume that the accelerating region is close to the Schwarzschild radius of a Black hole with a mass of $10^8 M_8$ solar masses. The shock velocity is taken to be a fraction α of the free-fall velocity $v_{ff} \approx c\sqrt{(R_g/r)}$. The typical loss-time t_6 in units of 10^6 seconds varies between 10^{-3} and 1 for different models [14.15]

in kinetic energy $T \equiv (\gamma - 1)mc^2$: $dN/dT \cdot \kappa T^{-2.7}$ for T in the range of $3 - 10^5$ GeV/nucleon. [35]. Above that range, the spectrum seems to flatten, and then steepen again. This power-law behaviour is followed by protons and other primary nuclei, but the elemental abundances at a given energy/nucleon differ significantly from Solar values. The global energetics of cosmic rays in the galaxy seems to support the idea of SNR as the primary sources. A supernova typically releases an energy $E_{snr} \approx 10^{51}$ erg into the interstellar medium, mostly in the form of kinetic energy of the ejecta and swept-up interstellar matter. The supernova rate Q_{snr} is estimated at 1 per 30-100 years ($Q_{snr} \approx 10^{-9} - 3 \times 10^{-10}$ s^{-1} in cgs units).

Cosmic ray propagation in the galaxy is usually described in terms of the so-called "leaky box model" [36], in which a typical cosmic ray is retained in the galaxy for about $t_r \approx 10^7$ years in a volume of about $V \approx 10^{65}$ cm^3. This corresponds to a dilute halo with a thickness at least 5 times that of the stellar disk , i.e. a scaleheight of about 1 kpc above the galactic plane. The cosmic-ray energy density measured at earth is $\varepsilon_{cr} \approx 1$ eV/cm$^3 \approx 2 \times 10^{-12}$ erg/cm^3. Taking this to be a representative value, one can calculate the fraction of the energy from supernovae which has to be chanelled into cosmic rays in order to maintain their current energy density in the galaxy:

$$E_{cr}/E_{snr} \approx \varepsilon_{cr}V/Q_{snr}E_{snr}t_r \approx 10^{-2} \qquad (14)$$

Stellar winds have been proposed as a source of cosmic rays, mainly on the ground that the isotopic abundances of cosmic rays resemble those observed in the coronae of the Sun and other active stars [37]. However, the power dissipated by stellar winds in the interstellar medium is about a factor 5 less than that put in by supernovae. The efficiency with which this power would have to be converted into cosmic rays is correspondingly larger, which seems unlikely.

It is also suggestive that the break in the cosmic-ray spectrum at about 10^{15} eV/nucleon (which corresponds to $\gamma \approx 10^6$) is intriguingly close to the value of γ_{max} estimated for the process of shock acceleration by SNR blastwaves (table 1). This suggests that primary cosmic rays below 10^{15} eV/nucleon originate at SNR, but that particles at higher energies have a different origin. Acceleration at Superbubbles could give $\gamma_{max} \approx 10^7$ if scattering is very efficient so that $\varepsilon \approx 1$. But this is not nearly enough to explain the UHE cosmic rays above 10^{18} eV [38]. When expansion losses dominate, the maximum attainable energy scales as $\gamma_{max} \approx \varepsilon L u_s/cr_o$. So if one believes that shock acceleration is the process responsible for the production of these UHE particles, one needs either a very large, or a very fast shock. This has lead to the suggestion that these particles are accelerated at the termination shock of a galactic wind with $L \approx 300$ kpc, $u_s \approx 400$ km/s and $B \approx 1$ µG [39] . For protons this gives a maximum Lorentz-factor equal to $\gamma_{max} \approx 5 \times 10^8 \varepsilon (u_s/400$ km s$^{-1})(L/300$ kpc)$B_{\mu G}$. Other nuclei with a mass $m = Am_p$ and charge Ze one can boost γ_{max} by a factor Z/A and the energy by a factor Z. It is therefore very important to measure the chemical composition of the UHE cosmic rays. The occurence of high-Z nuclei above 10^{15} eV such as members of the Iron group (Z = 26) would be much easier to account for than, say, ^{4}He with Z = 2. The low particle flux arriving at earth in this energy range has made composition measurements impractical until now.

6. "Realistic" shock acceleration: non-linear effects.

Most of the estimates derived in this paper have been based on a number of simplifying assumptions, such as stationarity and a planar shock. Shocks in astrophysical objects evolve and usually have a non-planar geometry. Even more important is the fact that the full theory of shock acceleration is *inherently non-linear* . So far, I have employed a test-particle approximation, in which particles are accelerated by a shock of prescribed strength in a simple, one-dimensional flow. This test-particle approximation, although usefull for making estimates, is not likely to apply in practice if shock acceleration is efficient in the sense that a significant fraction of the incoming momentum flux ρu_s^2 + P is converted into energy density of accelerated (i.e. non-thermal) particles at the shock.

The strong coupling between the accelerated particles (or photons) and the bulk flow mediated by scattering is a necessary ingredient of the theory, even though the *details* of the scattering do not enter directly into the resulting distribution of accelerated particles. This coupling allows the bulk fluid to feel the pressure-gradients in the gas of accelerated particles ("radiation pressure"). The precursor formed by the accelerated particles in the upstream region slows down the incoming fluid due to the associated pressure gradient $\nabla\Pi$, with $\Pi(\mathbf{x},t) \equiv \int dpF(p,\mathbf{x},t)(pv/3)$ the particle pressure [40,41].

A two-fluid approach has been used by several authors to study the influence of the accelerated particles on the shock structure [42,43]. It describes the accelerated particles as an inertialess fluid with pressure Π and energy density $W_{cr} \equiv \Pi/(\Gamma_{cr} - 1)$ which defines an effective adiabatic heat ratio Γ_{cr} for the accelerated particles. The pressure of the accelerated particles satisfies an equation which includes the effect of diffusive propagation:

$$\frac{\partial}{\partial t} \Pi \; + \mathbf{u}\cdot\nabla\Pi - \nabla\cdot\bar{\mathbf{D}}\cdot\nabla\Pi \; = \; - \Gamma_{cr}\Pi\nabla\cdot\mathbf{u} \; . \tag{15}$$

Here $\bar{\mathbf{D}}$ is a suitably defined average of the diffusion tensor.

The effect of an increasing particle pressure at the shock is twofold. First of all, the deceleration of the incoming flow leads to a weaker (viscous) subshock with compression $r < 4$. If enough energy is put into the accelerated particles, the viscous subshock is erased altogether, and one is left with a "cosmic-ray dominated" or "radiation dominated" shock in which the velocity changes smoothly over several precursor scales. Secondly, for flat spectra one expects the cosmic-ray gas to behave as an ideal relativistic gas with $\Gamma_{cr} \approx 4/3$. This makes the system more compressible, so that the *overall* compression across the precursor + viscous subshock (if present) may exceed the value $r = (\Gamma+1)/(\Gamma-1) = 4$ (for $\Gamma = 5/3$) for a pure hydrodynamic shock. The value of the total compression depends sensitively on the efficiency of the acceleration process.

This non-linear modification of the shock structure will effect the momentum distribution of the accelerated particles. The mean-free-path $\lambda(\gamma)$ is increases with particle energy. High-energy particles diffuse further into the decelerating precursor flow and sample a larger effective compression across the shock, leading to a larger energy gain per cycle and a flattening of the distribution F(p) at higher energies. This breaks the scale-free power-law behaviour of Eqn. (7). The deviation from the simple power-law becomes noticible at a momentum scale defined by $\left(D(p)/u^2\right)\left(\partial u/\partial x\right) \approx 1$, with $\partial u/\partial x$ the precursor velocity gradient.

To calculate this one must consider the full equation for $F(p,x,t)$ rather than Eqn. (15) for the particle pressure $\Pi(x,t)$. In the simple case of elastic scattering centers advected with the fluid one has:

$$\frac{\partial}{\partial t} F(p,x,t) + \nabla\cdot\Sigma(p,x,t) = \frac{\partial}{\partial p}\left(\frac{1}{3}\nabla\cdot upF(p,x,t)\right),$$ (16a)

$$\text{where:} \quad \Sigma(p,x,t) \equiv uF(p,x,t) - (D\cdot\nabla)F(p,x,t).$$ (16b)

Results of numerical solutions of the full set of equations for the fluid and Eqns. (16) have confirmed this intuitive picture [44,45]. When the total compression is large, the spectrum becomes quite flat at higher energies $(s<2)$. Most of the particle pressure then resides at the high-energy tail of the distribution, depending quite sensitively on the momentum cut-off $p_{max} \approx \gamma_{max} mc$. The problem then becomes time-dependent, since γ_{max} (in absense of losses) depends on the age T of the shock. c.f. Eqn. (12). For instance, for a scaling $\lambda(\gamma) \approx r_g/\varepsilon = r_0\gamma/\varepsilon$ one has $\gamma_{max} \approx \varepsilon u_s T/cr_0$. This means that shocks in which a significant fraction of the available mechanical momentum flux ρu^2 is converted into pressure of energetic particles are probably intinsically time-dependent structures. In these circumstances a two-fluid approach is of limited value. The detailed microscopic behaviour of $F(p,x,t)$ as described by Eqn. (16) and the hydrodynamics of the shock and precursor become strongly coupled. A meaningfull averaging procedure for finding the average diffusion coefficient $\bar{D}$ in Eqn. (15) does not really exist, since the largest contribution ususally comes from the very time-dependent high-energy tail of the distribution. A further problem with the two-fluid approach is that the effective adiabatic heat ratio Γ_{cr} of the accelerated particles is in effect a variable, rather than a constant parameter. Although generally $5/3 \geq \Gamma_{cr} \geq 4/3$, its actual value depends sensitively on the precise shape of the momentum distribution, which one does not know in a two-fluid approach.

7. Outstanding problems.

Although the basic principles of shock acceleration are well understood, there are still a number of unsolved problems in the theory which, at least in this authors view, stand in the way of a fully quantitative comparison of theory and observations. The most pressing of these problems are:

A. The injection problem.

Most astrophysical shocks are collisionless in the sense that the conversion of directed kinetic energy of the upstream flow into randomized motion ("heat") in the downstream flow proceeds through collective (plasma) effects rather than by ordinary Coulomb collisions. The typical random energy per proton downstream equals $m_p u_s^2/2$. Shock acceleration takes a few of these particles, and accelerates them to energies much larger than this value. How many particles are picked up from the "thermal" pool at the shock to be accelerated further is not really known. This question is probably intimately connected with the detailed microscopic structure of quasi-parallel collisionless shocks in a magnetised plasma. The overall efficiency of the acceleration process can not really be calculated without some knowledge of this injection process. The momentum- and energy conservation laws for cosmic ray mediated shocks (generalised Rankine-Hugoniot relations) in the steady state usually allow (at least) two solutions [42,43]: one with low efficiency and a strong viscous subshock, and a high-efficiency solution with a weak subshock, or no viscous subshock at all. Which solution is chosen will depend on the injection process, unless enough seed particles are already present in the upstream flow to force the system into forming a cosmic-ray dominated shock .

One might hope that numerical simulations will answer this question.

B   Electron acceleration .

Non-thermal continuous radio emission is almost always interpreted as synchrotron radiation of relativistic electrons in a magnetic field. Relativistic protons remain invisible due to their small radiation losses. It is not known how much of the energy put into accelerated particles at a shock goes into protons and ions, and how much goes into electrons. This makes it very difficult to interpret the data in terms of the energy requirements needed to power the source.

Electrons have to be accelerated to some threshold momentum before they can be picked up by the shock acceleration process. This threshold momentum usually exceeds the typical momentum $\sqrt{(m_e m_p)}u_s$ of a thermalised electron in the post-shock flow.

The first threshold is associated with the requirement of efficient scattering. Protons boiling of the shock generate low-frequency MHD waves . These waves have frequencies below the proton cyclotron frequency $\Omega_p = eB/m_p c$, and wavenumbers below $k_{max} \approx \Omega_p/v_A$ ($v_A \equiv B/\sqrt{(4\pi\rho)}$ is the Alfven velocity). Electrons do not interact resonantly with these waves until their gyroradius $r_g \approx pc/eB > k_{max}^{-1} \approx m_p v_A c/eB$. So in order to be scattered efficiently, the electron momentum must exceed $p > m_p v_A$.

If acceleration proceeds at a collisionless shock of width Δ , electrons will not see the shock as a discontinuity unless their scattering mean-free-path $\lambda_e > \Delta$. Both observationally, and in numerical simulations [46, 47] one finds that a quasi-parallel collisionless shock has a width of a few times c/ω_{pi}, where $\omega_{pi} = \sqrt{(4\pi e^2 n/m_p)}$ is the ion plasma-frequency. If scattering is due to strong magnetic fluctuations one has $\lambda_e \approx r_g \approx pc/eB$. Electrons will not see the shock as a discontinuity unless $p > eB/\omega_{pi} = m_p v_A$.   This is the same threshold as was derived previously.

When the subshock is erased altogether by the diffusive action of accelerated protons, the shockwidth will exceed $\Delta \approx m_p u_s c/eB$, which is the gyro-radius of a proton with momentum $p \approx m_p u_s$. Electrons need to have at least the same momentum before shock acceleration becomes efficient. In order for shock acceleration to operate a shock has to be super-Alfvenic ($u_s > v_A$). So this last requirement stronger than the first two.

In all cases some mechanism is needed to pre-accelerate electrons to energies exceeding the typical "thermal" energy per particle$c m_p u_s^2/2$ in the downstream flow. A mechanism has been proposed for electron acceleration near a shock using compressive MHD waves generated by supra-thermal ions which are reflected by a electrostatic potential hump in the shock [46]. Such potential humps are usually part of the shock structure of collisionless quasi-parallel shocks.

This momentum threshold will limit the operating range of diffusive shock acceleration for electrons. In particular this is true for electron acceleration in the nuclei of active galaxies and quasars, where radiation losses become very strong. Assuming a shock dominated by accelerated protons and combined synchrotron and Compton losses as the loss mechanism scaling as $(d\gamma/dt)_{loss} \cdot \gamma^2/\tau$, electrons can be accelerated provided their Lorentz factor lies in the range:

$$(m_p/m_e)(u_s/c) < \gamma < \left(\varepsilon u_s^2 \tau / cr_e\right)^{1/2}. \qquad (17)$$

Here $\tau \equiv 6\pi m_e c/\sigma_T B_e^2$ with $B_e \equiv (B^2 + 8\pi U_{ph})^{1/2}$ is an effective magnetic field defined in terms of the actual field B and the energy density U_{ph} of the radiation, and $r_e \equiv m_e c^2/eB$. Electron acceleration is possible provided $\tau > (m_p/m_e)(1/\varepsilon\Omega_e)$, independent of the shock velocity u_s. This condition can be written in terms of B_e as $B_e (G) < 6 \times 10^4 \left(\varepsilon B(G)\right)^{1/2}$. The central object in active galaxies and quasars is usually thought to be a black hole with a mass $M \approx 10^8$ solar masses, surrounded by an accretion disk or - torus. The typical luminosity is the Eddington luminosity $L_{edd} \cdot 4\pi GMm_p c/\sigma_T$. At a distance r the radiation density $U_{ph} \approx L_{edd}/(4\pi r^2 c)$ corresponds to an effective field $B_e \approx (m_e c^2/\sigma_T R_G)^{1/2} x^{-1} \approx 10^4 M_8^{-1/2} x^{-1} G$. Here $R_G \cdot 2GM/c^2$ is the Schwarzschild radius of the black hole, $x \equiv r/R_G$ and $M_8 \equiv M/10^8 M_\odot$. This shows that the operating range for electron acceleration is quite small in these conditions. Another source of relativistic electrons therefore seems likely, such as a pair cascade induced by decay of pions into γ-rays [14,15]. The pions are generated by proton-proton/proton-photon collisions of shock-accelerated protons.

8. Conclusions.

The theory of shock acceleration has firmly established itself in the astrophysical community as an important mechanism for the generation of energetic particles. In its simplest form, it is deceivingly straightforward. Practical application to astrophysical objects however is hindered by the fact that some of the details of the process such as injection, and the acceleration of electrons remain ill-understood. It seems therefore premature to use the proces as a panacea whenever efficient generation of energetic particles is required by the observations. Some of these problems are closely linked to our lack of understanding of all the details of quasi-parallel shock structure in astrophysical environments.

References.

1. Axford, W.I., Leer, E. & Skadron, G.: *Proc. 15th International Cosmic Ray Conference*, Plovdiv , 1977.
2. Krimsky, G.F.: *Dokladay Acad. Nauk. SSR*, **234**, 1306, 1977.
3. Bell, A.R.: *Mon. Not. R. astr. Soc.* **182** , 147, 1978.
4. Blandford, R.D. & Ostriker, J.P. : *Astrophys. J. (Letters)* **221**, L29, 1978.
5. Blandford, R.D. & Eichler, D.: *Physics Reports* **154**, 1, 1987.
6. Lagage, P.O. & Cesarsky, C.J.: in *Plasma Astrophysics* , ESA SP-161, p.317, 1982.
7. Pomraning, G.C. : *Radiation Hydrodynamics* , p.51, Pergamon Press, Oxford, 1973.

8.   Webb, G.M.: *Astron. Astrophys.* **124**, 163, 1983.
9.   Neufeld, D.A. & McKee, C.F.: *Astrophys. J. (Letters)* **331**, L87,1988.
10. Jokipii, J.R.: *Astrophys. J. (Letters)* **146**, L80, 1966.
11. Kulsrud, R. and Pearce, W.F.: *Astrophys. J.* **156**, 445, 1969.
12. Max, C. E. , Zachary, A.L. & Arons, J.: in *Proc. Joint Varenna-Abatsumani School and Workshop on Plasma Astrophysics* Vol. 1, ESA SP-285, p. 45, 1988.
13. Webb, G.M., Drury, L. O'C & Biermann, P. *Astron. Astrophys.* **137**, 185, 1984.
14. Kazanas, D. & Ellison, D.C.: *Astrophys. J.* **304**, 178, 1986.
15. Sikora, M. & Shlosman, I.: *Astrophys. J.* **336**, 593, 1989.
16. Achterberg, A.: *Mon. Not. R. astr. Soc.* **232**, 323, 1988.
17. Achterberg, A. & Norman, C.A.: *Astron. Astrophys.* **89**, 353,1980.
18. Ellison, D.C. & Ramaty, R.: *Astrophys. J.* **298**, 400, 1985.
19. Smith, D. F. & Brecht, S.H.: *J. Geophys. Res.* **90**, 205, 1985.
20. Teresawa, T. : *J. Geophys. Res.* **86**, 7595, 1981.
21. Cesarsky, C. J. & Montmerle, T.: *Space Science Rev.* **36**, 173, 1983.
22. McCray, R.: in *Spectroscopy of Astrophysical Plasmas* , A. Dalgarno & D. Layzer eds. Cambridge University Press, p. 274, 1987.
23. Mundt, R.: in *Formation and Evolution of Low Mass Stars*, A.K. Dupree & M.T.V.T. Lago, eds. NATO ASI , Vol. 241, p. 257, Kluwer Academic Publishers, Dordrecht, 1988.
24. Ögelman, H.: in *Astrophysical Jets and their Engines* , W Kundt, ed. , NATO ASI, Vol. 208, p. 67, Kluwer Academic Publishers, Dordrecht, 1987.
25. Begelman, M.C., Blandford, R.D. & Rees, M.J.: *Rev. Mod. Phys.* **56**, 255, 1984.
26. Spruit, H.J. : *Astron. Astrophys.* **194**, 319, 1988.
27. Schneider, P. & Bogdan, T.J.: *MPA Preprint 431* , 1989.
28. Cowsik, R. and Lee, M.A.: *Proc. Roy. Soc. London A*, **383**, 409, 1982.
29. Kennel , C.F. et. al. : *J Geophys. Res.* **89**, 5419, 1984.
30. Scholer, M. et al. : *J. Geophys. Res.* **88**, 1977, 1983.
31. Scholer, M.: in *Collisionless Shocks in the Heliosphere, Reviews of Current Research*, B. T. Tsurutani & R.G. Stone, eds. , Geophysical Monograph 35, p. 287, American Geophysical Union, Washington DC, 1985.
32. Blandford, R.D. & Ostriker, J.P.: *Astrophys. J.* **237**, 793, 1980.
33. Bogdan, T.J. & Völk, H.J.: *Astron. Astrophys.* **122**, 129, 1983.
34. Völk, H.J. & Biermann, P.L. : *Astrophys. J. (Letters)* **333**, L65, 1988.
35. Wefel, J.P.: in *Genesis and propagation of Cosmic Rays*, M.M. Shapiro & J.P. Wefel eds., NATO ASI Vol. 220, p. 1. D. Reidel Publ., Dordrecht, Holland, 1988.
36. Cesarky, C. J.: *Ann. Rev. Astron. Astrophys.* **18**, 289, 1980.
37. Montmerle, T.: *Adv. Space Res.* **4**, 357, 1984.
38. Hillas, A.M.: *Ann. Rev. Astron. Astrophys.* **22**, 425, 1984.
39. Jokipii, J.R. & Morfill, G.: *Astrophys. J.* **312**, 170, 1987.
40. Achterberg, A: *Astron Astrophys.* **98**, 195, 1981.
41. Eichler, D.: *Astrophys. J.* **229**, 419, 1979.
42. Drury, L. O'C & Völk, H.J. : *Astrophys. J.* **248**, 344, 1981.
43. Achterberg, A., Blandford, R.D. & Periwal, V.: *Astron. Astrophys.* **132**, 97, 1984.
44. Achterberg, A.: *Astron. Astrophys.* **174**, 329, 1987.
45. Falle, S. A.E.G. & Giddings, J. R. : *Mon. Not. R. astr. Soc.* **225**, 399, 1987.
46. Greenstadt, E.W.: in *Collisionless Shocks in the Heliosphere, Reviews of Current Research*, B.T. Tsurutani & R.G. Stone, eds. Geophysical Monograph Vol. 35, p.169, American Geophysical Union, Washington DC, 1985.
47. Quest, K.B. : ibid. p. 185.
48. Galeev, A.A.: in *Advances in Space Plasma Physics*, B. Buti, ed., p. 273 & p. 401, World Scientific Publishers, Singapore, 1985.

NUMERICAL SIMULATIONS OF COLLISIONLESS SHOCKS

B. LEMBEGE
C.R.P.E./C.N.E.T
38-40 rue du General Leclerc
92131 Issy-les-Moulineaux
France

ABSTRACT. This tutorial-style review presents recent developments of the numerical simulation activity devoted to the study of collisionless shock waves. First, background (theoretical and numerical) characteristics of non-linear waves and shock waves are reviewed. Second, a short description of the main features of shocks commonly observed in space plasmas is presented for both fields components and particles dynamics. Third, a review of the various appropriate simulation codes is presented with a particular emphasis of their respective advantages and disadvantages for the study of collisionless shock waves. Fourth, typical patterns of magnetosonic shock waves, such as the downstream (or upstream) "wavetrain", the "ramp", and the "foot" are shown to be well recovered in numerical results and are associated with particle acceleration mechanisms which are discussed in terms of dissipation sources. Fifth, a classification of the various plasma instabilities related to a magnetosonic shock is presented; these instabilities are strongly related to the particle dynamics presented in the previous section and are additional sources of energy dissipation; these are shown to affect strongly the overall structures of the shock and its nearest downstream and upstream neighborhood. Finally, some applications to astrophysical shocks will be presented. Numerical simulation appears to be a very powerful tool in order to interpret complex patterns of shock waves associated with some intricate wave-particle interaction mechanisms which are important sources of energy dissipation. In particular, such a tool is shown to represent, until now, the only way "to visualize" carefully the overall dynamics of particle species by the use of appropriate diagnostics.

Table of Contents

81

W. Brinkmann et al. (eds.), Physical Processes in Hot Cosmic Plasmas, 81–139.
© 1990 *Kluwer Academic Publishers.*

Glossary:

M_A	:	Mach number
M_A*	:	First critical Mach number
θ	:	Angle between the normal to the shock and the magnetostatic field $\vec{B}_0$ [$\equiv(\vec{k}, B_0)$]
β	:	Ratio of the plasma thermal energy to the magnetic field energy
MS wave	:	Magnetosonic wave
1 and 2	:	denotes respectively the upstream and downstream region of the shock
"i" and "e"	:	denotes respectively the ions and the electrons
"l" and "t"	:	denotes respectively longitudinal and transverse fields components
q-$\perp$	:	quasi-perpendicular shocks
q-‖	:	quasi-parallel shocks

1. INTRODUCTION

A large amount of theoretical, experimental, and numerical studies have already dealt with the features of a collisionless shock propagating in a magnetized plasma. The aim of this paper has been restricted to and mainly focused on the background features of magnetosonic shocks and on their associated mechanisms for energy dissipation (Section 2 - 6). These mechanisms strongly differ according to the shock parameters and may largely affect its overall structure; this explains the growing interest given to the numerical simulation activities devoted to the shock studies since almost 25 years. Subsequent numerical studies of other kinds of shocks have also been developed in association with both the improvement of numerical codes and space experiments which have been successfully launched and have gathered an impressive quantity of data.

This review has been written in a tutorial-course style rather than in reviewing a deep and delicate presentation of the top-level numerical results recently obtained for various kinds of shock waves. In the purpose of clarity, many phenomena are illustrated by sketchs and/or classifying tables. Readers who feel unsatisfied by this introductory presentation will find a large list of references which is provided at the end of this review in order "to satisfy their appetite" and to look for a deeper analysis. Preparing this review has been easier by previous monographs and reviews which are recommended reading (Tidman and Krall, 1971; Biskamp, 1973; Sagdeev, 1979 ; Kennel et al., 1985; Goodrich, 1985; Quest, 1985; Papadopoulos, 1985; Burgess, 1987). Detailed and comprehensive presentation of plasma simulations can be found in tutorial-style books (Birdsall and Langdon, 1985; Hockney and Eastwood, 1981), reviews (Dawson, 1983), and also in the various proceedings of the International School for Space Simulation (ISSS-1, 1985; ISSS-2, 1985; ISSS-3, 1988, 1989). For improving the presentation, a simplified glossary of terminology commonly used is presented at the beginning; in addition, a few appendices on more specific and technical computer aspects are presented at the end of the paper.

2. Nonlinear Waves and Shock Waves

Before dealing with a collisionless magnetosonic shock wave, we will progressively introduce the main characteristics of nonlinear waves by considering, first, the simplest example of an electrostatic nonlinear wave propagating in an isotropic and unmagnetized plasma ($|\vec{B}_0| = 0$), second, the same wave in magnetized plasma ($|\vec{B}_0| \neq 0$) and, third, nonlinear effects in both electrostatic and electromagnetic fields as in the case of a nonlinear magnetosonic wave. In the latter case, the difference between a so-called nonlinear magnetosonic wave and a magnetosonic shock wave will be reviewed. Let us start by considering when a wave can be called "nonlinear".

2.1 NONLINEAR ELECTROSTATIC WAVES ($|\vec{B}_0| = 0$)

Let us consider a simple example of an electrostatic wave (density wave) in an isotropic plasma. When the amplitude of the wave is not too large, linear theory applies and predicts that the phase velocity v_ϕ of this wave only depends on the frequency ω and the wavenumber k ($v_\phi = \omega/k$). However, when the amplitude E_0 of the wave is larger, there exists a critical value E_m of the amplitude above which nonlinear effects are important and v_ϕ becomes dependant on the wave amplitude. Coupled with the dispersion nature of the plasma, this leads to a unique amplitude and shape that a wave can have for amplitudes less than this critical value. In other words, the threshold E_m is due to the fact that a given wavelength λ_0 ($\equiv 2\pi/k_0$) can support a certain maximum wave amplitude without the waveform suffers any distortion and leads to higher harmonics excitation (Dawson, 1959; Dawson et al, 1973). For $E_o > E_m$, the phase velocity depends strongly on the local perturbation δE and in particular will be different at the locations of the troughs and at the crests of the E profile (respectively velocity v_1 and v_2 in Fig. 1b). This difference $\Delta \vec{v} = \vec{v}_2 - \vec{v}_1$ is the source of the steepening effect. As a consequence, the waveform distorts and a second harmonic is generated (what we call herein "wave steepening" or wave "overtaking".) Owing to dispersion effects, the second harmonic propagates at a different velocity from the fundamental and will gradually slip out of phase with it (Dawson et al., 1973). Generally, the various harmonics converge before an appreciable slip takes place provided that wave steepening is strong enough and occurs rapidly. A useful way to study this mechanism consists of analyzing the energies per mode versus time which indicates how the energy of the initial large amplitude wave (at $t = 0$) is distributed among the higher k-harmonics while the wave evolves or propagates in time. Then one can estimate the time range Δt_s, illustrating the time necessary for a second harmonic (for instance) to slip from the fundamental mode over a reasonable distance (say $\lambda_o/2$) as soon as it is excited.

Particle trapping is commonly associated with wave steepening and has been analyzed with a cold and warm plasma approach (Coffey, 1971; Kruer, 1979; Kono and Mulser, 1983). In the cold approximation, wave overtaking takes place at the same time as the trapping of particles and as the density perturbation δn becomes infinite. In a warm plasma, pressure effects oppose such high compression and the density perturbation is large but remains finite; in other words, wave steepening is limited and the maximum amplitude E_m is reduced because of the thermal effects. Particles are brought into resonance not from $v = 0$ but from v equal to a few times the thermal velocity v_{th} (i.e., from the extremity of the distribution tail). More generally, this trapping mechanism applies both to a propagating wave (phase velocity v_ϕ) and to a steady state wave (which can be considered as resulting from two waves propagating in opposite directions v_ϕ and $-v_\phi$). When the wave amplitude is large enough, it can accelerate and trap a noticeable number of particles as soon as this velocity reaches the value v_ϕ, or equivalently when their energy $m v^2/2$ becomes comparable to the wave potential $q\Phi$, where q is the particle electric charge.

The arising question is: *which phenomenon, wave steepening or particle trapping, occurs first?* This depends mainly on the relative distance of the resonant region (trapping area around v_ϕ) with respect to the thermal velocity v_{th} of the particle distribution at rest. In other words, the energetic particles of the Maxwellian tail will be trapped by the electrostatic wave toward the resonant region over a certain time range Δt_{trap} which has to be compared with the steepening time Δt_s mentioned before. In order to avoid any confusion of terminology, let us note that "steepening" (also "overtaking") of a wave is used in the absence of any particle trapping, while "wave breaking" is commonly used when particle trapping is included. Then, for a growing large amplitude wave (wavelength λ_o) and for $\Delta v = v_\phi - v_{th}$ large, the value λ_o will not support the maximum amplitude E_o of the wave and wave steepening occurs first; particle trapping may follow at later times ($\Delta t_s < \Delta t_{trap}$). In the opposite case ($\Delta t_s > \Delta t_{trap}$), for smaller range Δv, particle trapping may take place first before steepening occurs.

Another important result is that the trapping mechanism reinforces the wave steepening. Basically, this can be explained by the longitudinal nature of the electrostatic wave (electric field parallel to the propagation vector $\vec{k}$). While accelerated and getting trapped, particles accumulate strongly over a narrow space width around the resonant region as illustrated in Fig. 2a. Density of the trapped particles becomes higher and higher within the potential troughs at least before some particles lose their phase with respect to the wave and become detrapped, i.e., during the resonance time range $\Delta t = t_{dt} - t_{trap}$, where t_{trap} and t_{dt} are respectively the trapping and detrapping times. This means that pressure effects increase

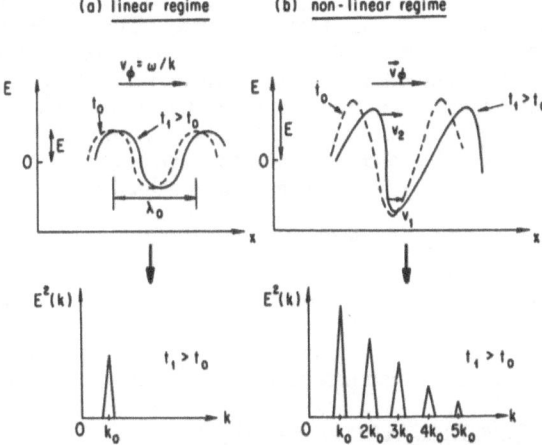

Figure 1: Sketch of a propagating electrostatic wave according that its initial amplitude E (defined at time t_0) is small (case a), or large (case b) i..e., when non-linear effects are negligible or important respectively. Higher harmonics of the initial wavenumber k_0 are excited in case (b).

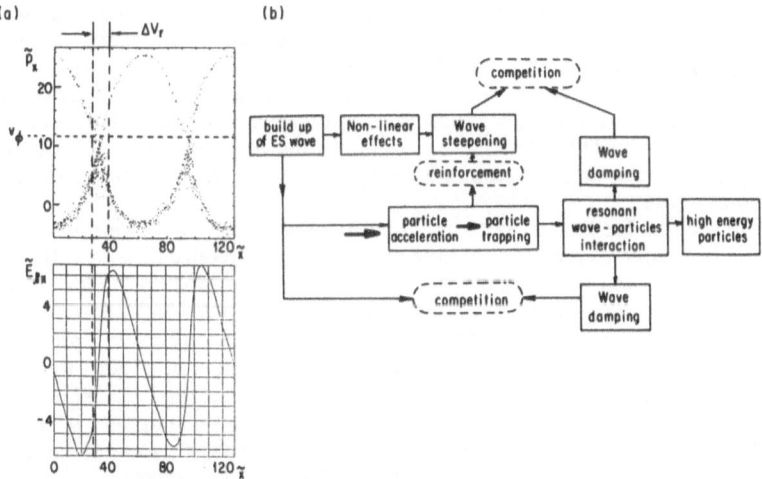

Figure 2: (a) Numerical simulation of a steepening (sawtooth shape) of an electrostatic (ES) wave in presence of particles trapping from a 1D, ES particle code; $|\vec{B}_0| = 0$; v_ϕ is the phase velocity of the wave (Lembege, B., unpublished material); (b) sketch illustrating the competition between wave steepening, wave damping, and particle trapping.

very locally, since regions out of the trapping locations have a low particles density and exhibit only a bulk motion of the background plasma. These effects push the waveform from its initial shape towards a sawtooth shape and force the potential trough to steepen.

This steepening reinforcement is well maintained as long as particles are continuously trapped in time. However, trapped particles are also a source of wave damping. Then, at later times when damping becomes important, the wave amplitude becomes reduced and so does wave steepening. On the same, the reduced wave amplitude may become not strong enough to trap any more particles. Trapping reduces, which means that wave damping also reduces. A competition results between the three processes: particle trapping, wave steepening, and wave damping (Fig. 2b). Then, in the case there is an external feeding mechanism allowing to maintain the presence of the large amplitude electrostatic wave (i.e., by space charge effects, or external applied $\vec{E}$ field), the growth of the nonlinear wave becomes dominant again and particle trapping plus wave steepening may restart. In the absence of any external feeding mechanism, resulting energetic particles have been formed but the efficiency of the acceleration process dies out in time.

2.2 NONLINEAR ELECTROSTATIC WAVE ($|\vec{B}_0| \neq 0$)

In the preceding case where $|\vec{B}_0| = 0$, the trapping loop has a scalelength equal to the wavelength λ_0. We consider now a large amplitude wave E_x propagating along the x-direction, i.e. $\vec{k} = (k_x, 0, 0)$) and perpendicular to a magnetostatic field $\vec{B}_0$ (= (0, 0, B_{0z})). As soon as a particle is getting trapped in the potential well, i.e. when $v_x \simeq v_{\phi x}$, it suffers an important y-acceleration parallel to the wave front by the $\vec{E}_x \times \vec{B}_z$ drift (also called "dynamo effect" or "non-relativistic surfatron effect"). As a consequence, there is a rapid velocity transfer from v_x to v_y; trapped particles have no time to describe a large loop over the complete wavelength λ_0 and a loop of reduced size λ_1 results. In fact, the width λ_1 is of the order of the particle Larmor radius ρ_c since λ_1 is only the x-projection (phase space v_x - x) of the large gyromotion perpendicular to $\vec{B}_0$ (velocity space v_x - v_y), as illustrated in the 3D plot of Fig. 2c. This y-acceleration becomes very large at the expense of the wave energy (wave damping); this dissipation process continues until the detrapping force (i.e., Lorentz force or $q\vec{v} \times \vec{B}$) becomes comparable to the trapping force ($q\vec{E}$) which is itself affected by the wave damping. Then, trapped particles are getting detrapped and describe a large Larmor radius.

We will see that for a quasi-perpendicular magnetosonic wave, ions are trapped by the electrostatic field (due to the space charge effects) and have only time to suffer one bounce within the potential well before being detrapped. Then, detrapped particles are ejected on a large Larmor orbit and start forming a ring distribution [(v_x - v_y) projection in Fig. 2c] provided that there is a continuous feeding of particles through the 3-steps mechanism "acceleration-trapping-detrapping". This overall mechanism is analogous to that of a surfer who tries to accelerate along the wave direction until he succeeds to be in phase with it. While the wave grows and propagates, it stays (or tries to stay!) in phase within the bottom of the wave, and suffers an important acceleration parallel to the wave front (i.e., perpendicular to the wave propagation). The longer it stays in phase, the longer it will the length of the displacement be (provided that the waves do not die out too quickly). This explains the so called "non-relativistic surfatron" terminology used sometimes in the literature; corresponding "relativistic surfatron effect" has been studied by Katsouleas (1984). In summary, more important is the y-acceleration for the particle, larger is the size of larmor orbit (i.e., of the corresponding ring radius) and more efficient is the acceleration process.

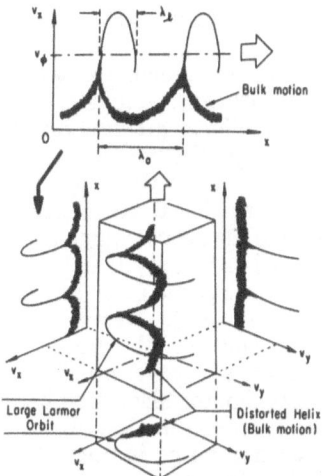

Figure 2c: Sketch of particle trapping by an ES wave when $|\vec{B}_0| \neq 0$; top panel: phase space v_x - x; bottom panel: 3D representation of the phase space (x, v_x, v_y) with corresponding 2D projections $(v_x - x)$, $(v_y - x)$ and $(v_x - v_y)$; $\vec{B}_0$ is assumed along z; the ES wave propagates from the bottom to the top of the figure.

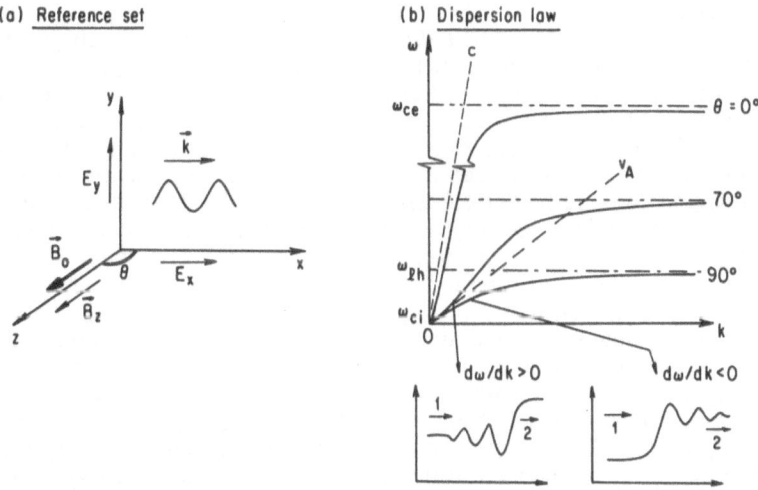

Figure 3: (a) Reference set and field components of MS wave propagating at $\theta = 90°$; (b) Sketch of the dispersion law of MS waves as a function of θ where $\theta = (\vec{k}, \vec{B}_0)$; the sketch is not to scale; ω_{ci}, ω_{lh}, and ω_{ce} denote respectively the ion cyclotron frequency, the lower hybrid frequency, and the electron cyclotron frequency; v_A and c are respectively the Alfven velocity and the light velocity.

2.3   NONLINEAR MAGNETOSONIC (MS) WAVE

Until now, we have focused our attention on the simple example of an electrostatic wave and on the associated particle mechanisms. A logical transition with magnetosonic shocks consists of considering, first, a nonlinear periodic magnetosonic wave (a magnetosonic shock is a particular case of a step profile magnetosonic wave). For simplicity, we will consider a propagation direction ($k = (k_x, 0, 0)$, perpendicular to the magnetostatic field B_0 = (0, 0, B_{0z}). In such a case, the wave has three noticeable components: one electrostatic field E_{lx} (longitudinal field due to the space charge effects and solution of the Poisson's equation); one electric field E_{ty} and one magnetic field B_{tz} (both are solutions of Maxwell's equations); both components E_{ty} and B_{tz} are transverse electromagnetic fields and are now in addition with respect to Section 2.2). Reference set and dispersion features are respectively sketched in Fig. 3a and Fig. 3b. From dispersion relation, the dominant sound-like linear plasma wave is the MS wave, which is a compression wave (Tidman and Krall, 1971). As indicated in Fig. 3b, the dispersion branch of the MS wave for θ = 90° concerns low frequency range (between the ion cyclotron frequency ω_{ci} and the lower hybrid frequency ω_{lh}). Since this range is much below any characteristic electron frequency, ions alone are mainly affected by the presence of MS waves. This means that preceding discussions (Sec. 2.2) concerning particle trapping and acceleration by a nonlinear electrostatic wave may apply to ions in the present case.

A few questions arise with respect to previous discussions and may be expressed as follows:

2.3.1. *How a nonlinear magnetosonic (electromagnetic) wave may steepen from an initial sinusoidal waveform?*
The answer is illustrated in Fig. 4 by numerical simulation results (Lembege and Dawson, 1986) obtained from a 1D electromagnetic, fully particle code and may be summarized as follows: in the presence of a growing MS wave, electrons and ions react quite differently and an electrostatic space charge field builds up [see (i) in Fig. 4]. This large amplitude field is strong enough to trigger a 3-steps mechanism "acceleration-trapping-detrapping" in ions dynamics [see (ii)], while electrons have only a quasi-static answer and suffer a poor adiabatic compression [see (iii) - (iv)]. This coherent (non-stochastic) ion motion leads to an important ion heating; the stochastic ion motion (such as "nonlinear particle trapping" by Karney, 1978) causes some diffusion in velocity space but brings a poor heating contribution in the present case and will not be considered herein. As emphasized in Sec. 2.1, ion trapping reinforces the steepening of the E_{lx} field which strongly distorts in time because of the continuous feeding of trapped ions; in other words, electrons follow the ion dynamics and are compressed at the locations where ion density is very rich, i.e., at the trapping loops, i.e., where the waveform steepens. However, since $v_{thi} \ll v_{the} \lesssim v_\phi$, electrons fall almost immediately into the resonant region ($v_{xe} \simeq v_\phi$) and suffer a rapid but large $E_{lx} \times B_z$ drift parallel to wave front; let us remember that now $B_z = B_{tz} + B_0$. This is at the origin of v_y - x profile [see (iv)] which is more distorted than v_x - x profile, and sawtooth-shape in j_{ye} current results. In contrast with ions, electrons do not stay in the resonance region (i.e., $\Delta t = t_{dt} - t_{trap} \simeq 0$) since their rapid associated cyclotron motion kicks them out of phase from the wave; finally, electrons only receive a kick when passing through $v_{xe} \simeq v_\phi$.

Peaked profiles of currents result, especially in $j_{ye} \gg j_{yi}$ [see (v) - (vi)] because of the light mass of electrons. These peaked profile currents induce corresponding distortions in the (transverse) electromagnetic components E_{ty} and B_{tz}; a steepening of the MS wave takes place at these same locations (trapping loops) and a sawtooth (shock-like) waveform

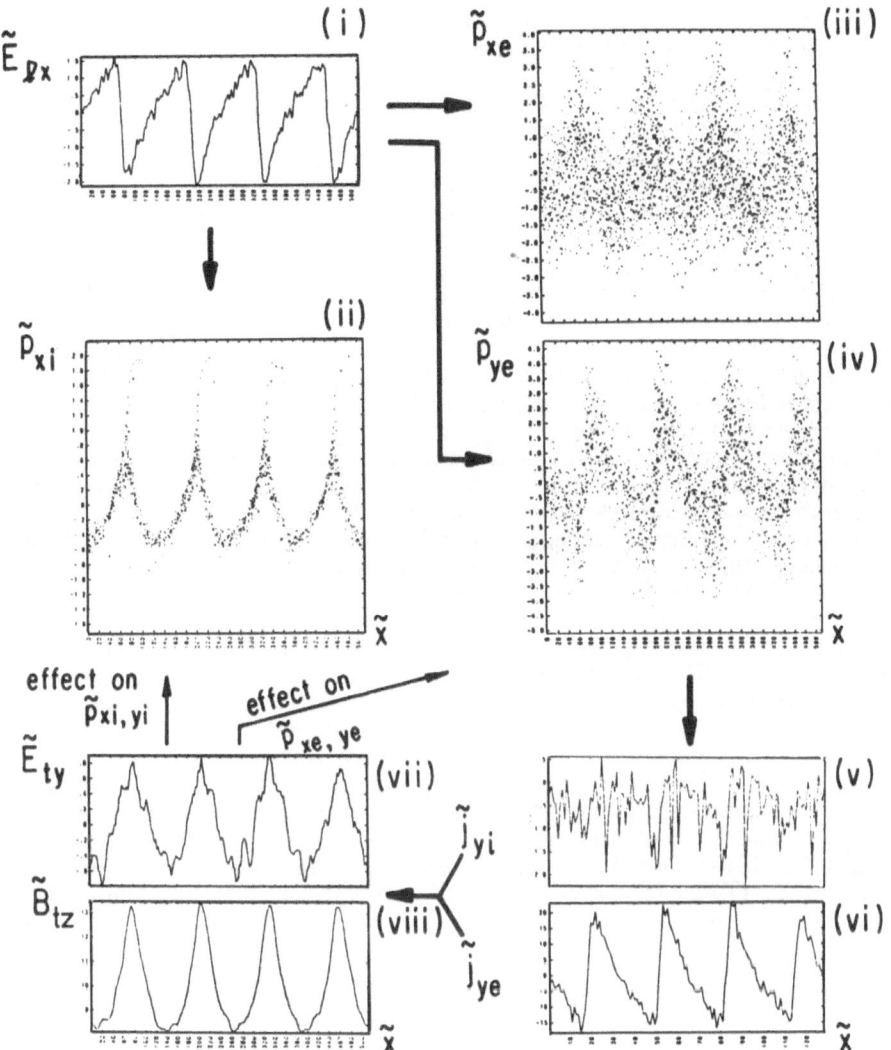

Figure 4: Numerical simulation illustrating the steepening of a large amplitude magnetosonic wave, E_{lx} is the longitudinal electrostatic component; E_{ty} and B_{tz} are the transverse electromagnetic components; results obtained from a 1D electromagnetic, fully particle code (Lembege and Dawson, 1986).

results in the MS wave [see (vii) - (viii)]. Analogous wave particle interactions will also be recovered in the magnetosonic shock wave.

2.3.2 How a nonlinear MS wave differs from a MS shock wave?

Both dispersion and dissipation effects need to be included in addition with the wave steepening. In a general case, wave steepening is characterized by an excitation of higher harmonics in both electrostatic and electromagnetic field components. This means that the characteristic wavelength λ reduces. Inside an idealistic dispersion-free plasma, i.e., where ω varies linearly with k, this process should continue indefinitely. For a more realistic plasma, the wave overtaking will continue until the wavelength λ reaches characteristic values over which dispersion and/or dissipation effects can be efficient enough and counteract the steepening effect. The question whether dispersion or dissipation effect occurs first depends mainly on the relative size of their characteristic scalelengths, respectively λ_{disp} and λ_{diss}, as discussed in details by Kennel et al (1985). We will see in Section 6 that λ_{diss} may vary differently according to the relevant source of energy dissipation, i.e., by viscosity ($\lambda_{diss} = \lambda_{vis}$) or by resistivity ($\lambda_{diss} = \lambda_{res}$).

In summary, dispersion effects are responsible for the formation of trailing or leading wavetrains according to the concavity and the convexity of the dispersion curves around the origin. A dispersive medium assumes at least a variation of $\omega(k)$ as $\omega = ak^3 + \ldots$. Then, concavity and convexity corresponds respectively to the case $d\omega/dk > 0$ and $d\omega/dk < 0$ (Tidman and Krall, 1971). In practice, these cases correspond to directions of propagation $\theta = 90°$ and $\theta \neq 90°$. This curvature effect explains why no precursor wave is expected in front of the shock at $\theta = 90°$.

On the other hand, dissipation effects can come from different sources according to the plasma parameters. Then, a competition between dispersion effects and dissipation effects results on the steepened waveform and can be summarized as follows: the oscillating steepened MS wave will transform (a) into a pure solitary wave (MS soliton) if the dispersion effects prevail, i.e., within a dissipation-free medium, or (b) into an oscillatory MS shock as soon as a small amount of dissipation is considered. In the latter case, an oscillatory wavetrain will appear in the upstream or downstream region (respectively in front of or behind the shock front) according to the angle θ [see Fig. 3b]. The evolution of a steepened profile into a soliton or a shock pattern has already been largely developed in previous studies (Tidman and Krall, 1971; Biskamp, 1973), and is beyond the scope of the present review.

3. Shock Waves in Space Plasma

Very intensive activity has already been involved on shock studies, in particular after the large amount of experimental data on magnetospheric shocks which have been gathered by the successful space missions ISEE and VOYAGER. The first mission was mainly devoted to the study of solar wind-earth magnetosphere interaction, while the second mission was oriented rather on planetary studies (other than the Earth). Extensive studies have been more recently performed on the so-called "cometary shocks" (the observation of a well defined shock pattern for comets is still a subject of controversy at the present time) in connection with the experimental observations issued from the encounters with the Vega, Giotto, Sakigake, and Suisei spacecrafts with comet Halley, as well as that of the International Cometary Explorer (ICE) with comet Giacobini-Zinner. As indicated previously, we had to restrict this review in order to introduce the reader to the elementary features of shocks. A lot of experimental, theoretical, and numerical studies related to

ISEE missions have allowed a detailed study of the magnetospheric bow shock. We will follow the same way in order to see what a "real" shock looks like.

Let us remember shortly the main features of the Earth magnetosphere as sketched in Fig. 5a. The solar wind (plasma continuously emitted from the sun) distorts the Earth magnetic field by "compressing" and "dilatating" the magnetic field lines respectively on the day side and on the night side of the Earth, and form the magnetospheric cavity surrounding our planet. The solar wind continuously flows on this cavity which plays the role of an obstacle. Because of the high values of this continuous flow speed, a shock interface forms and separates both media: a medium of high plasma density and of strong magnetic field (terrestrial magnetosphere) and a medium of low plasma density and of weak magnetic field (solar wind), where the interplanetary magnetic field is considered as being that of the sun (Fig. 5b).

As a first approach, MS shock waves that are observed in front of the Earth's magnetosphere are commonly classified as follows: (1) according to the angle θ between the normal $\vec{k}$ to the shock front and the magnetostatic field $\vec{B_0}$ which leads to distinguish the so-called quasi-perpendicular shocks (hereinafter named q-$\perp$ shocks, i.e., $90° > \theta > 45°$) and quasiparallel shocks (hereinafter named q-$\parallel$ shocks, i.e., $45° > \theta > 0°$); (2) according to their associated Mach number $M_A = v_{sw}/v_A$, which leads to subcritical ($M_A < M_A^*$) and supercritical ($M_A > M_A^*$) Mach number, where v_{sw} and M_A^* are respectively the drift velocity of the solar wind and the first critical Mach number which will be discussed later on. A third parameter, ß, the ratio of the plasma thermal energy over the magnetic field energy plays also an important role and will be taken into account when necessary ; a fourth parameter, T_e/T_i, can be also important in particular for ion-acoustic shocks. Let us mention that a detailed analysis of a typical crossing of the terrestrial bow by ISEE satellites has been performed by Scudder and al. (1986a, b, c).

A shock may be considered as a transition layer between a low entropy state (medium 1) and a high entropy state (medium 2). All macroscopic plasma quantities, such as density, temperature and pressure, defined on each side (1) and (2) of the transition layer are ruled out by the Rankine-Hugoniot fluid conditions. The dissipation mechanisms efficiency in this situation determines the structure of the shock. For a collisionless shock (such as the Earth bow shock), certain suitable mechanisms must be invoked to explain the small scale length of the shock (~ 100 Km) despite the large collision mean free path (~ 10^7 Km) of the solar wind. The main difficulty is to identify correctly the various dissipation sources since the associated mechanisms (i) vary considerably according to the shock parameters and in particular to the values θ, β and M_A of concern and, (ii) are not only localized at the transition layer but may also strongly affect the upstream and downstream regions far away from the transition layer. Typical experimental data obtained from ISEE satellites are represented in Fig. 6. Let us precise that the abscisse is the Universal Time (UT) which is equivalent to considering the distance travelled by the spacecraft when it crosses the shock environment, N_E and N_i are respectively the electron and ion density (cm^{-3}); T_p and T_e are the corresponding temperatures (K°); v_p is the proton bulk speed (Km sec^{-1}); P_E the electron pressure (10^{-9} N cm^{-2}); B is the magnetic field (nT) and $\theta_{Bn} = (\vec{k}, \vec{B})$ indicates a perpendicular shock ($\theta_{Bn} \simeq 90°$).

3.1 VARIATION VERSUS θ

A q-$\perp$ shock is characterized by a narrow, abrupt and well-defined jump in the magnetic field profile; similar remarks apply to the other quantities. In contrast, the transition region is hard to define in q-$\parallel$ shocks since the associated magnetic field is very turbulent and presents an oscillatory pattern over a much wider space range. From "in situ"

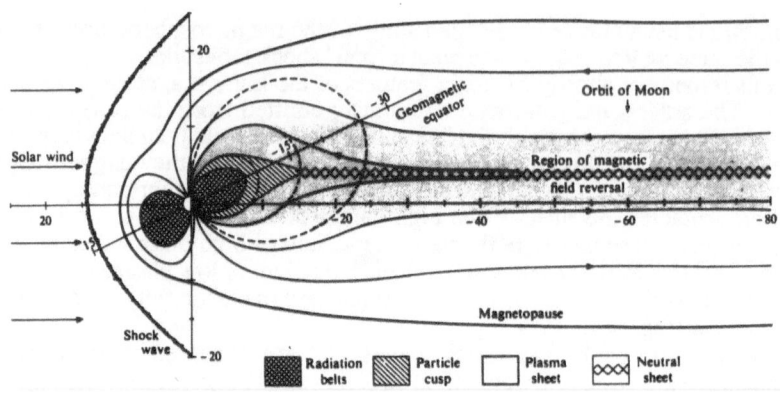

Figure 5 (a): Sketch of the Earth magnetosphere in the noon-midnight plane (from Speiser and Ness, 1967).

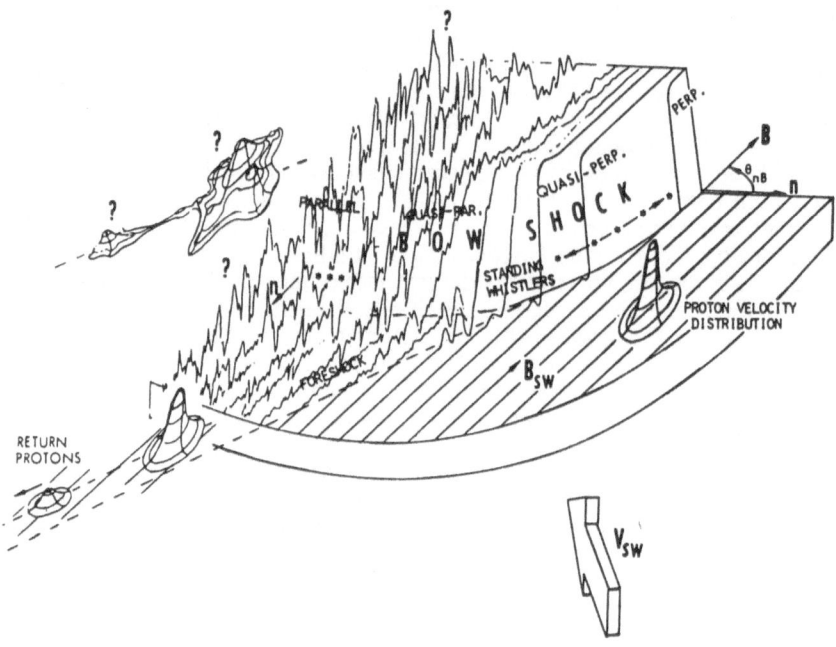

Figure 5 (b): Sketch of collisionless shocks as manifested in the earth's curved bow shock. Unshocked interplanetary field direction B_{sw} is indicated in the foreground field "platform". Field magnitude is plotted vertically. The superimposed 3D sketches represent solar wind proton thermal properties as distribution in velocity space; V_{sw} is the direction of the solar wind flow. The foreshock defines the region when particles are accelerated back from the shock into the solar wind (from Greenstadt and Fredricks, 1979).

observations, q-⊥ shocks appear "stable" in contrast with q-∥ shocks. However, this stability feature seems to be not strictly evidenced and is discussed in Section 5.

3.2   VARIATION VERSUS M_A

At low Mach (subcritical) numbers, shocks are characterized by a monotonic-like transition layer with an abrupt jump in B field. Sometimes an oscillatory wavetrain may occur downstream according to the value of β (mentioned earlier). The wavetrain is downstream only if $(\theta - \pi/2) \leq (m_e/m_i)^{1/2}$. Otherwise, wavetrain is upstream. The upstream phase-standing whistler wave has been observed for subcritical quasi perpendicular shocks.

At higher Mach (supercritical) numbers, shocks have a more rich pattern, including three distinct regions: a foot (with thickness ~ c/ω_{pi} where c/ω_{pi} is the ion inertial length) just ahead of a thin ramp (<< $c/\omega/_{pi}$) where the magnetic field rises rapidly, and a post-ramp overshoot (several c/ω_{pi}) which is the beginning of an oscillatory trailing wavetrain. However, since $c/\omega_{pi} \sim 1/(N)^{1/2}$ and the density N can jump by 8-10 times the upstream value, it is not clear that the ramp thickness is much less than the local value c/ω_{pi} (as measured in the overshoot). This point is important since it is commonly assumed that the ramp thickness is of the order of c/ω_{pe}. The wavetrain dies out further behind the ramp where the magnetic value achieves its final downstream value. These signatures appear both in magnetic field profiles (Livesey et al., 1982, 1984) and in ion observations (Schopke et al., 1983). The foot in the B field is an important signature of high-M_A shocks and is directly associated with an analogous ion density profile and a drastic jump in the ion temperature T_i; in contrast the variation in T_e is rather limited. This result is clearly apparent by comparing the different profiles located between the vertical dashed lines in the left hand panel of Fig. 6. Observations of the ion distribution function have shown that a certain percentage of incident ions are specularly reflected in the foot; these ions gyrate in the upstream magnetic foot and then re-encounter the shock where they pass downstream.

Deeper analysis of particles and fields observations obtained from ISEE satellites may be found in "ISEE Upstream Wave and Particles" (1981), which is recommended reading. We will see in Sections 5 and 6 how numerical simulations allow the recovery of most shock features. At once, let us consider in Section 4 what kinds of numerical methods are best appropriate to deal with collisionless shocks.

4. Numerical Methods

The main computational methods used for simulating collisionless shocks are briefly summarized. They are three basic types of codes: hydrodynamic, full particle, and hybrid; their main characteristics are summarized in Appendix 1, 2, and 3 respectively.

4.1   HYDRODYNAMIC CODES

They solve a set of fluid equations with assumptions about the plasma pressure in order to close the set of equations (e.g., MHD model). This kind of code is ideal for studying long time periods; however, they are of restricted use since important phenomena in collisionless shocks take place over scalelengths less than the fluid scale length. As a consequence, some kinetic effects need to be included.

As mentioned in Section 3, a shock wave can be defined as a transition layer causing a change of state through the plasma. In the analysis of Appendix 1, the shock is considered as a discontinuity with zero thickness, across which the fluid changes state. This is an idealistic case. In a more realistic description, the process involved in the build up of the

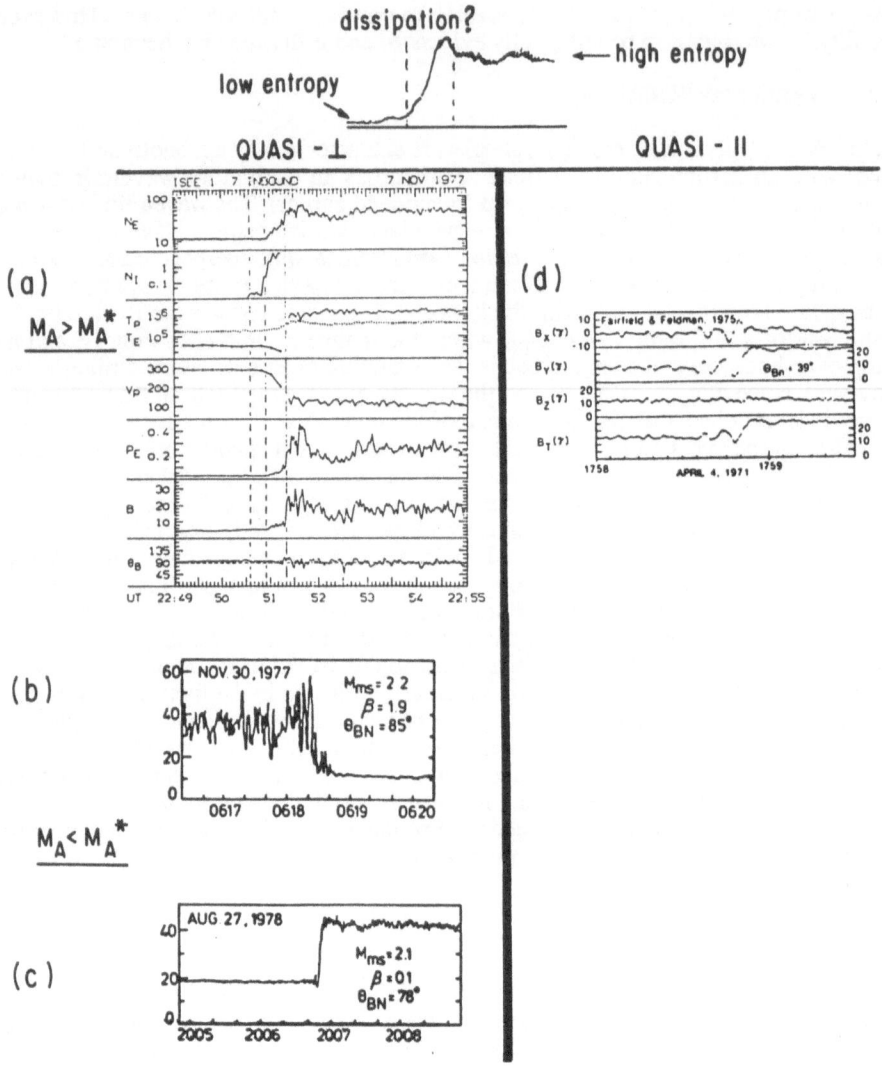

Figure 6: Space experimental data observed when crossing the terrestrial bow shock for various directions θ (q-⊥ and q-‖) and various Mach regimes; (a) from Sckopke et al (1983); (b) and (c) from Formisano (1981); (d) from Fairfield and Feldman (1975).

transition layer and its accompanying scalelength need to be included. Then the shock arises as a result of the non-linear steepening of propagating compressional waves, which is balanced by dissipation or dispersion effects. Both effects limit the steepening by damping out the short wavelength components of the shock region (Section 2). In MHD code, the steepening is contained in the $\vec{u} \cdot \nabla \vec{u}$ term while the dispersion is provided by the Hall current $\vec{j} \times \vec{B}$ term (Eq. 10 in Appendix 1); the dissipation may be issued from two sources: resistivity (term S_η) and/or viscosity (term S_μ). The characteristic transition length (i.e. the shock thickness Δ) for each one can be estimated by taking μ (Δ_μ) and η (Δ_η) constant. Finally, the observed shock structure depends on the ordering of the lengths Δ_μ, Δ_η, or L_H, where L_H is the dispersive wavetrain scalelength. In other words, the steepening stops until it reaches the longest wavelength (from dispersion or dissipation) which will define the shock thickness (Kennel et al, 1985).

A question arises: where is the entropy jump associated to shock wave coming from? A simple approach may be expressed as follows: in collisional shocks, the plasma state changes by binary encounters. For collisionless shocks, which are our main interest herein, the mechanisms are much more complex. The shock transition is accomplished by collective interactions between particles and self-consistently generated electric and magnetic fields. Further, electrons and ions are affected quite differently by field fluctuations. A basic tool for investigating collisionless plasma processes is the Vlasov equation, supplemented by Maxwell's equations for $\vec{E}$ and $\vec{B}$ in terms of self-consistent charge and current densities (Papadopoulos, 1985). The shock transition is accompanied by an entropy increase which, in the presence of dissipative processes such as resistivity or viscosity, is given (Kulikosky and Lubinov, 1962) by:

$$S = C_v \ Ln \ (P/N^\gamma)$$

On the other hand, the stationary solution of the Vlasov equation gives $\partial_x \int dv \ v \ f \ Ln \ (f) = 0$, which means that entropy flux is conserved. However, if the plasma is unstationary, i.e., turbulent or partially turbulent in time, the entropy flux is not conserved. In this case, laminar and turbulent fields can be formally distinguished by writing all quantities in terms of their true average values and their fluctuations about the mean value (Tidman and Krall, 1971):

$$f = <f> + \delta f, \quad \vec{E} = <\vec{E}> + \delta \vec{E}, \qquad \vec{B} = <\vec{B}> + \delta \vec{B}$$

When applied to the Vlasov equation, it appears that the origin of the entropy production by turbulence is the time behavior of f, where δf, δE, and δB are caused by microinstabilities. A deeper analysis may be found in Papadopoulos (1985), which presents also a generic class of shock transitions including (a) laminar dispersion dominated shocks, (b) quasi-turbulent subcritical and supercritical shock (Formisano, 1981), and (c) turbulent shock (Greenstadt, 1985).

4.2   FULL PARTICLE CODES

Full particle simulation codes (reviewed by Forslund, 1985) include a complete set of Poisson's and Maxwell's equations, associated with the equation for the motion of each particle which is followed individually for both electrons and ions (Appendix 2). No simplifying assumption is used in the basic equations, i.e., this kind of code is appropriate to approach the description of a real plasma. However, because of computer limits (both in memory size and in computer time), some numerical artifacts are necessary which means that this duplication cannot be exactly achieved. Then, a simulation run will include many fewer particles than in any real situation. Each particle has a finite size in order to deal with collisionless plasma (Birdsall and Langdon, 1985; Dawson, 1983). The time spent by the simulation will cover a few ion gyroperiods and the size of the simulation box is restricted to a few (or a few tens) of gyroradii. The ion to electron mass ratio are chosen to be artificially low in order to make simulation runs possible. Particle simulation codes tend to be best suitable to a study of oblique shocks, of microinstabilities expected within shocks, space charge effects, and of any other phenomena where electrons exhibit some dynamics and control the length and the time scales.

4.3   HYBRID CODES

A hybrid code represents a compromise between a full particle code and a MHD code (Appendix 3). In practice, fluid equations (Appendix 1) are used for electrons, while a particle description (Appendix 2) applies to ions which are followed individually. The main consequences may be summarized as follows:

advantages:

(i)   as shown later, ion dynamics play a dominant role on the shock pattern, in particular for supercritical shocks. Then, the ion-shock wave interaction may be followed easily over a large simulation box (a few hundreds of ion dissipation length c/ω_{pi}) and over long time runs (several tens of ω_{ci}^{-1}) since electrons time and space scales are neglected and do not need to be followed. This code also includes various ion distribution instabilities which affect the overall pattern of the shock.

(ii)   some electron behavior can be introduced artificially by using a phenomenological resistivity or changing the electron fluid equations of state. This resistivity allows us to maintain a coupling between the electron fluid equations [terms of ηj^2 and $\eta(\vec{j_e} + \vec{j_i})$ of Eqs. (a)] and the particle pusher [term $\vec{P}$ in Eqs. (b)] as shown in Appendix 3; it illustrates the anomalous resistivity, i.e. the "friction" created on ions by the electrons related to some cross-field current instabilities.

disadvantages:

(i)   all space and time electron scales are neglected since $m_e = 0$. Electron dynamics which can be relevant in some oblique shocks are forbidden.

(ii)   the exclusion of the displacement current leads to neglect the high frequency electromagnetic effects.

(iii)   quasi-neutrality condition imposes an electrostatic field which may differ in amplitude from that obtained from the self-consistent space charge field. As an example, space charge effects are important over Debye length scales, so ion acoustic subshocks cannot be modeled with hybrid codes.

(iv)   introducing an artificial resistivity η (external parameter defined by the user) leads to a breakdown of the self-consistency of the code since η is fixed to a constant value for each simulation. Self-consistency is also broken down by (i) and (iii), and this kind of code cannot be applied for shocks where electrons and ions have some individual dynamics, i.e., where electrons cannot be considered any longer as a charge neutralizing fluid.

4.4   GENERAL FEATURES OF SIMULATION CODES

4.4.1   *Particles of Finite Size.*
Hybrid and particle codes are based on the idea of "macroparticles" and on the "particle in cell" method (PIC) (Morse, 1970; Dawson, 1983; Birdsall and Langdon, 1985). At the present time it is impossible to include a real number of particles in a computer simulation. However, if the simulation particles were treated as point particles, i.e., equivalent to

calculating the particle moments as a sum over delta functions, the calculated particle moments would be irregular functions which would lead to some important numerical noise. In order to smooth the moments, and redistribute the spectrum of numerical noise, particles are treated with a finite size so that they contribute to more than one cell when moments are computed. Their contributions to any cell depends on some function (commonly the "shape" or the "size" of the particle) of their distance from that cell (Dawson, 1983; Birdsall and Langdon, 1985). A brief description of numerical algorithms is included in Appendices 2 and 3 for full particle and hybrid codes respectively.

4.4.2 *How to Launch a Shock Numerically ?*

There are three basic methods used to form a shock in numerical simulations (Burgess, 1987a) which are sketched in Fig. 7.

(a) Rigid boundary method: The plasma simulation box is open at one boundary (l.h.s.) where plasma is continuously injected into the box. At the other end (r.h.s.), a rigid wall is fixed and reflects particles. At early times, the inflowing plasma reflects on the r.h.s. wall and leads to a two counterstreaming plasma flow right at this rigid boundary. This initiates a wave which grows, steepens, and propagates (from the right to the left side in fig. 7a) into the incoming flow. One advantage of the method is that the magnetic piston is left behind by the shock it creates, for quasi-perpendicular shocks. However, for oblique shocks, downstream ions can bounce between the shock and the rigid wall, maintaining communication between them. This problem is particularly severe for slow shocks, when ion thermal speed is much faster than shock speed.

(b) Magnetic piston method: Initially the simulation box is occupied by a uniform magnetized plasma at rest. At early times, an electric field (or an external current) is applied at one boundary, forcing a magnetic flux into the simulation box ; this flux, equivalent to a piston, pushes the plasma (as a "snowplowing" machine) and initiates a shock (Fig. 7b). Particles (ions) pressure gradients resulting from the thermalization cause the leading edge of the disturbance to travel at a velocity higher than that of the mass center while the trailing edge moves slower. By this way, the shock (finite jump in the density and fields space profiles) separates from the piston (where the plasma density drops to zero). A certain care is necessary since, in some cases, the magnetic piston and shock separates slowly and a noticeable plasma heating may result at the piston itself which may provide disturbances in the structure of the shock wave.

(c) Injection method: Initial conditions are carefully chosen to be near to those expected of the shock one wants to create. Typically, both boundaries of the plasma simulation box are open (Fig. 3c) with an incoming (l.h.s.) and outcoming (r.h.s.) plasma flow at each boundary respectively. A smooth jump between upstream and downstream states is used with the downstream conditions calculated from the Rankine-Hugoniot jump conditions. The difficulty of the method consists in correctly anticipating the downstream conditions and keeping the shock within the simulation box by providing appropriate conditions of particle injection at both upstream and downstream boundaries.

In all methods, the shock regime (i.e., the values of M_A, or the amplitude of the shock itself), may vary according to the injected plasma velocity (in methods (a) and (c)), or the amplitude of the applied electric field or external current (which generates the piston in method (b)). A few other methods exist but can be considered only as derivations of the one of the techniques quoted above.

How to generate a shock numerically

(a) Rigid boundary method

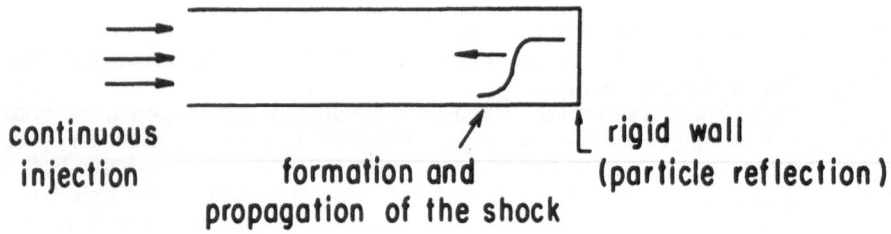

continuous
injection

formation and
propagation of the shock

rigid wall
(particle reflection)

(b) Magnetic piston method

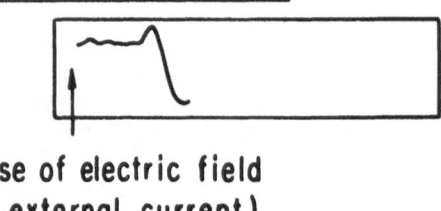

pulse of electric field
(or external current)

(c) Injection method

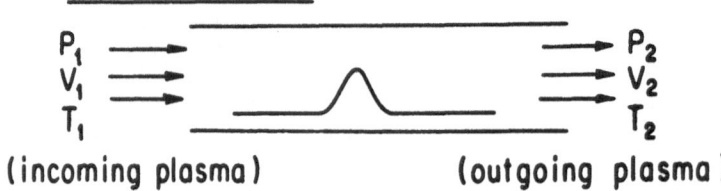

P_1
V_1
T_1

P_2
V_2
T_2

(incoming plasma)

(outgoing plasma)

Figure 7: Sketch of the different numerical methods used for launching a shock.

5. Quasi-Perpendicular Magnetosonic Shocks; Numerical Results

5.1   NECESSITY FOR A SYNTHETIC CLASSIFICATION

As mentioned in Section 3, the increasing interest for studying collisionless shocks has been particularly motivated by the observations made by ISEE spacecrafts on the terrestrial bow shock. Numerical simulation reveals to be a very appropriate tool in order to analyze the shock characteristics in despite of its complexity and of its dependence on a large number of plasma parameters. This deeper numerical investigation was particularly helpful for improving progressively a synthetic view of collisionless shocks through a detailed parametric study instead of considering several different shock models. We will focus herein on quasi-perpendicular shocks (q-⊥) which exhibit a rather well defined pattern. It results from space observations and discussion of Section 3, that ion kinetic dynamics needs at least to be included in details. This is made possible in either hybrid or full particle codes on which we will concentrate now. We will consider the two typical cases, subcritical and supercritical shocks.

As notified in Section 3, the main dissipation sources which are required to complete a shock turned out to vary considerably according to the associated Mach number M_A (also with angle θ and β values). Then, the electron and ion pressure may largely differ from each other depending on the dissipation mechanism of concern. In a first approach, a simplified overview of both subcritical (low M_A) and supercritical (high M_A) shock regimes is sketched in Figure 8 for different values of β_{i1} in order to clarify when ion thermal effects prevail over the magnetic effects in the shock pattern; β_i is the ratio of ion kinetic to the magnetic energy while subscripts 1 and 2 will denote respectively upstream and downstream regions. The β dependance needs to be considered with a particular caution and requires certainly a deeper analysis. The difficulty comes from the fact that varying β leads to a change of M_A^* (and then to a possible change of the Mach regime M_A/M_A^*); also, a certain lack of complete experimental and numerical analysis still persists on this question at the present time. As emphasized in Section 2.3, while the shock propagates, a competition takes place between the build up of the oscillatory wavetrain (dispersion effect) which occurs behind or before the jump of the compression wave (steepening effect) depending on the angle θ, and its damping by thermal effects (dissipation effect). This competition leads to a more or less steady-state shock pattern, and rules the shock structure according to the strength and the kind of the dissipation of concern.

5.2   SUBCRITICAL SHOCKS

At low M_A, a dissipation source is necessary to satisfy the jump conditions as required by the Rankine-Hugoniot equations. In such a regime, the dissipation is thought to be provided by resistive heating, primarily the electrons, which is associated with the strong electric currents required to produce the steep gradients in the magnetic field (Tidman and Krall, 1971; Biskamp, 1973). The subject of subcritical shocks has been studied extensively over the last several years (to see the review by Mellott, 1985). A noticeable ion heating as well as electron has been evidenced depending on parameters.

Sometimes, subcritical shocks are labelled laminar when $\beta_{i1} \ll 1$ and quasi-turbulent when $\beta_{i1} \simeq 1$. The transition from laminar to a turbulent shock transition is conveniently described by the growth of an instability which can be classified as macroscopic or microscopic according to the scalelength over which they develop (Biskamp, 1973). Basically, the shock is thought to be resistive characterized by a smooth monotonic transition layer which scales with the resistive length. For low β_{i1}, the dissipation effects

Low Mach shock ($M_A < M_A{}^*$)	High Mach shock ($M_A > M_A{}^*$)

a) Dissipation dominated by anomalous resistivity (?)
 - electron heating "dominant"
 - ion heating (low) created
   . by resistivity from electrons if B weak
   . by non linear Landau damping if B "strong"
	a) Dissipation dominated by anomalous viscosity
	 - ion heating dominant
	   (reflection and cyclotron motion)
	 - poor electron heating

b) No foot	b) Well defined foot
c) Limited viscous effect (ion reflection)	c) Strong ion reflection
$\beta_{i1} \ll 1$	
d) Weak dissipation effects	d) Noticeable dissipation effects
e) Trailing wavetrain quasi undamped → laminar <u>oscillatory</u> magnetosonic shock	e) Wavetrain strongly perturbed and damped
$\beta_{i1} \approx 1$	
f) Noticeable dissipation effects	f) Very large dissipation effects
g) Wavetrain partially damped by ion Landau damping	g) Wavetrain completely suppressed by more diffuse processes of ion damping → "<u>monotonic</u> magnetosonic shock"?

Figure 8: Sketched overview of low and high Mach number shocks characteristics according to the β_{i1} value. This applies to quasi-perpendicular shocks.

are negligible, i.e., dispersion prevails and a laminar oscillatory magnetosonic wavetrain is observed. At larger β_{i1}, ion Landau damping acts to reduce the trailing wavetrain as in Mason (1972); let us remember that Mason (1972) used a hybrid code that retains electrons inertial effects and pressure but neglects resistivity (i.e., $m_e\, dv_e/dt \neq 0$ and $\eta = 0$ in equations of Appendix 3). This study differs from that of Leroy et al (1981, 1982), where resistivity has been included ($\eta \neq 0$, but $m_e = 0$) and leads to a dominant effect of η in washing out the wavetrain. Which dominant process between ion Landau damping or resistivity is still an open question and certainly requires a detailed parametric study. On the other hand, the electrostatic jump at the shock front is just strong enough to reflect only a small fraction of upstream ions (Biskamp and Welter, 1972a; Ohsawa and Sakai, 1985; Lembege and Dawson, 1987a), meaning that the associated dissipation mechanism is limited.

Usually the invoked anomalous resistivity is provided by cross-field currents instabilities; in order to be included, this effect requires a careful choice in the simulation code (hybrid or fully particle) and in the orientation of $\overline{B}_0$ field (in 2D codes) as illustrated in Appendix 4. Identifying the dominant kinds of instabilities associated with subcritical as well as with supercritical shocks is not an easy task. The present section will try to introduce this problem while a rather broad overview of corresponding plasma instabilities will be presented in Section 6. Basically, the resistivity allows a momentum transfer from streaming electrons to ions to be heated and/or between different streamings of particles of the same species.

However, it is important to note that a relevant resistive electron heating has not been clearly evidenced both experimentally and numerically until now. The usual correspondence "low Mach shocks are resistive shocks" requires a deeper analysis for confirmation. Electron heating is not so "dominant", as expressed by quotes " in the l.h.s. panel of Figure 8. Its importance "seems" larger with respect to ions at low M_A in the sense that ion heating is very large for high M_A, and drops to lower values for low M_A regime. However, the amount of resistive electron heating (by ion acoustic instability) was found to be small and ion reflection still provides a major source of dissipation (Tokar et al., 1987) for low M_A regime. This results seems quite consistent with detailed and careful analysis of shock overshoots by Mellott and Livesey (1987), who found that most of "laminar" shocks observed by ISEE 1 and 2 satellites do exhibit overshoots, albeit small ones. The omnipresence of overshoots is consistent with laboratory data (Strokin, 1985) and spacecraft observations (Schopke and al., 1983; Thomsen and al., 1985; Greenstadt and Mellott, 1987); all of which have indicated the presence of a small but measurable number of reflected ions associated with even nominally subcritical shocks.

5.3   SUPERCRITICAL SHOCKS

At high Mach numbers, the dissipation source provided by resistivity is not strong enough to satisfy the Rankine-Hugoniot jump relations across the shock and an additional dissipation mechanism is needed. The reflection of a substantial fraction of upstream ions at the shock front was shown to be a good candidate to provide the additional dissipation. However, the consequences of ion reflection on the shock features vary according that ions are unmagnetized (Papadopoulos et al., 1971a; Forslund and Freidberg, 1971; Mason, 1972) or magnetized (Auer et al., 1962, 1971; Auer and Evers, 1971; Biskamp and Welter, 1972a; Chodura, 1975; Leroy et al., 1981, 1982; Leroy, 1983; Winske and Leroy, 1984a; Winske, 1985; Forslund et al., 1984; Lembege and Dawson, 1984a, 1987a; Ohsawa, 1985, 1986; Lee et al., 1986 a, b, 1987). In the first case ($|\overline{B}_0| = 0$), reflected ions can escape freely upstream as a beam and interact with the background plasma to develop some ion-ion instabilities. In the second case ($|\overline{B}_0| \neq 0$), the upstream ion escaping is prevented by the upstream $\overline{B}_0$; this corresponds to the case of the terrestrial magnetosphere where the upstream magnetostatic field (i.e. interplanetary field) is weak but finite. Reflected ions gain a large energy in suffering an upstream gyration, succeed to overcome the potential barrier (both electrostatic and magnetostatic) at the second encounter with the shock front and gyrate downstream where they thermalize.

In summary, ion heating required to satisfy the Rankine-Hugoniot equations for high M_A shocks requires extraction of energy. Since this extra-energy is in the upstream ion flow, an ion thermalization process is required. In a fluid description, this corresponds to ion viscosity. For a collisionless plasma, such anomalous viscosity can be attributed to counterstreaming ion instabilities (Papadopoulos et al., 1971a; Tidman and Krall, 1971). The counterstreaming can be produced either by the mixing of the upstream and downstream state or by the presence of ions reflected by the shock as shown in Section 5.4.

5.4 NUMERICAL RESULTS

Ion dynamics through high M_A shocks will be illustrated by results obtained from hybrid and full particle codes. Such comparisons will allow us to emphasize the differences obtained by both kinds of codes and to stress the consequences of their respective weakness on the simulation results.

5.4.1 *Hybrid Codes*.

The original code evolved out of earlier work by Chodura (1975) and numerical details can be found in Winske and Leroy (1984a). Simulations have been carried out by Leroy et al., (1981, 1982) with a 1-D hybrid code and including an artificial resistivity ($\eta \neq 0$, in Appendix 3). Shocks have been launched by the injection method (Section 4.4.2). Numerical results have clearly emphasized that the "self-consistent" shock structure which evolved in time was closely linked to the dynamics of the reflected ions. The simulations produced magnetic (and electric) field profiles which strongly reproduce the ISEE observations, namely, foot, ramp, and overshoot i.e., beginning of the downstream trailing wavetrain; in particular, connection between the reflected ions and the magnetic foot was quite clear (Figures 9 and 10). With increasing M_A, the reflected ions in the simulation appeared simultaneously with the overshoot in the magnetic field and has been analyzed analytically (Leroy, 1983). Thus, reflected ions behave essentially like a beam in the shock front, including the foot and the ramp region, whereas they tend to form a gyrating stream in the downstream region behind the overshoot; in addition, reflected ions are ejected on a large Larmor orbit during their gyration and tend to form a ring. All these various ion populations are ideal candidates for additional dissipation sources via plasma instabilities. It was found that the ramp thickness was mainly determined by the value of the "external" resistivity η; let us remember that the mechanisms defining self-consistently the value η are unknown in this case. The process of ion reflection has been described in terms of a combination of a potential jump (not self-consistent because of the quasi-neutrality condition and the use of a fixed value η) at the shock ramp, and the deceleration of the upstream flow in the magnetic foot. Appropriate numerical diagnosis have allowed a direct comparison between numerical results and experimental data obtained from ISEE satellites (Goodrich, 1985).

Simulations have been extended to oblique q-$\perp$ shocks ($90° > \theta > 45°$) by Leroy and Winske (1983). Similar shock structures (foot-ramp-overshoot) have been recovered. For strongly oblique shocks, it was found that some incident ions arriving on the shock could be turned around by the shock in such a way as to escape upstream (a process known as "direct reflection", Figure 11). These ions are regarded as one source of the ions in field aligned beams seen upstream of the Earth's bow shock (Thomsen, 1985; Burgess, 1987b). It was suggested that such incident ions which return upstream after one near specular reflection at the shock front (as observed numerically for $\theta = 30°$), are the reflected gyrating ions reported upstream of the q-$\parallel$ shock by Gosling et al. (1982). However, the electron dynamics for such cases has been completely neglected since $m_e = 0$ (Appendix 3). We will see that electron kinetic effects play an important role for oblique shocks.

In summary, hybrid codes have been very helpful in allowing a direct comparison with experimental results. Simulation results have also confirmed quantitatively previous ideas as (i) the importance of ion reflection representing the "breaking of laminar flow" and the

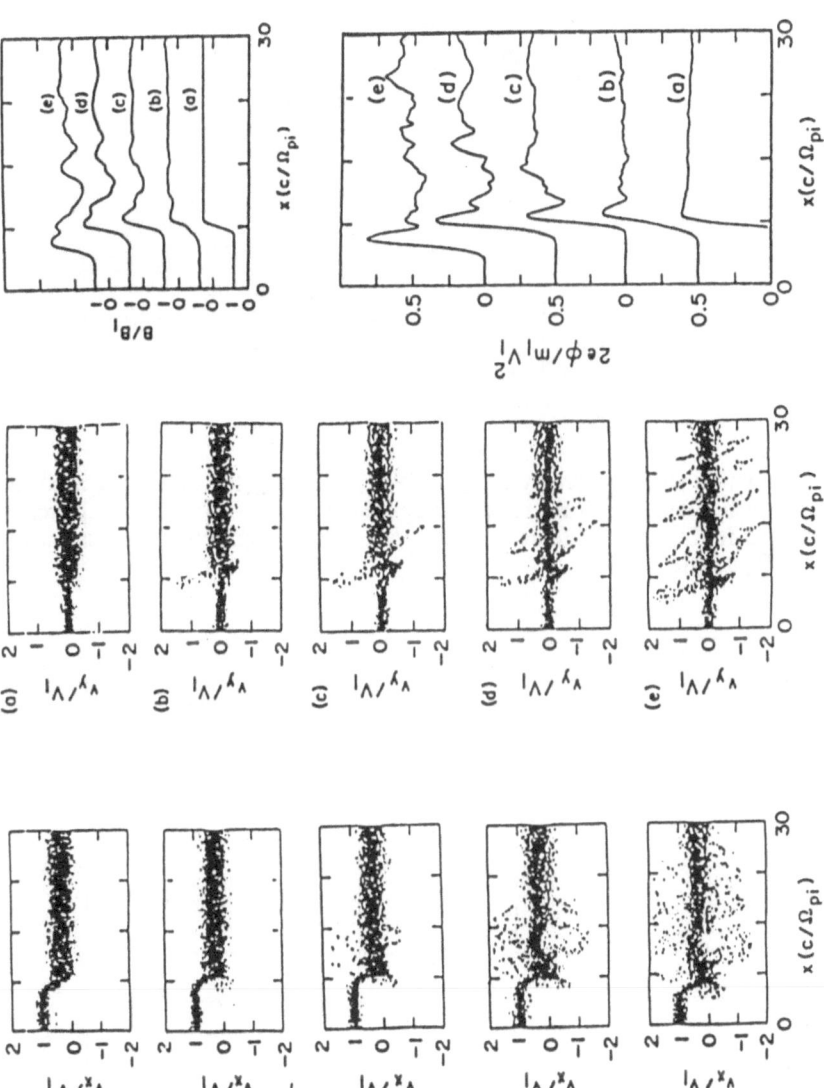

Figure 9: Simulation results from Leroy et al. (1982) showing developments of structure at five successive instants (a) - (e); ion phase space, $(v_x - x)$ and $(v_y - x)$; magnetic field (B/B_1) and electric potential $(2e\phi/m_1v_1^2)$. Note that $\theta = 90°$ and $\bar{B}_0 = (0, 0, B_0)$; $M_A = 6$, $\beta_e = \beta_i = 1$, (a) - (e): $t = 0$, $1.3\ \Omega_{ci}^{-1}$, $2.6\ \Omega_{ci}^{-1}$, $5.2\ \Omega_{ci}^{-1}$, and $9.6\ \Omega_{ci}^{-1}$ respectively.

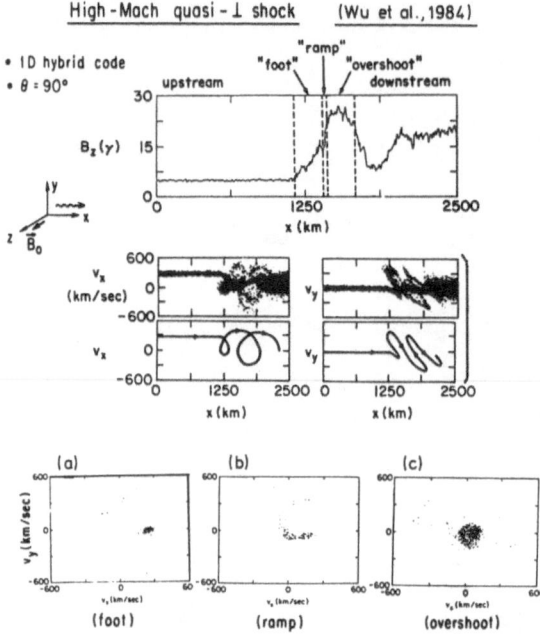

Figure 10: Simulation results from Wu et àl. (1984) showing the detailed connection between the magnetic field profile B_z, ion phase space (v_x - x, v_y - x) and the trajectory of a typical reflected ion. Bottom panels represent ion velocity-space (v_x - v_y) at various locations of the B_z profile.

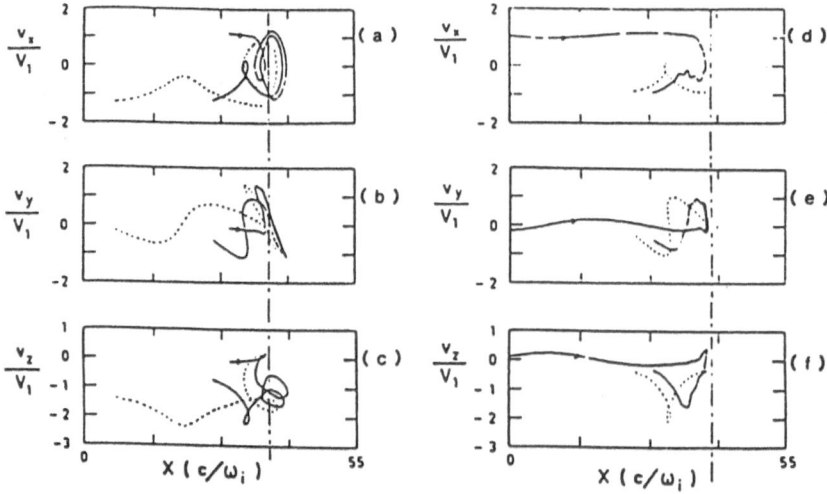

Figure 11: Trajectories of directly reflected ions at $\theta_{Bn} = 45°$ and $30°$ obtained from hybrid simulations. Solid lines are phase space trajectory; dashed lines are predictions from specular reflection model (from Leroy and Winske, 1983).

existence of a critical Mach number as discussed by Sagdeev (1962); (ii) the connection between the foot and reflected gyrating ions (Woods, 1969); (iii) Bimodal ion distributions as observed by Montgomery et al., (1970) downstream of the bow shock are related to ion stream gyrating in the downstream region (Auer et al., 1962, 1971). All explain the success met by hybrid codes in the analysis of collisionless MS shocks. However, it turns out that several effects have been neglected which can be taken into account by full particle codes only.

5.4.2 *Full Particle Codes.*

The main basic features of a collisionless MS shock propagating perpendicular to a static field $\overline{B}_0$ has also been recovered in early simulations of Biskamp and Welter (1972). At early times of the run, a potential jump is formed and becomes large enough to reflect sufficient ions to provide the necessary dissipation, as in Mason (1972). At latter times of the run (t > τ_{ci}, where τ_{ci} is the upstream ion gyroperiod), a large fraction of ions are reflected and gyrate in the upstream field structure. Thus, magnetic fields and potential largely grow at the location of the foot (upstream ion turning point); the foot evolves into a peak and forms a new-shock front. This peak then starts to reflect ions, thus forming another foot and peak. A periodic reformation of the shock results (Figure 12); this result largely differs from that observed with hybrid codes.

More detailed analysis (Lembege and Dawson,1984a, 1987a; Ohsawa, 1985) have confirmed and refined the earlier results as illustrated in Fig. 13. These emphasize that reflected ions are strongly accelerated and heated (in the plane perpendicular to $\overline{B}_0$) without invoking artificial electron resistivity or cross-field currents resistivity as a source of dissipation. Again, ion reflection provides important dissipation but is now expressed in terms of the "three-steps" process (acceleration, trapping, and detrapping as presented in Section 2) by Lembege and Dawson (1987a) or in terms of "resonant acceleration" by Ohsawa (1985). This difference is not only semantic since this viewpoint allows to estimate the resonant time range Δt (difference between detrapping and trapping time of ions by the electrostatic potential at the shock front); this time range provides a useful information on the field strengths at the front in terms of wave damping and on the consequences to the wave velocity. This detailed kinetic dissipation allows a comparison with corresponding theoretical analytical approaches. After being detrapped (or equivalently roughly after reflected ions have reached their turning point upstream), ions are ejected on a large Larmor orbit and initiate a large radius ion ring while penetrating the downstream region (p_x - p_y momentum space of Figure 13).

In summary, three different ion populations appear, as illustrated in Figure 14: (a) the unperturbed upstream drifted ions, i.e. incoming flow of solar wind plasma; (b) directly transmitted ions which successfully pass downstream through the shock ramp without reflection; and (c) reflected ions, i.e. a fraction of the incoming plasma flow which suffers one bounce against the shock front before succeeding to pass downstream through the shock front (Figure 10). The question of "how directly transmitted ions (DTI) and reflected ions (RI) are respectively responsible for the increase of ion temperature across a shock" has been dealt with by many authors. The answers may be expressed as follows:

(i) DTI population suffers heating under three ways: a local adiabatic heating (Biskamp and Welter, 1972a); a local "three-steps" mechanism (see Section 2.3) provided that the downstream electric field fluctuations are large enough to trigger ion trapping, and a local cyclotron heating (Lee et al., 1986a, b).

(ii) RI population suffers a noticeable ion heating by converting the ion bulk energy into large Larmor gyromotions (viscous heating), convected downstream after succeeding to cross the shock front. This leads to the formation of a beam distribution (Figure 10a,b)

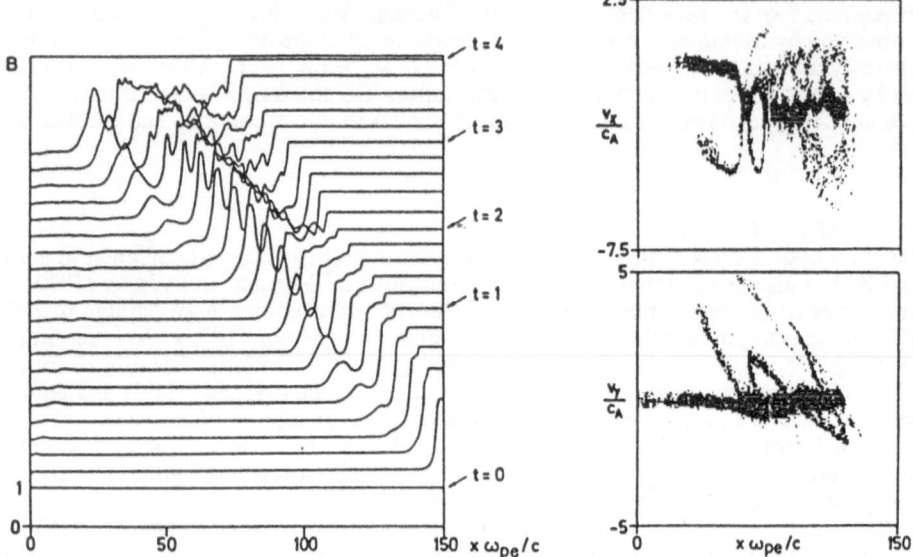

Figure 12: Time evolution of the space profile of the magnetic field and ion phase space at $T = 3\Omega_{ci}^{-1}$; $\vec{B}$ is along z axis (from Biskamp and Welter, 1972a).

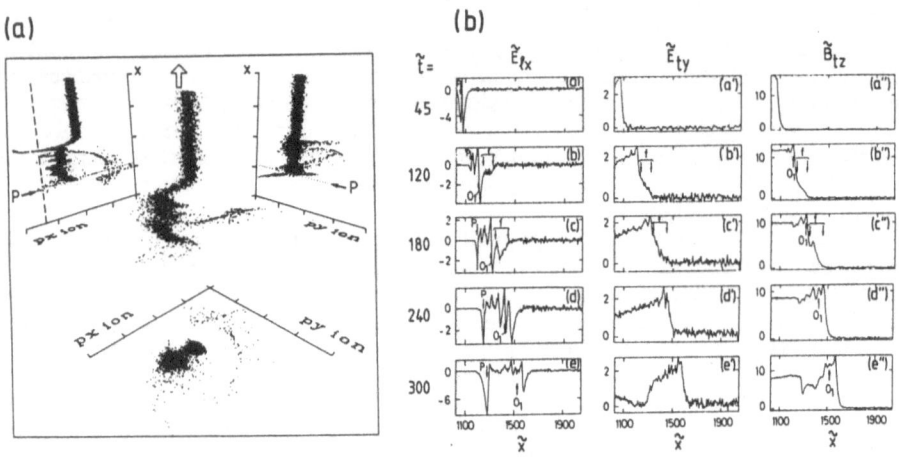

Figure 13: (a) Three dimensional plot of the ion phase space at $t \simeq \tau_{ci}/2$ and associated two-dimensional projection. The shock propagates from the bottom to the top of the picture; (b) time evolution of the longitudinal (E_{lx}) and magnetic field (B_{tz}) (from Lembege and Dawson, 1987a).

107

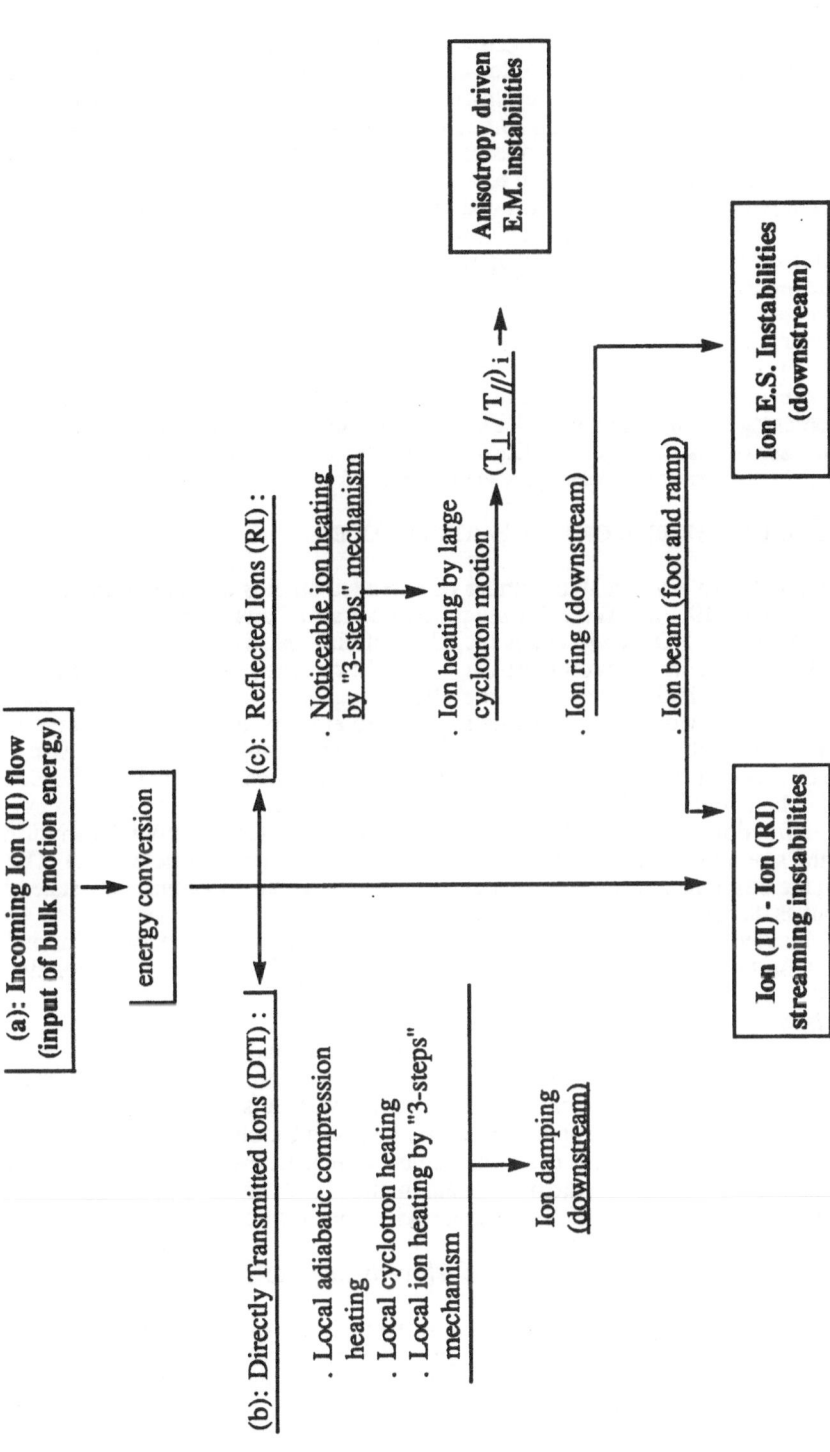

Figure 14: Sketch illustrating the various sources of ion heating coming, respectively, from directly transmitted ions, reflected ions, and plasma instabilities for a quasi-perpendicular shock. Other sources of energy dissipation related to electron dynamics and associated plasma instabilities are not included here.

which is a source of ion (II) - ion (RI) streaming instability (type 1) and to a downstream ion ring distribution which is the source of ion electrostatic instability (type 2). In addition, an important perpendicular ion heating results from both DTI and RI populations and leads to a strong ion anisotropy which is the source of driven anisotropy instability (type 3). Type 1-3 instabilities are additional sources of energy dissipation, and illustrate the three classes of instabilities presented in Section 6.

Finally, simulation results have shown that in the absence of resistivity, reflected ions are observed for both supercritical and subcritical shocks (Biskamp and Welter, 1972a; Ohsawa, 1985; Ohsawa and Sakai, 1985; Lembege and Dawson, 1987a; Lee et al., 1986b). In the presence of artificial resistivity, the dissipation mechanism is dominated by both the reflected and transmitted ions and the resistive electron heating becomes unimportant as resistivity decreases. More exactly, the (non adiabatic) local cyclotron heating of DTI population, i.e. a large percentage of incoming ions convected over a "small" downstream Larmor radius, is of the same order of that created by RI population, i.e. a small percentage of incoming particles convected over a "large" downstream Larmor radius. In the case of resistiveless subcritical shocks, non-adiabatic cyclotron motion of DTI dominates the ion heating process (Lee et al., 1987).

5.5   WHAT TO USE: HYBRID OR FULL PARTICLE CODE?

This question has led to a noticeable number of extended studies and has been already discussed by Burgess (1987a). We will only present a draft of this discussion illustrated by Figure 15. Standard 1-D hybrid codes include artificial resistivity η and assume the charge neutrality and a zero electron mass ($m_e = 0$). In contrast, 1-D full particle codes exclude resistivity, but include space charge effects (no quasi-neutrality) and a finite electron mass. In other words, an electrostatic jump $\Delta\phi$ appears at the shock in order to satisfy the quasi-neutrality condition in hybrid codes, while it results from self-consistent space charge effects in full particle codes. The consequence appears on the ion reflection which occurs in both hybrid and full particle codes but the source of the dominant forces is different. The fraction α of reflected ions depends mainly on the resistivity η (hybrid code) or on the strength of the electric field at the shock front (full particle code). This could explain the difference observed respectively between the steady or unsteady state of the shock front as discussed by Leroy et al (1982).

Results obtained from the hybrid code appear to be issued from two "forced" situations: one on the jump $\Delta\phi$ at the shock, the other on η. Then, two corresponding questions arise:

(i) *When is the quasi-neutrality assumption justified?*
Using arguments based on charge separation fields in low amplitude magnetosonic waves, Quest (1986a) has concluded that this validity depends on M_A and ω_{pe}/ω_{ce}. It is not valid for large M_A and ω_{pe}/ω_{ce} small (use of full particle code), while it is for low M_A and larger ω_{pe}/ω_{ce} (use of hybrid code). A simple picture may be given for clarity by reminding the reader that high ω_{pe} case corresponds to a large number of plasma oscillations "forcing" the plasma to a quasi-neutral state. On the other hand, inertia effects and charge accumulation can be stronger for low ω_{pe} case, since plasma oscillations do not have enough time to neutralize them and space charge effects result.

(ii) *What should the consequences be of a self-consistent resistivity on the fraction α of reflected ions?*
It is clear that no 1D code (of any kind) can answer to this question since a second dimension is necessary in real space. For so doing, Forslund et al (1984) have made use of a 2D full particle code; since the code was also implicit, it excludes high frequency electron plasma oscillations because of the use of a large timestep Δt; an oblique shock (θ

= 78,7°) was considered in order to allow some electrons motion along $\vec{B}_0$. Self-consistent resistivity from cross-field current-driven instabilities is excluded or included, by placing respectively the magnetic field in the simulation plane or perpendicular to it (see Appendix 4). It was found that the resistivity both reduces the fraction α of reflected ions and changes the magnetic ramp scalelength. The lower hybrid instability (LH) seemed to be responsible for this resistivity; an electron heat flux was seen coming from the shock and seemed to be due to the escape of the high energy tail of electrons heated at the shock, in a good agreement with experimental data on Earth's bow shock. Winske et al (1985) made a deeper analysis and identified the modified two-stream instability (MTS) as responsible for

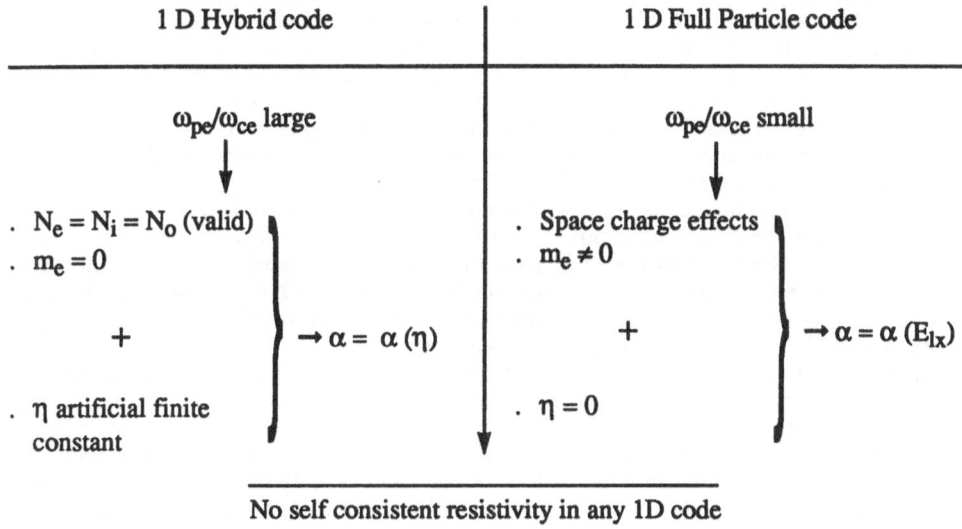

2 D Full particle code : Resistivity caused by lower hybrid (LH), modified two-streams (MTS), or ion acoustic (IA) instabilities.

- Substantial magnetic fluctuations at and behind the shock (LH and/or IA), or at the foot and ramp (MTS).
- Reduction of α.
- Electron heat flux coming from the shock
- Small component of the magnetic field in the simulation plane
  (in MTS case) → electron dynamics // $\vec{B}_0$ curtailed→ no flattop electron distribution)

Figure 15: Sketch illustrating the possible reconciliation between hybrid and full particle simulations; α is the fraction of reflected ions and η is the resistivity.

the turbulence at the foot and ramp. In both cases (LH and MTS), no flat-topped electron distributions, as observed in the experimental data (Feldman et al., 1982, 1983; Feldman, 1985) have been evidenced in the simulation results yet.

Before closing this section, it is important to state that dynamics of both electrons and ions may be important parallel to $\vec{B}_0$, for oblique shocks provided that the direction is below a first critical angle ($90° > \theta_{te} > \theta$) for electrons or above a second critical angle ($90° > \theta > \theta_{ti}$) for ions. These results have been observed for oblique nonlinear MS waves (Lembege and Dawson, 1984b), and have been applied to oblique MS shock waves for ions (Ohsawa, 1986) and for both particle species (Lembege and Dawson, 1987b). Let us mention that parallel ion trapping has been also observed in earlier work (Biskamp and Welter, 1972b). The parallel electron acceleration gives birth to a strong parallel current which can generate a whistler wave propagating ahead of the shock front and forms an upstream wavetrain provided that $\theta < \theta_{te} < 90°$; this can be easily explained by considering the discussion on dispersion curves in Figure 3b and Section 2. On the other hand, the strong acceleration of electrons parallel to $\vec{B}_0$ gives rise to trapping loops around the shock front, of which width is narrower than that of the ion trapping loop. A large potential drop builds up where trapped electrons accumulate; a "double layer" structure forms and stays more or less steady in time because of the continuous feeding of such trapped electrons and ions as observed by Lembege and Dawson (1989a). Such results have been obtained for "pseudo-oblique" shocks, i.e., using a 1-D code (Appendix 2) and for plasma parameters not exactly covering those of the Earth's bow shock; then, an extended parametric study and the use of 2-D codes are required for confirmation.

Acceleration and heating mechanisms of electrons need to be studied in more details in order to estimate the exact quantitative role of the associated resistivity for perpendicular (2D code), and oblique (2D code) or pseudo-oblique (1D code) shocks. Quantitative contributions in energy dissipation, via plasma instabilities which trigger in 2D codes (i.e., resistive effects) or via "direct" wave-electron energy transfer, as, for instance, in parallel trapping loop (i.e., viscous effects) need a deeper analysis.

In summary, hybrid codes have revealed to be very helpful tools for simulating the overall structure and ion dynamics at a reasonable computer cost, provided that assumptions of the model are justified. However, two main points restrict the use of such codes: different plasma conditions may require the inclusion of self-consistent effects and electron dynamics may play an important role. One way to satisfy both points are in the use of a 2D full particle code (even using an unrealistic mass ratio). Such 2D simulations are expected to be quite costly but necessary since one wants to include self-consistently the main plasma instabilities. The question concerning the identification of the most relevant plasma instabilities is now considered in the next section.

6. Instabilities in Collisionless Magnetosonic Shocks

6.1 NECESSITY FOR A SYNTHETIC CLASSIFICATION

Previous sections have emphasized the various particles dynamics according to values of M_A and θ, and the resulting associated "direct" heating (anomalous viscosity), i.e., without involving any plasma instabilities. However, additional mechanisms of energy dissipation may take place self-consistently via plasma instabilities as mentioned in Section 5.4.2 and Figure 14. Instabilities are usually classified as macroscopic or microscopic, and, within the latter category, a distinction is made between resonant and non-resonant. Let us briefly clarify this distinction (Biskamp, 1973). Resonant instabilities directly give rise to collisionless dissipation. They are driven by the interaction (resonant) of a small fraction of particles with a wave, which in the presence of a broad spectrum of waves leads to turbulence in phase space, i.e. to collisionless dissipation. The ion acoustic instability is of this kind, and will be discussed in Section 6.2. On the other hand, in a non-resonant instability, a wave first interacts coherently with the whole particle distribution. But, often the non-linear saturation is again a resonant process, some particles being reflected or "trapped". An example is the instability between two counterstreaming ion beams (for instance two-stream instability discussed in Section 6.2.2.). Since these non-resonant instabilities often grow to substantial amplitudes until they are stabilized, they have a large effect on the plasma (heating, friction). While microscopic instabilities playing a role in collisionless shocks have mainly some (full or partial) electrostatic nature with $k \lambda_D \sim 1$, macroscopic instabilities have much longer wavelengths and are therefore associated with magnetic oscillations. Such macroturbulence may lead to an increase of entropy by phase mixing of waves and may also induce microinstabilities, i.e. by creating steep gradients and thus enhance dissipation. A special kind of macroscopic instability is the decay instability of a finite-amplitude wave; this type has been claimed to cause damping of non linear wavetrains (for instance of the whistler precursor in oblique shocks) in some dispersive shocks (Wright, 1971).

Herein, we will mainly focus on microinstabilities which have already been analyzed in details by many authors; the major difficulty is in determining: (i) which instabilities are dominant in collisionless magnetosonic shocks according to the simple classification of M_A and θ introduced before, and (ii) how to include them in appropriate simulation codes and for suitable box and plasma parameters. Two helpful reviews respectively by Wu et al. (1984) and Papadopoulos (1985) deal with the main physical criteria of plasma microinstabilities and are recommended reading. Herein, our first step consists in identifying the free energy sources associated with collisionless shocks and which might excite these microinstabilities (Figure 16).

Previous Section 5 has already emphasized that at low M_A regime, a dissipation source is necessary in order to satisfy the jump conditions (i.e., the entropy jump) across the shock as required by the Rankine-Hugoniot equations. In such a region, the dissipation is mainly provided by anomalous resistive heating (often thought of primarily the electrons) which is associated with strong electric currents. Then, such resistivity is caused by the relative drift between electrons ($\vec{u}_e$) and ions ($\vec{u}_i$) and thereby contributes a frictional force, which is associated with the magnetic transition. The net effect is to reduce the available free energy, which is typically manifested through reduction of $\vec{u}_d$ ($= \vec{u}_{de} - \vec{u}_{di}$) and particles heating. The reduction in $\vec{u}_d$ results in decrase of the magnetic field gradient (increase of the gradient length). This illustrates the physical process through which resistivity "controls" the thickness L of the shock ramp as shown in numerical simulations (L decreases with η, as confirmed in hybrid simulations of Leroy and al., 1982). In a first kinetic approach, the transition length L is such that $\rho_e \ll L \ll \rho_i$ so that the ions follow almost "unmagnetized" (or large scalelengths) orbits while the electrons suffer $\vec{E} \times \vec{B}$, $\vec{\nabla}B$, $\vec{\nabla}N_e$ and $\vec{\nabla}T_e$ drifts where ρ_e and ρ_i denote the electron and ion Larmor radius respectively. These drifts are the basic free energy sources for the excitation of various (resistive) instabilities. These instabilities excite a wave spectrum which acts back on the current thus reducing the available free energy while heating the plasma.

On the other hand, these current-driven instabilities seem to play a minor role in the Mach regime where the additional source of energy dissipation brought by ion reflection is required; this can take place in low M_A cases (poor but finite ion reflection) or high M_A case (larger ion reflection). However, reflected ions are themselves free-energy sources for the excitation of other instabilities; in contrast with the current-driven ones, these instabilities are not related with any frictional (i.e., resistive) effects and constitute an additional source of anomalous heating. They include two subclasses: the anisotropy-driven instabilities and the ion instabilities. In summary, three groups can be defined as illustrated in Figure 16. Correspondingly, we retrieve the three instabilities of type 1, 2, and 3 mentioned in Section 5.4.2 and in Figure 14.

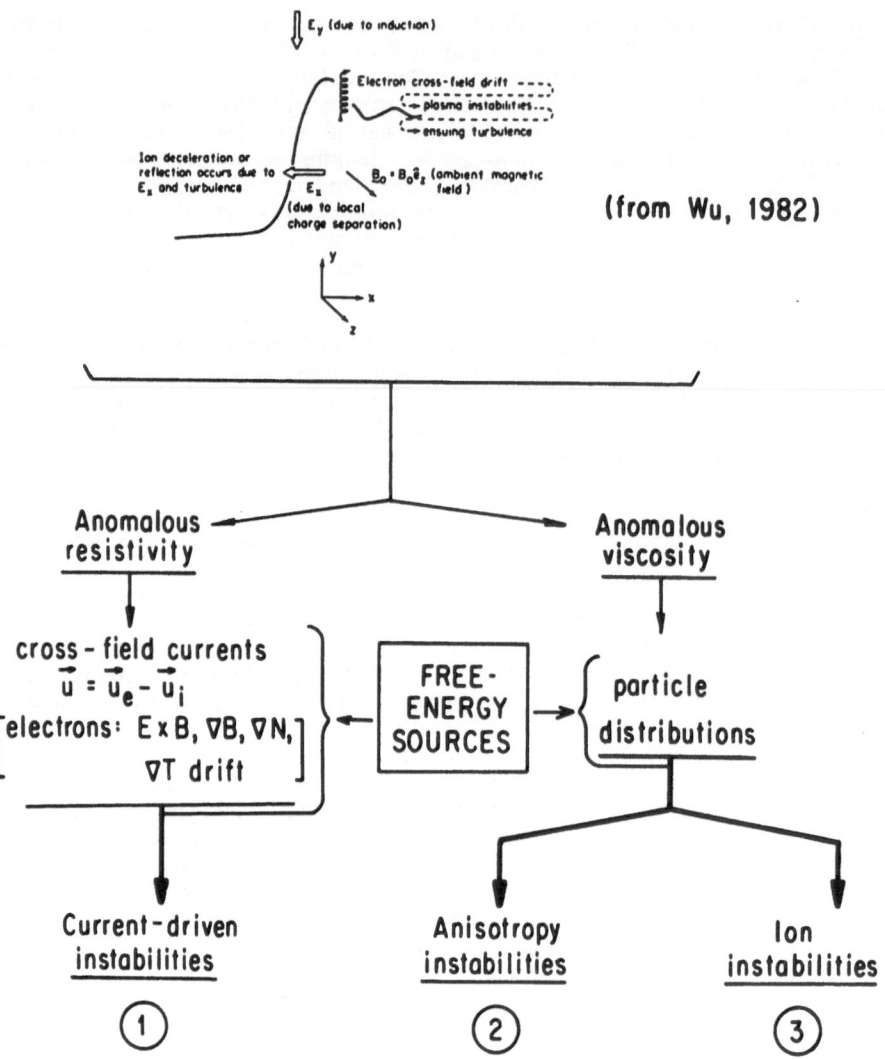

Figure 16: Table of the various-free energy sources involved in the excitation of microinstabilities through a collisionless shock.

6.2 CURRENT-DRIVEN INSTABILITIES (RESISTIVE HEATING)

The different instabilities potentially concerned are summarized in Table 1 according to the associated source of free-energy, to their nature (electromagnetic: EM or electrostatic: ES), to the propagation direction ϕ, and their triggering threshold. In short, the major instabilities investigated and discussed in the literature are: (i) ion acoustic instability (Gary, 1970; Lashmore-Davies and Martin, 1973; Arefev, 1970; Tokar et al., 1987); (ii) electron-cyclotron-drift instability (Wong, 1970; Gary and Sanderson, 1970; Lashmore-Davies, 1971; Forslund et al., 1970; Gary, 1972); (iii) lower-hybrid-drift instability (Davidson and Gladd, 1975; Huba and Wu, 1976; Davidson et al., 1977; Gladd, 1976; Krall and Liewer, 1971), (iv) ion-ion streaming instability (Auer et al., 1971; Papadopoulos et al., 1971b). In addition to these flute modes (i.e., characterized by $k_\parallel = 0$), there is the so-called modified two-stream instability (Gladd, 1976; Krall and Liewer, 1971; McBride et al., 1972), which can be considered as a particular case of the kinetic cross-field streaming instability (Wu et al., 1983). Let us remember that three conditions must be satisfied in order that any instabilities produce significant stochastic electron heating within a perpendicular shocks: (i) the threshold for exciting the instabilities must be exceeded, (ii) the electrons must have a non-adiabatic response to the fluctuating fields of the instability and (iii) the instability must be able to grow to significant amplitude within the time period for plasma to convect across the shock (Tokar et al., 1987).

The ion-sound (or ion acoustic) instability is well adapted when $T_e/T_i \gg 1$; this instability turns out to be the so-called Buneman instability for $u_e = v_{th}$ (Buneman, 1959) or the lower hybrid instability when $T_e/T_i \leq 1$ (Lemons and Gary, 1978). As indicated in Table 1, the Buneman instability is not expected since its threshold condition is too strong for classical subcritical or supercritical shock; however it may play a certain role for super high Mach number (SHMN) shocks as those expected in astrophysical plasmas as for instance around supernovae remnants, as discussed in Section 7. The electron cyclotron drift instability is minor because of its very small associated resistivity; indeed, this instability can be easily suppressed by the effect of the magnetic field gradient in the perpendicular shock. Then, most relevant current-driven instabilities expected in collisionless magnetosonic shocks are:

6.2.1. *Ion acoustic (IA) instability* which has been largely studied in particular within a self-consistent shock simulation where shock parameters have been chosen to maximize its growth rate (Tokar et al., 1987). An explicit code has been used which allows the authors to resolve electrostatic waves in the ion acoustic regime in contrast with an implicit code (Forslund et al., 1984). Electron heating by ion acoustic turbulence is observed at the shocks at rates in agreement with second-order Vlasov theory predictions. However, a striking feature is that the amount of resistive electron heating is still small and ion reflection provides the major source of dissipation even at low M_A regime. Let us note that strictly resistive shocks cannot be simulated for the parameters suitable for explicit particle codes running on present time supercomputers because the plasma convects through these shocks so quickly that current driven instabilities have little time to be amplified and to heat the electrons resistively.

6.2.2 *Modified two-stream (MTS) instability* which allows the transfer of ion cross-field drift energy to parallel electron energy when the electron drift speed u_d is lower than the local Alfven speed v_A. This instability may be suppressed by electromagnetic effects when $u_d > v_A$. Previous studies mentioned above have stressed that the electron energy transfer results in bulk electron heating. However, the modified two-stream instability has been shown to result in electron acceleration instead of electron heating, i.e., formation of

Table 1 : Current Driven Instabilities ($\vec{u}_d = \vec{u}_{de} - \vec{u}_{di}$)

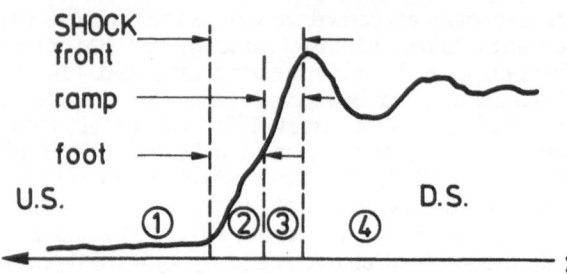

$u_d = u_d(x) \rightarrow$ different instabilities may dominate at different locations

Instability	Source of energy	Mode	Propagation	Condition	Expected?
Buneman	Cold é-ions Two streams	E.S.	$(\vec{k}.\vec{B}_o) = 90°$	$u_d \geq 1.8 \, v_{the}$	No
Ion acoustic ③	$\neq$ streaming between ions species and el.	E.S.	$(\vec{k}.\vec{B}_o) < 90°$ out of the coplanarity plane $(\vec{n}, \vec{B}_o)$	$\dfrac{u_d}{v_{the}} \geq \dfrac{T_i}{T_e}$	often observed
Beam cyclo. (= el. cyclo. drift)	el. drift relative to solar wind ions	E.S.	$90°$	$u_d > \max \left(c_s, \dfrac{\omega_{ce} \, v_{the}}{\omega_{pe}} \right)$	No
Lower hybrid drift ②, ③	$\nabla N, \nabla T, \nabla B$ drift	mixed ES-EM	$90°$	$L_n < \left(\dfrac{m_i}{m_e} \right)^{1/2} \rho_i$ where $\nabla B \approx B/L_n$	Yes
Kinetic cross field streaming ②, ③	Cross-field drift between el. and transmitted /reflected ions	• mixed ES-EM • E.M.	$\cos \theta \geq \left(\dfrac{m_e}{m_i} \right)^{1/2}$ $\cos \theta \leq \left(\dfrac{m_e}{m_i} \right)^{1/2}$	$\left. \begin{array}{l} u_d > v_a \\ \beta_e = \beta_i \geq 1 \end{array} \right\}$	Yes
Modified two-stream ②, ③	Cross-field drift between el. and transmitted /reflected ions	E.M.	$\cos \theta \leq \left(\dfrac{m_e}{m_i} \right)^{1/2}$	$\beta_e = \beta_i \ll 1$ $u_d < v_a$	Yes

suprathermal electron tails (Tanaka and Papadopoulos, 1983). Ion trapping controls the instability saturation while a small fraction of electrons are accelerated in the direction parallel to the magnetic field by a systematic nonlinear increase in the phase velocity caused by a nonadiabatic transition of the plasma mode.

6.2.3. *Kinetic cross-field streaming (KCFS) instability.*

The MTS instability has been shown to be a particular case, for small β and large θ, of the so-called kinetic cross-field streaming instability (Wu et al., 1983). Unlike the modified two-stream instability, the kinetic mode is not stabilized by electromagnetic effects when $u_d > v_A$ (see Table 1). For weak deviation θ, i.e., $k_\parallel / k = \cos \theta > (m_e/m_i)^{1/2}$, the mode has both electrostatic and electromagnetic components while for large deviation θ, i.e., for $\cos \theta < (m_e/m_i)^{1/2}$, the unstable mode is predominantly electromagnetic. The excited waves are whistler as shown from an approximate stability theory; for $\cos \theta \leq (m_e/m_i)^{1/2}$, the instable waves are basically magnetosonic waves. Applications of this mode have been made by Winske et al., (1985) for estimating the heating rates at collisionless shocks. Quasi-linear theory and comparison with results from particle simulations have shown that heating for both the electrons and ions increases with β. Results suggest that electron dynamics determines the saturation level of the instability which is manifested by the formation of a flattop electron distribution parallel to the magnetic field. As a result, both the saturation levels of the fluctuations and the heating rates decrease sharply with β. Two cases may be considered when applied to the collisionless bow shock: (i) for supercritical q-⊥ shocks, waves due to the interaction of the electrons with both the reflected and solar wind incoming ions in the foot region are possible and can rise to an appreciable electron heating if β is small; in contrast, only the solar wind ion-electron interaction is significant at higher β and the heating is much less. On the other hand, (ii) at laminar shocks (low M_A and low β, to see Section 5.1), the heating, especially of ions, may be significant ($\Delta T/T \sim$ 10) and is accompanied by flattop electron distributions, as typically observed (Thomsen et al., 1985).

6.2.4. *Lower hybrid drift (LHD) instability* which includes relative drifts between

electrons and ions due to gradients (∇N, ∇T, ∇B). Simulation results and ISEE data have indicated that reflected ions may attain high velocities relative to the transmitted ions in the direction of the diamagnetic drift; thus waves propagating along the drift direction may couple to the drift mode. This coupling drastically changes the nature of the ion-ion streaming instability. Let us remember that two-stream instability and LHD instability are rather similar in nature with the exception that the latter requires a density gradient.

Then a question arises: *which instability is dominant?*

The lower hybrid modes, particularly the lower hybrid drift instability and the modified two-stream instability, are attractive because they heat both electrons and ions, are not suppressed when $T_e \sim T_i$, and have lower thresholds than the ion acoustic instability (Lemons and Gary, 1978). Nevertheless, their small growth rates and longer wavelengths may one wonder whether they can heat the plasma to the required downstream temperature over the relatively narrow current layer of the shock; however, calculations by Revathy and Lakina (1977) indicate that significant heating is possible. The problem is complicated by a lack of space experimental data concerning the waves in the lower hybrid frequency range. Winske et al. (1987) initiated a first approach of the problem by performing a comparative theoretical study of plasma heating by IA and MTS instabilities at q-⊥ subcritical shocks and have concluded that the dissipation at these shocks is most likely due to the lower frequency MTS instability. Indeed, the wave intensities at higher frequencies are about 4 orders of magnitude smaller than those predicted for the IA instability at saturation; this is consistent with the fact that the measured shock widths imply cross-fields drift speeds that are below threshold for this instability. However, let use note that some effects, such as electron temperature gradients which drastically increase the growth rate of IA instability (Priest and Sanderson, 1972), has been neglected in the study of Winscke and al. (1987).

In summary, the principal instabilities which are expected in magnetosonic shocks are the ion acoustic (IA) and "lower hybrid" (LH) modes:

(i) The IA instability is a short wavelength (approximately the Debye length), high frequency (approximately the ion plasma frequency or the electron gyrofrequency) electrostatic mode (Biskamp, 1973; Galeev, 1976; Boyd, 1977; Papadopoulos, 1977). Evidence for the existence of ion acoustic modes at shocks derives from the observed electrostatic noise. Moreover, it is interesting to note that the marginal stability condition (growth rate $\gamma = 0$) for ion acoustic turbulence has often been used to estimate the shock width (Morse and Greenstadt, 1976; Greenstadt and al., 1978; Russell and al., 1982) based on the argument that the instability relaxes nonlinearly to marginal stability (Manheimer and Boris, 1972). The threshold condition has also been used to distinguish between resistive and diffusive laminar shocks (Mellot and Greenstadt, 1984). However, the validity of this method, when applied to both IA and MTS instabilities, has been critized by Winscke and al. (1987).

(ii) on the other hand, when conditions of the excitation of IA mode are not verified, the principal heating and resistivity are thought to come from the lower hybrid modes. The term "lower hybrid" (LH) is used to describe various instabilities: the modified two-stream instability, the kinetic cross-field streaming instability, and the lower hybrid drift instability. All of these modes have frequencies and growth rates in the lower hybrid frequency range with wavelengths which are longer than the electron but shorter than the ion gyroradius and propagate nearly perpendicular to the magnetic field.

In fact, the distinction between IA and LH instabilities is somewhat superficial since both join together smoothly for $T_e \simeq T_i$ as shown by Lashmore-Davies and Martin (1973) in the electrostatic limit. More exactly, these instabilities differ primarily in their electron behaviour (respectively, unmagnetized for $\omega \approx \omega_{pi}$, magnetized for $\omega = \omega_{lh}$), but both modes lie on the same branch of the dispersion equation in the electrostatic limit. Most of these IA and/or LH instabilities have been mainly studied in subcritical shock regimes where their relevant features are particularly expected. Indeed, such microscopic processes are also likely to occur in supercritical shocks but are masked by the effects due to ion reflection. All current-driven instabilities lead to energy transfer mainly along the drift direction, i.e., involve 1D heating processes; on the opposite, microinstabilities presented in Sections 6.3. and 6.4. allow to redistribute the energy over the other degrees of freedom. In the present case, since the drift current strongly varies according to distance x, different instabilities may dominate at different locations. Then, LHD and KCFS (and MTS) instabilities are expected in both the foot (2) and the ramp (3), while IA instability should develop in the ramp region (3) (see Table 1); similar comments apply to the overshoot region. However, IA instability would not always be expected because of the large required value T_e/T_i; its excitation in terrestrial shocks (where $T_e \approx T_i$) is still not completely understood at the present time, and will remain a mystery until a more complete non linear analysis can reduce this requirement.

6.3   ANISOTROPY DRIVEN ELECTROMAGNETIC INSTABILITIES

These instabilities may occur respectively for electrons and ions and are summarized in Table 2; all of these have an electromagnetic nature. Electron whistler mode is expected along the direction of B_o; let us note that whistler waves may also be directly excited by field-aligned electron beams, such as observed in oblique shocks. Non-linear theories (Davidson and Hammer, 1972; Hamasaki and Krall, 1973) and computer simulations

(Ossakow et al., 1972) indicate very fast isotropization ($T_{e\parallel} > T_{e\perp}$) by the electron whistler instability, which seems to agree with observations.

On the other hand, ion anisotropy created by both reflected and directly transmitted ions (Section V) may excite two types of instabilities: the Alfven ion cyclotron, or AIC, instability (Davidson and Ogden, 1975; Tanaka et al., 1983; Tanaka, 1985; Otani, 1986; Ambrosiano and Brecht, 1987) and the mirror drift, or MD, instability (Lee et al., 1988); AIC and MD propagate respectively at $\theta = 90°$ and $\theta \neq 90°$. They are expected to be dominant sources of dissipation for high M_A regimes where ion anisotropy is very large (in particular at the shock front). Thomas and Brecht (1986) have performed a self-consistent study of AIC instability by the use of a 2D-hybrid code for a perpendicular shock; AIC mode develops with an additional non-linear perturbed magnetic field which allows the system to redistribute, through the shock structure, the perpendicular ion anisotropy created by the reflected ions.

Figure 17 illustrates the growth of AIC instability until it reaches a saturation level (for a typical run), and the significant modifications to the plasma shock front for different Mach regimes. It appears clearly that low M_A shocks keep their planarity features (for $M_A = 3.5$) in contrast with high M_A shocks where a noticeable departure from planarity takes place. For oblique shocks, two electromagnetic instabilities have been recovered (Thomas and Brecht, 1987). The AIC instability mentioned before, which is present for q-$\perp$ interaction and the electromagnetic counterstreaming ion instability, which is present for q-$\parallel$ interactions. In short, AIC instability leads to a significant structuring of the shock region and causes isotropization in the ion velocity space for $\theta > \pi/3$; on the other hand, the electromagnetic counterstreaming instability leads to velocity space isotropization but does not cause much two-dimensional spatial structuring (at least for plasma parameters used in the study).

It has been confirmed by further studies that AIC instability behind the shock overshoot causes waves which propagate parallel to the magnetic field and which cause ion isotropization by pitch angle scattering. However, a discrepancy appears at the ramp between the theoretically predicted wavelength and that observed in the simulation. This can be accounted for by obliquely propagating modes due to the AIC instability rather than the MD instability. This point is important with respect to the 1D simulation study of Lee et al (1988), who found evidence for the generation of mirror waves well downstream of the shock. However, the use of a 1D (hybrid) code cannot include the completion of the various wave modes with different propagation angles for the free energy of the ion distribution; then, a 2D code is necessary. For so doing, a more complete approach has been performed by Winske and Quest (1988) with a 2D hybrid code, in order to determine the relative importance between AIC and MD instabilities. Distinguishing between both possible wave modes, by using polarization and frequency analysis, as well as their respective compressibility, has revealed to be a difficult task. Simulation results seem to indicate a slight predominant effect of AIC instability with respect to MD instability; however, a more complete explanation awaits the derivation and analysis of an appropriate dispersion relation describing the growth and coupling of low frequency modes in the inhomogeneous high beta environment of the shock.

Finally, the formation of backstreaming particles has been initiated by Tanaka et al., (1983), who discussed the problem of isotropization of the anisotropic ion distribution due to the electromagnetic ion cyclotron instability. The isotropization occurs over 4-5 ω_{ci}^{-1}, where ω_{ci} is the ion cyclotron frequency. This has led to an interesting mechanism accounting for the ion foreshock phenomena by using the leakage of energetic ions upstream following this isotropization, as sketched in Figure 18. Many other studies have dealt with "direct" or "indirect" particle leakage mechanisms, which lead to ion (and electron) backstreaming from the shock into the solar wind (Burgess and Luhmann, 1986;

118

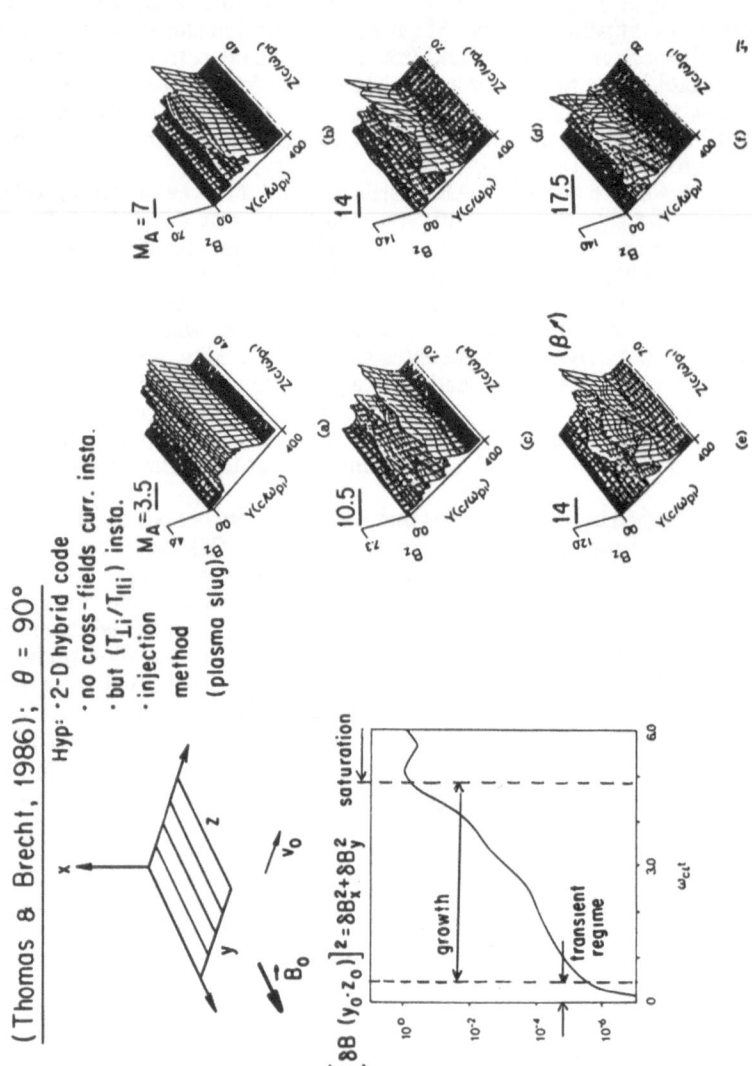

Figure 17: 2D hybrid simulation illustrating the deformation of the shock front (departure from planarity) according to M_A regime (from Thomas and Brecht, 1986).

Table 2 : Anisotropy-driven e.m. Instabilities

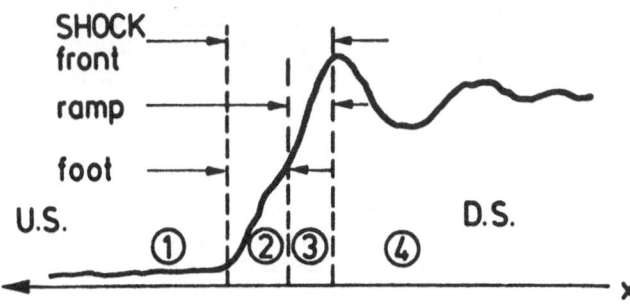

Instability	Free energy source	Propagation	Observed ?
Whistler	$(T_{\perp e}/T_{/\!\!/e}) > 1$	$(\vec{k} \cdot \vec{B_o}) = 0°$	Yes
Alfven ion cyclotron ②,③,④	$(T_{\perp i}/T_{/\!\!/i}) > 1$ of both reflected and transmitted ions	$\theta = 0°$	Yes
Mirror drift ②,③,④	$\dfrac{\beta_\perp}{\beta_{/\!\!/}} > 1 + \dfrac{1}{\beta_\perp}$ where high $\beta = \dfrac{8\pi nkT_i}{B^2}$	θ oblique $(\gamma = 0$ for $\theta = 90°)$	Yes

Table 3 : Ion Instabilities

Instability	Free energy source	Mode	Propagation	Charateristics
Lower hybrid ring ②	ion ring & électrons	E.M.	$q - \perp$ $\left[\cos\theta = \left(\dfrac{m_e}{m_i}\right)^{1/2}\right]$	$f\left[\theta, \dfrac{v_\perp}{v_A}\right]$
Ion acoustic ring ④	. transmitted ions $(T_\perp/T_{/\!\!/}) > 1$. ion ring . & électrons	E.S.	$q - /\!\!/$	$f\left[\theta, \dfrac{v_\perp}{v_A}, \dfrac{T_e}{T_{i\perp}}\right]$
Ion-Ion ring ④	. ion ring & . core ions	E.S.	$90°$	$f(n_r/n_c)$

Burgess, 1987b,c; Leroy and Winske, 1983). These are beyond the scope of the present review and will not be developed herein. However, let us mention one relevant consequence, i.e., backstreaming particles may form field-aligned streams which interact with the solar wind ions in order to excite ion-beam electromagnetic instabilities (Gary et al., 1981; Winske and Leroy, 1984b; Gary et al., 1984b; Gary and Tokar, 1985; Winske and Quest, 1986) as illustrated in Fig. 18.

6.4   ION INSTABILITIES (VISCOUS HEATING)

As emphasized in Section 5, reflected ions behave as a beam in the foot-ramp region (Figure 10) and form a ring as they gyrate downstream after crossing the shock front. The ring distribution is characterized by a large Larmor radius ($v_\perp \gg v_A$). The various ion instabilities expected by the mixing of various ion distributions are summarized in Table 3. In addition to the ion anisotropy instabilities (Section 6.3), related to both transmitted and reflected ions, an important issue in the downstream region is the thermalization of the free energy available in the ring distribution of the reflected ions. In 1D simulations, the ion ring distribution is stable; however, instabilities are expected in higher dimensions (Akimoto and al., 1985). This configuration is unstable to several types of instabilities : two instabilities involving the ring ions with the electrons, a lower hybrid ring (LHR) mode and an ion acoustic ring (IAR) mode, and a low frequency mode involving the two ion species (IIR). These instabilities mainly differ from each other according to their nature (electromagnetic : EM or electrostatic : ES), and their direction θ of propagation. They are mainly dependant on θ and the size $v_\perp$ of the ring (LHR); on θ, $v_\perp$ and the ratio $T_e/T_{i\perp}$ (IAR); and on the ratio of ion ring over core ions densities (IIR), respectively. Basically, two cases are relevant:

(i)   for $\sin \theta < (m_e/m_i)^{1/2}$, the interaction is basically an ion-ion interaction. This applies mostly downstream. A quasi-linear computation of this instability by Kulygin et al. (1971) shows that the thermalization occurs by the formation of a strong non-Maxwellian tail which extends up to energies $E \simeq 4\, E_{ring}$.

(ii)   for $\sin\theta > (m_e/m_i)^{1/2}$, the ion ring is expected to interact with the whistler. This interaction is much weaker than the previous one and results in electron heating but with little ion thermalization. In the case of the strongest interaction ($\sin \theta \simeq (m_e/m_i)^{1/2}$, the energy goes to the generation of strong field-aligned electron tails. This process has been proposed as a possible electron acceleration mechanism at the foot of supercritical shocks (Papadopoulos, 1981). Another possibility for ring thermalization is the onset of stochasticity in the ion orbits due to magnetosonic and ion cyclotron waves (Varvoglis and Papadopoulos, 1982).

In summary, such results illustrate the mutual interaction between downstream and upstream regions which are strongly dependant one each other. Defining which instability is dominant is still a difficult task within the frame of this mutual interaction. At the present time, only a "chain" of various phenomena may be defined in order to understand their inter-dependence, although some parts of the chain need further analysis.

It is important to stress that most studies on microinstabilities have been performed for homogeneous plasma as an initial value problem; moreover, the non-linear development of the instability is characterized by a gradual relaxation process that can be described in terms of quasi-linear theory. However, for shocks, processes occur differently as newly reflected ions at the shock front are continuous sources of free energy and are isotropized as they gyrate downstream; then we have a time quasi-stationary but spatially inhomogeneous situation (Winske and Quest, 1988). Consequently, the spatially homogeneous quasi-linear theory is not expected to be strictly valid, although the basic picture of quasi-linear relaxation with spatial rather than temporal variation should apply.

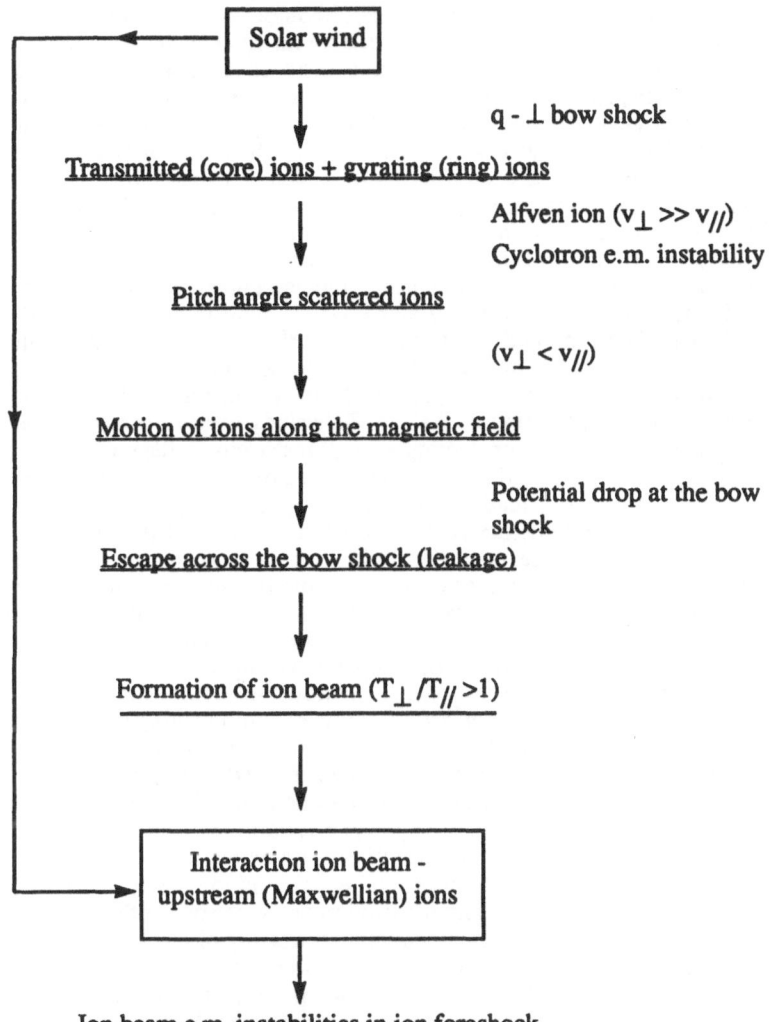

**Escape of ions from the bow shock
and consequences in ion foreshock**

Figure 18: Sketch of ions leakage and consequences with upstream solar wind

7. Applications to Super High Mach Number Shocks and Relativistic Shocks

This tutorial-style review was mainly focused on collisionless magnetosonic q-$\perp$ shocks, which correspond to cases where the physics of the shock, although being already intricate, is the simplest when a kinetic description needs to be taken into account (at least for the ions). In fact, many other important problems related to shocks, such as particle leakage, Fermi-acceleration mechanisms, shocks colliding, quasi-parallel shocks, electron and ion foreshocks, self-consistent effects of the shocks curvature and cometary shocks are not discussed herein. A special attention must be given to the so called "Super High Mach Number", or SHMN, shocks as those expected in astrophysical plasma and/or around remnants of supernovae; indeed these are of direct interest in the view of the present book. These astrophysical shocks generally have much higher Mach numbers than the terrestrial bow shock, of the order of 1000 for supernovae shocks. Already Leroy et al (1982) have discussed high Mach number shocks and have found that the percentage of reflected ions saturates at $\alpha = 0.2 - 0.25$ when M_A increases. Later on, two main features of SHMN shocks were observed in simulations and are particularly relevant; one concerns the unsteady state, the other concerns a strong non-adiabatic electron heating.

Quest (1985) found that for a $M_A = 22$ shock, the structure was not steady but instead periodically breaks and reforms, with a period 2 - 3 Ω_{ci}^{-1}. This study was performed by use of a 1D hybrid code (Appendix 3) and by considering a small resistivity in the code so that the electron response was approximately adiabatic. The effects of varying the resistivity were examined by Quest (1986b) and lead to the following results: if the resistive scalelength $\lambda_{res} \ll c/\omega_{pi}$, the shock structure is stable for $M_A = 5\text{-}60$ and $\beta_i = 1$; on the other hand if $\lambda_{res} > 0.4\,c\,/\,\omega_{pi}$, then the magnetic field overshoot is damped and the imbalance in the electron momentum equation results in a periodic fluctuation of the fraction α of reflected ions. For $M_A > 10$, this percentage α peaks at 40% and the field overshoot increases more slowly. At the present time, the extension of these results to astrophysical shocks is not clear, since the structure of the shock depends strongly on the (artificial) resistivity η and the self-consistent electrons response is unknown in SHMN shocks.

For so doing, a deeper analysis has been performed on the electron heating and has led to two kinds of results. By using a full particle code, Tokar et al (1986) have found that (i) the shock structure exhibits periodic reformation as at lower M_A (Biskamp and Welter, 1972a; Lembege and Dawson, 1987a), and (ii) electrons can be non-adiabatically heated without the need of cross-field current-driven resistivity, provided that the shock ramp becomes sufficiently thin. On the other hand, Cargill and Papadopoulos (1988) have shown that a sequence of plasma instabilities (Buneman, or BI, and ion acoustic, or IAI) between the reflected and/or transmitted ions and the background electrons at the foot and the shock front (see Table 1), can give rise to rapid anomalous heating of electrons; hybrid codes have been used which means that the strict self-consistent electron answer is not included. The striking result is that the threshold conditions of both BI and IAI are quite hard (in particular BI) when applied to plasma conditions of "classical" shocks, such as the terrestrial bow shock, but seem to be well appropriate for astrophysical plasmas; in such cases, electron heating may dominate (Papadopoulos, 1988). However, these recent results suggest again the need for a deeper comparison between hybrid and full particle results.

Furthermore, much attention has been already brought on the acceleration mechanisms of both electrons and ions to high energy in studies of solar flares, astrophysical plasma and fusion devices. Ultrarelativistic waves have been simulated in overdense electron-positron unmagnetized plasma with different initial fields waveforms (Lebœuf and al., 1982). Studies have been also performed with electromagnetic pulses in underdense electron-positron plasmas in order to determine the physical processes involved in the acceleration of particles to very high energies (Ashour-Abdalla and al., 1981). High amplitude electromagnetic waves may occur in the outer magnetosphere of pulsars which are thought to be electron-positron plasmas; in such a case, relativistic effects in particles dynamics are relevant. Recent simulation studies have been performed on the formation of relativistic particles in strongly magnetized plasma ($\omega_{ce} > \omega_{pe}$) by Ohsawa and Sakai (1988). Although a SHMN shock regime has not been considered ($M_A = 2$), these numerical studies based on full particle codes have shown that the time needed for acceleration of both particles species is very short, more exactly of the order of Ω_{ci}^{-1} for ions (i.e., less than 1 second for solar plasma) and less than one Ω_{ci}^{-1} for electrons respectively. Moreover, the transition from a sinusoidal large amplitude magnetosonic wave to a shocklike wave has been studied in relativistic regime by Lembege and Dawson (1989) in similar strongly magnetized plasmas ($\omega_{ce}/\omega_{pe} \simeq 3$); full particle codes have also been used there. Both relativistic ions and electrons have been observed, accompanied by the formation of electrons solitary waves propagating at velocities much larger than the initial magnetosonic wave.

On the light of these separate results, it seems that more complete studies are necessary on the formation of relativistic particles by SHMN shocks and on the relative importance of the dissipation sources (i) by direct particle acceleration mechanisms, or (ii) by plasma instabilities as those mentioned in Section 6. As an example of point (i), two known mechanisms can accelerate charged particles to high energy (Decker and Vlahos, 1986) at fast-mode shocks:

a) the shock drift mechanism (i.e., $\vec{E} \times \vec{B}$ drift, also called throughout this review "dynamo effect", or "non-relativistic surfatron", or as being a part of the "3-steps mechanism").

b) the first-order Fermi or diffuse shock acceleration mechanism, which is a statistical process in which particles undergo spatial diffusion along field lines and are accelerated as they scatter back and forth across the shock; thereby particles are being compressed between scattering centers fixed in the converging upstream and downstream flows.

Then, for oblique shocks, the drift or first-order Fermi processes may be simultaneously operative if a sufficient level of turbulence exists in the shock's vicinity. Which of both processes is dominant in SHMN shocks, and for what set of parameters (M_A, θ, and β) is still an open question, as confirmed by some discussions at NATO meeting at Vulcano.

8. Conclusions

The purpose of this tutorial-style review was double : to introduce the reader, on one hand, with the basic processes involved in the physics of collisionless shocks, and on the other hand, with numerical simulation (applied to shocks physics in the present case).

Shocks problem has been gradually introduced herein by considering, first, the steepening of a simple electrostatic wave and the mutual competition between dissipation, dispersion, and steepening effects on this wave, and, second, a similar approach applied to a large amplitude magnetosonic wave. Such an approach allows to apply to shock waves all the helpful informations (time range of the resonant waves-particles interaction, consequences and estimate of wave damping, etc.) already analyzed for ES or MS non linear waves, and to remind that a MS shock wave can be considered as a particular case (step profile) of a MS wave. Previous studies have shown that a shock wave largely differs (i.e. may be classified) according to the value ß, the ratio T_e/T_i, the angle θ and in particular to the Mach number M_A. The M_A dependance may be expressed in terms of

124

relative importance of ion reflection, i.e. ion heating, with respect to electron heating and not only according to the simple classification : no ion reflection, no foot and no overshoot for $M_A < M_A*$ and the reverse for $M_A > M_A*$. Such distinction, often accepted in a first approach, reveals to be too strict since numerical results as well as revisited experimental data clearly show some (residual) ion reflection and the presence of some overshoots for $M_A < M_A*$. A smoother consideration is then required.

A certain class of microinstabilities, -cross-field current instabilities (Table1) -, seem to be always present since they are directly related to the strong currents which support the shock profile. Although they seem to play an important role in subcritical shocks, they are masked by ion reflection effects in supercritical shocks. "Ion reflection effects" include not only the reflection itself which is a source of viscous heating, but also all microinstabilities (Tables 2 and 3) which are "consequences" of this reflection and which are additional sources of energy dissipation. Then, instead of distinguishing subcritical to supercritical shocks by focusing mainly on the absence or the presence of ion reflection itself, one deeper approach would consist in determining:

(i) which kinds of microinstabilities (among Tables 2 and 3) are dominant in supercritical regime according to ß, T_e/T_i and θ,

(ii) how these instabilities decrease in importance from a supercritical to a subcritical shock,

(iii) what is their relative importance with respect to the microinstabilities of Table 1, when approaching the subcritical regime. A similar approach could also apply to astrophysical or relativistic shocks. However, one needs to take into account two additional features for such shocks : (i) the electrons do not behave adiabatically and have their own dynamics, and (ii) some instabilities which play a minor role in "classical" MS shocks may become dominant in SHMN shocks. A few studies have been already performed at least in subcritical and supercritical regimes separately; they have clearly emphasized the difficulties of the problem due to either computer limitations, or the difficulty for distinguishing accurately different microinstabilities in a given M_A regime. New promising hardwares in near-future computers could satisfy the first point, while a more extensive parametric study (often a problem of computer ressources) could answer to the second point.

Indeed, numerical simulation reveals to be a very powerful tool for the progressive understanding of phenomena associated with shocks (as well as in other areas of physics) for four reasons :

(i) it allows a direct comparison with experimental results with the help of appropriate numerical diagnosis.

(ii) it represents, sometimes, a unique way of diagnosis, even taking into account the numerical restrictions of the simulation of concern (ex : visualizing accurately particles trajectories through self-consistent fields).

(iii) it strengthens the validity of some results via extended parametric studies and/or reproducibilities of these results by different kinds of numerical codes.

(iv) it makes easier the approach of a difficult but necessary task, combined with experimental and theoretical studies : a synthetic overview of the various kinds of shocks. As illustrated in Section 6, various studies have shown the relationship existing between some microinstabilities which have been analysed separately at the origin. A similar approach is now necessary in shock studies in general.

In summary, a shock pattern is supported self-consistently by a particular dynamics of electrons and of ions which constitute various sources of energy dissipation; by a feed-back effect, these affect strongly the overall structure of the shock and its downstream and upstream neighbourhood. This mutual interaction can be expressed in terms of "phenomena-in-chain"; a typical (but incomplete) example is illustrated in Fig. 18. Some parts of this chain require further analysis, in order to improve the understanding of the overall shock-particles interaction mechanisms.

Acknowledgements

Preparation of this review was carried out while the author was at the Physics Department of UCLA (Los Angeles, USA) which is thanked for its nice hospitality. I am also particularly grateful to Dr. K.B. Quest from San Diego U.C. for his helpful comments and proofreading over the manuscript.

APPENDIX 1 : Magnetohydrodynamic code : two fluids, (electrons and ions)

This two fluid MHD approach (Tidman and Krall, 1971, Papadopoulos, 1985) is based on the following assumptions :
   a) quasi neutrality
   b) scalar pressure
   c) negligible electron and heat flux
   d) dissipation provided by resistivity which heats the electrons and viscosity which heats the ions
   e) neglects the displacement current ($\partial E / \partial t = 0$) and includes the set of equations :
   1) quasi neutrality : $N = N_i = N_e$

2) continuity : $dN / dt = - N \vec{\nabla}.\vec{u_i} = - N \vec{\nabla}.\vec{u_e}$

3) electron momentum :

$$m_e N \frac{d\vec{u_e}}{dt} = - \vec{\nabla}P_e - Ne\left(\vec{E} + \frac{\vec{u_e} \times \vec{B}}{c}\right) - \vec{P_{ei}}$$

4) electron energy :

$$\frac{dP_e}{dt} + \gamma P_e \vec{\nabla}.\vec{u_e} = (\gamma - 1) S_\eta$$

5) ion momentum :

$$m_i N \frac{d\vec{u_i}}{dt} = - \vec{\nabla}P_i - \vec{\nabla}\pi + Ne\left(\vec{E} + \frac{\vec{u_i} \times \vec{B}}{c}\right) + \vec{P_{ei}}$$

6) ion energy :

$$\frac{dP_i}{dt} + \gamma P_i \vec{\nabla}.\vec{u_i} = (\gamma - 1) S_\mu$$

7) Maxwell's equations :

$$\vec{\nabla} \times \vec{B} = 4\pi \ \vec{j} / c$$

$$\vec{\nabla} \times \vec{E} = -\frac{1}{c} \frac{\partial \vec{B}}{\partial t}$$

$$\vec{\nabla} \cdot \vec{B} = 0$$

The terms π, S_μ and S_η represent, respectively, the viscous stresses, viscous dissipation and resistive dissipation. A fluid description assumes that the pressure stays isotropic (or at least gyrotropic) ; then, the pressure and viscous correction terms can be related to the other transport variables. In the case of collisionless plasma, this is often justified for electrons but not for the ions.

Steady-state description

An important consequence of equations (1) - (7) for steady state conditions is a set of relations which connect the low entropy medium (upstream or unshocked plasma) to the high entropy medium (downstream or shocked plasma). This transition is mainly function of the Mach number M_A, the angle θ and the value of the upstream β (ratio of thermal energy over magnetic energy). For so doing, one assumes :

(8a) time derivative neglected, i.e. $\partial / \partial t = 0$
(8b) gradient in the x- direction, i.e. $\vec{\nabla} = (d/dx, 0, 0)$
(8c) the fluid velocity : $\vec{u} = (u, \vec{v_t}) \approx \vec{u_i}$ within m_e/m_i
(8d) the electron-ion momentum exchange:
$$P_{e,i} = N e \eta j \text{ which gives } S_\eta = \eta j^2 ;$$

η is the resistivity.

8e) The viscous term : $\pi = \mu \ du/dx$ which gives $S_\mu = \mu (du/dx)^2$;
μ is the viscosity.
(8f) electron inertia neglected i.e. $m_e = 0$
(8g) electron fluid velocity
$$\vec{u_e} = \vec{u} - \frac{\vec{j}}{Ne}$$

which leads to :

(9) reduced Maxwell's equations :
$$\vec{e_x} \times \frac{d\vec{B}}{dx} = 4\pi \ \vec{j} / c$$
E_t = constant ; B_x = constant
where the subscript "t" refers to components transverse to the flow.

(10) the momentum conservation equation by summing (3) et (5) :

$$N\,m_i\,u\,\frac{d\vec{u}}{dx} = -\,\vec{e}_x\,\frac{d}{dx}\,(\,P_e + P_i - \mu\frac{d\mu}{dx}\,) + \frac{\vec{j}\times\vec{B}}{c}$$

(11) the energy conservation equation by summing (4) and (6).

All the equations allow to compute the downstream fluid quantities in terms of the upstream ones and give the shock jump conditions. The simplest way for deriving the upstream to downstream relations, usually called "Rankine-Hugoniot" equations, is to use equations (9) - (11) for a uniform plasma outside the shock region. For more complete analysis, see Tidman and Krall (1971).

APPENDIX 2 : Electromagnetic, full particle code (both electrons and ions)

This kind of code (1 D or 2 D) includes basically :

(1) the Poisson's equation :

$$\vec{\nabla}.\vec{E} = \rho/\varepsilon_o$$

(2) the full set of Maxwell's equations :

$$\vec{\nabla}\times\vec{B} = \mu_o\vec{j} + \frac{1}{c}\frac{\partial\vec{E}}{\partial t} \qquad \text{where } \vec{j} = \vec{j}_e + \vec{j}_i$$

$$\vec{\nabla}\times\vec{E} = -\,\frac{1}{c}\frac{\partial\vec{B}}{\partial t}$$

(3) the equation of particle motion :

$$\frac{d\vec{p}_\alpha}{dt} = q_\alpha\,(\,\vec{E} + \vec{v}_\alpha\times\vec{B}\,)$$

$$\vec{v}_\alpha = \frac{d\vec{r}_\alpha}{dt}\,;\,\vec{p}_\alpha = m_\alpha\vec{v}_\alpha$$

(4) relativistic effects (in option) :

$$\vec{p}_\alpha = \gamma_\alpha\vec{v}_\alpha\,;\,\text{with }\gamma_\alpha = (1 - v_\alpha^2/c^2)^{1/2}$$

Comments :

Equation (1) includes the real space charge effects and replaces the quasi-neutrality equation used in MHD and hybrid codes ($\rho = 0 \rightarrow N_e = N_i$)

Equations (2) are the so-called "fields pusher". Pushing fields can be performed either in the real space $\vec{r}$ by finite-difference methods, or by using FFT (Fast Fourier Transform) techniques in the associated $\vec{k}$-space. The latter method allows to provide also direct diagnosis of wave spectrum (fields energy versus the wave number k) at any time of the run.

Equations (3) are the so-called "particles pusher" and are applied along each direction x, y, z of the velocity $\vec{v}$ and the space $\vec{r}$ variables for each component (electrons and ions) of the plasma. The particle pusher is the most expensive part of the code where vectorisation computing techniques need to be the most efficient.

1D (one dimensional) and 2D (two dimensional) codes refer mainly to the dimension of the real -or k- space used in the code ; hence 1D → [$\vec{k}$ =(k$_x$,0, 0)], and 2D - > [$\vec{k}$ = (k$_x$, k$_y$, 0)] for instance.

The code dimension changes the number of projected equations of Eqs 1-3; as a consequence, it drastically changes the requested size of the computer memory and the cost of the computer time. This can be illustrated by the following remarks :

a)   1D code provides five fields components; namely :

E_{lx} :   longitudinal electrostatic field (Eq. 1) ; subscript "l"
E_{ty}, E_{tz} $\Big\}$:   electromagnetic
B_{ty}, B_{tz} transverse fields (Eqs.2) ; subscript "t".

a) perpendicular case (θ = 90°)

→ (E_{lx}, E_{ty}, B_{tz})

b) "pseudo-oblique" case (θ ≠ 90°)

→ all fields

In the case θ = 90°, only three fields components are relevant namely E_{lx}, E_{ty} and B_{tz}. For θ ≠ 90° (called "pseudo-oblique" case), all fields are relevant provided that θ is below some critical angle. This is due to the fact that an important particle dynamics takes place along $\vec{B}_o$ (i.e. electron current along z) which induces both additional components E_{tz} and B_{ty}.

b)   2D code provides eight components, namely:

E_{lx}, E_{ly} :   longitudinal fields
E_{tx}, E_{ty}, E_{tz} $\Big\}$: transverse fields
B_{tx}, B_{ty}, B_{tz}

In the case θ = 90°, we recover again only three relevant components E_{lx}, E_{ty} and B_{tz} (assuming $\vec{B}_o$ = (0, 0, B$_o$)). the situation is more complex for a "real" oblique shock propagating within the plane (k_x, k_y) and requires a more complete set of fields components.

Algorithms : In brief, various numerical techniques apply to fully particle codes : explicit, spectral and implicit methods. In the first standard method, particles and fields are calculated at time t from their values at previous time. A standard "leap-frog" scheme is used to advance the particles and fields which means that particle velocities are calculated at the half time steps while the particle positions and the fields are known at the even time steps. This is due to the fact that velocities and positions of particles cannot be defined at the same time (Eq. 3). In addition, explicit methods are mainly characterized by the restriction required by dispersion effects (high frequencies to be included) on the maximum size of the time step used in the fields and particles pushers. In spectral methods, a harmonic representation for the fields is used (see the review by Dawson, 1983). In implicit methods (Brackbill and Forslund, 1982; Langdon, 1985), fields quantities are calculated from moments of the particle distribution in such a way to reduce the constraints in temporal and spatial step size. In principle, this method allows computer simulation for long time periods (large Δ t) but cannot apply to situations including electron inertia effects i.e. where high frequency plasma effects, or short wavelenghts, have to be taken account.

APPENDIX 3: Hybrid codes (electrons : fluid ; ions : particles)

a) Electrons: fluid   Conséquences:

- $m_e = 0$ ———————————(1)———————————→ neglect time and space scales of electrons

- $\vec{E} = - \dfrac{1}{N_e q} \left(\vec{J}_e \times \vec{B} \right) - \dfrac{1}{N_e q} \vec{\nabla} P_e - \eta \left(\vec{J}_e + \vec{J}_i \right) \longrightarrow$ no high frequency electromagnetic effects

- $\dfrac{3}{2} \dfrac{\partial P_e}{\partial t} + \vec{\nabla}.\left(\dfrac{3}{2} P_e \vec{u}_e \right) + P_e \vec{\nabla}.\vec{u}_e = \eta \, j^2$

where: $\vec{u}_e = - \vec{J}_e / q N_e$

η = resistivity

b) Ions: particles

- $m_i \dfrac{d\vec{v}_i}{dt} = q_i \left(\vec{E} + \dfrac{\vec{v}_i \times \vec{B}}{c} \right) + \vec{P}$

where: $\vec{P} = - q \eta \, \vec{j}$

c) <u>Electromagnetic equations</u>

- $\vec{\nabla} \times \vec{E} = -\dfrac{1}{c} \dfrac{\partial \vec{B}}{\partial t}$

- $\vec{\nabla} \times \vec{B} = \dfrac{4\pi}{c} (\vec{j}_e + \vec{j}_i)$ $\longrightarrow$ $\partial E / \partial t = 0$ (no displacement current)

d) <u>Quasi-neutrality</u>

$N_e = N_i = N$ $\xrightarrow{\quad + (1) \quad}$ no electron dispersion effects

<u>Algorithm</u> : In brief, three numerical techniques apply to hybrid codes : resistive Ohm's law, predictor-corrector and Hamiltonian method. In all these methods, a standard "leap-frog" scheme is used (to see algorithms and references of Appendix 2). Details of these methods may be found in Winske (1985) and Quest (1989).

APPENDIX 4: How to include or exclude resistivity effects in 1D and 2D simulation codes ?

We will separately distinguish 1D, 2D, fully particles and hybrid codes. Strictly perpendicular shock ($\theta = 90°$) will be considered for simplicity.

<u>1) 1D code</u> [$\vec{k} = (k_x, 0,0)$]

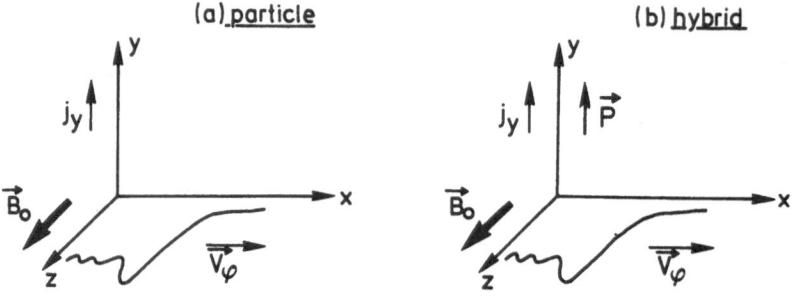

(a) particle   (b) hybrid

<u>Fully particle code</u> (case a) : the code is totally self-consistent in the sense that time and space behavior of the shock do not depend on any external plasma parameter included artificially. However, because of the one-dimensionality, any resistivity effect arising from cross-fields currents instabilities is excluded. This effect can be taken into account either by an artificial inclusion of η in 1D (case b), or by adding a second dimension in real space (2D code; case b).

Hybrid code (case b) : an artificial resistivity η is included through the ηj^2 and $\eta(\vec{j}_e + \vec{j}_i)$ terms and the friction force term $\vec{P}$, which couples fluid electron equations and ion particle pusher (Appendix 3). However, since η is fixed as constant throughout the run, the code is not strictly self-consistent.

b) 2D code $[\vec{k} = (k_x, 0, k_z)]$

We will consider a fully particle code for which two situations are possible as proposed by Forslund and al. (1984) :

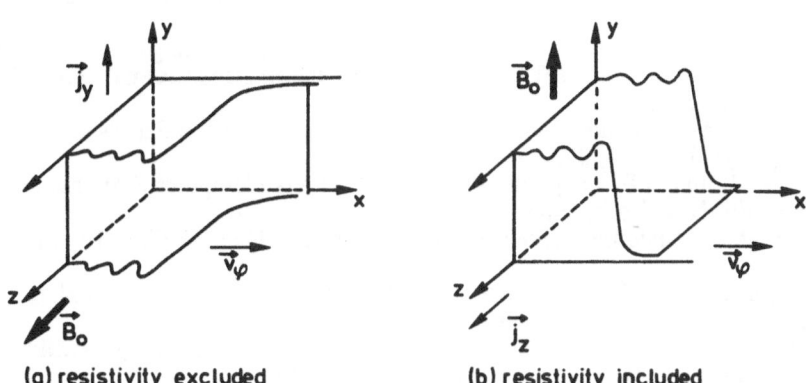

(a) resistivity excluded   (b) resistivity included

a) In the left hand side panel, the magnetostatic field $\vec{B}_o$ is inside the propagation plane (x,z). The diamagnetic cross field current is in the y direction (j_y) and out of the simulation plane. Thus, cross fields currents instabilities are not allowed (resistivity excluded) but pitch angle scattering instabilities (also called anisotropy-driven instabilities) are (to see Sec. 6).

b) In the right hand side panel, the field $\vec{B}_o$ is rotated out of the plane $[\vec{B}_o = (0, B_{oy}, 0]$; the currents ($j_z$) move into the propagation plane allowing the cross-field instabilities to occur (resistivity included).

Then, by running a simulation code for identical plasma parameters, but by rotating or not the static field $\vec{B}_o$ out of the plane (x, z) one may illustrate a transition from essentially zero cross-field resistivity to the finite level resistivity, self-consistently produced by cross-field currents instabilities. An overview of possible plasma instabilities is provided in Sec 6.

132

References :

Akimoto, K., Papadopoulos, K. and Winske, D., 'Lower hybrid instabilities driven by an ion velocity ring', J. Plasma Phys. 34, 445-479, 1985.

Ambrosiano, J. and Brecht, S. H., 'A simulation study of the alfven ion cyclotron instability in high beta plasma', Phys. Fluids, 30, 108-114, 1987.

Arefev, V.I., 'Instability of a current-carrying homogeneous plasma', Sov. Phys. Tech. Phys. 14, 1487-1491, 1970

Ashour-Abdalla M., J.N. Lebœuf, T. Tajima, J.M. Dawson and C.F. Kennel, 'Ultrarelativistic electromagnetic pulses in plasmas', Phys. Rev., 23, 1906-1914, 1981.

Auer, P.L. and W.H. Evers, 'Collision free shock formation in finite temperature plasmas', Phys. Fluids, 14, 1177-1182, 1971.

Auer, P.L., H. Hurwitz, Jr. and R. W. Kilb, 'Large amplitude magnetic compression of a collision-free plasma,II,Development of a thermalized plasma', Phys. Fluids, 5, 298-316, 1962.

Auer, P.L., R.W. Kilb and W.F. Crevier, 'Thermalization of the Earth's bow shock' , J. Geophys. Res., 76, 2927-2939, 1971.

Birdsall, C.K. and A.B. Langdon, 'Plasma Physics via Computer Simulations', Mc. Graw-Hill, New-York, 1985.

Biskamp, D., 'Collisionless shock waves in plasmas', Nuclear Fus., 13, 719- 740, 1973.

Biskamp, D. and H. Welter, 'Numerical studies of magnetosonic collisionless shock waves' , Nucl. Fusion, 12, 663-666, 1972a.

Biskamp, D. and H. Welter, 'Ion heating in high-Mach number oblique collisionless shock waves , Phys. Rev. Lett., 28, 410-413,1972b.

Boyd, T.J.M., 'Shock waves in collisionless plasmas', in Proceedings of the Eleventh International Symposium on Shock Tubes and Waves, edited by B. Ahlborn, A. Hertzberg and D. Russel, p. 156, University of Washington Press, Seattle, 1977.

Brackbill, J.U. and Forslund, D.W., 'An implicit method for electromagnetic plasma simulation in two dimensions', J. Comp. Phys., 46, 271, 1982 .

Buneman, O., 'Dissipation of currents in ionized media', Phys. Rev., 115, 303, 1959.

Burgess, D., 'Numerical simulation of collisionless shocks', in Proceedings of International Conference on Collisionless Shocks, Bulatonfured, Hungary , pp. 89-111, ed. by Szego, 1987a.

Burgess, D., 'Simulations of backstreaming ion beams formed at oblique shocks by direct reflection', Annales Geophysicae, 5, 133-145, 1987b .

Burgess, D. , 'Shock drift acceleration at low energies', J. Geophys. Res. , 92, 1119-1130, 1987c .

Burgess, D. and Luhmann J.G., 'Scatter-free propagation of low-energy protons in the magnetosheath : implications for the production of field aligned ion beams by non-thermal plasma', J. Geophys. Res., 91, 1439-1449, 1986.

Cargill, P.J. and K. Papadopoulos, 'A mechanism for strong electron heating in supernova remnants', The Astrophys. J., 329, L29-L32, 1988.

Chodura, R., 'A hybrid fluid-particle model of ion heating in high-Mach number shock waves', Nucl. Fusion, 15, 55-61, 1975 .

Coffey, T.P., 'Breaking of large amplitude plasma oscillations', Phys. Fluids, 14, 1402-1406, 1971.

Davidson, R.C., Gladd, N.T., Wu, C.S. and Huba, J.B., 'Effect of finite plasma beta on the lower-hybrid-drift instability', Phys. Fluids, 20, 301-310, 1977.

Davidson, R.C. and Hammer, D.A., 'Nonequilibrium energy constants associated with large amplitude electron whistlers', Phys. Fluids, 15, 1282-1284, 1972.

Davidson, R.C. and Ogden, J.M., 'Electromagnetic ion cyclotron instability driven by ion energy anisotropy in high beta plasmas', Phys. Fluids, 18, 1045-1050, 1975.

Davidson, N. C. and Gladd, N.T. 'Anomalous transport properties associated with the lower-hybrid-drift instability', Phys. Fluids, 18, 1327-1335, 1975.

Dawson, J.M., 'Nonlinear electron oscillations in a cold plasma', Phys. Rev., 113, 383-387, 1959.

Dawson, J.M., 'Particle simulation of plasmas', Rev. Mod. Phys., 55, 403, 1983.

Dawson, J.M., W.L. Kruer and B. Rosen, 'Investigation of ion waves', in dymanics of ionized gases, ed. by M.J. Lighthill, I. Imai and H.Sato , University of Tokyo Press, Tokyo, Japan, pp. 47-61, 1973.

Decker, R.B. and L. Vlahos, 'Numerical studies of particle acceleration at turbulent, oblique shocks with an application to prompt ion acceleration during solar flares', 306, 710-729, 1986.

Fairfield, D.H. and W.C. Feldman, 'Standing waves at low Mach number bow shocks', J. Geophys. Res., 80, 515-522, 1975.

Feldman, W.C., 'Electron velocity distributions near collisionless shocks', in Collisionless Shocks in the Heliosphere : Reviews of Current Research, Geophys. Monogr. Ser., Vol. 35, pp. 195-206, ed. by B.T. Tsurutani and R.G . Stone, AGU, Washington, D.C., 1985.

Feldman, W.C., S.J. Bame, S.P. Gary, J.T. Gosling, D. Mc. Thomas, M.F. Thomsen, G. Paschmann, N. Sckopke, M.M. Hoppe and C.T. Russell, 'Electron heating within the Earth's bow shock', Phys. Rev. Lett., 49, 199-201, 1982.

Feldman, W.C., R.C. Anderson, S.J. Bame, S.P. Gary, J.T. Gosling, D.J. Mc Comas, M.F. Thomsen, G. Paschmann and M.M. Hoppe, 'Electron velocity distributions near the Earth's bow shock', J. Geophys. Res., 88, 96-110, 1983.

Formisano, V., 'Plasma processes at collisionless shock waves', in Plasma Astrophysics Course and Workshop, Rep. ESA-SP-161, pp. 145-165, 1981.

Forslund, D.W., 'Fundamentals of plasma simulations', Space Sci. Rev., 42, 3-16, 1985.

Forslund, D.W., Morse, R.L. and Nielson, C.W., 'Electron cyclotron drift instability', Phys. Rev. Lett. 25, 1266-1270, 1970.

Forslund, D.W., and J.P. Freidberg, 'Theory of laminar collisionless shocks' , Phys. Rev. Lett., 27, 1189-1192, 1971 .

Forslund, D.W., K.B. Quest, J.U. Brackbill and K. Lee, 'Collisionless dissipation in quasi-perpendicular shocks', J. Geophys. Res., 89, 2142-2150, 1984.

Galeev, A.A., 'Collisionless shocks', in Physics of solar planetary relationships, edited by D.J. Williams, p. 464, AGU, Washington DC, 1976.

Gary. S.P. 'Longitudinal waves in a perpendicular collisionless plasma shock ; I, Vlasov ions' J. Plasma Phys., 4, 753-760, 1970.

Gary, S.P., 'Longitudinal waves in a perpendicular collisionless plasma shock, Part 4, Gradient B', J. Plasma Phys., 4, 417-425, 1972.

Gary, S.P. and J.J. Sanderson, 'Longitudinal waves in a perpendicular collisionless plasma shock, I, Cold ions', J. Plasma Phys., 4, 739-751, 1970.

Gary, S.P., J.T. Gosling and D.W. Forslund, 'The electromagnetic ion beam instability upstream of the Earth`s bow shock', J. Geophys. Res., 86, 6691 -6696, 1981.

Gary, S.P., C.W. Smith, M.A. Lee, M.L. Goldstein and D.W. Forslund, 'Electromagnetic ion beam instabilities', Phys. Fluids, 27, 1852-1862, 1984.

Gary, S.P. and R.L. Tokar, 'The second order theory of electromagnetic hot ion beam instabilities', J. Geophys. Res., 90, 65-72, 1985.

Gladd, N.T., 'The lower hybrid-drift instability and the modified two stream instability in high density theta pinch environment', Plasma Phys. 18, 27-40, 1976.

Goodrich, C.C., 'Numerical simulations of quasi-perpendicular collisionless shocks', in Collisionless Shocks in the Heliosphere : Reviews of Current Research, Geophys. Monogr. Ser., Vol. 35, pp 153-168, ed. by B.T. Tsurutani and R.G. Stone, AGU, Washington, D.C. , 1985.

Gosling, J.T., M.F. Thomsen, S.J. Bame,W.C. Feldman, G. Paschmann, and N., Sckopke, 'Evidence for specularly reflected ions upstream from the quasi- parallel bow shock', Geophys. Res. Lett., 9, 1333-1336, 1982.

Greenstadt, E.W., 'Oblique, parallel and quasi-parallel morphology of collisionless shocks', in Collisionless Shocks in the Heliosphere : Reviews of Current Research, Geophys. Monograph. Ser., Vol. 35, pp.169-184 , ed. by B.T. Tsurutani and R.G. Stone, AGU, Washington, DC, 1985.

Greenstadt, E.W., V. Formisano, C.T. Russell, M. Neugebauer and F.L. Scarf, 'Ion acoustic stability analysis of the earth's bow shock', Geophys. Res. Lett., 5, 399-402, 1978.

Greenstadt, E.W. and R.W. Fredricks, 'Shock systems in collisionless space plasmas', in Solar System Plasma Physics III, ed. by C.K. Kennel, L.J. Lanzerotti and E.N. Parker, pp. 3, North-Holland, Amsterdam, 1979.

Greenstadt, E.W. and Mellott, M.M., 'Plasma wave evidence for reflected ions in front of subcritical shocks, ISEE-1 and 2 observations', J. Geophys. Res., 92, 4730-4734, 1987.

Hamasaki, S. and N.A. Krall, 'Relaxation of anisotropic collisionless plasma', Phys. Fluids, 16, 145-149, 1973.

Hockney, R.W. and J.W. Eastwood, Computer Simulation Using Particles , Mc. Graw-Hill, New York, 1981.

Huba, J.D. and Wu, C.S., 'Effects of a magnetic field gradient on the lower hybrid-dirft instability', Phys. Fluids, 19, 988-994, 1976.

ISEE Upstream Waves and Particles, Special Issue of J. Geophys. Res., 86, 4319-4536, 1981.

ISSS-I, 'Computer Simulation of Space Plasmas', Proceedings of the First International School for Space Simulations, ed. H. Matsumoto and T. Sato, D. Reidel Publishing Company, Dordrecht, Holland, 1985.

ISSS-2, 'Space Plasma Simulations, Proceedings of the Second International School for Space Simulations, ed. M. Ashour-Abdalla and D.A. Dutton, Reidel Publishing Company Dordrecht/Boston, 1985 ; (reprinted from Space Science Rev., Vol. 42, Nos. 1-4, 1985).

ISSS-3, 'Numerical Simulation of Space Plasmas', Proceedings of the Third International School for Space Simulations, Part 1, ed. by B. Lembege and J.W. Eastwood, North-Holland, Amsterdam, 1988; (reprinted from Computer Physics Communications, Vol. 49, n° 1, 1988).

ISSS-3, 'Tutorial courses', Proceedings of the Third International School for Space Simulations, Part 1, CRPE, Cepadues Editions, Toulouse, France, 1989.

Karney, C.F.F., 'Stochastic ion heating by a lower hybrid wave', Phys. Fluids, 21, 1584-1599, 1978

Katsouleas, T., 'Cross magnetic field plasma phenomena', Ph.D., Thesis, University of California at Los Angeles, 1984.

Kennel, C.F., Edminston, J.P. and Hada, T., 'A quarter century of collisionless shock research', in Collisionless Shocks in the Heliosphere, Reviews of Current Research, Geophys. Monograph. Ser., Vol. 35, pp. 1-36, ed. by B.T. Tsurutani and R.G. Stone, AGU, Washington, DC, 1985.

Kono, M. and P. Mulser, 'Effect of trapping on breaking large-amplitude plasma oscillations', Phys. Fluids, 26, 3004-3007, 1983.

Krall, N.A. and Liewer, P.C., 'Low frequency instability in magnetic pulses', Phys. Rev. A4, 2094-2103, 1971.

Kruer. W.L., 'Wave breaking amplitude in warm inhomogeneous plasma', Phys. Fluids, 22, 1111-1114, 1979.

Kulikovski, A.G. and Lubinov, G.A., 'Magnetohydrodynamics', pp. 9-31, Addison-Wesley, Reading, Mass., 1962.

Kulygin, V.M., Mikhailovskii, A.B. and Tsapelkin, E.S., 'Quasilinear relaxation of fast ions moving transverse to a magnetic field', Plasma Phys., 13, 1111-1116, 1971.

Langdon, A.B., 'Implicit plasma simulation', in Space Plasma Simulations, ed. by M. Ashour-Abdalla and D.A. Dutton, D. Reidel Publishing Company / Dordrecht / Boston, pp. 67-83, 1985.

Lashmore-Davies, C.N., 'Instability in a perpendicular collisionless shock wave for arbitrary ion temperatures' Phys. Fluids, 14, 1481-1484, 1971.

Lashmore-Davies and C.N., Martin, T.J., 'Electrostatic instabilities driven by an electric current perpendicular to a magnetic field', Nuclear Fus. 13, 193-203, 1973.

Lebœuf J.N., M. Ashour-Abdalla, T. Tajima, C.F. Kennel, F.V. Coroniti and J.M. Dawson, Phys. Rev. 25, 1023-1039, 1982.

Lee, L.C., Wu, C.S. and Hu, X.W., 'Increase of ion kinetic temperature across a collisionless shock: 1- A new mecanism', Geophys. Res. Lett., 13, 209-213, 1986a.

Lee, L.C., Mandt, M.E. and Wu, C.S., 'Increase of ion kinetic temperature across a collisionless shock: 2 - A simulation study', J. Geophys. Res., 92, 13438-13446, 1986b.

Lee, L.C., C.P. Price, C.S. Wu, and M.E. Mandt, 'A study of mirror waves generated downstream of a quasi-perpendicular shock', J. Geophys. Res., 93, 247-250, 1988.

Lembege, B. and J.M. Dawson, 'Kinetic perpendicular collisionless shocks', UCLA report No PPG-882, 1984a.

Lembege, B. and J.M. Dawson, 'Plasma heating and acceleration by strong magnetosonic waves propagating obliquely to a magnetostatic field', Phys. Rev. Lett., 53, 1053-1056, 1984b.

Lembege, B. and J.M. Dawson, 'Self-consistent plasma heating and acceleration by strong magnetosonic waves for theta=90, Part I: Basic mechanisms', Phys. Fluids, 29, 821-836, 1986.

Lembege, B. and J.M. Dawson, 'Self-consistent study of a perpendicular collisionless and non-resistive shock', Phys. Fluids, 30, 1767-1788, 1987a.

Lembege, B. and J.M. Dawson, 'Plasma heating through a supercritical oblique collisionless shock', Phys. Fluids, 30, 1110-1114, 1987b.

Lembege, B. and J.M. Dawson, 'Formation of double layers within an oblique collisionless shock', Phys. Rev. Lett., 62, 2683-2686, 1989a.

Lembege, B.L. and J.M. Dawson, 'Relativistic particle dynamics in a steepening magnetosonic wave', Phys, Fluids, 1, 1001-1010, 1989b.

Lemons, D.S. and Gary, S.P., J. Geophys. Res. 83, 1625-1631, 1978.

Leroy, M.M., 'Structure of perpendicular shocks in collisionless plasma', Phys. Fluids, 26, 2742-2753, 1983.

Leroy, M.M., C.C. Goodrich, D. Winscke, C.S. Wu, and K. Papadopoulos, 'Simulations of a perpendicular bow shock', Geophys. Res. Lett., 8, 1269-1272, 1981.

Leroy, M.M., D. Winscke, C.C. Goodrich, C.S. Wu, and K. Papadopoulos, 'The structure of perpendicular bow shocks', J. Geophys. Res., 87, 5081-5094, 1982.

Leroy,M.M. and D. Winscke, 'Backstreaming ions from oblique Earth bow shocks ',Annales Geophysicae., 1, 527-536, 1983.

Livesey, W.A., C.F., Kennel and C.T. Russell, 'ISEE-1 and -2 observations of magnetic field strengh overshoots in quasi-perpendicular bow shocks', Geophys. Res. Lett., 9, 1037-1040, 1982.

Livesey, W.A., C.T. Russell and C.F. Kennel, 'A comparison of specularly reflected gyrating ion orbits with observed foot thickness', J. Geophys. Res., 89, 6824-6828, 1984.

Manheimer, W.M. and J.P. Boris, 'Self consistent theory of a collisionless resistive shock', Phys. Rev. Lett., 28, 659-662, 1972.

Mason, R.J., 'Ion and electron pressure effects on magnetosonic shock formation', Phys. Fluids, 15, 1082-1089, 1972.

McBride, J.B., Ott, E., Boris, J.P. and Orens, J.H., 'Anomalous resistance due to cross field electron-ion streaming instabilities', Phys. Fluids, 15, 2356-2362, 1972.

Mellot, M.M., 'Subcritical collisionless shock wave in Collisionless Shocks in the Heliosphere : Reviews of current research', Geophys. Monogr. Ser., vol. 35, pp. 131-140, ed. by B.T. Tsurutani and R.G. Stone, AGU, Washington DC, 1985.

Mellot, M.M. and Greenstadt, E.W., 'The structure of oblique subcritical bow shocks: ISEE1 and 2 observations', J. Geophys. Res. 89, 2151, 1984.

Mellot, M.M. and W.A. Livesey, 'Shocks overshoots revisited', J. Geophys. Res', 92, 13661-13665, 1987.

Montgomery, M.D., J.R. Asbridge and S.J. Bame, 'Vela 4 plasma observations near the Earth's bow shock', J. Geophys. Res., 75, 1217-1231, 1970.

Morse, R.L., 'Multidimensional plasma simulation by the particle in cell method', in Methods of Computional Physics, Vol.9, ed. by B. Alder, S. Fernbach, and M. Rotenberg, p.213, Academic Press, New York, 1976.

Morse, D.L. and Greenstadt, E.W., 'Thickness of magnetic structures associated with the earth's bow shock', J. Geophys. Res., 81, 1791-1793,1976.

Ohsawa, Y., 'Strong ion acceleration by a collisionless magnetosonic shock wave propagating perpendicularly to a magnetic field', Phys. Fluids, 28, 2130-2136, 1985.

Ohsawa, Y., 'Resonant ion acceleration by oblique magnetosonic schok waves in a collisionless plasma', Phys. Fluids, 29, 773-781, 1986.

Ohsawa, Y. and J. Sakai, 'Ion acceleration in quasi-perpendicular collisionless magnetosonic shock waves with subcritical Mach number', Geophys. Res. Lett., 12, 617-619, 1985.

Ohsawa,Y. and J.I. Sakai, 'Prompt simultaneous acceleration of protons and electrons to relativistic energies by shock waves in solar flares', The Astrophy. J., 332, 439-446, 1988.

Ossakow, S.L., Ott, E. and Haber, I., 'Nonlinear evolution of whistler instabilities', Phys. Fluids, 15, 2314-2326, 1972.

Otani, N.F., 'Saturation and post-saturation behavior of the Alfven ion cyclotron instability; a simulation study', Memo. Rep. M86/15, University of California at Berkeley, Berkeley, California, 1986.

Papadopoulos, K., 'A review of anomalous resistivity for the ionosphere', Rev. Geophys. Space. Phys., 15, 113, 1977.

Papadopoulos, K., 'Electron acceleration in magnetosonic shock fronts', in Plasma Astrophysics, Spec. Publ. ESA SP-161, pp. 313-315, European Space Agency; Neuilly, France, 1981.

Papadopoulos, K., 'Microinstabilities and anomalous transport', in Collisionless Shocks in the Heliosphere', Reviews of Current Research, Geophys. Monograph.

Ser., Vol. 35, pp. 59-88, ed. by B.T. Tsurutani and R.C. Stone, AGU, Washington, DC, 1985.

Papadopoulos, K., 'Electron heating in supercritical Mach number shocks', Astrophys. and Space Sci., 144, 535-547, 1988.

Papadopoulos, K., C.E. Wagner and I.Haber, 'High mach number turbulent magnetosonic shocks', Phys. Rev. Lett., 27, 982-986, 1971a.

Papadopoulos, K., Davidson, R.C., Dawson, J. M., Haber, I., Hammer, D. A., Krall. N. A., and Shanny, R. 'Heating of countersteaming ion beams in an external magnetic field', Phys. Fluids, 14, 849-857, 1971b.

Priest, E.R. and J.J. Sanderson, 'Ion acoustic instability in collisionless shocks', Plasma Phys. 14, 951-958, 1972.

Quest, K.B., 'Simulations of high Mach number collisionless perpendicular shocks in astrophysical plasmas', Phys. Rev, Lett., 54, 1872-1874, 1985.

Quest, K.B., 'Very high Mach number shocks', Adv. Space Res., 6, 33-39, 1986a.

Quest, K.B., 'Simulations of high Mach number perpendicular shocks with resistive electrons', J. Geophys. Res., 91, 8805-8815, 1986b.

Quest, K.B., 'Hybrid simulation', in Numerical Simulation of Space Plasmas, Proceedings of ISSS-3, Part 1, pp. 177-182, Cepadues Editions, Toulouse, France, 1989.

Revathy, P. and G.S. Lakina, 'Ion and electron heating in the Earth's bow shock region', J. Plasma Phys., 17, 133-138, 1977

Russell, C. T., M. M. Hoppe, W. A. Livesey, J. T. Gosling, and S. J. Bame, 'ISEE 1 and 2 observations of laminar bow shocks : Velocity and thickness', Geophys. Res. Lett., 9, 1171-1174, 1982.

Sagdeev, R.Z., 'Collisionless shocks in a rarified plasma', Sov. Phys. Tech. Phys., 6, 867, 1962.

Sagdeev, R.Z., 'The 1976 Oppenheimer lectures : critical problems in plasma astrophysics,II, Singular layers and reconnection', Rev. Mod. Phys., 51, 11, 1979.

Sckopke, N., G. Paschmann, S.J. Bame, J.T. Gosling and C.T. Russell, ' Evolution of ion distributions across the nearly perpendicular bow shock : Specularly and nonspecularly reflected gyrating ions', J. Geophys. Res., 88, 6121-6136, 1983.

Scudder, J.D., Mangeney, A., Lacombe, C., Harvey, C.C., Aggson, T.L., Anderson, R., Gosling, J.T., Paschmann, G. and Russell, C.T., 'The resolved layer of a collisionless, high ß, super-critical, quasi-perpendicular shock wave ; 1: Rankine Hugoniot Geometry, currents and stationarity', J. Geophys. Res., 91, 11019-11052, 1986a.

Scudder, J.D., Mangeney, A., Lacombe, C., Harvey, C.C. and Aggson, T.L., 'The resolved layer of a collisionless, high ß, super-critical, quasi-perpendicular shock wave ; 2 : Dissipative fluid electrodynamics', J. Geophys. Res., 91, 11053-11073, 1986b.

Scudder, J.D., Mangeney, A., Lacombe, C., Harvey, Wu, C.S. and Anderson, R., 'The resolved layer of a collisionless, high ß, super-critical, quasi-perpendicular shock wave ; 3: Vlasov electrodynamics',J. Geophys. Res., 91, 11075-11097, 1986c.

Speiser T.W. and N.F. Ness, 'The neutral sheet in the geomagnetic tail : its motion, equivalent currents, and field line connection through it', J. Geophys. Res., 72, 131-141, 1967.

Strokin, N.A., 'Ion heating and energy redistribution in a collisionless shock wave', Sov. Phys. JETP, 61, 1187, 1985.

138

Tanaka, M., 'Simulations of heavy ions heating by electromagnetic ion cyclotron waves driven by proton temperature anisotropies', J. Geophys. Res., 90, 6459-6468, 1985.

Tanaka, M. and Papadopoulos, K., 'Creation of high-energy electron tails by means of the modified two-stream instability', Phys. Fluids, 26, 1697-1699, 1983.

Tanaka, M., C.C. Goodrich, D. Winske and K. Papadopoulos, 'A source of backstreaming ion beams in the foreshock region', J. Geophys. Res., 88, 3046-3054, 1983.

Thomas, V.A. and S.H. Brecht, 'Two dimensional simulation of high Mach number plasma interactions', Phys. Fluids, 29, 2444-2454, 1986.

Thomas, V.A. and S.H. Brecht, 'Angular dependence of high Mach number plasma interactions', J. Geophys. Res., 92, 3175-3186, 1987.

Thomsen, M.F., 'Upstream suprathermal ions', in Collisionless Shocks in the Heliosphere : Reviews of Current Research, Geophys. Monogr. Ser., Vol. 35, pp 253-270, ed. by B.T. Tsurutani and R.G. Stone, AGU, Washington, D.C., 1985.

Thomsen, M.F., J.T. Gosling, S.J. Bame and M.M. Mellott, 'Ion and electron heating at collisionless shocks near the critical Mach number', J. Geophys . Res., 90, 137-148, 1985.

Tidman, D.A. and Krall, N.A., 'Shock waves in collisionless plasmas', Wiley-Interscience, 1971.

Tokar, R.L., C.H. Aldrich, D.W. Forslund and K.B. Quest, 'Nonadiabatic electron heating at high Mach number perpendicular shocks', Phys. Rev. Lett., 56, 1059-1062, 1986.

Tokar, R.L., S.P. Gary and K.B. Quest, 'Electron heating by ion acoustic turbulence in simulated low Mach number shocks', Phys. Fluids, 30, 2569-2575, 1987.

Varvoglis, H. and Papadopoulos, K., 'Stochastic ion acceleration by electrostatic ion cyclotron waves', Rep. AP 82-051, University of Maryland, College Park, 1982.

Winske, D., 'Hybrid simulation codes with application to shocks and upstream waves ', Space Sci. Rev., 42, 53-66, 1985.

Winske, D. and M.M. Leroy, 'Hybrid simulation techniques applied to the Earth's bow shock', in Computer Simulation of Space Plasmas- Selected Lectures from the First ISSS, ed. by H. Matsumoto and T. Sato, pp. 255-278 , Kluvier Academic, Hingham, Mass., 1984a.

Winske, D. and M.M. Leroy, 'Diffuse ions produced by electromagnetic ion beam instabilities', J. Geophys. Res., 89, 2673-2688, 1984b.

Winske, D., M. Tanaka, C.S. Wu and K.B. Quest, 'Plasma heating at collisionless shocks due to the kinetic cross-field streaming instability', J. Geophys. Res., 90, 123-136, 1985.

Winske, D. and Quest, K.B., 'Electromagnetic ion beam instabilities; comparison of one and two dimensional simulations', J. Geophys. Res., 91, 8789-8797, 1986.

Winske, D., J. Giacalone, M.F. Thomsen and M.M. Mellott, 'A comparative study of plasma heating by ion acoustic and modified two stream instabilities at subcritical quasi-perpendicular shocks', J. Geophys. Res. , 92, 4411-4422, 1987.

Winske, D. and Quest, K.B., 'Magnetic field and density fluctuations at perpendicular supercritical collisionless shocks', 49, 9681-9693, 1988.

Wong, H.V., 'Electrostatic electron-ion streaming instability', Phys. Fluids, 13, 757-760, 1970.

Woods, L.C., ' On the structure of collisionless magnetoplasma shock waves at supercritical Alfven Mach numbers', J. Plasma Phys., 3, 435-447, 1969.

Wright, Th. P., 'Instability of one-dimensional, oblique, magnetic shock structures', Phys. Fluids, 14, 2337-2340, 1971.

Wu, C.S., 'Physical mechanisms for turbulent dissipation in collisionless shock waves', Space Sci. Rev., 32, 83-97, 1982.

Wu, C.S., Zhou, Y.M., Tsai, S.T., Guo, S.C., Winske, D. and Papadopoulos, K., 'A kinetic cross field streaming instability', Phys. Fluids, 26, 1259-1267, 1983.

Wu, C.S., Winske, D., Zhou, Y.M., Tsai, S.T., Rodriguez, P., Tanaka, M., Papadopoulos, K., Akimoto, K., Lin, C.S., Leroy, and M.M., Goodrich C.C., 'Microinstabilities associated with a high Mach-number, perpendicular shock', Space Sci. Rev. 37, 63-109, 1984.

NUMERICAL SIMULATIONS OF VLASOV EQUILIBRIA

L. DEMEIO
Center for Transport Theory and Mathematical Physics
Virginia Polytechnic Institute and State University
Blacksburg, Virgina, 24061
United States

ABSTRACT. Solutions of the Vlasov–Poisson system which correspond to undamped travelling waves near a maxwellian equilibrium are analysed numerically. The results are strongly in favour of a recently developed theory which predicts the existence of BGK modes arbitrarily close to any spatially homogeneous equilibrium

1. INTRODUCTION

Very recently, a theoretical result has been obtained concerning the Vlasov–Poisson system for a one-dimensional unmagnetized plasma confined in a box of length L upon which periodic boundary conditions are imposed [1,2]; with fixed ions, the equations governing the time evolution of the electron distribution function $f(x,v,t)$ and the self–consistent electric field $E(x,t)$ are:

$$\frac{\partial f}{\partial t} + v \frac{\partial f}{\partial x} - E \frac{\partial f}{\partial v} = 0$$

$$\frac{\partial E}{\partial x} = 1 - \int f dv \qquad \text{(VP)}$$

with initial condition $f(x,v,0)=g(x,v)$ and boundary conditions $f(0,v,t)=f(L,v,t)$ and $E(0,t)=E(L,t)$. Here, $t\geq 0$ is time, $x\in[0,L]$ the space and $v\in R$ the velocity variable (dimensionless units have been used; lengths are measured in Debye lengths, time in plasma periods and velocities in thermal units). The result found in [1,2] predicts the existence of undamped travelling waves (BGK equilibria [3]) arbitrarily near any spatially homogeneous equilibrium. The subject has already received some attention, although from a heuristic point of view only [3]. In [1,2] the statement is proved rigorously and precise conditions are given for the existence of such BGK modes. In particular, it is found that the phase velocity v_φ and the wave number k of such waves obey the Vlasov dispersion relation [4]

W. Brinkmann et al. (eds.), Physical Processes in Hot Cosmic Plasmas, 141–155.
© 1990 *Kluwer Academic Publishers.*

$$1 + \frac{1}{k^2} P \int dv \, \frac{\partial f_0 / \partial v}{v - v_\phi} = 0$$

where P denotes the principal value. The importance of this theory lies in the fact that one should be able to start, for example, from a maxwellian distribution, perturb it slightly, and obtain solutions whose time behaviour contradicts the predictions of the well established linear theory [5,6,7]. As is well known, according to linear theory any small perturbation to a maxwellian equilibrium should exhibit damped oscillations, at a frequency and with a damping rate given by a linear dispersion relation which depends upon the equilibrium distribution only (Landau damping). As explained in [1,2], the discrepancy is only apparent, since the kind of perturbation that one has to add to the equilibrium distribution in order to obtain undamped travelling waves would probably make the initial data undescribable by linear theory. In any case, a test of the theory against experimental and numerical evidence is in order.

In this paper, we present our first attempt to look for undamped travelling waves near a maxwellian equilibrium, comparing the time evolution of the system in a typical Landau damping case with the one that occurs when the initial data are chosen near a BGK equilibrium. In Section II we briefly recall the theory of linear Landau damping and of BGK modes; in Section III we present a typical Landau damping simulation and in Section IV we follow the time evolution of the system when starting with an initial distribution which differs from a maxwellian by an arbitrarily small amount (uniformly) and is very close to a BGK equilibrium as well; in Section V we state our conclusions.

The numerical results have been obtained with the code, based on the well known splitting–scheme algorithm [8], which we have written and implemented on the Cray X–MP (and subsequently on the Y–MP) of the Pittsburgh Supercomputing Center.

2. THEORY OF LANDAU DAMPING AND OF BGK MODES

Here, we briefly recall the results given by linear theory about the stability of Vlasov equilibria and then give a description of BGK modes.

2.1 Linear theory

A function of velocity only, say $f_0(v)$, is an equilibrium solution of (VP) with $E(x,t)=0$. For small deviations from the equilibrium, the equations can be linearized; with $f(x,v,t) = f_0(v) + \tilde{f}(x,v,t)$ the equations become, after Fourier transforming,

$$\frac{\partial \tilde{f}_k}{\partial t} + ikv\tilde{f}_k - E_k(t)\frac{\partial f_0}{\partial v} = 0$$

$$ikE_k(t) = -\int \tilde{f}_k dv$$

(LVP)

which is called linearized Vlasov–Poisson system. Note that, in obtaining the linearized equations, the term containing the velocity gradient of the perturbation has been discarded. The linearized equations have been solved in at least three different ways [5,6,7] all of them giving the same answer about the stability of the various equilibria. Except for other unimportant contributions, the solution can be written in the form

$$\tilde{f}_k(v,t) = \frac{1}{k^2}\sum_{j=1}^{\infty} q_j \frac{\partial f_0/\partial v}{v-\nu_j} e^{-ik\nu_j t}$$

where the ν_j's are the complex roots of the Landau dispersion function

$$\Lambda_k(\nu) \equiv 1 + \frac{1}{k^2}\int_{\mathscr{L}} \frac{\partial f_0/\partial v}{v-\nu}\, dv \tag{1}$$

where $\mathscr{L}$ is the well–known Landau contour, and the q_j's are known coefficients. The function Λ_k is analytic in the complex ν plane and its roots determine the linear stability properties of the solution. The terms corresponding to roots with positive imaginary part grow exponentially in time giving rise to instability, while those corresponding to roots with negative imaginary part are exponentially damped in time. This is called Landau damping. For any k and for any equilibrium distribution $f_0(v)$ (1) has infinitely many roots in the lower half plane (i.e., there are infinitely many damped modes) and only a finite number in the upper half plane, possibly none, according to the nature of the equilibrium distribution. When, in a reference frame in which the plasma is at rest, $f_0(v)$ is a monotonically decreasing function of the energy (such as a maxwellian), there are no unstable roots and the plasma is stable against perturbations with any wave number. When $f_0(v)$ has a rising portion, then there exists a range of wave numbers for which there are roots in the upper half plane whose real part falls in the rising portion of the distribution, so the plasma is unstable against perturbations with those wave numbers. For that particular value of k for which the real part of ν falls exactly on the minimum of the distribution linear theory predicts the onset of steady–state oscillations.

2.2 BGK modes

While the linearized equations have been extensively studied and their solution has been known for many years, very few analytical results have been obtained for the full nonlinear system. BGK modes [3] are solutions of the full nonlinear equations which correspond to stationary states in some reference frame. Namely, consider the stationary Vlasov–Poisson system

$$v\frac{\partial f}{\partial x} - E\frac{\partial f}{\partial v} = 0$$

(SVP)

$$\frac{\partial E}{\partial x} = 1 - \int fdv$$

where now $f = f(x,v)$ and $E = E(x)$. Due to the invariance of (VP) with respect to any Galilean transformation, $g(x,v,t) = f(x-Vt,v-V)$ and $H(x,t) = E(x-Vt)$ are solutions of (VP) if $f(x,v)$ and $E(x)$ are solutions of (SVP). BGK modes are solutions of (VP) which can be cast in the form $f(x-Vt,v)$ and $E(x-Vt)$ for some real V, i.e., if there exists a reference frame in which the distribution function and the electric field are time independent. In the stationary problem, the single–particle energy

$$\mathcal{E} = \frac{v^2}{2} - \Phi(x)$$

is a constant of motion. Here, $\Phi(x)$ is the electric potential, so that $E(x) = -d\Phi/dx$. The phase space trajectories will occur on curves with constant $\mathcal{E}$ if $\Phi_{min} \leq \Phi(x) \leq \Phi_{max}$, then electrons with $-\Phi_{max} \leq \mathcal{E} \leq -\Phi_{min}$ are trapped and follow closed phase–space orbits, while electrons with $\mathcal{E} > -\Phi_{min}$ are untrapped and follow open curves. It can be shown that one can prescribe, for example, the potential $\Phi(x)$ and the distribution of the untrapped electrons and then solve for the trapped electron distribution. Alternatively, one can solve for the distribution of the untrapped electrons or for the potential. The class of BGK equilibria is quite large, since almost any potential wave form with continuous second derivative can be achieved, by properly prescribing the distribution of the untrapped electrons.

3. LANDAU DAMPING

When studying numerically the evolution of the system starting from a perturbed maxwellian equilibrium, the initial distribution is usually chosen of the form

$$g(x,v) = f_0(v)(1 + \epsilon \cos(kx))$$

(2)

with $\epsilon \ll 1$, $k = 2\pi m/L$ (m is the mode number) and

$$f_0(v)^2 = \frac{1}{\sqrt{(2\pi)}} e^{-v^2/2}.$$

Here, we choose $k = 0.3$ and $m = 1$, in which case the linear dispersion relation (in the Landau form) gives $\omega = 1.16$ and $\gamma = 0.013$ for the frequency of the oscillations and the damping rate.

Since in this case f(L-x,-v,0) = f(x,v,0), according to a well-known symmetry property of the solution [10,11] we expect the electric field to behave like a standing wave and not like a travelling wave. In other words, due to the symmetry of the equilibrium distribution, both waves at $\pm v_\varphi \equiv \pm\omega/k$ are excited, so when integrating over v to obtain the density they are superposed yielding a standing wave.

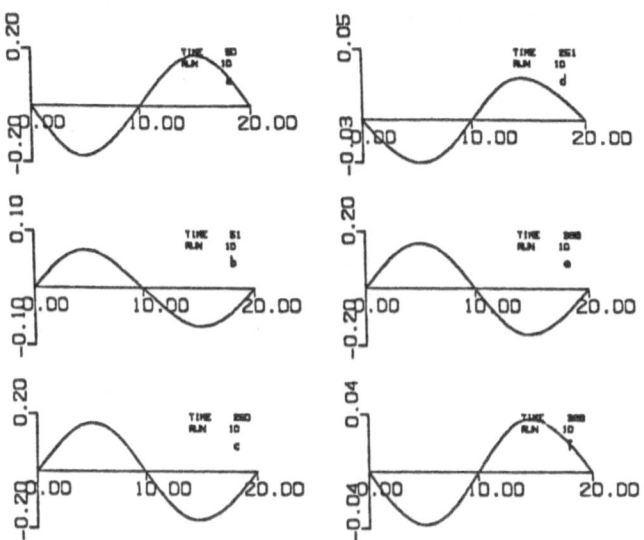

Figures 1a–f. Electric field E(x,t) as function of space at selected times for the linear Landau damping case.

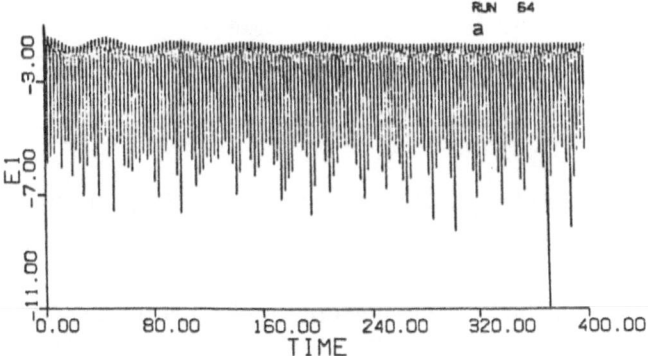

Figure 2. Amplitude of the m=1 mode of the electric field as function of time for the linear Landau damping case.

146

In figs. 1a–f we show the electric field as a function of space at selected times, while in fig. 2 we show the amplitude of the fundamental mode, i.e. $|E_1|$ when the electric field is expanded as $E(x) = \sum_m E_m e^{ikx}$. At early times, Landau damping is observed, in qualitative and quantitative agreement with linear theory, followed by the amplitude oscillations predicted by O'Neil [9]. Asymptotically in time, the electric field is seen to perform steady–state oscillations at the frequency given by the linear dispersion relation (the frequency has never changed during the whole simulation).

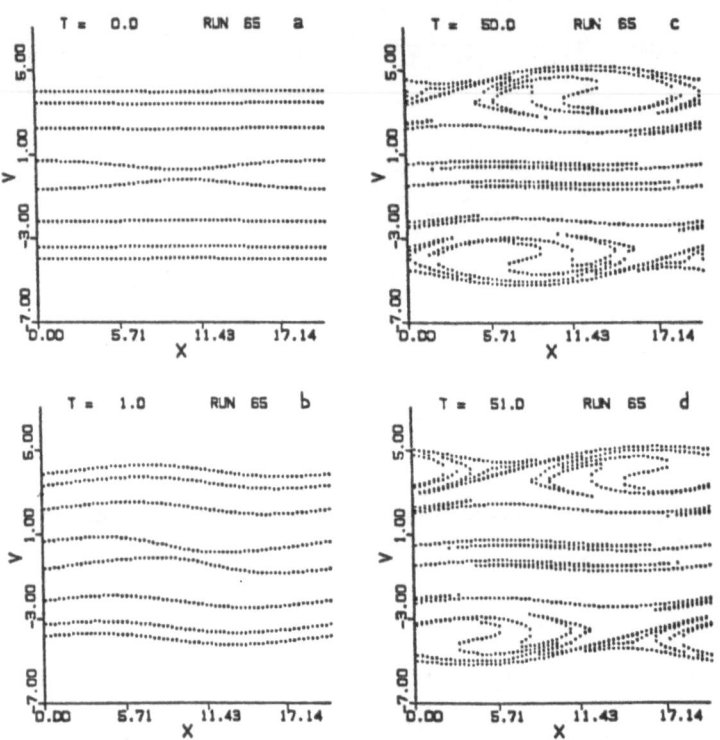

Figures 3a–d. Level curves of $f(x,v,t)$ in phase space at $t=0, 1, 50, 51$ for the linear Landau damping case.

Looking at the level curves of the distribution function in phase space (shown in figs. 3a–h) we see that two vortices have formed, centered at $\pm v_\varphi$; they travel, parallel to the space axis, in opposite directions and equal speed v_φ. Note the finer and finer structure, typical of Vlasov solutions, which is formed in the region of the vortices as time proceeds. Asymptotically in time, the solution seems to be well represented by a superposition of two BGK modes (see [10,11] and references cited therein for more details). Obviously this is not a BGK equilibrium.

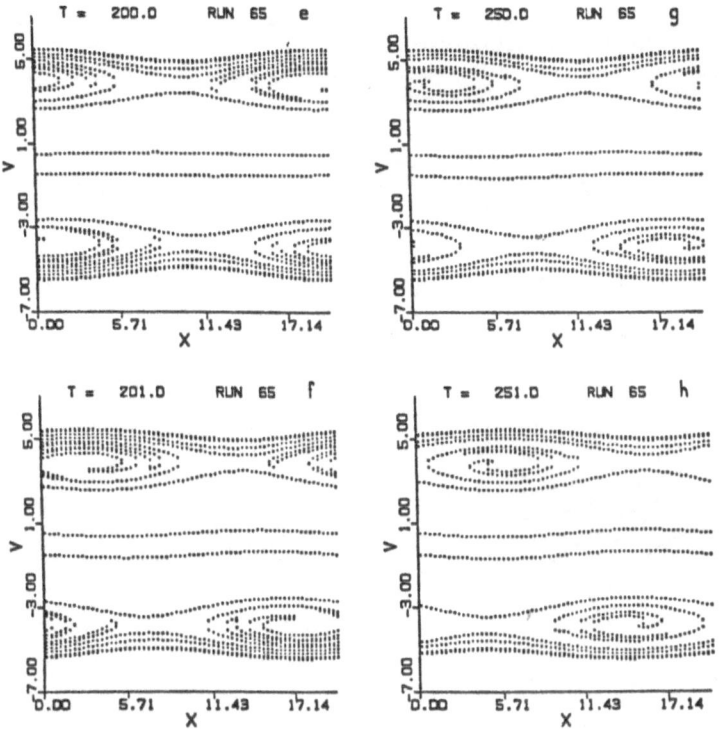

Figures 3e–h. Level curves of f(x,v,t) in phase space at t=200, 201, 250, 251 for the linear Landau damping case

4. BGK MODES NEAR A MAXWELLIAN

Constructing a BGK equilibrium near a maxwellian is not an easy task. It is easier to write a function which is very close to a BGK equilibrium and differs uniformly from a maxwellian by an arbitrarily small amount and then, initializing the system with this function, to follow the time evolution. If the initial data were exactly a BGK equilibrium, the electric field amplitude and the electric field energy would have to remain constant in time; when starting the system only near a BGK equilibrium, as we do here, we expect to see some evolution, but only a very slow one, and hopefully not one that linear theory can describe. Moreover, we anticipate that, in relating the wave phase velocity v_φ and the wave number k of the travelling waves we shall make use of Landau's dispersion relation [5]

$$1 + \frac{1}{k^2} \int_{\mathscr{L}} dv \, \frac{\partial f_0 / \partial v}{v - \nu} = 0,$$

where $\mathscr{L}$ is the well–known Landau contour [5], and not of Vlasov's. As mentioned in [1], however, the roots of the two dispersion relations are approximately the same in this particular case.

Consider the initial data (2) chosen in the previous section. Then we have $E(x,0) = -(\epsilon/k)\sin(kx)$ for the electric field and

$$\Phi(x,0) = -\int_0^x E(y,0)dy = -\frac{\epsilon}{k^2}\cos(kx) \qquad (3)$$

for the electric potential. The single–particle energy is $\mathscr{E} = \frac{v^2}{2} + \frac{\epsilon}{k^2}\cos(kx)$; the curves of constant energy (phase space characteristics) are shown in fig. 4.

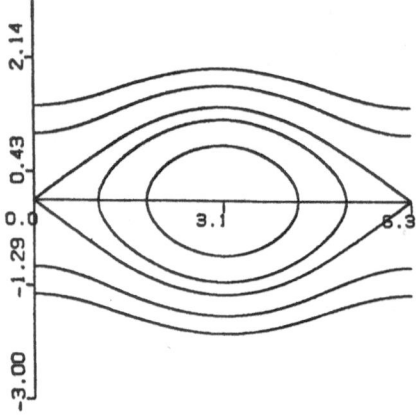

Figure 4. Phase space characteristics for a cosine potential.

We distinguish three families of curves; open curves (when $\mathscr{E} > \frac{\epsilon}{k^2}$), closed curves (when $\mathscr{E} < \frac{\epsilon}{k^2}$) and the two separatrices (when $\mathscr{E} = \frac{\epsilon}{k^2}$). The open curves are the trajectories of the untrapped electrons and the closed curves the trajectories of the trapped ones. Due to the particular shape of the closed orbits and the separatrices, this particular pattern is called "cat's eye". Of course, these curves are the same as the ones for the nonlinear harmonic oscillator. Our aim is to construct a function which is constant along the phase space characteristics, i.e. a function of the single–particle energy. Before doing so, however, we have to account for the fact that in the case of a BGK equilibrium $\mathscr{E}$ is constant in the wave frame, not in the laboratory frame; the equations of the two separatrices $v_{\pm}(x)$ in the laboratory frame are then:

$$v_{\pm}(x) = v_{\varphi} \pm \sqrt{2\epsilon \left(1 - \frac{1}{k^2} \cos(kx)\right)}\,.$$

In the numerical calculations, we shall use $k = 0.3$ and $v_{\varphi} = 3.866$, i.e. the same values as in Section 3 for the Landau damping case, which corresponds to using the Landau dispersion relation to relate k and v_{φ}. The single–particle energy in the wave frame is now

$$\mathcal{E} = \frac{1}{2}(v-v_{\varphi})^2 + \frac{\epsilon}{k^2} \cos(kx)$$

and the equation of the characteristics in phase space is

$$v_{\mathcal{E}}^{\pm}(x) = v_{\varphi} \pm \sqrt{2\left(\mathcal{E} - \frac{\epsilon}{k^2} \cos(kx)\right)}\,.$$

Note that, in transforming from the wave frame in the laboratory frame, we have never replaced x with $x-v_{\varphi}t$ since we make use of the characteristics at $t=0$ only. We define the following function:

$$g(0,v) = g(L,v) = f_0(v)$$
$$g(x,v) = f_0(v_{\varphi}), \quad v_-(x) \le v \le v_+(x)$$
$$g(x,v) = f_0(v_{\mathcal{E}}^+(0)), \quad v > v_+(x)$$
$$g(x,v) = f_0(v_{\mathcal{E}}^-(0)), \quad v < v_-(x)$$

where $\mathcal{E}$ is the single–particle energy (in the wave frame) that pertains to the characteristic passing through the point (x,v) in phase space. The function $g(x,v)$ is uniformly arbitrarily close to the equilibrium distribution f_0 which we have chosen to be maxwellian (as $\epsilon \to 0$, $g(x,v) \to f_0(v)$ uniformly) and is continuous but not differentiable on the separatrices, which makes it undescribable by linear theory. It is obviously a function of the single–particle energy (in the wave frame) for the potential Φ given by (3); however, it easily seen that when $g(x,v)$ is substituted into Poisson's equation the potential Φ is not recovered, i.e. the function thus constructed is not self–consistent. But, when assigning $g(x,v)$ as the initial condition to the numerical code, the program calculates the actual electric potential given by $g(x,v)$; therefore, the problem is self–consistent, although what we start with is not an exact BGK equilibrium. Since, as it appears from our numerical results, the actual potential sustained by $g(x,v)$ is very close to the expression given in (3), our initial data are very close to a BGK equilibrium.
The time evolution of the system when starting with $f(x,v,0) = g(x,v)$ with $\epsilon = 0.1$ is shown in figs. 5a–l (electric field as function of space at selected times),

150

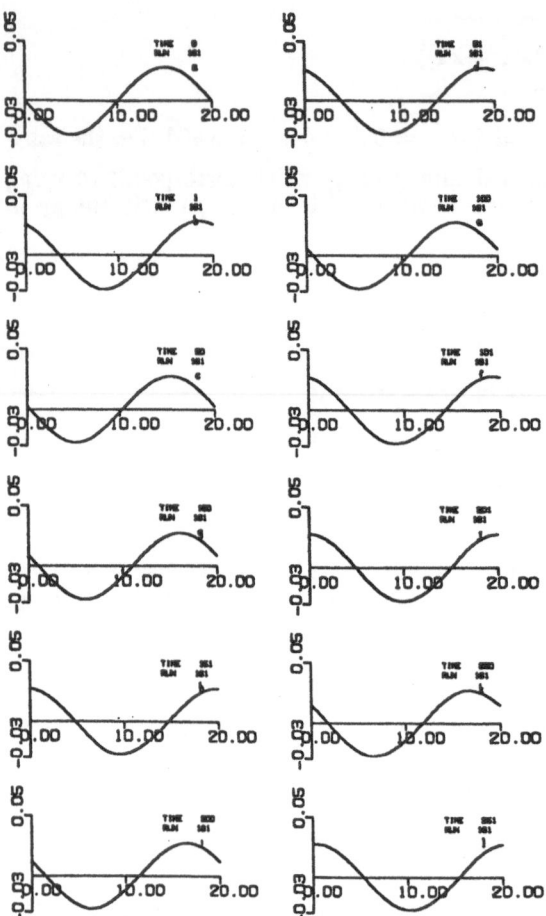

Figures 5a–l. Electric field $E(x,t)$ as function of space at selected times for the BGK case.

6a–d (amplitude of the fundamental mode of the electric field, first and second harmonic and electric field energy) and 7a–h (level curves of the distribution function in phase space at selected times). The electric field behaves like a travelling wave, with phase velocity $v_\varphi = 3.866$, and remains sinusoidal with excellent approximation, since the harmonics are at least two orders of magnitude smaller than the fundamental for all times. In the Landau damping simulation, the harmonics were zero (to machine precision) at t=0, growing to higher values during the time evolution, but always remaining about two orders of

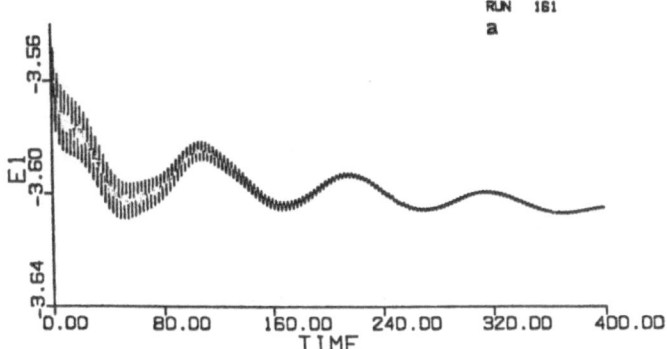

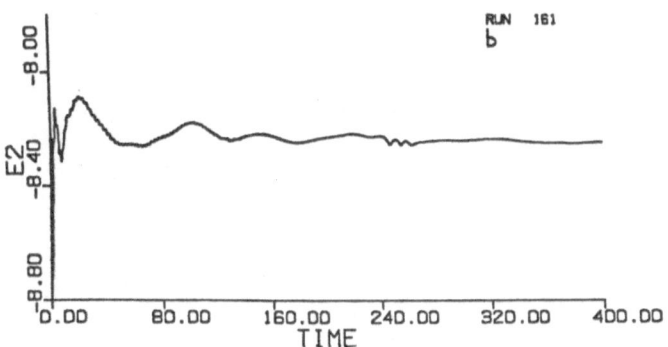

Figures 6a–b. Amplitude of the m=1 mode (a) and first harmonic (b) of the electric field as functions of time for the BGK case.

magnitude smaller than the fundamental [10,11]. In the case considered here, instead, they are non–zero at t=0, because of the above mentioned reason concerning the self–consistency of the initial distribution, but they don't grow significantly in time. No variation in the amplitude of the electric field can be noticed from figs. 5a–l, but from figs. 6a–d we note that a small evolution is present. There are oscillations at the frequency given by the linear dispersion relation and oscillations on the slower scale characterized by the trapping period are also present. In phase space, we see that the cat's eye structure which was placed at v_φ at t=0 is travelling to the right at a speed equal to v_φ. Little fine

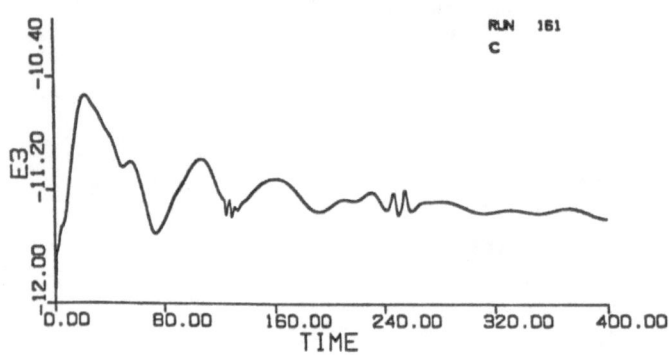

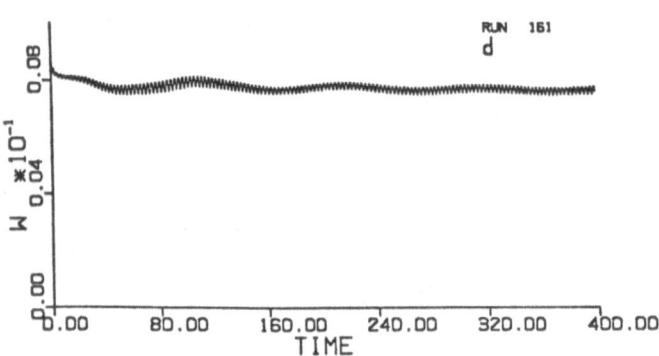

Figures 6c–d. Second harmonic of the electric field (c) and electric field energy (d) as functions of time for the BGK case.

structure is formed inside the cat's eye and, what is probably most important, the distribution function has remained unperturbed away from the cat's eye; in particular, no formation of vortices has occurred at $-v_\varphi$.

No qualitative changes are noticed when ϵ is decreased from 0.1 to 0.01, making our initial data closer to the maxwellian equilibrium. Again, a small evolution is present (smaller than in the $\epsilon=0.1$ case).

For an exact BGK equilibrium we expect the electric field to be a travelling wave with constant amplitude, so we expect a constant (in time) electric energy, while the cat's eye should only travel at the phase velocity without any formation of fine structure. The small time evolution which is observed is

153

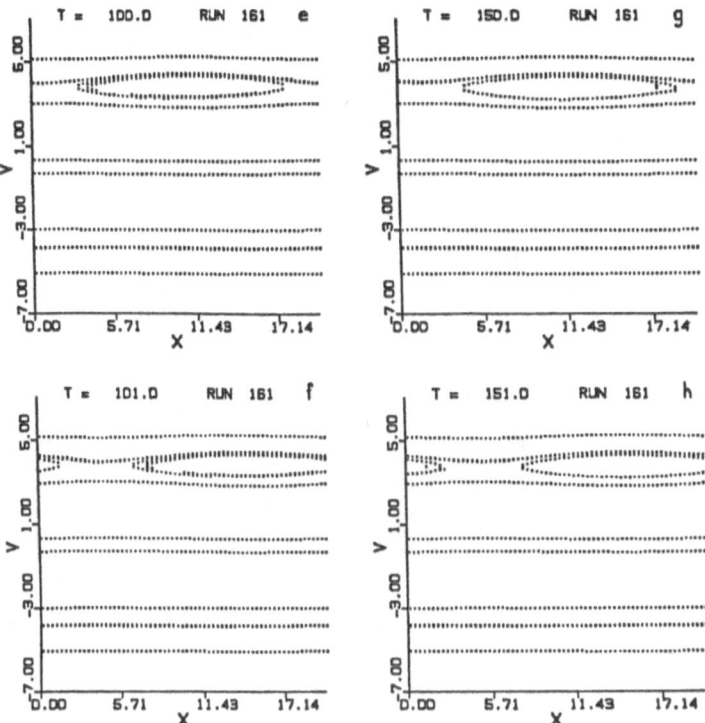

Figures 7a–d. Level curves of f(x,v,t) in phase space at t=0, 1, 50, 51 for the BGK case.

certainly due to the fact that the system has been initialized only very near a BGK equilibrium but not exactly on it; also, we have used the Landau version of the linear dispersion relation and not the Vlasov one. Finally, it should be mentioned that the discontinuity in the gradients of the initial distribution might have induced some small spurious numerical effect.

5. CONCLUSIONS

We have presented a first attempt to look numerically for undamped travelling waves near a spatially homogeneous equilibrium, in a one–dimensional, collisionless, unmagnetized plasma with fixed ions, whose existence has recently been predicted [1,2]. We have considered a maxwellian equilibrium and constructed a funcion which is as close as wished to the equilibrium distribution and is close to a BGK equilibrium as well. Using this function as the initial distribution, we have followed numerically the time evolution of the system, finding that it cannot be described by linearizing the equations about the maxwellian equilibrium. The electric field and the distribution function behave

154

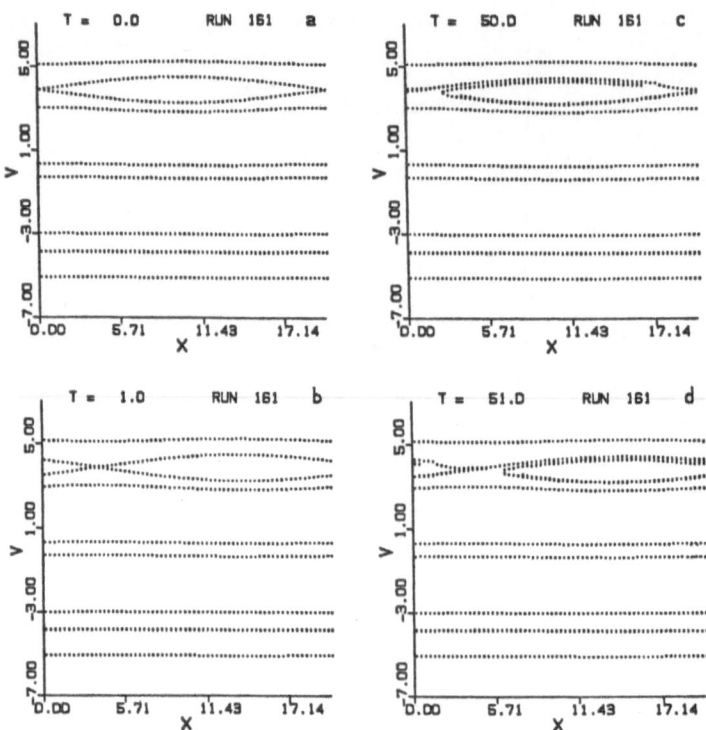

Figures 7e–h. Level curves of f(x,v,t) in phase space at t=100, 101, 150, 151 for the BGK case.

like travelling waves, whose amplitude undergoes only very small variations in the course of time. This small evolution is to be attributed to the fact that we did not start the system on an exact BGK equilibrium but only very close to it, and to the use of the Landau dispersion relation instead of the Vlasov one.

These results are strongly in favour of the theory developed in [1,2], although further and more accurate tests should be carried out.

ACKNOWLEDGEMENT

The author wishes to thank Dr. J.P. Holloway, Prof. P.F. Zweifel and Prof. J.D. Dorning for many helpful discussions and comments on the subject.
This work was supported by the Center for Transport Theory and Mathematical Physics through DOE grant DE–FG05–87ER25033 and NSF grant DMS8701050.

REFERENCES

[1] J.P. Holloway, Ph.D. Thesis, University of Virginia, Charlottesville, Virginia, January 1989.
[2] J.P. Holloway, this conference.
[3] I.B. Bernstein, J.M. Green, M.D. Kruskal, Phys. Rev. 108 (1957) 546.
[4] A. Vlasov, J. of Physics 9 (1945) 25.
[5] L.D. Landau, J. of physics 10 (1946) 25
[6] K.M. Case, Ann. Phys. 7 (1959) 349
[7] M.D. Arthur, W. Greenberg, P.F. Zweifel, Phys. Fluids 20 (1977) 1926
[8] C.G. Cheng, G. Knorr J. Comput. Phys. 22 (1976) 330
[9] T. O'Neil, Phys. Fluids 8 (1965) 2255
[10] L. Demeio, P.F. Zweifel, to appear.
[11] L. Demeio, Ph.D. Thesis, Virginia Polytechnic Institute and State University, Blacksburg, Virginia, April 1989.

CLASSICAL TRANSPORT PROPERTIES OF PLASMAS

C. T. DUM
Max-Planck-Institut für Physik und Astrophysik
Institut für extraterrestrische Physik
D-8046 Garching, Federal Republic of Germany

ABSTRACT. Classical transport theory is discussed, with special emphasis on recent developments necessitated by the particular properties of Coulomb collisions. These properties are contrasted with scattering by neutral particles and by plasma turbulence. The rapid decrease of the scattering cross section with energy usually requires extensions of the Chapman-Enskog method. A method applicable to isotropization by Coulomb collisions, or any other (turbulence) scattering mechanism, is outlined. The structure of transport relations is exhibited for this more general case. The potential and the problems associated with various truncation schemes for an expansion of the distribution function or its moments is discussed. Recent methods for dealing with the breakdown of collision dominated transport are illustrated for electron heat flux.

1. Introduction

Flows couple different regions of space by mass, momentum, or energy transfer. Polar, solar or stellar winds are outstanding examples. The simplest description is given by

$$\frac{1}{r^2}\frac{\partial}{\partial r}r^2\rho u(\frac{u^2}{2} + \frac{5}{2}\frac{p}{\rho} - \frac{GM_o}{r}) = -\frac{1}{r^2}\frac{\partial}{\partial r}r^2 q \tag{1}$$

where u is the radial wind speed, ρ the density, p the pressure, q the heat flux, and a gravity term is included. Heat conduction plays the crucial role in this energy balance. In contrast to convection, described by the terms on the left hand side of (1), it is a microscopic process which may be controlled by collisions of plasma particles. Transport theory is supposed to relate heat flux to the temperature gradient (and other gradients). Transport theory should also describe dissipation processes, such as viscosity and electrical resistivity, which play an important role in shock wave phenomena, magnetic reconnection and magnetic field generation. The essential aim of a transport theory is to provide a simplified description, relating a set of macroscopic variables, rather than having to rely on the detailed microscopic description which is provided by kinetic theories.

The basis of *classical transport theory* is the universal relaxation of particle distribution functions to a Maxwellian by collisions. If collisions are sufficiently frequent then deviations from Maxwellians remain small and can be found by (first order) perturbation theory. By taking moments of the perturbed distribution functions one obtains the desired transport

157

W. Brinkmann et al. (eds.), Physical Processes in Hot Cosmic Plasmas, 157–180.
© 1990 *Kluwer Academic Publishers.*

relations. A perturbation theory, known as the Chapman-Enskog method (*Chapman and Cowling* 1970), was first developed for neutral gases. The method was subsequently also applied to plasmas. However, in contrast to molecular forces, the Coulomb force, just like gravity, is a long range force. Astrophysicists pointed out that as a result, the cumulative effect of small deflections is more important than occasional large angle deflections by close encounters (*Jeans* 1929; *Chandrasekhar* 1942; *Cohen et al.* 1950). A Fokker-Planck term is therefore more appropriate for the small angle deflections than the Boltzmann collision term. Particles undergo diffusion in the selfconsistent electric field associated with thermal fluctuations.

The large difference in mass of the plasma constituents also has important consequences. Collisions between electrons and ions are nearly elastic. It is therefore more useful to develop a perturbation theory in which already to lowest order the plasma constituents have separate temperatures and drifts, rather than trying to maintain a single fluid description. An excellent presentation of this modified classical transport theory has been given by *Braginskii* (1967).

The present paper is devoted to more recent developments in classical transport theory. An outstanding problem is the breakdown of collision dominated classical transport for surprisingly weak gradients, especially for heat conduction. The reason is that the Coulomb scattering cross section is a rapidly decreasing function of energy (*Spitzer, Jr.* 1962). The problem arises for the solar wind flow and nearly every other application, see e.g. *Cowie and McKee* (1977) for a problem related to interstellar clouds or *Craig and Davys* (1984) for heat flux limitation in solar flare plasmas. A large number of remedies have been proposed in order to salvage classical transport theory. The only truly successful method, however, appears to a direct solution of the kinetic equations. Especially experiments on the interaction of lasers with a dense plasma, which create strong temperature gradients and large heat flux, have significantly contributed to these recent developments.

We need to mention another way nature may handle the problem of large gradients, *anomalous transport*. Non-equilibrium features in the distribution functions associated with sufficiently strong heat flux may give rise to microinstabilities, as may e.g. sufficiently large currents. The resulting enhanced small scale fluctuations can scatter particles, similar to the scattering by thermal fluctuations, but with a greatly increased effective collision frequency. The problem, unfortunately is very complicated. Unlike the thermal spectrum corresponding to Coulomb collisions, which is described by a known functional of the distribution functions, the wave spectrum is evolving separately in time and space, and may be limited by nonlinear effects which so far have not been fully explored. Computer simulation has been most successful in handling this problem. Even when the spectrum is known and a corresponding Fokker-Planck collision term can be determined, the problem is still difficult. There is no universal relaxation process that would be analogous to the relaxation to a Maxwellian by Coulomb collisions and which formed the basis of classical transport theory. As a result, anomalous transport has been mostly confined to rather elementary estimates of effective collision frequencies. So far, only for ion acoustic turbulence has it been possible to develop a perturbation theory which leads to a complete set of transport relations, similar to classical transport theory (*Dum* 1978a, *b*). The theory is based on the fact that ion acoustic fluctuations cause isotropization, similar to electron-ion collisions. The dominant isotropic part of the distribution function, however, is not Maxwellian in general and must be determined from a reduced kinetic equation. It is interesting to note that the same methods can be applied to classical transport, especially for plasmas with high ion charge numbers, such as have been frequently used in laser-plasma experiments. High Z corresponds to the dominance of elastic electron-ion collisions over electron-electron collisions. The method can also be applied to cosmic ray transport, where scattering is due to Alfvèn wave turbulence.

In the next section, some essential properties of the Coulomb collision term are reviewed. In Section 3 we discuss the hierarchy of equations which results from taking moments of the kinetic equations. Many attempts have been made at arriving at a transport theory by some truncation of this infinite set of coupled equations, e.g. after 13 moments, or 16 moments in case of a magnetic field. This is very tempting, especially as potentially the method is more general than the perturbation theory for collision dominated transport. The method can be applied even when the plasma is no longer collision dominated, or when it is subject to anomalous transport. We discuss the basics of this method and the problems associated with it.

For charged particles the magnetic field plays an important role in transport. If Larmor radii are small compared to gradient scales, distribution functions will remain nearly gyrotropic. Simplifications in the hierarchy of moment equations and in the solution of the kinetic equations result. Simplifications also result if the plasma is collision dominated. In Section 4 we review the Chapman-Enskog method and then describe the more general method which is based upon the dominance of isotropization. The structure of the set of transport relations is shown for this more general case, allowing for non-Maxwellian energy distributions. The self-consistent determination of the energy distribution is also discussed.

Section 5 is devoted to a discussion of the breakdown of classical collision dominated transport. Taking heat flux as the most important example, we discuss recent developments in the solution of this problem and some experimental results. Section 6 summarizes our conclusions.

2. Kinetic equation, Collision term, Collision times

The kinetic equation for the distribution function $f(\mathbf{x}, \mathbf{v}, t)$ has the form

$$\frac{\partial f}{\partial t} + \mathbf{v} \cdot \frac{\partial f}{\partial \mathbf{x}} + \frac{e_j}{m_j}(\mathbf{E} + \frac{1}{c}\mathbf{u} \times \mathbf{B}) \cdot \frac{\partial f}{\partial \mathbf{v}} = Cf \tag{2}$$

For short range forces the Boltzmann collision term is given by

$$Cf_a = \sum_b \int d\mathbf{v}' d\Omega g \sigma(g, \Omega)[f_j' f_k' - f_j f_k] \tag{3}$$

where the primes indicate that the distributions have the argument $\mathbf{v}'$, the velocity after the collision and $g = |\mathbf{v} - \mathbf{v}'|$. The Coulomb force like the gravitational force, however, is a long range force and as a result the cumulative effect of small deflections is dominant. Expanding (3) for small deflections gives the Landau collision integral

$$C_{ab}f_a = -\frac{\partial}{\partial \mathbf{v}} \cdot \int d\mathbf{v}' \mathcal{Q}_{ab}(\mathbf{v}, \mathbf{v}') \cdot [\frac{m_a}{m_b}\frac{\partial f_b}{\partial \mathbf{v}'}f_a(\mathbf{v}) - f_b(\mathbf{v}')\frac{\partial f_a}{\partial \mathbf{v}}] \tag{4}$$

$$\mathcal{Q}_{ab}(\mathbf{v}, \mathbf{v}') = \Gamma_{ab}\frac{1}{g}(\mathcal{I} - \frac{\mathbf{g}\mathbf{g}}{g^2}) \tag{5}$$

where $\Gamma_{ab} = (2\pi e_a^2 e_b^2/m_a^2)ln\Lambda$. $\Lambda = \lambda_D/b_0 = 12\pi n\lambda_D^3$ is the Coulomb logarithm, which is determined by the cutoff of the impact parameter at the Debye shielding distance

$$\frac{1}{\lambda_D^2} = \sum \frac{1}{\lambda_j^2} \tag{6}$$

where $\lambda_j = v_j/\omega_j$, $v_j = \sqrt{T_j/m_j}$, $\omega_j^2 = 4\pi n_j e_j^2/m_j$. The minimum impact parameter for a particle of speed v has been replaced by its thermal average $b_0 = e^2/3T_e$. It is implied that $n\lambda_D^3 \gg 1$, i.e. that the interparticle distance is small compared to the Debye length. The Coulomb logarithm is typically of order 10. For smaller values, i.e. very dense plasmas, many body collisions (correlations) become important.

It is possible to derive a more general form of the collision term from kinetic theory (*Balescu* 1960; *Lenard* 1960). The kernel Q_{ab} accounts then for the dielectric properties of the plasma. The cutoff at the Debye length is no longer necessary, but a cutoff at a minimum impact parameter is now required because higher order correlations are neglected.

The kinetic theory allows also to account for the effects of very strong magnetic fields on collisions. Basically, it amounts to replacing the Debye length in the Coulomb logarithm by the electron Larmor radius if it is smaller, but details depend on the velocity of the particle and its direction relative to the magnetic field (*Dum and Pfirsch* 1972; *Hassan and Watson* 1977a; *Baldwin and Watson* 1977; *Hassan and Watson* 1977b)

Electric fields with wavelengths longer than the Debye length can be excited to significant levels if the plasma is unstable. These fields are no longer a functional of the distribution functions, as assumed (adiabatic hypothesis) in the kinetic theory of stable plasmas, but must be determined from an additional kinetic equation for waves. Anomalous transport can then be determined from a kinetic equation for particles, with a diffusion term describing the wave-particle interaction. In some applications it may also be of importance to combine classical and anomalous transport (*Dum* 1978a).

2.1. PROPERTIES OF THE LANDAU COLLISION TERM

The collision term (4) can also be written as

$$Cf = \frac{\partial}{\partial \mathbf{v}} \cdot (-\mathbf{A} + \mathcal{D} \cdot \frac{\partial}{\partial \mathbf{v}})f \tag{7}$$

where

$$\mathcal{D}_{ab} = \Gamma_{ab} \int d\mathbf{v}' Q_{ab}(\mathbf{v}, \mathbf{v}') f_b(\mathbf{v}') \tag{8}$$

$$\mathbf{A}_{ab} = \frac{m_a}{m_b} \frac{\partial}{\partial \mathbf{v}} \cdot \mathcal{D}_{ab} \tag{9}$$

The usual form of the Fokker-Planck term is then

$$Cf = -\frac{\partial}{\partial \mathbf{v}} \cdot \mathbf{r}f + \frac{\partial^2}{\partial \mathbf{v} \partial \mathbf{v}} : \tilde{\mathcal{D}}f \tag{10}$$

where the force $r^i = A^i + \frac{\partial D^{ik}}{\partial v_k}$ consists of the polarization term and a dynamical friction term, and the diffusion tensor is symmetrized, $\tilde{D}^{ik} = (1/2)(D^{ik} + D^{ki})$.

2.1.1. *Diffusion and friction terms* Rosenbluth et al. (1957) have shown that polarization term and diffusion term are conveniently evaluated in terms of the potentials

$$H_b = \int d\mathbf{v}' \frac{f_b(\mathbf{v}')}{|\mathbf{v} - \mathbf{v}'|}; \qquad G_b = \int d\mathbf{v}' f_b(\mathbf{v}')|\mathbf{v} - \mathbf{v}'| \tag{11}$$

$$\mathcal{D}_{ab} = \Gamma_{ab} \frac{\partial^2}{\partial \mathbf{v} \partial \mathbf{v}} G_b \tag{12}$$

$$D_{ab} = 2\Gamma_{ab} H_b \tag{13}$$

$$\mathbf{A}_{ab} = \frac{m_a}{m_b} \frac{\partial}{\partial \mathbf{v}} D_{ab} \tag{14}$$

For an isotropic field particle distribution the expressions reduce to

$$\mathbf{A}_{ab} = -\frac{m_a}{m_b} \nu_{ab}(v) \mathbf{v} \tag{15}$$

$$\mathcal{D} = \mathcal{D}^l + \mathcal{D}^t = D^l \frac{\mathbf{vv}}{v^2} + D^t \frac{v^2 \mathcal{I} - \mathbf{vv}}{v^2} \tag{16}$$

$$\nu_{ab}(v) = 2\Gamma_{ab} v^{-3} N_b(v) \tag{17}$$

where $N_b = \int_0^v dv' 4\pi v'^2 f_b(v')$ is the number density of particles with speed $< v$ and the trace of the diffusion tensor as well as the magnitudes of its longitudinal and transverse components are determined by (*Dum* 1978a)

$$D_{ab}(v) = \nu_{ab} v^2 + 2\Gamma_{ab} \int_v^\infty dv' 4\pi v' f_b(v') = D^l + 2D^t \tag{18}$$

$$D^l(v) = \frac{2}{3} \Gamma_{ab} [v^{-3} \int_0^v dv' 4\pi v'^4 f_b(v') + \int_v^\infty dv' 4\pi v' f_b(v)] \tag{19}$$

This formulation is convenient for determining the high and low speed limits. For $v \gg v_b$, e.g. in e-i collisions, one obtains

$$\nu_{ab}(v) = 2\Gamma_{ab} n_b v^{-3} \tag{20}$$

$$D^t_{ab} = \nu_{ab}(v) v^2 \tag{21}$$

$$D^l_{ab} = \nu_{ab}(v) v_b^2 \tag{22}$$

where v_b is the thermal velocity of field particles. If the test particles a have a speed $v \gg v_b$ collisions are nearly elastic.

If $v \ll v_b$, as in i-e collisions, one obtains (inelastic) Brownian motion

$$\nu_{ab} = 2\Gamma_{ab} \frac{4\pi}{3} f_b(0) \tag{23}$$

$$D^l = D^t = \frac{D}{3} = \frac{2}{3} \Gamma_{ab} \int_0^\infty dv' 4\pi v' f_b(v') \tag{24}$$

Spitzer, Jr. (1962) has given a detailed discussion of the slowing down and deflection times for a test particle of arbitrary speed in a Maxwellian field particle distribution. Numerical results for other distributions have been presented by *Marsch and Livi* (1985), showing the dependence on the shape of the distribution function, which follows from (17-19).

2.1.2. *Momentum transfer* For the rate of momentum transfer from species b to a

$$\mathbf{R}_{ab} = m_a \int d\mathbf{v} \mathbf{v} C_{ab} f_a \tag{25}$$

we obtain

$$\mathbf{R}_{ab} = -m_a \int d\mathbf{v} \mathbf{v} (1 + \frac{m_a}{m_b}) \nu_{ab}(v) f_a(\mathbf{v}) \tag{26}$$

for an isotropic field distribution. Using (20), the ion-electron transfer rate is given by

$$\mathbf{R}_{ei} = -n_e m \mathbf{u}_{ei} \frac{1}{\tau_{ei}} \tag{27}$$

for a drifting isotropic electron distribution. The effective collision time is determined by

$$\frac{1}{\tau_{ei}} = 2\Gamma_{ei} \frac{4\pi n_i}{3n_e} f_e(0) \tag{28}$$

or

$$\frac{1}{\tau_{ei}} = \nu_{ei}(v_e) \frac{1}{3} (\frac{2}{\pi})^{1/2} = \frac{4\pi e^4 Z ln\lambda}{3T_e^{3/2} m^{1/2}} (\frac{2}{\pi})^{1/2} \tag{29}$$

for a Maxwellian electron distribution and ions of density $n_i = n_e/Z$. This expression provides a convenient characterization of collision times. The electron-electron collision time, in the low speed limit (23), is $\tau_{ee} = Z\tau_{ei}$, and the ions may be characterized by a collision time which is obtained by replacing the electron charge, mass and temperature in (29) with corresponding ion terms (*Braginskii* 1967). It is important, however, to be aware of the rapid increase of collision times and mean free paths with speed, $\lambda(v) = v/\nu(v) \propto v^4$, for energetic particles. Electron runaway for a plasma in an electric field has its origin in this speed dependence. Since heat flux is predominantly carried by more energetic particles, conditions on the ratio of (average) mean free path to gradient scale length are especially critical. This behavior is in strong contrast to collisions with neutrals. For the hard sphere model the mean free path is independent of speed, but actually it may be a strongly decreasing function of speed.

2.1.3. *Energy transfer* The rate of energy transfer may be determined from

$$K_{ab} = \int d\mathbf{v} m_a \frac{v^2}{2} C_{ab} f_a = \int d\mathbf{v} m_a (\mathbf{r}_{ab} \cdot \mathbf{v} + D_{ab}) f_a \tag{30}$$

Using the high temperature limit for the electrons, we obtain the rate of energy transfer from electrons to ions

$$K_{ie} = 3 \frac{Zm}{M \tau_{ei}} (T_e - T_i) \tag{31}$$

The transfer rates satisfy the conservation laws

$$\mathbf{R}_{ab} + \mathbf{R}_{ba} = 0; \quad K_{ab} + K_{ba} = 0$$

It is important to note that for electrons the collision times for momentum transfer from ions, self collisions, and energy transfer from ions have the ratios

$$\tau_{ei} : \tau_{ee} : \tau_{ei}^E = 1 : Z : M/Zm$$

For ions, on the other hand, the time for self-collisions is much shorter than the momentum exchange time with electrons, $\tau_{ii} \ll \tau_{ie}$. The modified transport theory of *Braginskii* (1967) accounts for these facts by allowing differences in drift and speed between electrons and ions to lowest order in the perturbation expansion.

3. Moment equations and Expansions of the Distribution Functions

Transforming velocities by $\mathbf{w} = \mathbf{v} - \mathbf{u}$, where $\mathbf{u} = \mathbf{u}(\mathbf{x}, t)$ is space and time dependent, the kinetic equation (2) takes the form

$$\frac{df}{dt} + \mathbf{w} \cdot \frac{\partial f}{\partial \mathbf{x}} + \mathbf{w} \times \Omega \cdot \frac{\partial f}{\partial \mathbf{w}} - \mathbf{w} \cdot \frac{\partial \mathbf{u}}{\partial \mathbf{x}} \cdot \frac{\partial f}{\partial \mathbf{w}} + \mathbf{a} \cdot \frac{\partial f}{\partial \mathbf{w}} = Cf \tag{32}$$

where $\frac{d}{dt} = \frac{\partial}{\partial t} + \mathbf{u} \cdot \frac{\partial}{\partial \mathbf{x}}$, $\Omega = e_j \mathbf{B}/m_j c$, $\mathbf{a} = -\frac{d\mathbf{u}}{dt}$. Moments of the distribution function are defined by

$$n\langle \Phi \rangle = \int d\mathbf{w} \Phi(\mathbf{w}) f(\mathbf{w}) \tag{33}$$

where n is the density ($\Phi = 1$). Taking moments of (32) gives

$$\frac{dn\langle \Phi \rangle}{dt} + n\langle \Phi \rangle \frac{\partial}{\partial \mathbf{x}} \cdot \mathbf{u} + \frac{\partial}{\partial \mathbf{x}} \cdot n\langle \mathbf{w} \Phi \rangle + n\langle \mathbf{w} \frac{\partial}{\partial \mathbf{w}} \Phi \rangle : \frac{\partial \mathbf{u}}{\partial \mathbf{x}}$$
$$- n\mathbf{a} \cdot \langle \frac{\partial}{\partial \mathbf{w}} \Phi \rangle - n\langle (\mathbf{w} \times \Omega) \cdot \frac{\partial}{\partial \mathbf{w}} \Phi \rangle = n \frac{\delta \langle \Phi \rangle}{\delta t} \tag{34}$$

where the collisional transfer rate is given by

$$n \frac{\delta \langle \Phi \rangle}{\delta t} = \int d\mathbf{w} \Phi Cf \tag{35}$$

An especially convenient choice for $\mathbf{u}$ is the particle mean velocity. In this frame $\langle \mathbf{w} \rangle = 0$ and $\Phi = 1$ gives the continuity equation

$$\frac{dn}{dt} + n \frac{\partial}{\partial \mathbf{x}} \cdot \mathbf{u} = 0 \tag{36}$$

The next moment determines acceleration

$$nm\mathbf{a} = \frac{\partial}{\partial \mathbf{x}} \cdot \mathcal{P} - \mathbf{R} \tag{37}$$

which is equivalent to the equation of motion

$$nm \frac{d\mathbf{u}}{dt} + \frac{\partial}{\partial \mathbf{x}} \cdot \mathcal{P} = nm(e_j/m_j)[\mathbf{E} + (\mathbf{u}/c) \times \mathbf{B}] + \mathbf{R} \tag{38}$$

The equation for the temperature is obtained by taking a moment with $\Phi = w^2/2$, noting that the pressure tensor $\mathcal{P} = nm\langle \mathbf{ww} \rangle$ can be decomposed as $\Pi = \mathcal{P} - p\mathcal{I}$, $p = nT$.

$$\frac{3}{2} n \frac{dT}{dt} + p \frac{\partial}{\partial \mathbf{x}} \cdot \mathbf{u} + \Pi : \frac{\partial \mathbf{u}}{\partial \mathbf{x}} + \frac{\partial}{\partial \mathbf{x}} \cdot \mathbf{q} = Q \tag{39}$$

where the heat flux vector is determined by the next order moment

$$\mathbf{q} = nm\langle \mathbf{w} w^2/2 \rangle \tag{40}$$

The continuity equation was used to rewrite to corresponding first two terms of (34) as a convective derivative

$$\frac{dn\langle\Phi\rangle}{dt} + n\langle\Phi\rangle\frac{\partial}{\partial\mathbf{x}}\cdot\mathbf{u} = n\frac{d\langle\Phi\rangle}{dt} \tag{41}$$

The equations for the density n, drift u and temperature T are the basic fluid equations for a plasma component. Transport theory is supposed to provide expressions for the heat flux, viscous stress and the transfer rates $\mathbf{R}$, Q in terms of the fluid variables, in order to close this set of equations. Applications assume that the plasma components are nearly isotropic, with the viscous stress tensor Π being a small correction to the isotropic pressure p. More generally, it may be useful to consider distributions which are gyrotropic to lowest order, $f(w_\parallel, w_\perp)$, where $w_\parallel$ is the velocity component parallel to the magnetic field direction $\tau_0 = \mathbf{B}/B$ and $w_\perp$ is the magnitude of $\mathbf{w}_\perp$. In this case we have to lowest order pressure components $p_\parallel, p_\perp$, with the remaining 4 independent components contained in the modified viscous stress tensor $\Pi = \mathcal{P} - p_\perp\mathcal{I} - (p_\parallel - p_\perp)\tau_0\tau_0$. Instead of a single heat flux vector we must now consider the heat flux vectors $\mathbf{q}^\parallel = nm\langle\mathbf{w}w_\parallel^2/2\rangle$, $\mathbf{q}^\perp = nm\langle\mathbf{w}w_\perp^2/2\rangle$ for parallel and perpendicular thermal energy.

3.1. HIERARCHY OF MOMENT EQUATIONS

It is straightforward to derive moment equations also for the higher order moments which appear in the fluid equations. The set contains a total of 13 moments, or 16 moments in case of a gyrotropic component. However, if the magnetic field is nonuniform the other 4 components of the heat flux tensor $Q = nm\langle\mathbf{www}\rangle$ also appear explicitly in the equations for $p_\parallel$ and $p_\perp$. These components are needed in any case in the moment equations for Π. In the equations for $\mathbf{q}^\parallel$, $\mathbf{q}^\perp$ and the other components of Q we need in addition the next order moment

$$S = nm\langle\mathbf{wwww}\rangle \tag{42}$$

which has 15 components, 3 of which arise for the lowest order gyrotropic distribution. We have an infinite chain of coupled moment equations, with greatly increasing complexity, especially if the magnetic field is nonuniform.

3.1.1. *Finite Larmor radius expansion* Some reductions in the chain of moment equations can be achieved by making use of symmetries in the problem. For transport across a magnetic field, with gradient lengths much larger than the Larmor radius one may also make use of a finite Larmor radius expansion. A well known example is the double adiabatic approximation of *Chew et al.* (1956)

$$\frac{d(p_\parallel B^2/n^3)}{dt} = 0, \quad \frac{d(p_\perp/nB)}{dt} = 0 \tag{43}$$

for a collisionless plasma in which heat flux across the magnetic field is negligible by virtue of the small Larmor radius, and longitudinal heat flux is ignored, assuming homogeneity along the magnetic field. Corrections to first order in the Larmor radius have been obtained by *Kennel and Greene* (1966) and by *Frieman et al.* (1966). A direct expansion of the chain of moment equations, to first order in the Larmor radius, has been carried out by *MacMahon* (1965). Expressions for collisionless heat flux and viscosity are derived. The equations contain, however, the 3 gyrotropic components of (42) for which one needs to make some assumptions in order to close the set of equations. These components may be written as

$$S_\parallel^\parallel = nm[w_\parallel^4] + \frac{3p_\parallel^2}{nm} \qquad (44a)$$

$$S_\perp^\parallel = nm[w_\parallel^2 w_\perp^2] + \frac{2p_\parallel p_\perp}{nm} \qquad (44b)$$

$$S_\perp^\perp = nm[w_\perp^4] + \frac{8p_\perp^2}{nm} \qquad (44c)$$

A possible ad hoc procedure consists in neglecting the semi-invariants characterized by the square brackets. They vanish for a bi-Maxwellian distribution, but in a collisionless plasma there is no particular reason that the lowest order distribution function should be of this form. Only the last moment, (44c), is needed if gradients along the magnetic field and the curvature of magnetic field lines are neglected. *Bowers and Haines* (1968) have derived a corresponding set of fluid equations, starting from a finite Larmor radius expansion of the distribution functions.

3.2. CLOSURE OF THE HIERARCHY: 13 AND 16 MOMENT APPROXIMATIONS

Even if one finds some arguments for truncating the chain of moment equations, the distribution function is still needed in general to evaluate the collision terms. A frequently employed procedure consists in taking the simplest possible form of the distribution function, consistent with the set of moments to be considered. For the closure of the set of 13 moment equations one would approximate a nearly isotropic distribution function by

$$f(\mathbf{w}) = F_0 + \mathbf{w} \cdot \mathbf{F}_1(w) + \mathbf{ww} : \mathcal{F}_2(w) + \ldots \qquad (45)$$

If F_0 is a Maxwellian with density n, drift $\mathbf{u}$ and temperature T, then, consistent with the heat flux and viscous stress, one may choose

$$\mathbf{F}_1 = \hat{\mathbf{q}} \frac{2}{5v_j} \left(\frac{\hat{w}^2}{2} - \frac{5}{2} \right) F_0, \qquad \mathcal{F}_2 = \frac{1}{2v_j^2} \hat{\Pi} F_0 \qquad (46)$$

where $\hat{w} = w/v_j$, $\hat{\mathbf{q}} = \mathbf{q}/nmv_j^3$, etc. In the case of the 16 moment approximation for a plasma in a magnetic field one may choose

$$f(\mathbf{w}) = F_0 + \mathbf{w} \cdot \mathbf{F}_1(w_\parallel, w_\perp) + \mathbf{ww} : \mathcal{F}_2(w_\parallel, w_\perp) + \ldots \qquad (47)$$

with the simplest possible dependence $\mathbf{F}_1 = [\mathbf{a} + \mathbf{b}w_\parallel^2 + \mathbf{c}w_\perp^2]F_0$, $\mathcal{F}_2 = \mathcal{D}F_0$. After choosing F_0 e.g. to be a bi-Maxwellian, the other coefficients are determined by $\langle \mathbf{w} \rangle = 0$, $\mathbf{q}^\parallel$, $\mathbf{q}^\perp$, and Π (*Oraevskii et al.* 1968). A set of 14 moment equations is obtained by including $\zeta = \langle w_r^4 \rangle - 3v_{thr}^4$ for the radial velocity, cf. (44a), in the moments and the assumed expression for $f(\mathbf{w})$. This set was used by *Larson* (1970) in treating the 'gravothermal catastrophe' in the evolution of star clusters and by *Cuperman et al.* (1980) for radial expansion of stellar winds. Fourteen moment equations for a plasma were also derived by *Belyi et al.* (1989), with an application to ion-acoustic wave excitation.

By these methods one obtains a set of moment equations which is closed, as higher order moments and the collisional transfer rates may be directly evaluated from the assumed distribution function. The latter are very complicated and are usually evaluated by linearizing the collision term in terms of the relative drifts, heat fluxes, and viscosity (*Chodura and Pohl* 1971), although expressions with arbitrary flow velocities have also been derived (*Burgers* 1969; *Salat* 1975). Various limits of the 16 moment approximation are listed in the review by *Barakat and Schunk* (1982) and compared with previous calculations. For a collision dominated plasma, in particular, the set of equations reduces to the first or second approximation of *Chapman and Cowling* (1970), depending on the particular transport coefficient, see below.

The comparison with transport theory for a collision dominated plasma shows that the simple expressions (45) or (47), however, are often not adequate to represent the speed dependence introduced by the collision terms. This is especially true for heat flux. If the plasma is not collision dominated and the time derivatives are important also in the moment equations for heat flux, then a new problem arises (*Palmadesso et al.* 1988). The equations in this case describe thermal waves, which actually may become unstable. By contrast, an exact solution of the problem in a collisionless plasma shows that such thermal waves decay exponentially, simply by phase mixing.

Many anomalous transport "theories" confine themselves to evaluating transfer rates between waves and drifting Maxwellians (5 moment approximation) or drifting bi-Maxwellians (6 moments), which is straightforward, if the wave spectrum were known. But with strong anomalous heating present, it is obvious that one must at least also include heat flux. *Ganguli and Palmadesso* (1987) in an application to the auroral zone solved a set of six equations which, in addition to phenomenological terms for anomalous resistivity and heating, includes also moment equations for the heat fluxes of parallel and perpendicular energy. The reduction from the 16 moments equations is due to neglecting viscosity and assuming one-dimensional space dependence. A two-dimensional extension of the code, which allows for the curvature of magnetic field lines, has been developed recently (*Ganguli and Palmadesso* 1989).

3.3. EXPANSION OF THE DISTRIBUTION FUNCTION IN ORTHOGONAL POLYNOMIALS

A systematic procedure for the solution of the kinetic equations is obtained from an expansion of the distribution function in orthogonal polynomials. Closure is achieved by truncating the chain of equations for the coefficients at some numerically determined order. For a distribution function which is Maxwellian to lowest order an expansion in Hermite or Laguerre (Sonine) is especially convenient. A generalization of these schemes by choosing the lowest order distribution function as weight function in the orthogonality relations has been proposed by *Mintzer* (1965).

3.3.1. *Hermite expansion* Grad (1949, 1958) discussed a formal solution of the kinetic equations which is based on an expansion in the infinite set of Hermite polynomials with the orthogonality property $\int_0^\infty dx e^{-x} H_n H_m = \delta_{nm} N_n$. The distribution function can be expanded as

$$
\begin{aligned}
f(\mathbf{x}) &= e^{-x^2} \sum C_{p_1 p_2 \ldots p_m} \prod H_{p_i}(x_i) \\
&= \sum (-1)^p C_{p_1 p_2 \ldots p_m} \frac{\partial^p}{\partial x_1 \partial x_2 \ldots \partial x_m} e^{-x^2}
\end{aligned}
\tag{48}
$$

where $p = \sum p_i$, $x^2 = \sum x_i^2$, $x = w/\sqrt{2}v_j$ and moments as

$$\langle \Phi(x_1, x_2 \ldots x_m) \rangle = \sum C_{p_1 p_2 \ldots p_m} \langle \frac{\partial^p}{\partial x_1 \partial x_2 \ldots \partial x_m} \Phi \rangle_0 \tag{49}$$

where the subscript 0 indicates that the moment is to be taken over the lowest order distribution $exp[-x^2]$. With the conventional normalization N_n the first few polynomials are given by $H_0 = 1$, $H_1 = 2x$, $H_2 = 2(2x^2 - 1)$, $H_3 = 4x(2x^2 - 3)$, $H_4 = 4(4x^4 - 12x^2 + 3)$, etc. The coefficients of these terms in the expansion (48) correspond to easily recognized moments (49). An expansion in one-sided Hermite polynomials has been considered by *Gross et al.* (1959) for the solution of a one-dimensional boundary value problem in a weakly collisional plasma. The expansion is applied to the two components of $f(w) = f^+(w) + f^-(w)$, which are restricted to positive and negative w respectively. Ad hoc models for these components have been discussed by *Bond* (1981) and by *Shirazian and Steinhauer* (1981). However, there is a large discrepancy between these models and the known collision dominated limit.

3.3.2. *Laguerre expansion*   Laguerre polynomials have the orthogonality property

$$\int_0^\infty dt e^{-t} t^\alpha L_n^\alpha L_m^\alpha = \delta_{nm} N_n \tag{50}$$

hence for a lowest order Maxwellian are convenient for an expansion in energy. More generally, for an isotropic component given by

$$F_0(w) = (nA_s/v_s^3) exp[-(w/v_s)^s] \tag{51}$$

where A_s is a normalization constant, one would use the expansions (*Dum* 1978b)

$$\mathbf{F}_1(w) = \frac{1}{v_s} F_0 \sum b_k L_k^{(5/s)-1}(t) \tag{52}$$

$$\mathcal{F}_2(w) = \frac{1}{v_s^2} F_0 \sum C_k L_k^{(7/s)-1}(t) \tag{53}$$

for the vector and tensor terms in (45), where $t = (w/v_s)^s$. The first two polynomials are $L_0^\alpha = 1$, $L_1^\alpha = \alpha + 1 - x$. The requirement $n\langle w \rangle = n\langle (w^2/3)\mathbf{F}_1 \rangle = 0$ (rest frame) amounts then to $b_0 = 0$. Heat flux $\mathbf{q} = nm\langle (w^2/3)(w^2/2)\mathbf{F}_1 \rangle$ is determined by b_1 in case of a Maxwellian (s=2), but involves all other terms for other values of the coefficient s. Viscous stress $\Pi = nm\langle (2w^4/15)\mathcal{F}_2 \rangle$ is determined by the first tensor coefficient in (53), C_0. Polynomials are not eigen functions of the collision term, however, thus one has in any case an infinite set of coupled equations which must be truncated at some order. In the Chapman-Enskog procedure (see next section) one usually uses the first or second order approximation. Especially for heat flux and intermediate values of the magnetic field errors are in the 100% range. Expansions up to 6th order have been carried out for this reason (*Landshoff* 1951; *Kaneko* 1960).

168

3.3.3. *Expansion in Spherical Harmonics*

In contrast to the expansions we have considered so far, the expansion

$$f(\mathbf{x}, \mathbf{w}, t) = \sum f_l^m(\mathbf{x}, w, t) Y_l^m(\theta, \phi) \tag{54}$$

in spherical harmonics with the orthogonality property $\int d\Omega Y_l^m Y_o^n = \delta_{lnmo}$ applies only to the angle dependence, but leaves the energy dependence open. The l=1 term corresponds to F_1 in (45) and $l = 2$ to $\mathcal{F}_2$. If we assume a slab geometry with space dependence only along z and no magnetic field (or along z), then (54) reduces to an expansion in Legendre polynomials

$$f(z, \mathbf{v}, t) = \sum_{l=0}^{N} f_l(z, v, t) P_l(\mu) \tag{55}$$

where $\mu = cos\theta$, with θ the angle between the z axis and $\mathbf{v}$. The kinetic equation (2) may be expanded as

$$\begin{aligned}
\frac{\partial f_l}{\partial t} + v\frac{\partial}{\partial z}(\frac{l}{2l-1}f_{l-1} + \frac{l+1}{2l+3}f_{l+1}) \\
- \frac{eE}{m}[\frac{l}{2l-1}(\frac{\partial f_{l-1}}{\partial v} - \frac{l-1}{v}f_{l-1}) \\
+ \frac{l+1}{2l+3}(\frac{\partial f_{l+1}}{\partial v} + \frac{l+2}{v}f_{l+1})] = (Cf)_l
\end{aligned} \tag{56}$$

The great advantage of this method is that the speed dependence is not specified ad hoc, but may be determined self-consistently. We see that the various orders are coupled by the gradient and the applied electric field. They are also coupled in general by the collision term, unless field particles are isotropic, where this term reduces to $C_0 f_l$. Again, one can use an ad hoc cutoff or a numerically determined cutoff (small residual error) for a truncation of this chain. Only the lowest order terms will be needed if collisions are sufficiently frequent to maintain a nearly isotropic distribution. In some applications it is possible to solve the coupled chain of equations by a numerical algorithm which is efficient also for very weak collisionality (*Dum* 1975).

The linearized collision operator with isotropic lowest order distributions

$$C_{ab}^1(f_b)f_a = C(f_b^0)f_a^1 + C(f_b^1)f_a^0 \tag{57}$$

is rotationally invariant, hence the spherical harmonics are eigen functions, where according to the Wigner Eckart theorem

$$C^1 f_l^m(w)Y_l^m = -\nu_l(w)\frac{l(l+1)}{2}f_l^m(w)Y_l^m \tag{58}$$

For like particle collisions, however, $\nu_l(w)$ is an operator on $f_l^m(w)$.

4. Transport theory for a collision dominated plasma

In the previous section we have discussed approximate methods for the solution of the kinetic equations which, formally at least, can be applied for any degree of collisionality. We now consider special cases in which perturbation theory can be used to simplify the solution of the kinetic equations. These cases also provide tests of the methods in the previous section.

4.1. CHAPMAN-ENSKOG METHOD

If the plasma component is collision dominated we can introduce a small parameter $\epsilon = O(\lambda/L) = O(\tau/T)$, expressing that the (average) mean free path and the collision time are small compared to the macroscopic length and time scales, respectively. The kinetic equation (32) has the structure

$$\frac{df}{dt} + Lf + (\mathbf{w} \times \Omega) \cdot \frac{\partial f}{\partial \mathbf{w}} = Cf \tag{59}$$

where L is a linear operator. Collision dominance is is assumed by setting $C = O(1/\epsilon)$, but one also allows for the influence of the magnetic field, $\Omega = O(1/\epsilon)$. The distribution function is expanded as $f = f^{(0)} + \epsilon f^{(1)} + \dots$. The two lowest order equations are

$$\mathbf{w} \times \Omega \cdot \frac{\partial f^{(0)}}{\partial \mathbf{w}} = C f^{(0)} \tag{60}$$

$$C(f^{(0)})f^{(1)} + C(f^{(1)})f^{(0)} + (\mathbf{w} \times \Omega) \cdot \frac{\partial f^{(1)}}{\partial \mathbf{w}} = \frac{df^{(0)}}{dt} + Lf^{(0)} \tag{61}$$

It has been shown by Hilbert in 1912 that this perturbation scheme has only normal solutions $f[(n(\mathbf{x}, t), \mathbf{u}(\mathbf{x}, t), T(\mathbf{x}, t), \mathbf{w}]$ with an implicit time dependence

$$\frac{d}{dt} = \frac{\partial f}{\partial n}\frac{dn}{dt} + \frac{\partial f}{\partial \mathbf{u}}\frac{d\mathbf{u}}{dt} + \frac{\partial f}{\partial T}\frac{dT}{dt}$$

through the lowest order moments. These moments determine the local Maxwellian which is, according to the H-theorem, the solution of the lowest order equation (60). The moment equations of appropriate order in ϵ can be used to determine $\frac{d}{dt}$ on the right hand side of (61) and higher order perturbations. In addition one has the solubility conditions that the perturbation terms make no contributions to the moments which are already contained in $f^{(0)}$. The equation (61) is linear in the perturbed distribution functions $f^{(1)}$ and in the perturbing forces, but still represents a rather complicated integro-differential equation, or, if several species are present, a system of such equations. This may be seen from the explicit expression for the collision terms, given in Section 2. The collision operator acting on the unperturbed distribution in (61) is a differential operator, but depends on integrals of the perturbed distributions.

The Chapman-Enskog method (*Chapman and Cowling* 1970) of solution consists of a low order truncated Laguerre expansion of $\mathbf{F}_1(w)$ and $\mathcal{F}_2(w)$. This expansion was discussed in the previous section, but now is inserted into the perturbation equation (61) rather than the full kinetic equation. As we have already mentioned, a reasonably accurate representation of the speed dependence, especially for heat flux, requires, however, at least 4-6 terms. The order can be recognized by noting that the magnetic field dependence becomes a rational function of the same order. A direct numerical solution of the integro-differential equations (without magnetic field) has been carried out by *Cohen et al.* (1950) for electrical conductivity, by *Spitzer,Jr. and Härm* (1953) for heat conductivity and by *Roussel-Duprè* (1981) for ion diffusion. These solutions, as the explicit solutions for a Lorentz gas given below, confirm the need for a more accurate representation of the speed dependence than is provided by the low order truncated expansions (moment approximations) of the previous section. At least it can be shown that the transport coefficients satisfy mini-max principles. They can be determined approximately by applying a variational method for maximum entropy production to trial distribution functions, see e.g. *Robinson and Bernstein* (1962)

The classical method of expansion (60–61) implies for a multi-component gas that to lowest order all distributions are isotropic Maxwellians with the same temperature and no relative drift. According to the H-theorem, these are the equilibrium solutions. In a plasma, however, the approach to the equilibrium between the electron and ion components is very slow, as has been shown in the previous section. A more practical perturbation scheme starts therefore with separate drifts and temperatures for electrons and ions, and treats some cross collision terms as a small quantity, by virtue of the small electron to ion mass ratio. Elastic scattering of electrons by ions is not a small term, but to lowest order in the mass ratio, only dependent on ion density and charge state (*Braginskii* 1967).

4.2. NEARLY ISOTROPIC DISTRIBUTIONS

A much more general scheme which not only includes the classical expansion we just discussed, but can also be applied to turbulent plasmas, is based on assuming dominance of isotropization by elastic scattering (*Dum* 1978a, b). This scattering can be due to electron-ion collisions or enhanced ion fluctuations connected to ion acoustic turbulence, but could also arise for cosmic ray particles in the magnetic fluctuations connected with Alfvèn wave turbulence. The distribution function can be written in the form

$$f(\mathbf{w}) = F(w) + \hat{f}(\mathbf{w}) \tag{62}$$

where the isotropic part (energy distribution) is not necessarily Maxwellian, but determined by the reduced kinetic equation, which is obtained by averaging (32) over angles

$$\frac{dF}{dt} - \frac{w}{3}\nabla \cdot \mathbf{u} \cdot \frac{\partial F}{\partial w} + \frac{\partial}{\partial \mathbf{x}} \cdot \langle w\hat{f}\rangle + \frac{1}{w^2}\frac{\partial}{\partial w}w\langle(\mathbf{a} \cdot \mathbf{w} - \mathcal{U} : \mathcal{W})\hat{f}\rangle - < \hat{C}\hat{f} > \tag{63}$$
$$= < C > F$$

where

$$U_{ik} = \frac{1}{2}\left(\frac{\partial u_i}{\partial x_k} + \frac{\partial u_k}{\partial x_i}\right) - \frac{1}{3}\frac{\partial}{\partial \mathbf{x}} \cdot \mathbf{u}\delta_{ik} \tag{64}$$

is the shear tensor and

$$\mathcal{W} = \mathbf{ww} - \frac{w^2}{3}\mathcal{I}$$

The anisotropic part satisfies

$$\mathbf{w} \cdot \frac{\partial F}{\partial x} + (\mathbf{a} \cdot \mathbf{w} - \mathcal{U} : \mathcal{W})\frac{1}{w}\frac{\partial F}{\partial w} - \hat{C}F = (\mathbf{w} \times \Omega) \cdot \frac{\partial \hat{f}}{\partial \mathbf{w}} + C\hat{f} - < \hat{C}\hat{f} > \tag{65}$$

where the time dependence is neglected to lowest order, by virtue of the assumed dominance of isotropization. In a strong magnetic field it is useful to make the further decomposition

$$\hat{f}(\mathbf{w}) = \bar{f}(w_{\parallel}, w_{\perp}) + \tilde{f}(w_{\parallel}, w_{\perp}, \phi) \tag{66}$$

and split (65) by taking a ϕ average

$$w_{\parallel} \frac{\partial F}{\partial w_{\parallel}} + (a_{\parallel} w_{\parallel} - \mathcal{U}_0 : \mathcal{W}_0) \frac{1}{w} \frac{\partial F}{\partial w} - \bar{C}F = \bar{C}\bar{f} - <\bar{C}\tilde{f}>$$

$$\mathbf{w}_{\perp} \cdot \frac{\partial F}{\partial \mathbf{x}} + (\mathbf{a}_{\perp} \cdot \mathbf{w}_{\perp} - \mathcal{U}_{\perp} : \mathcal{W}_{\perp}) \frac{1}{w} \frac{\partial F}{\partial w} - \tilde{C}F = -\Omega \frac{\partial \tilde{f}}{\partial \phi} + \tilde{C}\bar{f} + C\tilde{f} - \overline{C\tilde{f}} \tag{67}$$

A finite Larmor radius expansion may then be applied to (67).

For classical transport of electrons, the perturbing collision term on the left hand side of (65) arises from a relative drift of electrons and ions.

$$\hat{C}_{ei}F = -\mathbf{r}_{ei} \cdot \mathbf{w} \frac{1}{w} \frac{\partial F}{\partial w} \tag{68}$$

where

$$\mathbf{r}_{ei} = -\nu_{ei}(w)(\mathbf{u} - \mathbf{u}_i) \tag{69}$$

and, cf. Sect. 2,

$$\nu_{ei}(v) = 3Z(ln\Lambda/\Lambda)\omega_e(v_e/v)^3 \tag{70}$$

is the collision frequency. It is best to choose the frame frame $\mathbf{u} = \mathbf{u}_e$ for the electron kinetic equations, else one needs to determine Π from the second order perturbation equation. In this frame $< \mathbf{w} >= 0$ which gives the equation of motion as solubility condition

$$nm\mathbf{a} = \frac{\partial p}{\partial \mathbf{x}} - \mathbf{R} \tag{71}$$

see (37).

For ions with a high charge state $Z \gg 1$, elastic scattering of electrons $\nu(w) \propto w^{-3}$ is dominant. In this Lorentz gas approximation (65) has the explicit solution

$$F_1^j = -(\nu_1 - ij\Omega)^{-1}[\frac{\partial}{\partial \mathbf{x}} + (\mathbf{a} + \mathbf{r})\frac{1}{w}\frac{\partial}{\partial w}]_j F, \qquad j = 0, \pm 1 \tag{72}$$

$$F_2^j = (3\nu_2 - ij\Omega)^{-1}U_j \frac{1}{w}\frac{\partial F}{\partial w} \qquad j = 0, \pm 1, \pm 2 \tag{73}$$

where $F_1^0 = \tau_0 \cdot \mathbf{F}_1$, $F_1^{\pm 1} = (\tau_1 \pm i\tau_2) \cdot \mathbf{F}_1$, $F_2^0 = \tau_0\tau_0 : \mathcal{F}_2$, $F_0^{\pm 1} = \tau_0(\tau_1 \pm i\tau_2) : \mathcal{F}_2$, $F_2^{\pm 2} = (\tau_1 \pm i\tau_2)(\tau_1 \pm i\tau_2) : \mathcal{F}_2$. If electron-electron collisions cannot be neglected one must consider the collision frequencies ν_1, ν_2 for the vectorial ($l = 1$) and tensorial ($l = 2$) perturbations as operators. The equations become integro-differential equations, as noted above.

Inserting the solution for the anisotropic part of the distribution function into (63) completes the equation for the evolution of the energy distribution. The resulting equation is quite complicated in general and will not be given here, see Dum (1978b). The perturbation $\hat{f}$ introduces diffusion terms which depend on gradients or applied fields, in addition to the energy diffusion which is represented by the collision term on the right hand side of

(63). Only if inelastic collisions are also very frequent, as assumed in the Chapman-Enskog method, can one use the lowest order approximation (60), which in the case of Coulomb collisions gives a Maxwellian, or a distribution (51) with $s = 5$, in the case of scattering by ion sound waves.

A power law is obtained for the energy distribution of elastically scattered electrons in a strong static electric field. The quiver motion in a high frequency electric field, on the other hand, produces an effective diffusion term which gives rise to the same flattened distribution as was obtained for ion sound turbulence (*Langdon* 1980). Inserting the anisotropy connected with shear into the kinetic equation for F also gives energy diffusion terms, even if scattering is fully elastic. A simplified form of (63) which includes the effects of shear has been obtained by *Earl et al.* (1988) for cosmic ray propagation. Isotropization by Alfvèn waves was represented by a simple phenomenogical collision term. Green's functions for the solution of the kinetic equations may be found for a large class of effective diffusion terms (*Dum* 1978a).

If effective energy diffusion is not dominant, then a nonlocal solution of the kinetic equation (63) for the energy distribution function must be obtained. Some analytic results have been found for the laser-plasma interaction with strong gradients (*Lindman and Swartz* 1986), but usually a numerical solution is required in this case (*Kho and Haines* 1986).

4.3. TRANSPORT RELATIONS

Once the distribution functions are known, transport relations are obtained by carrying out the integrations required in the moment equations. For electrons the following set of relations is obtained for the rate of momentum transfer, the heat flux and the viscous stress tensor (*Dum* 1978b).

$$\mathbf{R}_u = \mathbf{R}_u^0 + \mathbf{R}_u^1 = -\frac{nm}{\tau_e}[(1 - \rho_u^{\parallel})\mathbf{u}_0 + (1 - \rho_u^{\perp})\mathbf{u}_\perp - \rho_u^\Lambda(\tau_0 \times \mathbf{u})] \tag{74}$$

$$\mathbf{R}_T = -n[\rho_T^{\parallel}\nabla_{\parallel}T_e + \rho_T^{\perp}\nabla_\perp T_e + \rho_T^\Lambda(\tau_0 \times \nabla T_e)] \tag{75}$$

$$\mathbf{R}_n = T_e[\rho_n^{\parallel}\nabla_{\parallel}n + \rho_n^{\perp}\nabla_\perp n + \rho_n^\Lambda(\tau_0 \times \nabla n)] \tag{76}$$

$$\mathbf{q}_u = nT_e[\kappa_u^{\parallel}\mathbf{u}_{\parallel} + \kappa_u^{\perp}\mathbf{u}_\perp + \kappa_u^\Lambda(\tau_0 \times \mathbf{u})] \tag{77}$$

$$\mathbf{q}_T = -\frac{nT_e\tau_e}{m}[\kappa_T^{\parallel}\nabla_{\parallel}T_e + \kappa_T^{\perp}\nabla_\perp T_e + \kappa_T^\Lambda(\tau_0 \times \nabla T_e)] \tag{78}$$

$$\mathbf{q}_n = \frac{T_e^2\tau_e}{m}[\kappa_n^{\parallel}\nabla_{\parallel}n + \kappa_n^{\perp}\nabla_\perp n + \kappa_n^\Lambda(\tau_0 \times \nabla n)] \tag{79}$$

$$\Pi = -nT_e\tau_e[\eta_0\mathcal{U}_0 + \eta_1\mathcal{U}_1 + \eta_2\mathcal{U}_2 + \eta_{1,\Lambda}\mathcal{U}_{1,\Lambda} + \eta_{2,\Lambda}\mathcal{U}_{2,\Lambda}] \tag{80}$$

$$\mathcal{U}_0 = (3/2)U_0(\tau_0\tau_0 - \mathcal{I}/3)$$

$$\mathcal{U}_1 + \mathcal{U}_{1,\Lambda} = U_{-1}[\tau_0(\tau_1 + i\tau_2) + (\tau_1 + i\tau_2)\tau_0]$$

$$\mathcal{U}_2 + \mathcal{U}_{2,\Lambda} = U_{-2}(\tau_1 + i\tau_2)(\tau_1 + i\tau_2)$$

$$\mathcal{U} = \mathcal{U}_0 + \mathcal{U}_1 + \mathcal{U}_2 = \frac{1}{2}(\nabla u)^s - \frac{1}{3}\nabla \cdot \mathbf{u}$$

In these equations τ_e is the effective collision time for elastic scattering of electrons, $\tau_0 = B/B$ is the direction of the magnetic field, and τ_1, τ_2 are unit vectors which complete a righthanded orthogonal set. We notice that the magnetic field introduces an anisotropy with different transport coefficients for flows along the magnetic field and across the field. There is also transport in the direction perpendicular to magnetic field and gradients. The relation (80) between stress tensor and shear tensor, evidently is more complicated. The structure of the transport relations follows from rather general considerations. It applies as long as the only anisotropy is introduced by the magnetic field, i.e. scattering is isotropic.

The magnitude of the transport coefficients depends strongly on the shape of the energy distribution, which need not be Maxwellian. Explicit expressions for the transport coefficients in terms of $F(w)$, e.g. the model distributions of the form (51) have been derived by *Dum* (1978b), demonstrating this strong dependence. The energy distribution will be Maxwellian, only if inelastic scattering by electron-electron collisions is sufficiently frequent. Only in this case vanish the transport terms (76), (79) which are directly related to the density gradient, and one also has the Onsager relation $\rho_T = \kappa_u$. If, on the other hand, electrons are subject to scattering by an (isotropic) ion sound turbulence spectrum, the weak inelastic component would relax the energy distribution function to a flattened distribution of the form (51) with $s = 5$. In this case one has the symmetry relations $\rho_n = \rho_u$, $\kappa_n = \kappa_u$. The interaction of the plasma with a high frequency electric field also produces this energy distribution, hence the transport coefficients are also applicable to the laser-plasma interaction. As usually ions of high charge state Z are produced in this interaction, elastic scattering by electron-ion collisions dominates transport. In the numerical model of *Kho and Haines* (1986) modifications of the energy distribution by nonlocal transport are also included. The transport coefficients, including some which are important for magnetic field generation, are modified correspondingly.

If electron-electron collisions are important (small Z), one must, as mentioned above, solve integro-differential equations to obtain the transport coefficients, which become Z dependent. Approximate values based on a truncated expansion with two Laguerre polynomials may be found e.g. in *Braginskii* (1967) whose notation we have followed here. Ion transport coefficients are also listed.

5. Heat Flux and breakdown of collision dominated transport

Of the transport relations appearing in the fluid equations (36–39), heat flux (40) is the highest velocity moment of the distribution function. High speed particles contribute thus significantly, which in Coulomb collisions, however, have the largest mean free path, cf. Section 2. This is the reason why collision dominated classical transport theory is most likely to break down for electron heat flux. Using (72) for $f_1 = wF_1^0$ we obtain

$$\frac{f_1}{f_0} = -\frac{1}{16}\left(\frac{\pi}{2}\right)^{1/2}x^4(4 - x^2)\hat{q} \tag{81}$$

where $x = w/v_e\sqrt{2}$. The transport relation (78), in the absence of a magnetic field or a gradient along the field was used in expressing the temperature gradient in terms of the normalized heat flux $\hat{q} = q/nmv_e^3$. The heat conductivity coefficient is $\kappa_T^{\parallel} = 128/3\pi$ for a Lorentz gas, $Z \rightarrow \infty$, with Maxwellian energy distribution. Zero electric current requires an electric field which is given by the acceleration term (71). The temperature gradient in the pressure term can be added to the gradient term arising from the first term in (72). The density gradient terms cancel for a Maxwellian. The momentum transfer term in (71) reduces to the thermal force, which is given by (75) with the coefficient $\rho_T^{\parallel} = 3/2$. This friction force arises from the distortion of the distribution function and the speed dependence of scattering, even in the absence of a net drift (current).

We see that f_1/f_0 is a rapidly increasing function of x. The assumption $|f_1/f_0| \ll 1$ made in the derivation of this equation will always break down for sufficiently large x. The same is true if, instead of considering a Lorentz gas, we allow for the effect of electron-electron inelastic collisions in trying to maintain a Maxwellian. The corresponding numerical solutions of the integro-differential equation for f_1 have been tabulated by *Spitzer,Jr. and Härm* (1953) for various Z.

If we are very lenient and determine a critical velocity by the condition $|f_1/f_0| < 1$ and require that these particles carry 90% of the heat flux, then we obtain for $Z = 1$, $x_c = 9$, and $\hat{q} = 0.03$, corresponding to the small ratio $\lambda_e/L = 0.01$, and not unity as often stated, for the breakdown of collision dominated transport. The result depends not only on Z, but even more strongly on the shape of the energy distribution, which as we have seen, need not be Maxwellian.

A number of cures haves been proposed for this nearly universal problem in applying classical transport theory. The most frequent and simplest proposal is to introduce an ad hoc limitation of heat flux in the transport relation (78) by arguing that heat flux must be limited to some fraction of the free streaming value, $\hat{q} \leq \alpha < 1$. A common argument is that heat flux instabilities limit effective streaming velocities to some fraction of the electron thermal velocity. A seemingly more sophisticated proposal is to limit $|f_1/f_0|$ to some value less than unity, arguing that $f(w) = f_0 + f_1 cos\theta$ would otherwise assume negative values (*Shvarts et al.* 1981). A somewhat related method is implemented in numerical codes such as LASNEX (*Zimmerman and Kruer* 1975) which are used to model the laser-plasma interaction. In these codes electrons are divided into several groups, depending on the energy, and flux limiters are applied to heat fluxes of individual groups. Although a large range of values for the flux limiter has been considered, agreement with detailed measurements is poor, see e.g. *Rogers et al.* (1989). In their experiments it was not only possible to carry out a comparison between measured temperature profiles and code results, but electron distribution functions could also be measured. Many such measurements have also been carried out for the solar wind (*Pilipp et al.* 1987). The similarity of results is remarkable. An analysis of heat flow typically leads to results shown schematically in Figure 1.

The energy distribution f_0 is non-Maxwellian, and often may be considered of consisting of a cool core and a more energetic halo. The $l = 1$ component f_1 differs even more radically from the predictions of classical collision dominated transport theory. Instead of the predicted rapid increase with energy, f_1/f_0 remains limited. In the solar wind this ratio often reaches a plateau $|f_1/f_0| < 1$, as illustrated in Figure 1. This need not be the case, however. There is no general reason that $|f_1/f_0|$ should be limited to values less than unity. From the definition by the series (55) it follows only that

$$f_1(w) = \frac{3}{2} \int_{-1}^{1} dcos\theta f(w,\theta)cos\theta \leq 3f_0$$

For $f(w,\theta) = f(w)$, $\theta \leq \theta_m$ we obtain e.g. $f_1/f_0 = (3/2)(1 + cos\theta_m)$. For scatterfree expansion of the solar wind such a restriction of the pitch angle θ would follow from con-

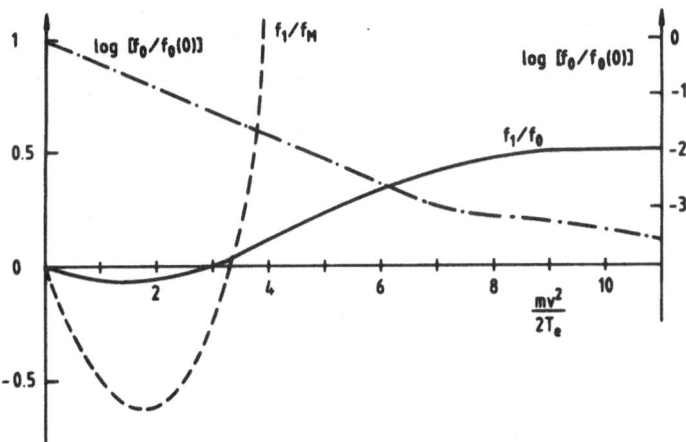

Figure 1. *Breakdown of the collision dominated approximation for heat flux. Shown are schematically the isotropic part f_0 of the distribution function (dot-dashed), and the $l=1$ anisotropic component f_1 which determines the heat flux (solid line). Also shown is the prediction of classical collision dominated transport theory (dashed line) with a Maxwellian energy distribution, for the ratio $\lambda/L_T = 0.32$ of mean free path and temperature gradient scale, demonstrating the total breakdown of this theory.*

servation of magnetic moment in the decreasing interplanetary field. Observed high energy electron distributions, indeed, are often restricted in this way, with a maximum pitch angle as small as 10° inside magnetic sectors. Near isotropy in the vicinity of sector boundaries, however, suggests strongly enhanced scattering. Ion acoustic turbulence provides a reasonable explanation for these observations, as a detailed comparison with the theory outlined in Section 4.2 for nearly isotropic non-Maxwellian distributions shows (*Dum* 1983).

In the measurements of *Rogers et al.* (1989) f_1/f_0 reached a maximum of approximately 1.2 and then decayed with energy. The ratios for the next two terms in (55) reached maxima of about 0.4, with $f_3 < 0$ in the entire energy range. These terms guarantee that the total distribution $f(\mathbf{w})$ remains positive. The measurements apparently can be explained by Coulomb collisions and collisions with neutrals, as a comparison with the predictions from a solution of the coupled equations (56) shows. This chain was truncated after $l = 3$. The ratio of mean free path to scale length was about 1/3. Temperature profile predicted by the numerical solutions of the kinetic equation also agreed well with measurements, in contrast to predictions using flux limiters. The collisions with neutrals play an important role in this experiment. Code results without electron-neutral collisions gave rise to a plateau in f_1/f_0 at 0.9, similar to Figure 1, but in strong contrast to the experiment. The decrease in anisotropy at large energies can be understood from the energy dependence of the electron-neutral collision frequency

$$\nu_{en} \propto E^{1.71}$$

where E is the electron energy. This dependence is also in strong contrast to the energy dependence $E^{-1.5}$ in Coulomb collisions. With Coulomb collisions alone, solutions of the coupled kinetic equations (56) at higher l have also been carried out (*Bell et al.* 1981; *Matte and Virmont* 1982). The most important factor in determining electron heat flux, however, is the non-Maxwellian energy distribution which is obtained in such calculations.

6. Conclusions

We have contrasted classical collision dominated transport theory, as it has been developed for neutral gases, with transport theories appropriate for a plasma. Whereas in collisions with neutrals the mean free path may be decreasing with energy, it is strongly increasing in Coulomb collisions, $\lambda \propto E^2$. As a result, collision dominated transport theory frequently breaks down, especially for electron heat flux which is carried predominantly by more energetic particles. Collisions may be still frequent enough to maintain a nearly isotropic distribution. Classical transport theory may then be extended by a perturbation theory which calculates the anisotropic part as a functional of the perturbing forces and an energy distribution which no longer is Maxwellian but is determined self-consistently from a reduced kinetic equation. The same methods can be applied to other isotropizing scattering mechanism such as the interaction of electrons with ion acoustic turbulence or the scattering of cosmic rays by Alfvèn wave turbulence. The shape of the energy distribution, generally non-Maxwellian, has a strong effect on the structure and magnitude of the transport coefficients. If gradients are stronger it may be necessary to include higher order anisotropies in the system of coupled equations. We have compared these solutions of the kinetic equations with truncation procedures for moments or ad hoc assumptions for the distribution function. Although such procedures are much easier to carry out, their success is limited, mainly by the failure to model the intrinsic speed dependence of Coulomb collisions.

Another outstanding characteristic of plasma transport is the important role of ambient magnetic and electric fields. Heat flux always involves also an electric field which is required to maintain zero electric current and charge neutrality. It too must be determined self-consistently. Effects connected with the topology of electric and magnetic fields can only be briefly mentioned here. A diverging magnetic field has a strong influence on the particle distribution functions in polar, solar, or stellar winds. Because of the speed dependence of Coulomb collisions, the more energetic particles are generally only weakly collisional and hence sample a large region of space, making transport non-local, and dependent on the structure of electric and magnetic fields. *Scudder and Olbert* (1979) considered this problem for solar wind electrons. Numerical solutions for the distribution of solar wind ions were obtained by *Griffel and Davis* (1969) and by *Livi and Marsch* (1986), using a highly simplified collision term.

For photoelectrons in polar magnetic flux tubes, distributions have been found by a Monte Carlo method for modeling Coulomb collisions (*Yaseen et al.* 1989). Electron heat flux has also been studied by this method (*Khan and Rognlien* 1981), as has cosmic ray transport in the presence of shear (*Earl et al.* 1988). *Jokipii et al.* (1989) and *Jokipii and Morfill* (1989) used the Monte Carlo method for cosmic rays also at discontinuities in the flow, where reduced kinetic equations of the form (63) are no longer applicable. The Monte Carlo method requires the calculation of a large number of particle orbits in the ambient

177

fields and of their random perturbation by collisions. The method can be competitive with solutions of the kinetic equations, if field topology and boundary conditions are very complex. In any case, this Lagrangian description may be intuitively more appealing than the Eulerian description by a kinetic equation.

Although this paper concentrated on classical transport, we have outlined parallels with anomalous transport. We feel that that a familiarity with the richness of phenomena in classical transport and the modern methods for dealing with them is an absolute prerequisite for advances towards more quantitative theories of anomalous transport.

References

Baldwin, D. E. and Watson, C. J. H. (1977) 'Magnetized plasma kinetic theory–II. Derivation of the Rosenbluth potentials for a uniform magnetized plasma', Plasma Phys., 19, 517–528.

Balescu, R. (1960) 'Irreversible processes in ionized gases', Phys. Fluids, 3, 52–63.

Barakat, A. R. and Schunk, R. W. (1982) 'Transport equations for multicomponent anisotropic space plasmas: A review', Plasma Phys., 24, 389–418.

Bell, A. R., Evans, R. G., and Nicholas, D. J. (1981) 'Electron energy transport in steep temperature gradients in laser-produced plasmas', Phys. Rev. Lett., 46, 243–246.

Belyi, V. V., Dewulf, D., and Paiva-Veretennicoff, I. (1989) 'Anomalous transport in strongly inhomogenous systems. II. The generailzed hydrdynamics of a two-component plasma', Phys. Fluids B, 1, 317–324.

Bond, D. J. (1981) 'Approximate calculation of the thermal conductivity of a plasma with an arbitrary temperature gradient', J. Phys. D, 14, L43–L46.

Bowers, E. and Haines, M. G. (1968) 'Fluid equations for a collisionless plasma including finite ion Larmor radius and finite β effects', Phys. Fluids, 11, 2695–2708.

Braginskii, S. I. (1967) 'Transport processes in a plasma', in Reviews of Plasma Physics, vol. I, M. A. Leontovich (ed.), Consultants Bureau New York, pp. 205–311.

Burgers, J. M. (1969) Flow Equations for Composite Gases, Academic Press, New York.

Chandrasekhar, S. (1942) Principles of Stellar Dynamics, Chapter II, Univ. Chicago Press, Chicago.

Chapman, S. and Cowling, T. G. (1970) The Mathematical Theory of Non-Uniform Gases, 3rd edition, Cambridge Univ. Press, London.

Chew, G. F., Goldberger, M. L., and Low, F. E. (1956) 'The Boltzmann equation and the one-fluid hydromagnetic equations in the absence of particle collisions', Proc. Roy. Soc., A 236, 112–118.

Chodura, R. and Pohl, F. (1971) 'Hydrodynamic equations for anisotropic plasmas in magnetic fields–II Transport equations including collisions', Plasma Phys., 13, 645–658.

Cohen, R. S., Spitzer,Jr., L., and Routly, P. McR. (1950) 'The electrical conductivity of an ionized gas', Phys. Rev., 80, 230–238.

Cowie, L. L. and McKee, C. F. (1977) 'The evaporation of spherical clouds in a hot gas. I. Classical and saturated mass loss rates', Astrophys. J., 211, 135–146.

Craig, I. J. D. and Davys, J. W. (1984) 'Heat flux saturation in hydrodynamic soft x-ray solar flare plasmas', Sol. Phys., 90, 343–356.

Cuperman, S., Weiss, I., and Dryer, M. (1980) 'Higher order fluid equations for multicomponent nonequilibrium stellar (plasma) atmospheres and star clusters', Astrophys. J., 239, 345–359.

Dum, C. T. (1975) 'Strong-turbulence theory and the transition from Landau to collisional damping', Phys. Rev. Lett., 35, 947–950.

Dum, C. T. (1978a) 'Anomalous heating by ion sound turbulence', Phys. Fluids, 21, 945–955.

Dum, C. T. (1978b) 'Anomalous electron transport equations for ion sound and related turbulent spectra', Phys. Fluids, 21, 956–969.

Dum, C. T. (1983) 'Electrostatic waves and anomalous transport in the solar wind', in Proc. Solar Wind V, NASA Conf. Publ. 2280, M. Neugebauer, NASA, Washington,D. C., pp. 369–376.

Dum, C. T. and Pfirsch, D. (1972) 'Neoclassical transport in a strong magnetic field', Fifth European Conference on Controlled Fusion and Plasma Physics, vol.I, 13, Euratom CEA, Grenoble.

Earl, J. A., Jokipii, J. R., and Morfill, G. (1988) 'Cosmic-ray viscosity', Astrophys. J., 331, L91–L94.

Frieman, E., Davidson, R., and Langdon, B. (1966) 'Higher order corrections to the Chew-Goldberger-Low theory', Phys. Fluids, 9, 1475–1482.

Ganguli, S. B. and Palmadesso, P. J. (1987) 'Plasma transport in the auroral return current region', J. Geophys. Res., 92, 8673–8690.

Ganguli, S. B. and Palmadesso, P. J. (1989) Private communication.

Grad, H. (1949) 'On the kinetic theory of rarefied gases', Comm. Pure a. Appl. Math., 2, 331–407.

Grad, H. (1958) 'Principles of the kinetic theory of gases', in Handbuch der Physik Vol. 12, S. Flügge (ed.), Springer, Berlin, pp. 205–294.

Griffel, D. H. and Davis, L. (1969) 'The anisotropy of the solar wind', Plan. Space Sci., 17, 1009–1020.

Gross, E. P., Jackson, E. A., and Ziering, S. (1959) 'Boundary value problem in kinetic theory of gases', Ann. Phys. N.Y., 1, 141–167.

Hassan, M. H. A. and Watson, C. J. H. (1977a) 'Magnetized plasma kinetic theory–I. Derivation of the kinetic equation for a uniform magnetized plasma', Plasma Phys., 19, 237–247.

Hassan, M. H. A. and Watson, C. J. H. (1977b) 'Magnetized plasma kinetic theory–III. Fokker-Planck coefficients for a uniform magnetized plasma', Plasma Phys., 19, 627–649.

Jeans, J. H. (1929) Astronomy and Cosmogony, Cambridge Univ. Press, Cambridge.

Jokipii, J. R., Kóta, J., and Morfill, G. (1989) 'Cosmic Rays at fluid discontinuities', to be published.

Jokipii, J. R. and Morfill, G. (1989) 'Particle acceleration in step function shear flows: A microscopic analysis', to be published.

Kaneko, S. (1960) 'Transport coefficients of plasmas in a magnetic field', J. Phys. Soc. Japan, 15, 1685–1696.

Kennel, C. F. and Greene, J. M. (1966) 'Finite Larmor radius hydromagnetics', Ann. Phys., 38, 63–94.

Khan, S. A. and Rognlien, T. D. (1981) 'Thermal heat flux for arbitrary collisionality', Phys. Fluids, 24, 1442–1446.

Kho, T. H. and Haines, M. G. (1986) 'Nonlinear electron transport in magnetized laser plasmas', Phys. Fluids, 29, 2665–2671.

Landshoff, R. (1951) 'Convergence of the Chapman-Enskog method', Phys. Rev., 82, 442.

Langdon, A. B. (1980) 'Nonlinear inverse Bremsstrahlung and heated-electron distributions', Phys. Rev. Lett., 44, 575–579.

Larson, R. B. (1970) 'A method for computing the evolution of star clusters', M. N. R. A. S., 147, 323–337.

Lenard, A. (1960) 'On Bogoliubov's kinetic equation for a spatially homogeneous plasma', Ann. Phys. N. Y., 3, 390.

Lindman, E. L. and Swartz, K. (1986) 'Analytic studies of hot electron transport', Phys. Fluids, 29, 2657–2664.

Livi, S. and Marsch, E. (1986) 'Generation of solar wind proton tails and double beams by Coulomb collisions', J. Geophys. Res., 92, 7255–7261.

MacMahon, A. (1965) 'Finite gyro-radius corrections to the hydromagnetic equations for a Vlasov Plasma', Phys. Fluids, 8, 1840–1845.

Marsch, E. and Livi, S. (1985) 'Coulomb self-collision frequencies for self-similar and kappa distributions', Phys. Fluids, 28, 1379–1386.

Matte, J. P. and Virmont, J. (1982) 'Electron heat transport down steep temperature gradients', Phys. Rev. Lett., 49, 1936–1939.

Mintzer, D. (1965) 'General orthogonal polynomial solution of the Boltzmann equation', Phys. Fluids, 8, 1976–1990.

Oraevskii, V., Chodura, R., and Feneberg, W. (1968) 'Hydrodynamic equations for plasmas in strong magnetic fields-I Collisionless approximation', Plasma Phys., 10, 819–828.

Palmadesso, P. J., Ganguli, S. B., and Mitchell,Jr., H. G. (1988) 'Multimoment fluid simulations of transport processes in the auroral zones', in Modelling Magetospheric Plasma, AGU Monograph, vol.44, T. E. Moore and J. H. Waite, AGU, Washington, p. 133.

Pilipp, W. G., Miggenrieder, H., Montgomery, M. D., Mühlhäuser, K.-H., Rosenbauer, H., and Schwenn, R. (1987) 'Characteristics of electron velocity distribution functions in the solar wind derived from the Helios Plasma Experiment', J. Geophys. Res., 92, 1075–1092.

Robinson, B. B. and Bernstein, I. B. (1962) 'A variational description of transport phenomena in a plasma', Ann. Physics, 18, 110–169.

Rogers, J. H., De Groot, J. S., Abou-Assaleh, Z., Matte, J. P., Johnston, T. W., and Rosen, M. D. (1989) 'Electron heat transport in a steep temperature gradient', Phys. Fluids B, 1, 741–749.

Rosenbluth, M. N., MacDonald, W. M., and Judd, D. L. (1957) 'Fokker-Planck equation for an inverse-square force', Phys. Rev., 107, 1–6.

Roussel-Dupré, R. (1981) 'Computations of ion diffusion coefficients from the Boltzmann-Fokker-Planck equation', Astrophys. J., 243, 329–343.

Salat, A. (1975) 'Non-linear plasma transport equations for high flow velocity', Plasma Phys., 17, 589–607.

Scudder, J. D. and Olbert, S. (1979) 'A theory of local and global processes which affect solar wind electrons, I. The origin of typical 1 AU velocity distribution functions–steady state theory', J. Geophys. Res., 84, 2755–2772.

Shirazian, M. H. and Steinhauer, L. (1981) 'Kinetic-theory description of electron heat transfer in a plasma', Phys. Fluids, 24, 843–850.

Shvarts, D., Delettrez, J., McCrory, R. L., and Verdon, C. P. (1981) 'Self-consistent reduction of the Spitzer-Härm electron thermal heat flux in steep temperature gradients in

180

laser-produced plasmas', Phys. Rev. Lett., 47, 247–250.

Spitzer,Jr., L. I. (1962) The Physics of Fully Ionized Gases, Interscience, New York.

Spitzer,Jr., L. I. and Härm, R. (1953) 'Transport theory in a completely ionized gas', Phys. Rev., 89, 977–981.

Yaseen, F., Retterer, J. M., Chang, T., and Winningham, J. D. (1989) 'Monte-Carlo modeling of polar wind photoelectron distribution with anomalous heat flux', Geophys. Res. Lett., 16, 1023–1026.

Zimmerman, G. B. and Kruer, W. L. (1975) 'Numerical simulation of laser-initiated fusion', Comments Plasma Phys., 2, 85.

PROGRESS IN LTE AND NON-LTE RADIATIVE TRANSPORT PROPERTIES

J. J. KEADY, W. F. HUEBNER*, J. ABDALLAH, JR., N. H. MAGEE, JR.
Los Alamos National Laboratory, Los Alamos, NM 87545
* and Southwest Research Institute, San Antonio, TX 78228-0510*

ABSTRACT. Atomic processes in ionized gases are reviewed along with recent calculations. Opacity calculations in progress at Los Alamos are described. Several of the open issues in the calculations relevant to opacities are discussed and preliminary results of the new advances are illustrated.

1. Introduction

Progress in the calculation of absorption coefficients and opacities proceeds along several fronts. The atomic physics of an ion having several open subshells can result in spectra containing tens of millions of lines. Until very recently, computer limitations made it very difficult to systematically generate these data, and next to impossible to handle and manipulate them. The advent of the present generation of supercomputers has considerably eased (though by no means eliminated) this burden. We shall describe some ongoing efforts at Los Alamos on more comprehensive and internally consistent opacity calculations.

By and large, most atomic structure data and cross section calculations assume an isolated atom. Plasma (density) corrections are usually incorporated perturbatively (cf. Cox, 1965). At some point, the perturbative approach cannot work, requiring a self-consistent many-body quantum mechanics and statistical mechanics approach to calculating energy levels, transition moments, etc. There is much to be done in this area.

Even for densities much less than solid, the plasma environment has an important role perturbing the excited states (cf. Hummer and Mihalas, 1988), especially when they are radiating. Much work has been done on spectral line shapes, (Griem, 1974), but the sheer number of spectral lines has made it impossible to incorporate most of these detailed results except for certain special cases (i.e., Stark broadening of hydrogenic and helium-like systems). One relies on simplified results that merit some scrutiny as to their effects on opacity calculations.

2. Atomic Structure

The basic atomic structure is calculated using Cowan's atomic structure code (Cowan, 1981). For our multi-electron system described by a Hamiltonian, the important interactions include the electron – nucleus interaction, the electron – electron interaction, and the spin – orbit interaction (the coupling of the electronic magnetic moment to the induced magnetic field).

With one overwhelmingly dominant interaction (Coulomb vs. spin – orbit) various "pure coupling" representations are achieved. For light elements, a commonly though not universally occurring

181

W. Brinkmann et al. (eds.), Physical Processes in Hot Cosmic Plasmas, 181–195.
© 1990 *Kluwer Academic Publishers.*

situation, first analyzed by Russell and Saunders, is when the Coulomb interaction dominates the spin – orbit interaction, so that this latter interaction can be either ignored or treated as a perturbation. This is the familiar case of LS (or Russell-Saunders) coupling.

For an atomic state well described by LS coupling, the behavior along an isoelectronic sequence with increasing nuclear charge Z takes the state into intermediate coupling, where no one interaction is necessarily a perturbation on the others. At the other extreme, where the spin – orbit interaction dominates, lies the case of so-called j-j coupling. Intermediate coupling occurs more commonly than the extreme case of j-j coupling.

For a particular atomic state, k, the wavefunction, Ψ_k, is written as an expansion involving a set of basis functions, ψ_b. From a mathematical point of view, the choice of a set of basis functions is arbitrary, but of course subject to such considerations as completeness, orthogonality, etc. However, the observations above suggest that a choice of basis functions that are possibly realizable physical states of the system may provide physical insight and simplify the calculations. The basis functions utilized in the calculations described here are LS coupled wavefunctions.

In the central field approximation, the one-electron spin orbitals are given by

$$\psi_i(r_i) = \frac{1}{r} P_{n_i l_i}(r) Y_{l_i m_{l_i}}(\theta_i, \phi_i)\, \sigma_{m_{s_i}}(s_{i_z}),\tag{1}$$

where r_i represents the position of electron i. $Y_{l_i m_{l_i}}$ is the usual spherical harmonic, s_{i_z} is the z component of the spin vector s_i, and $\sigma_{m_{s_i}}(s_{i_z})$ is the one-electron spin eigenfunction. The composite atomic state with N electrons characterized by

$$(n_1 l_1)^{w_1}(n_2 l_2)^{w_2}...(n_q l_q)^{w_q}, \qquad \sum_{i=1}^{q} w_i = N,\tag{2}$$

is constructed from angular momentum coupling of antisymmetrized products of the above spin-orbitals.

The radial functions $P_{nl}(r)$ are obtained from a solution of the Hartree-Fock equations, or usually a more stable approximation to them, using a local density approximation for the exchange contribution to potential energy operator (the so-called Hartree plus statistical exchange or Hartree-X).

The elements of the Hamiltonian matrix between all basis vectors b and b' are written

$$\begin{aligned}(H)_{bb'} = (\delta_{bb'})E_{av} &+ \sum_{j=1}^{q}\left[\sum_{k>0}(f_k(l_j l_j))F^k(l_{jj}) + (d_j)\,\zeta_j\right]\\ &+ \sum_{i=1}^{q-1}\sum_{j=i+1}^{q}\left[\sum_{k>0}(f_k(l_i l_j))F^k(l_i l_j) + \sum_{k}(g_k(l_i l_j))G^k(l_i l_j)\right].\end{aligned}\tag{3}$$

The so-called Slater integrals, F^k and G^k are given by

$$F^k(ij) = \int_0^\infty \int_0^\infty \frac{2r_<^k}{r_>^{k+1}}|P_i(r_1)|^2|P_j(r_2)|^2 dr_1 dr_2,\tag{4a}$$

$$G^k(ij) = \int_0^\infty \int_0^\infty \frac{2r_<^k}{r_>^{k+1}}P_i^*(r_1)P_j^*(r_2)P_j(r_1)P_i(r_2)dr_1 dr_2,\tag{4b}$$

where ζ_j is the spin-orbit parameter

$$\zeta_j(r) = \frac{\alpha^2}{2}\int \frac{1}{r}\left(\frac{dV^j}{dr}\right)|P_j(r)|^2 dr.\tag{5}$$

The f_k, g_k, and d_j containing the angular coupling are computed within the program using Racah algebra.

Diagonalization of the intermediate coupling Hamiltonian leads to the eigenvalues. Transformation matrices between various pure coupling representations are computed so that the coupling conditions can be investigated.

Once the intermediate coupling calculations are performed, the fine structure contributions to the energy levels (e.g., $^3P_{0,1,2}$) are appropriately averaged, resulting in a kind of post-facto LS coupling approximation. An example of the energy level structure is given in Fig. 1, for four times ionized sulfur. For most of the temperatures kT of interest, averaging the fine structure energies is justified as the splitting is small compared to kT. Additionally, the splitting is also small compared to the the the multi-frequency group widths currently feasible in multi-frequency group radiation transport calculations. For spectral synthesis calculations – not the intended use of these averaged data – more detail is usually needed.

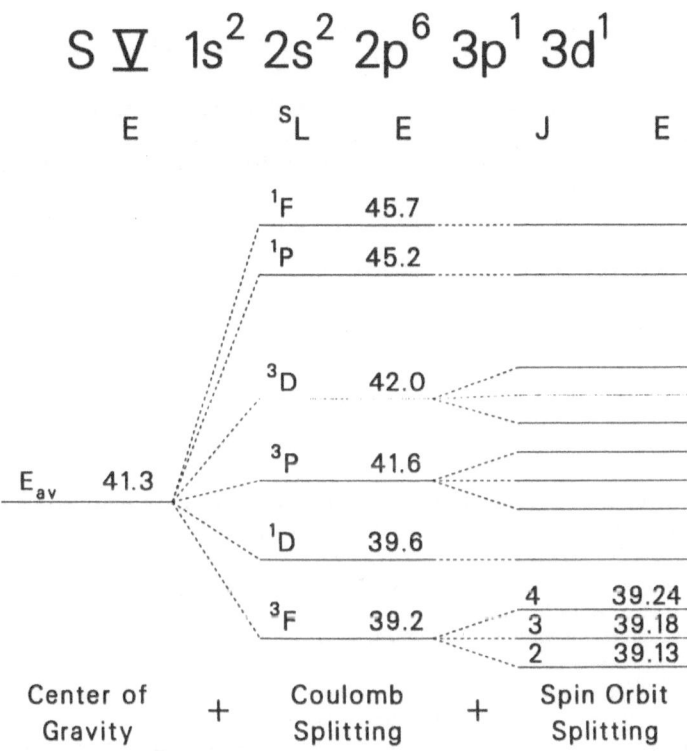

Fig. 1. Typical energy level structure for a configuration in S V showing the spin-orbit splitting. This is typically much smaller than the relevant temperature kT of interest.

Labeling the eigenvalues with the dominant LS coupling basis function component in the eigenfunction expansion allows the calculation of the LS coupling oscillator strength using the appropriate one-electron radial integral.

Configurations lying near each other in energy can perturb each other. The computations can include this configuration interaction effect, although it is not now included. The so-called single configuration approximation is now used. By incorporating configuration interaction effects, the basis function expansion for an atomic configuration will involve basis functions from other configurations. Configuration interaction effects generally decrease with increasing ionization.

Despite the simplifications discussed above, the atomic structure calculations still generate voluminous sets of data. Generation and manipulation of these very large data sets has been greatly facilitated by taking specific advantage of the Cray supercomputer environment (Abdallah et al., 1988). The intrinsic nature of the data sets, largely calculated along Rydberg series, admits the further simplification of quantum defect theory (Seaton, 1966).

In the simplest case of one highly excited electron outside some core configuration, the field experienced by the excited electron may deviate only slightly from a Coulomb field, becoming even more Coulombic for more highly excited states. The deviations are caused by the electron exchange interaction between the excited electron and the core electrons as well as polarization of the core by the excited electron, and are of course more noticeable for less highly excited states. In the simplest level of approximation, an atomic Rydberg level, E_{nl} , can be characterized

$$E_{nl} = -\left(\frac{Z_c}{n^*}\right)^2 = -\left(\frac{Z+N-1}{n-\delta_{nl}}\right)^2 , \tag{6}$$

where Z is the nuclear charge and N is the number of bound electrons and $n^* = n - \delta_{nl}$ is the effective principal quantum number of the excited electron. Historically, about the first systematic application of this Ritz-type formula was to the analysis of alkali spectra (Condon and Shortly, 1935).

In applying quantum defect theory to LS terms along Rydberg series, the general form used is

$$\delta_{nl} = c_0 + \frac{c_1}{n} + \frac{c_2}{n^2}, \tag{7}$$

where the c_i are fit coefficients (Clark and Merts, 1987).

The oscillator strengths are calculated from a similar type of fit to the radial dipole matrix element (Clark and Merts, 1987) via

$$f_{ij} = \frac{\Theta_{ij}\,\Delta E_{ij}\,T_{ij}^2}{12g_i(Z-N+1)^2} , \tag{8}$$

where Θ_{ij} is the angular factor between the two levels, ΔE_{ij} is the transition energy, T_{ij} is the radial dipole matrix element and g_i is the statistical weight of the initial level.

The fits use level data up to principal quantum number $n = 20$, even though bound – bound transitions are only calculated up to $n = 10$, and have orbital quantum numbers up to $l = 4$ (g states). Generally, transitions going into as many as five open subshells are allowed.

As an example, for all ionization stages of oxygen for the conditions described above, approximately 3000 fit coefficients and about 1400 angular matrix elements will generate about 65000 lines (Magee, 1987). For very dilute plasmas, the fits to data up through $n = 20$ will accurately generate many more energy levels for much more highly excited states.

For ionized systems, the fits are typically good to a few parts in a thousand. This usually exceeds the absolute accuracy of most of the atomic structure data.

3. Line Broadening

The perturbation of an ensemble of radiating ions by other ions and electrons is a complicated time dependent quantum mechanical collision problem with an enormous literature (Griem, 1974 and references therein). In general, the power spectrum (frequency dependence) is the Fourier transform of the autocorrelation function of the time dependent dipole moment. Most results in line broadening theory result when various simplifications are applied to the general quantum mechanical problem, or from enhancements of the classical oscillator picture. The main difference in the way electrons and ions interact with a radiating ion results from their differing masses. A multipole expansion of this interaction, through first order, includes a Coulomb interaction between the ion and the perturber, and a first order term involving the ionic dipole moment and the electric field at the ion resulting from the plasma background. Thus pressure broadening by ions and electrons is often called Stark broadening, and the calculation of the electric micro-field distribution, and suitable averages over it, are the central issues.

The most basic assumptions are that the relative motion of the radiator and perturber are quasi-classical. The trajectories are assumed to be rectilinear, and the densities low enough for binary collisions to dominate. Except in certain special cases (Baranger, 1958; Kolb and Griem, 1958), internal excitation of the radiating ion is neglected. This is the so-called adiabatic approximation. Time averages are usually replaced with ensemble averages via the ergodic hypothesis.

From considerations of collision timescales relative to the coherence time of the emitted wave, one of two simplifying assumptions is usually made. Suppose the collision duration time is short compared to the coherence time. Furthermore, suppose the disruption time is also short compared to the mean time between collisions, then the radiator is regarded as unperturbed between collisions, but suffering phase disruptions during the collision. This is the impact or interruption approximation. As considered by Weisskopf (1932), where the phase shift is determined from a power law interaction with distance, a Lorentz profile is obtained having a damping width determined by the interaction constant. In this treatment, only collisions able to induce phase shifts in excess of some prescribed (but arbitrary) amount are included. Relaxation of this last assumption (Lindholm, 1941; Foley 1946), incorporating the weak collisions, leads to a shifted Lorentzian profile.

Now suppose the coherence time is short compared to the collision duration time. Then the perturber may not have moved very much while the radiator is perturbed, and may be regarded as stationary. Then the frequency distribution arising from an ensemble of radiators is evaluated from averaging the probability distribution of stationary perturbers. This is the quasi-static (statistical) approximation. The simplest implementation would be to consider the perturbing effect of the nearest neighbor, since it will tend to be the strongest individual perturbation. However, in a low density ionized gas, the existence of long range potentials suggests that the many distant perturbers have a cumulative and continuous effect on the radiating ion. This of course is in addition to the strong perturbations of the nearest collision partners. This statistical theory was first elucidated by Holtzmark (1919).

Electron broadening is usually treated in the impact approximation (cf. Cox, 1965; Huebner, 1985) and ion broadening in the quasi-static approximation. In a situation where impact broadening holds in the line core and quasi-static broadening in the wings, there is then a domain between the line core and wing where neither the impact nor the quasi-static approximation is valid. There have been some efforts to formulate a more generalized theory (Vidal et al., 1970) and apply it to astrophysical situations (Schöning and Butler, 1989 and references therein).

There are plasma screening effects on the long range ionic potentials mentioned above, that will modify the Holtzmark distribution. More general treatments incorporating a Debye type screening have been developed (Hooper, 1968) and elaborated upon (Tighe and Hooper, 1977; Iglesias et al., 1983; Lee, 1988).

In addition to collision broadening, thermal broadening and natural broadening are always present. The convolution of the natural broadening and electron impact broadening Lorentzians also yields a Lorentz profile. A subsequent convolution with the Doppler profile yields the Voigt profile. If ion broadening contributes, still another convolution must be performed to yield the final profile.

Given that the absorption spectrum can contain millions of spectral lines, one of the biggest challenges for the future is the systematic incorporation of more sophisticated results for line shapes.

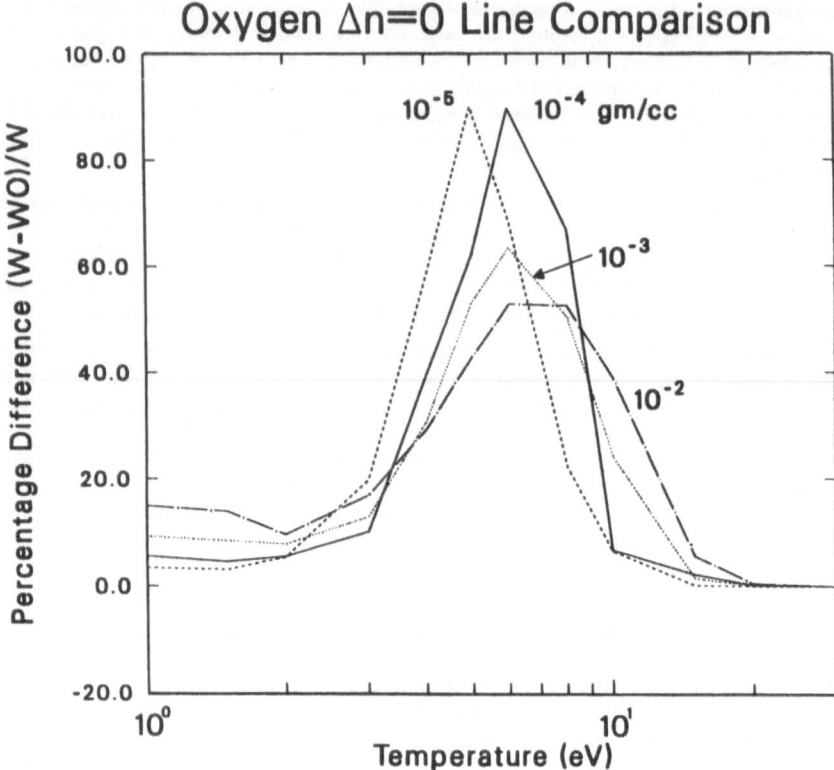

Fig. 2. Contribution of the $\Delta n = 0$ transitions to the oxygen opacity. The percentage difference of the calculations, with (w) and without (wo) their contribution is shown vs. temperature, for various densities.

4. Ongoing Calculations

Calculation of the new atomic data base is in progress. Inclusion of the new Hartree-X cross section data introduces an effective oscillator strength redistribution. This makes it difficult to quantitatively and systematically predict how the new opacities will differ from the old.

However, it is possible to identify how some features of the newer data can conceivably influence the outcome. For example, previous use of hydrogenic oscillator strengths precluded the presence of $\Delta n = 0$ transitions. Independent calculations (Iglesias et al., 1987) indicate that for an astrophysical mixture at a temperature of $2.3 \, 10^5$ K and density of $1.0 \, 10^{-5} \, \text{g/cm}^3$, the presence of the $\Delta n = 0$ transitions can increase the opacity by a factor of 2.3. The effect is maximized when the $\Delta n = 0$ transition frequencies occur under the peak in the Rosseland mean weighting function (at about $4 \, kT$). For pure oxygen, Fig. 2 shows the per cent change in the opacity, using the new oscillator strength data, when the contribution of the $\Delta n = 0$ transitions are simply excluded. The transition energies are such that the bulk of the effect occurs for temperatures kT between 4 and 10 eV. At $6.9 \, 10^5$ K, the independent calculations discussed above result in an insignificant difference when compared with the Astrophysical Opacity Library results (Huebner et al., 1977).

The oscillator strength redistribution also involves the bound – free cross sections. For excited state photoionization, the use of hydrogenic cross sections sometimes considerably overestimated the near threshold values, and vice versa at some distance above the edge. The photoionization cross sections are now calculated using Hartree potentials (Clark, 1989). In the single configuration approximation, this assures continuity in the absorption strength across the ionization limit into the continuum. This behavior in Fe XXI along a configuration averaged Rydberg sequence ($1s^2\ 2s^2\ 2p^1\ 3s^1$, $3s \rightarrow np$) is shown in Fig. 3 for the bound states, up to $n = 30$. The dashed curve on the positive energy side is directly proportional to the bound – free cross section ($\sigma = 8.067\ 10^{-18}df/dE\ \mathrm{cm}^2$).

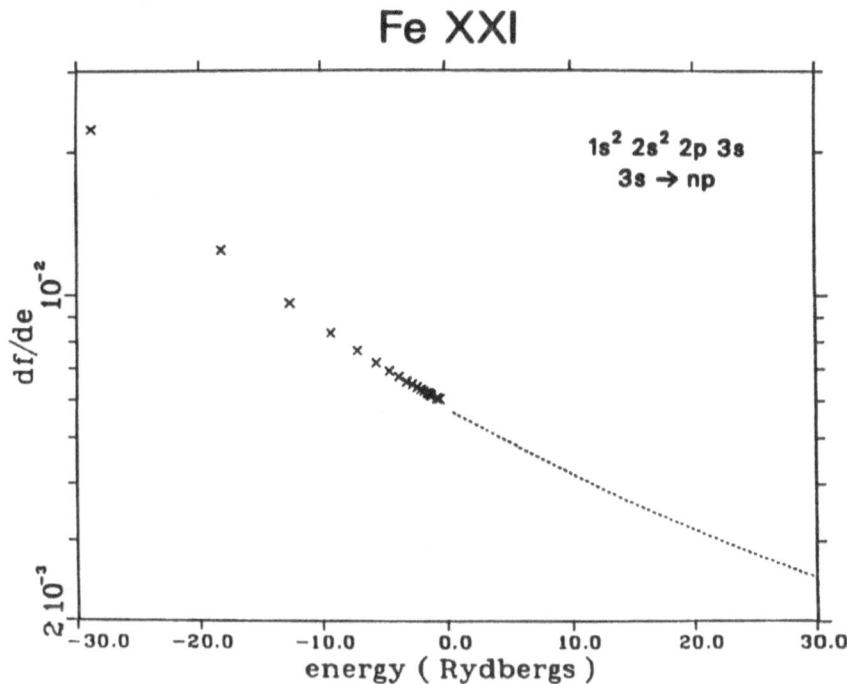

Fig. 3. The differential oscillator strength along a configuration averaged Rydberg sequence in Fe XXI across the ionization threshold into the continuum (positive energy side).

A preliminary result for Aluminum at $kT = 10$ eV and density of $1.0\ 10^{-5}\ \mathrm{g/cm}^3$ is shown in Fig. 4a. For ease of comparison, the next figure (Fig. 4b) shows the monochromatic data compacted into 50 Rosseland group means (dotted line) compared to the older Astrophysical Opacity Library data using the same group mean structure. At the peak in the Rosseland weighting function, $h\nu \simeq 40$ eV, the decrease in the absorption in the preliminary result comes about from a decrease in the bound – free cross sections relative to the older data. The Rosseland mean opacity calculated using the preliminary monochromatic data decreases by about 20% relative to the older data.

5. Non-LTE Effects

For photon mean free paths large compared to some characteristic length scale, specification of the population distribution becomes an immensely more complicated problem. The local radiation

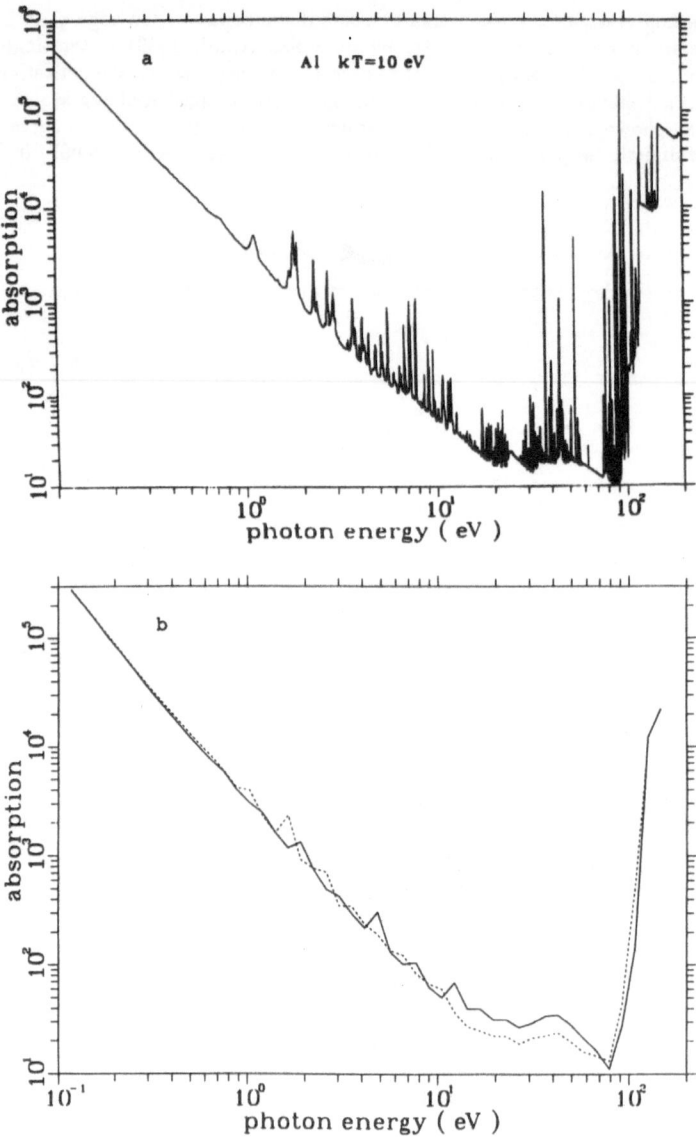

Fig. 4. (a) A preliminary result for absorption (in cm^2/g) by Aluminum at $kT = 10$ eV, and density of 1.0 10^{-5} g/cm^3. (b) The same result (dotted line) compacted into 50 Rosseland group means, compared to the older Astrophysical Opacity Library results (solid line). A decrease in the bound free absorption results in about a 20% decrease in the Rosseland mean opacity as calculated from the monochromatic data.

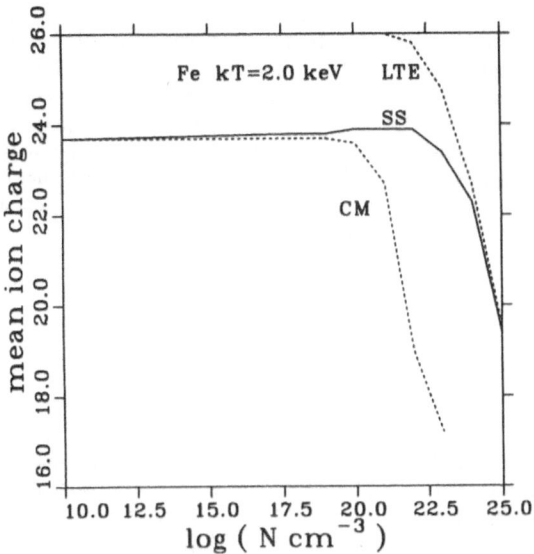

Fig. 5. For iron at an electron temperature of 2 keV, the mean ion charge vs. ion number density is shown for local thermodynamic equilibrium (LTE) the coronal model (CM), and a steady state (SS) calculation (with no radiation field), providing an approximate indication of the domains of validity for the CM and LTE models.

field depends on the global properties, and the radiative rates are competitive with the collisional rates. In general, a simultaneous solution of the radiation transport and statistical equilibrium equations is required. This is a a highly nonlinear system of equations. For the radiation transport, geometry usually matters, and many astrophysical media are moving as well. In some cases, this latter situation may permit some simplifications; i.e., escape probability methods when there are large velocity gradients.

With very tenuous media, the so-called coronal model (CM) provides a very useful simplification. Ionization is via collisions only, while the radiative decay rates are assumed so large that essentially all the population resides in the ionic ground states. With the Saha equation, the population ratios of adjacent ionic species depend inversely on the electron density. In contrast, the system of rate equations in the CM involve collisional ionization and radiative and dielectronic recombination, and do not depend on the electron density.

With an electron temperature of 2.0 keV, Fig. 5 shows the mean ion charge for iron vs. ion number density for the CM, LTE, and steady state (SS) conditions. The steady state calculations include from 30 to 80 configurations per ion. Processes that can be included in the SS calculation are radiative excitation and de-excitation, electron collisional excitation and de-excitation, photoionization, radiative recombination, collisional ionization, three body recombination and dielectronic recombination. Most cross sections are calculated using the Hartree-X potentials. Collisional excitation uses a plane-wave Born approximation, although distorted wave cross sections are available (Clark and Csanak, 1989). The collisional ionization cross sections use a simple fit to scaled hydrogenic cross sections.

At 2 keV, the SS calculations indicate that for ion densities below about 10^{20} cm^{-3}, the CM should be a good approximation, while LTE holds above about 10^{24} cm^{-3}.

The steady state calculations have no radiation field included. As the density increases in the steady state calculations, collisional excitation eventually competes effectively against radiative decay. Thus the excited states eventually have significant population, and collisional ionization out of these excited states modifies the ionization balance. Since spontaneous radiative decay and three-body recombination are included, this SS system of equations does depend on the electron density.

Most treatments of the non-LTE problem assume that the free electrons follow a Fermi-Dirac (Maxwellian) distribution. In the solar transition region this may not to be the case (Keenan et al., 1989). Perturbations of the high energy tail may have a profound effect on the optical appearance of the plasma.

6. Density Effects

6.1 Scattering

With increasing ionization, scattering processes become increasingly more important, particularly at low densities. At the simplest level of approximation, the scattering of photons off electrons, is given by the Thomson cross section

$$\sigma_{\rm T} = \frac{8\pi}{3} \left(\frac{e^2}{mc^2} \right)^2.$$ (9)

For sufficiently energetic photons (the Compton region), the Klein-Nishina formula is appropriate. This provides the cross section in the electron rest frame. For high enough temperature, electron thermal motion in the laboratory frame must be considered (Sampson, 1959).

When the product of the photon wavenumber and Debye length is less than unity, then correlation effects can influence the scattering. In this case, the densities are high enough so that the photon scatters coherently off at least two electrons (Diesendorf and Ninham, 1969; Watson, 1969; and Huebner, 1985).

The correction to the Thomson cross section when collective effects are important can be written (Huebner, 1985)

$$f(\delta) = \frac{3\delta}{8} \left(\frac{r_{\rm D}}{r_{\rm De}} \right)^2 \left[(\delta^3 + 2\delta^2 + 2\delta) \ln \left(\frac{\delta}{2+\delta} \right) + 2\delta^2 + 2\delta + \frac{8}{3} \right],$$ (10a)

with

$$\delta = \frac{1}{2} \left(\frac{c}{2\pi\nu r_{\rm D}} \right)^2,$$ (10b)

where the electron Debye radius (including degeneracy) is

$$r_{\rm De} = \left[\frac{kT(\bar{r}/a_o)^3}{6\bar{N}_f I'_{1/2}(\eta)/I_{1/2}(\eta)} \right]^{1/2} a_o,$$ (11)

and the total Debye radius (including degeneracy) is

$$r_{\rm D} = \left\{ \frac{kT(\bar{r}/a_o)^3}{6 \left[\bar{N}_f^2 + \bar{N}_f I'_{1/2}(\eta)/I_{1/2}(\eta) \right]} \right\}^{1/2} a_o,$$ (12)

where $\bar{r}$ is the ion sphere radius.

In Fig. 6, the high temperature – high energy, and correlation corrections to the Thomson cross section are shown. The dashed line (labelled a) shows the ratio of the effective Compton cross section to the Thomson cross section for a temperature $kT = 1.5\,\text{keV}$. Curves b, c, and d show the correlation correction for electron densities of $1.6\,10^{26}$, $4.5\,10^{25}$, and $1.7\,10^{25}\,\text{cm}^{-3}$, respectively, at $kT = 1.5\,\text{keV}$. These calculations ignore exchange correlations. Boercker (1987) provides a convenient correction formula to Eq. (10a). Without it, for condition c discussed above in Fig. 6, the total scattering is overestimated by about 4%. At low densities and temperatures above about $10\,\text{keV}$, scattering off electron – positron pairs starts to contribute (Sampson, 1959).

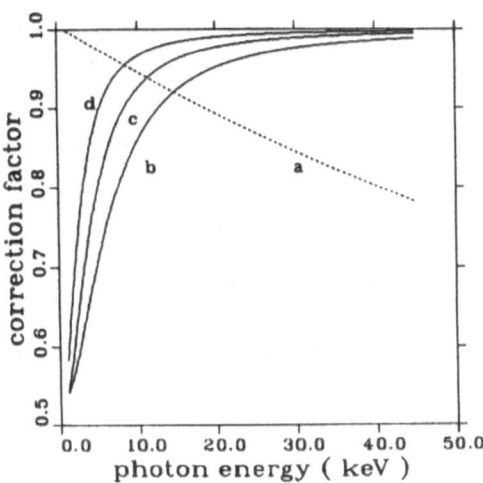

Fig. 6. Modifications of the Thomson scattering cross section caused by high energy and high temperature effects (labelled a), at $kT = 1.5\,\text{keV}$. Correlation effects at high densities are also shown (labelled b, c, d at electron densities of $1.6\,10^{26}$, $4.5\,10^{25}$, and $1.7\,10^{25}\,\text{cm}^{-3}$, respectively) at $kT = 1.5\,\text{keV}$. Both corrections to the Thomson cross section are multiplicative.

The binding energy of an electron to a plasma is the plasmon energy. When an energetic photon interacts with a plasma, the recoil energy that an electron receives may be sufficient to remove it from the collective modes of the plasma, generating density fluctuations in the plasma and creating plasmons. This will further influence the Compton scattering. Energy conservation requires

$$h\nu + \gamma mc^2 = h\nu' + h\nu_p + \gamma' mc^2, \tag{13}$$

where

$$\gamma = \left[1 - \left(\frac{v}{c}\right)^2\right]^{-1/2}, \tag{14}$$

ν is the frequency of the incoming photon, v is the initial velocity of the electron (in case of a high temperature plasma, v is the thermal velocity), m is the rest mass of the electron in the plasma, and ν_p is the plasma frequency

$$\nu_p = \left(\frac{e^2 N_e}{\pi m}\right)^{1/2}, \text{[Hz]}. \tag{15}$$

Primed quantities refer to the scattered photon and electron. The electron density is N_e.

Momentum conservation requires

$$\frac{h\nu}{c} + \gamma mv\cos(\theta) = \frac{h\nu'}{c}\cos(\phi) + \gamma'mv'\cos(\theta'), \tag{16a}$$

$$\gamma mv\sin(\theta) = -\frac{h\nu'}{c}\sin(\phi) + \gamma'mv'\sin(\theta'). \tag{16b}$$

Angles are measured relative to the direction of the incident photon. The momentum transfer to the plasma has been neglected. Eliminating θ' and v' from the above three equations yields

$$h\nu' = \frac{(h\nu_p)^2 - 2\gamma h\nu_p mc^2 - 2h\nu h\nu_p + 2h\nu\gamma mc^2 - 2h\nu\gamma mcv\cos(\theta)}{2h\nu - 2\gamma mcv\cos(\theta+\phi) - 2h\nu\cos(\phi) - 2h\nu_p + 2\gamma mc^2}. \tag{17}$$

The differential cross section for such an excitation is

$$\frac{d\sigma}{d\omega} = \left(\frac{e^2}{mc^2}\right)^2 \left[1 - \frac{1}{2}\sin^2(\phi)\right] S(k), \tag{18}$$

where

$$S(k) = \frac{hk^2}{8\pi^2 m\nu_p}. \tag{19}$$

The quantity $hk/(2\pi)$ is the momentum transferred to the electron by the incident photon.

$$\frac{d\sigma}{d\omega} = \left(\frac{e^2}{mc^2}\right)^2 \frac{[1 - \frac{1}{2}\sin(\phi)][(h\nu)^2 + (h\nu')^2 - 2h\nu h\nu'\cos(\phi)]}{2mc^2 h\nu_p}. \tag{20}$$

Substituting for $h\nu'$ and integrating gives the total cross section. For $\nu \gg \nu_p$ the total plasma interaction cross section is

$$\sigma = \frac{8\pi}{3}\left(\frac{e^2}{mc^2}\right)^2 \frac{(h\nu)^2}{2mc^2 h\nu_p}. \tag{21}$$

For photon energies in the keV range and typical plasma conditions this cross section is of about the same order of magnitude as the Compton cross section.

6.2 Bremsstrahlung

Another important process at high temperatures is the free – free or bremsstrahlung process. This is a three-body interaction involving a free electron in both the initial and final states, an ion, and an emitted or absorbed (inverse bremsstrahlung) photon. Internal excitation of the ion during the collision is usually ignored. Relaxation of this adiabatic assumption could possibly introduce resonance structures into the cross section. However, since a thermally averaged cross section is needed in opacity calculations, any resonance structure will be diffused.

Under nonrelativistic conditions, the bremsstrahlung cross section in a Coulomb field can be calculated exactly (Sommerfeld, 1953), while numerical calculations are required to incorporate the effects of electron degeneracy (Green, 1960; Karzas and Latter, 1961). At very high temperatures, relativistic effects are important. Extensive tables have been calculated these conditions (Nakagawa et al., 1987).

For very high densities, ion – ion correlations exist (Dagdeviren and Koonin, 1987) and can decrease the effective cross section. For conditions appropriate to the center of the Sun (condition c in Fig. 6) classical correlations decrease the absorption cross section by a few per cent. For larger density or lower temperature, such that the product of photon wavenumber and Debye length are less than unity, the effect will be larger.

6.3 Bound – Bound Transitions

The atomic structure is calculated for atoms and ions in isolation. The plasma environment perturbs the potential and in principle influences the energy levels, oscillator strengths, etc. Plasma effects on the energy levels are usually incorporated by a perturbation procedure (cf. Cox, 1965; Huebner, 1985), involving a 'continuum lowering' correction. There is no such simple correction for pressure effects on the transition moments, and isolated atom values are usually used.

A more consistent approach would be to directly incorporate the pressure effect on the potential and wavefunctions in a self-consistent field calculation. One such model, an adaptation from quantum mechanical solid state band theory, is the so-called atom in jellium model (Liberman, 1979). The discrete ion charge distribution around a particular ion is replaced with a constant positive charge distribution outside the atomic sphere. There is also an electronic charge distribution to guarantee charge neutrality. The system free energy and various constraints (electrical neutrality, orbital function normalization, and a fixed number of electrons) are used to to construct a free energy functional to be minimized variationally. A crucial assumption is that the energy of exchange and correlation are assumed to be a local function of the atomic electronic charge density (the local density approximation, Kohn and Sham, 1965). A mean field (average atom) approximation allows consideration of finite temperatures. Another and perhaps the simplest example of this class of Density Functional Theories (DFT) is probably Thomas-Fermi Theory.

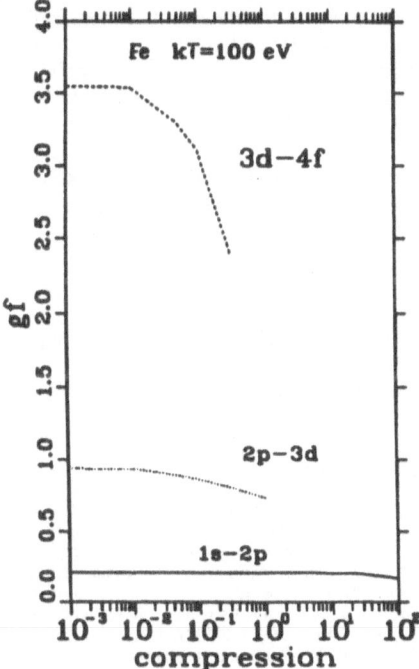

Fig. 7. The effects of compression of the plasma on selected oscillator strengths calculated with the average atom approximation for iron at $kT =100\,$eV.

An application of the atom in jellium model (Bennett and Liberman, 1985) is to calculate the average atom oscillator strengths for iron at a temperature $kT = 100\,$eV, as a function of compression

194

(in units of normal solid density). These can be seen in Fig. 7. Since the orbital functions for the 1s and 2p eigenstates are relatively more localized around the nucleus compared to the d and f states, the 1s to 2p transition requires relatively high compressions before being affected. As the pressure increases, the eigenvalues are raised toward and pushed into the continuum. Eventually the (upper) bound states cease to exist as such, as does the bound – bound oscillator strength.

This is meant to qualitatively illustrate possible density effects on the transition moments. The electronic charge distributions involved in these transitions will differ from the average atom distribution. Nor, strictly speaking, are the eigenvalues calculated from the single particle Dirac equation binding energies, so that the transition energies are not just simple differences of the eigenvalues (Perrot and Dharma-Wardana, 1985).

7. Summary

It is now possible to use modern atomic structure codes to systematically generate large quantities of energy level and cross section data for isolated atoms and ions. These data are presently being applied to the calculation of internally consistent absorption coefficient and opacity data.

It is also possible to use these data in limited non-LTE opacity calculations. However, compared to the non-LTE problem in its full generality (arbitrary radiation fields and electron velocity distributions), the models solved are very idealistic.

Prescriptions and models exist for incorporating background effects of plasma perturbations, although it may be that more sophisticated treatments are required (Adcock and Griem, 1983), even for relatively low densities ($N_e < 10^{17}$ cm^{-3}; Däppen et al., 1987). Achieving a capability of performing routine and internally self-consistent atomic structure calculations in plasmas is perhaps the most important challenge for the future.

Acknowledgements. We are indebted to R. E. H. Clark for making his photoionization code available to us. Part of this work was performed under the auspices of the U.S. Department of Energy.

References

1. Abdallah Jr., J., Clark, R. E. H., and Cowan, R. D. (1988): 'CATS, the Cowan Atomic Structure Code.' Los Alamos National Laboratory report LA-11436-M, Vol. I.
2. Adcock Jr., J. C., and Griem, H. R. (1983): Phys. Rev. Let. 50, 1369.
3. Baranger, M. (1958): Phys. Rev. 112, 855.
4. Bennett, B. I., and Liberman, D. A. (1985): 'Inferno.' Los Alamos National Laboratory report LA-10309-M.
5. Boercker, D. B. (1987): Astrophys. J. 316, L95.
6. Clark, R. E. H. (1989): Personal communication.
7. Clark, R. E. H., and Csanak, G. (1989): Los Alamos National Laboratory report LA-UR-89-2341.
8. Clark, R. E. H., and Merts, A. L. (1987): J. Quant. Spectrosc. Rad. Transfer 38, 287.
9. Condon, E. U., and Shortley, G. H. (1935): *Principles of Atomic Spectra*, Cambridge University Press.
10. Cowan, R. G. (1981): *The Theory of Atomic Structure and Spectra*, University of California Press.
11. Cox, A. N. (1965): 'Stellar Absorption Coefficients and Opacities.' In *Stars and Stellar Systems, Vol. 8; Stellar Structure*, Eds. L. H. Aller and D. B. McLaughlin, University of Chicago Press.
12. Dagdeviren, N. R., and Koonin, S. E. (1987): Astrophys. J. 319, 192.
13. Däppen, W., Anderson, L., and Mihalas, D., (1987): Astrophys. J. 319, 195.
14. Diesendorf, M. O., and Ninham, B. W. (1969): Astrophys. J. 156, 1069
15. Foley, H. M. (1946): Phys. Rev. 69, 616.
16. Green, J. (1960): 'Fermi-Dirac Averages of the Free – Free Hydrogenic Gaunt Factor.' Rand Corp. report RM-2580-AEC.
17. Griem, H. R. (1974): *Spectral Line Broadening by Plasmas*, Acedemic Press, N.Y.
18. Holtzmark, J. (1919): Z. Physik. 20, 162.
19. Huebner, W. F. (1985): 'Atomic and Radiative Processes in the Solar Interior.' In *Physics of the Sun*, Ed. P. Sturrock, D. Reidel.

20. Huebner, W. F., Merts, A. L., Magee Jr., N. H., and Argo, M. F. (1977): 'Astrophysical Opacity Library.' Los Alamos Scientific Laboratory report, LA-6760-M.
21. Hummer, D. G., and Mihalas, D. (1988): Astrophys. J. **331**, 794.
22. Iglesias, C. A., Hooper, C. F., and Dewitt, H. E. (1983): Phys. Rev. A **28**, 361.
23. Iglesias, C. A., Rogers, F. J., and Wilson, B. G., (1987): Astrophys. J. **322**, L45.
24. Karzas, W. J., and Latter, R. (1961): Astrophys. J. Suppl. **6**, 167.
25. Keenan, F. P., Cook, J. W., Dufton, P. L., and Kingston, A. E. (1989): Astrophys. J. **340**, 1135
26. Kohn, W., and Sham, L. J. (1965): Phys. Rev. **140**, A 1133.
27. Kolb, A. C., and Griem, H. R. (1958): Phys. Rev. **111**, 514.
28. Liberman, D. A. (1979): Phys. Rev. **B20**, 4981
29. Lindholm, E. (1946): Ark. Mat. Astron. Phys. **32a**, No. 17.
30. Lee, R. W. (1988): J. Quant. Spectr. Rad. Transfer **40**, 561.
31. Magee Jr., N. H. (1988): 'Opacity Calculations for Laser Plasmas,' In Proceedings of Int. Conf. Lasers: 1986, Ed. R. W. McMillan, STS Press; McLean, Va.
32. Nakagawa, M., Kohyama, Y., Itoh, N. (1987): Astrophys. J. Suppl. **63**, 61.
33. Perrot, F. and Dharma-Wardana, M. W. C. (1985): 'Density Functional and Many-Body Theories of Hydrogen Plasmas.' In *Radiative Properties of Hot Dense Matter*, Eds. J. Davis, C. Hooper, R. Lee, A. Merts, and B. Rozsnyai, World Scientific Publishing Co.; Singapore.
34. Sampson, D. H. (1959), Astrophys. J. **129**, 734.
35. Schöning, T., and Butler, K. (1989): Astron. Astrophys. **219**, 326.
36. Seaton, M. J. (1966): Proc. Phys. Soc. (London) **88**, 801.
37. Sommerfeld, A. (1953): *Atombau und Spektrallinien*, Vieweg; Braunschweig, Vol. 2.
38. Tighe, R., and Hooper Jr., C. F. (1977): Phys. Rev. **31A**, 1044.
39. Vidal, C. R., Cooper, J., and Smith, W. E. (1970): J. Quant. Spectr. Rad. Transfer **10**, 1011.
40. Watson, W. D. (1969): Astrophys. J. **157**, 375.
41. Weisskopf, V. (1932): Z. Physik **75**, 287.

TRANSPORT PROCESSES IN HOT DENSE PLASMAS: GENERAL RESULTS AND THE EVOLUTION OF PULSAR MAGNETIC FIELDS

NAOKI ITOH
Department of Physics
Sophia University
7-1, Kioi-cho, Chiyoda-ku, Tokyo 102
Japan

ABSTRACT. Recent developments in the studies of the transport processes in hot dense plasmas are reviewed. Special emphasis is placed upon the accuracy of the calculations. Ionic correlation effects play an essential role in the transport processes. Next the evolution of pulsar magnetic fields is discussed. It is found that the pulsar magnetic field decays inversely proportional to a quarter power of the pulsar age.

1. INTRODUCTION

Transport processes play an essential role in the evolution of dense stars. Recent developments in plasma physics have made accurate calculations of the transport processes possible. In the first part of this paper we review the recent developments in the studies of the transport processes in the interior of dense stars.

In the second part of this paper we discuss the evolution of the pulsar magnetic fields. This problem is closely related to the electrical conductivity of the pulsar material. It turns out that the pulsar magnetic field decays as $B \propto \tau^{-1/4}$, where τ is the age of the pulsar.

2. TRANSPORT PROCESSES

Recent papers on the transport processes in the interior of dense stars include Flowers and Itoh (1,2,3), Yakovlev and Urpin (4), Raikh and Yakovlev (5), Itoh et al. (6), Mitake, Ichimaru, and Itoh (7), Itoh et al. (8), Nandkumar and Pethick (9), Itoh, Kohyama, and Takeuchi (10).

2.1 Electrical and thermal conductivities of dense matter in the liquid metal phase

Essential ingredients that go into accurate calculations of the transport properties of the dense matter include the inter-ionic correlations brought about by the strong Coulomb cou-

W. Brinkmann et al. (eds.), Physical Processes in Hot Cosmic Plasmas, 197–220.
© 1990 *Kluwer Academic Publishers.*

pling and the electron-ion interaction represented by the screening function of the electrons. Our understanding of such many-particle effects in the Coulomb system has progressed remarkably during the period of those developments due mainly to the advancement in the Monte Calro method and other theoretical means (see, e.g., Ichimaru (11)). In this section we take account of what we consider to be the most reliable results currently available on the description of those many-particle effects, and thereby present an accurate calculation of the electrical and thermal conductivities of dense matter limited by electron-ion scattering in the liquid metal phase.

We shall consider the case that the atoms are completely pressure-ionized. We further restrict ourselves to the density-temperature region in which electrons are strongly degenerate. This condition is expressed as

$$T \ll T_F = 5.930 \times 10^9 \left\{ [1 + 1.018(\frac{Z}{A})^{2/3} \rho_6^{2/3}]^{1/2} - 1 \right\} \quad [K], \tag{2.1}$$

where T_F is the Fermi temperature, Z the atomic number of the nucleus, and ρ_6 the mass density in units of $10^6 \, g \, cm^{-3}$. For the ionic system we consider the case that it is in the liquid state. The latest criterion corresponding to this condition is given by (Slattery, Doolen, and DeWitt (12)).

$$\Gamma \equiv \frac{Z^2 e^2}{a k_B T} = 2.275 \times 10^{-1} \frac{Z^2}{T_8} (\frac{\rho_6}{A})^{1/3} < 178, \tag{2.2}$$

where $a = [3/(4\pi n_i)]^{1/3}$ is the ion-sphere radius, and T_8 the temperature in units of $10^8 \, K$.

In the present calculation we restrict ourselves to the cases where the high-temperature classical limit is applicable to the description of the ionic system. Specifically we assume that the parameter

$$y \equiv \frac{\hbar^2 k_F^2}{2 M k_B T} = 1.656 \times 10^{-2} \frac{1}{A T_8} (\frac{Z}{A})^{2/3} \rho_6^{2/3} \tag{2.3}$$

is much less than unity, where k_F is the Fermi wave number of the electrons and M is the mass of an ion. In Figure 1, we show the parameter domain for the validity of the present calculation in the case of ^{56}Fe plasma.

For the calculation of the electrical and thermal conductivities we use the Ziman formula (13) as is extended to the relativistically degenarate electrons (1). On deriving the formula we retain the dielectric screening function due to the degenerate electrons. As to the explicit expressions for the dielectric function, we use the relativistic formula worked out by Jancovici (14). The use of the relativistic dielectric function is an essential difference between the present work and that of Yakovlev and Urpin (4). Yakovlev and Urpin set the dielectric function due to electrons equal to unity: this assumption is valid only in the high-density limit.

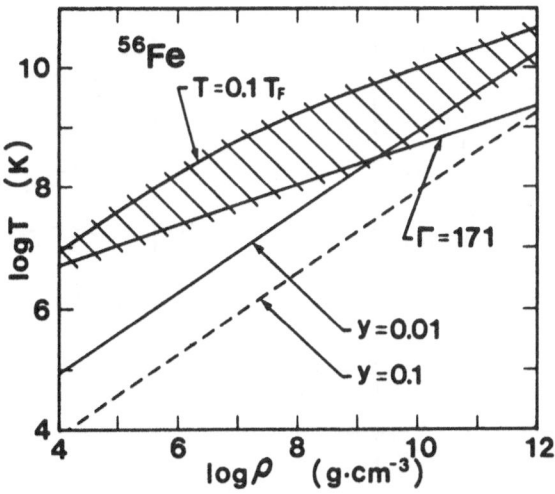

FIG. 1. Parameter domain (Shaded area) for the validity of the present calculation in the case of ^{56}Fe plasma.

Working on the transport theory for relativistic electrons given by Flowers and Itoh (1), we obtain the expression for the electrical conductivity σ:

$$\sigma = 8.693 \times 10^{21} \frac{\rho_6}{A} \frac{1}{[1 + 1.018(Z/A)^{2/3}\rho_6^{2/3}] < S >} \quad [s^{-1}]. \tag{2.4}$$

Here the scattering integral $< S >$ is evaluated for $y \ll 1$ as

$$< S > = \int_0^1 d(\frac{k}{2k_F})(\frac{k}{2k_F})^3 \frac{S(k/2k_F)}{[(k/2k_F)^2 \epsilon(k/2k_F,0)]^2}$$

$$- \frac{1.018(\frac{Z}{A})^{2/3}\rho_6^{2/3}}{1 + 1.018(\frac{Z}{A})^{2/3}\rho_6^{2/3}} \int_0^1 d(\frac{k}{2k_F})(\frac{k}{2k_F})^5 \frac{S(k/2k_F)}{[(k/2k_F)^2 \epsilon(k/2k_F,0)]^2}$$

$$\equiv < S_{-1} > - \frac{1.018(\frac{Z}{A})^{2/3}\rho_6^{2/3}}{1 + 1.018(\frac{Z}{A})^{2/3}\rho_6^{2/3}} < S_{+1} >, \tag{2.5}$$

where $\hbar k$ is the momentum transferred from the ionic system to an electron, $S(k/2k_F)$ the ionic structure factor, and $\epsilon(k/2k_F,0)$ the static dielectric screening function due to degenerate electrons. The first term in equation (2.5) corresponds to the ordinary Coulomb logarithmic term, and the second term is a relativistic correction term. For the ionic liquid structure factor we use the results of the improved hypernetted chain (IHNC) theory for the classical one-component plasma (15).

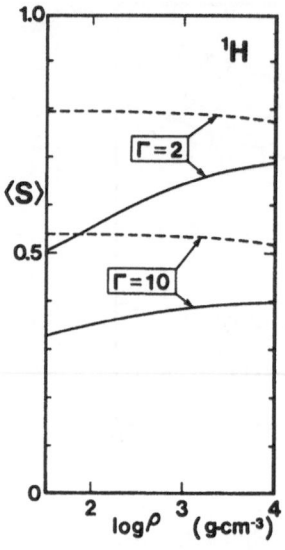
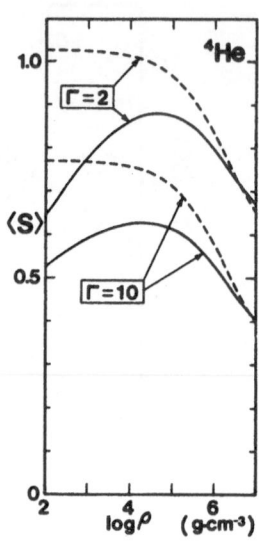

FIG. 2. Comparison of Yakovlev and Urpin's results   (dashed curves)
        with the present results (solid curves) for the ^{1}H matter.
FIG. 3. Same as FIG.2. for the ^{4}He matter.

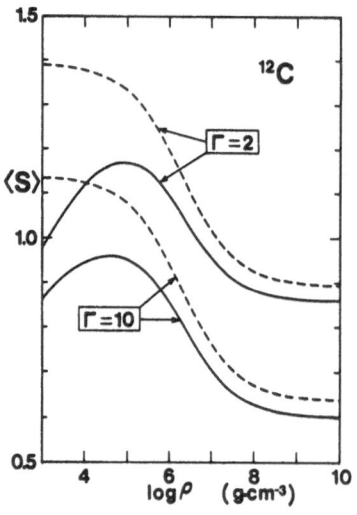
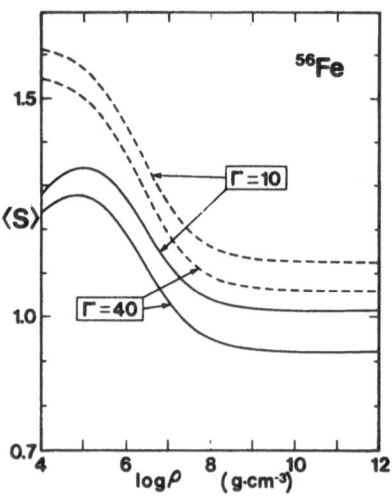

FIG. 4. Same as FIG.2. for the ^{12}C matter.
FIG. 5. Same as FIG.2. for the ^{56}Fe matter.

For the thermal conductivity κ for the relativistically degenerate electrons we analogously obtain the expression:

$$\kappa = 2.363 \times 10^{17} \frac{\rho_6 T_8}{A} \frac{1}{[1 + 1.018(\frac{Z}{A})^{2/3}\rho_6^{2/3}] < S >}$$
$$[ergs\, cm^{-1}\, s^{-1}\, K^{-1}], \qquad (2.6)$$

where $< S >$ is the same as that for electrical conductivity.

We have carried out the integrations in equation (2.5) numerically by using the IHNC structure factor of the classical one-component plasma and Jancovici's (14) relativistic dielectric functions for degenerate electrons. We have made calculations for the parameter ranges $2 \cong \Gamma \cong 160$, $10^{-4} \cong r_s \cong 0.5$, which cover most of the density-temperature region of the dense matter in the liquid metal phase of astrophysical importance.

In Figures 2,3,4, and 5 we compare the results of the calculation of $< S >$ by Yakovlev and Urpin (4) (dashed curves) with the present results (solid curves). For the ^1H matter and the ^4He matter Yakovlev and Urpin's results amount to an overestimation of $< S >$ by 60% at low densities. For the ^{12}C matter their overestimation of $< S >$ amounts to 40% at low densities. For the ^{56}Fe matter their overestimation is nearby 30% at low densities. At high densities Yakovlev and Urpin's results are reasonably close to the present ones. The large amount of the overestimation of $< S >$ at low densities by Yakovlev and Urpin is due to their neglect of electron screening. At high densities, however, the effect of the screening due to electrons is relatively small. This is the main reason for their overestimation of the resistivity (underestimation of the conductitity) at low densities.

2.2 Electrical and thermal conductivities of dense matter in the crystalline lattice phase

In this section we deal with the electrical and thermal conductivities of dense matter in the crystalline lattice phase $\Gamma > 178$. The electrical conductivity σ and thermal conductivity κ are related to the effective electron collision frequencies ν_σ and ν_κ by

$$\sigma = \frac{e^2 n_e}{m^* \nu_\sigma} = 1.525 \times 10^{20} \frac{Z}{A}\rho_6[1 + 1.018(\frac{Z}{A}\rho_6)^{2/3}]^{-1/2}$$
$$\times \frac{10^{18} s^{-1}}{\nu_\sigma} s^{-1}, \qquad (2.7)$$

$$\kappa = \frac{\pi^2 k_B^2 T n_e}{3 m^* \nu_\kappa} = 4.146 \times 10^{15} \frac{Z}{A}\rho_6[1 + 1.018(\frac{Z}{A}\rho_6)^{2/3}]^{-1/2}$$
$$\times T_8 \frac{10^{18} s^{-1}}{\nu_\kappa} ergs\, cm^{-1}\, s^{-1}\, K^{-1}, \qquad (2.8)$$

where n_e is the number density of electrons and m^* is the relativistic effective mass of an electron at the Fermi surface. In this section we are interested in the scattering of electrons by phonons. The collision frequencies ν_σ and ν_κ due to one-phonon processes can be calculated by the variational method (1,4,5) as

$$\nu_{\sigma,\kappa} = \frac{e^2}{\hbar v_F} \frac{k_B T}{\hbar} F_{\sigma,\kappa} = 9.554 \times 10^{16} T_8 \left\{ 1 + \frac{1}{1.018[(\frac{Z}{A})\rho_6]^{2/3}} \right\}^{1/2}$$

$$\times F_{\sigma,\kappa} \ s^{-1}, \tag{2.9}$$

$$F_{\sigma,\kappa} = \frac{2\gamma^2}{S^2} \int \frac{dSdS'}{k^4 |\epsilon(k,0)|^2} [1 - (\frac{\beta k}{2k_F})^2] e^{-2W(k)} |f(k)|^2$$

$$\times \sum_{s=1}^{3} [k \cdot \hat{\epsilon}_s(p)]^2 (e^{Z_s} - 1)^{-2} e^{Z_s} g_{\sigma,\kappa} . \tag{2.10}$$

In the above the integral is over the areas of the Fermi surface, k is the momentum transfer, $\hat{\epsilon}_S(p)$ the polarization unit vector of a phonon with momentum p and polarization s, and

$$\gamma \equiv \frac{\hbar \omega_p}{k_B T} = 7.832 \times 10^{-2} \frac{Z}{(AA')^{1/2}} \frac{\rho_6^{1/2}}{T_8}$$

$$= 0.3443 \frac{\rho_6^{1/6}}{A^{1/6}(A')^{1/2}Z} \Gamma , \tag{2.11}$$

$$\beta \equiv \frac{\hbar k_F c}{E_F} = \left\{ 1 + \frac{1}{1.018[(\frac{Z}{A})\rho_6]^{2/3}} \right\}^{-1/2}, \tag{2.12}$$

$$z_s \equiv \frac{\hbar \omega_s(p)}{k_B T} , \tag{2.13}$$

$$g_\sigma = k^2, \tag{2.14}$$

$$g_\kappa = k^2 - \frac{k^2 z_s^2}{2\pi^2} + \frac{3k_F z_s^2}{\pi^2} , \tag{2.15}$$

ω_p being the ionic plasma frequency. The momentum conservation requires $k = \pm p + K$, where K is the reciprocal-lattice vector for the Brillouin zone to which k is confined. In equation (2.10) we have included the dielectric screening function due to relativistically degenerate electrons $\epsilon(k,0)$, the Debye-Waller factor $e^{-2W(k)}$, and the atomic form factor $f(k)$. Yakovlev and Urpin (4) and Raikh and Yakovlev (5) have used the Thomas-Fermi screening and set $e^{-2W(k)} = 1$, $f(k) = 1$.

The phonon spectra are modified by the screening due to electrons. The longitudial optical phonon turns into an acoustic phonon in the long-wavelength limit, whereas the original transverse acoustic phonons are little affected by the electron screening (16). Because the low-frequency transverse phonons play dominant roles in the resistivity of dense stellar matter, we neglect the effects of the electron screening on the phonon spectra and use the frequency moment sum rules for the pure Coulomb lattice.

As we consider the case in which the Fermi sphere is much larger than the Debye sphere, $(k_F/k_D)^3 = Z/2 \gg 1$, Umklapp processes contribute to the scattering dominantly, and the vector k in equation (10) most probably falls in a Brillouin zone distant from the first zone. When we perform an integration within a single distant zone corresponding to the reciprocal-lattice vector K, we can make an approximation $k = K$ in the integrand and carry out an integration over p within the first zone only.

Here we follow the semianalytical approach adopted by Yakovlev and Urpin (4) and also by Raikh and Yakovlev (5). We write

$$\sum_{s=1}^{3}[k \cdot \hat{\epsilon}_s(p)]^2 z_s^n (e^{z_s} - 1)^{-2} e^{z_s} \approx \frac{\pi^n k^2}{\gamma^2} G^{(n)}(\gamma), \tag{2.16}$$

$$G^{(n)}(\gamma) = \frac{\gamma^2}{3V_B \pi^n} \sum_{s=1}^{3} \int dp z_s^n (e^{z_s} - 1)^{-2} e^{z_s}, \tag{2.17}$$

where $n = 0$ or 2, and integration is carried out over the first Brilloun zone, whose volume is V_B. By the use of this approximation F_σ and F_κ in equation (2.10) are expressed as

$$F_\sigma = I_\sigma G^{(0)}(\gamma), \tag{2.18}$$

$$F_\kappa = I_\sigma G^{(0)}(\gamma) + I_\kappa^{(2)} G^{(2)}(\gamma), \tag{2.19}$$

$$I_\sigma = \int_{-1}^{\mu_{max}} d\mu \frac{e^{-2W(q)}|f(q)|^2}{|\epsilon(q,0)|^2}(1 - \beta^2 q^2), \tag{2.20}$$

$$I_\kappa^{(2)} = \int_{-1}^{\mu_{max}} d\mu \frac{e^{-2W(q)}|f(q)|^2}{q^2|\epsilon(q,0)|^2}(1 - \beta^2 q^2)(-\frac{1}{2}q^2 + \frac{3}{4}), \tag{2.21}$$

$$q = (\frac{1-\mu}{2})^{1/2}, \tag{2.22}$$

$$q_{min} = (\frac{1-\mu_{max}}{2})^{1/2}, \tag{2.23}$$

$$\mu_{max} = 1 - 0.3575 Z^{-2/3}. \tag{2.24}$$

Here we have introduced a small momentum transfer cutoff q_{min} corresponding to the unavailability of Umklapp processes for $q < q_{min}$. The contributions of the normal processes are very much smaller than those of the Umklapp processes. For the choice of q_{min} we follow Raikh and Yakovlev (5). Yakovlev and Urpin (4) derived the asymptotic expressions of $G^{(0)}(\gamma)$ and $G^{(2)}(\gamma)$ for $\gamma \ll 1$ and $\gamma \gg 1$, and proposed the following analytic formulae for arbitrary γ, which fit the main terms of the asymptotic expressions:

$$G^{(0)}(\gamma) = u_{-2}[1 + (\frac{3u - 2\gamma}{\pi^2 c_2})^2]^{-1/2} \approx 13.00(1 + 0.0174\gamma^2)^{-1/2}, \tag{2.25}$$

$$G^{(2)}(\gamma) = \frac{\gamma^2}{\pi^2}[1 + (\frac{15}{4\pi^4 c_2})^{2/3}]^{-3/2} = \frac{\gamma^2}{\pi^2}(1 + 0.0118\gamma^2)^{-3/2}, \tag{2.26}$$

where $u_{-2} \approx 13.00$ (16) and $c_2 = 29.98$ (17) are the numerical constants that are characteristic of the phonon spectrum of the bcc Coulomb lattice. Raikh and Yakovlev (5) calculated $G^{(0)}(\gamma)$ and $G^{(2)}(\gamma)$ numerically with the exact spectrum of phonons for $\gamma < 100$. It has been confirmed that the fitting formulae (2.25) and (2.26) have an accuracy better than 10% even at $\gamma \sim 1$.

We have carried out the numerical integrations of equations (2.20) and (2.21) for ^{4}He, ^{12}C, ^{16}O, ^{20}Ne, ^{24}Mg, ^{28}Si, ^{32}S, ^{40}Ca, ^{56}Fe. Some of the results are presented in

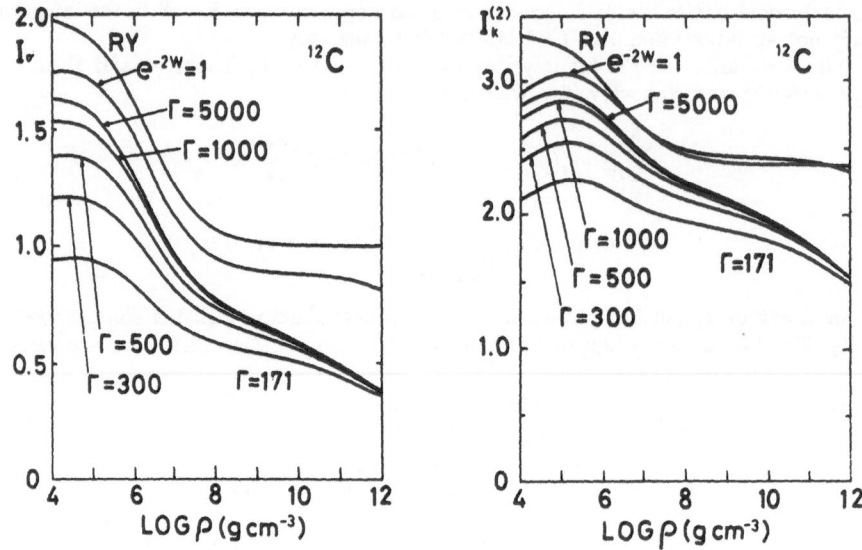

FIG. 6. I_σ for the ^{12}C matter.   RY stands for the results of Raikh and
Yakovlev (5).

FIG. 7. $I_\kappa^{(2)}$ for the ^{12}C matter.

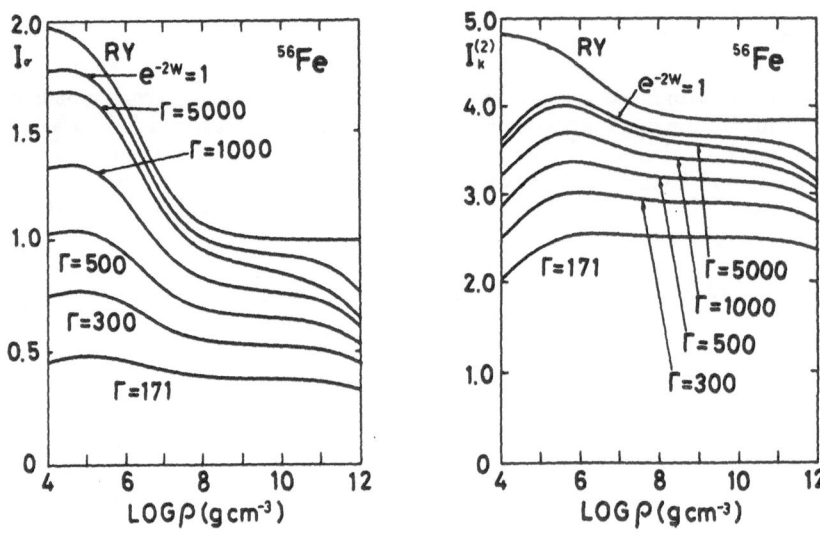

FIG. 8. I_σ for the ^{56}Fe matter.

FIG. 9. $I_\kappa^{(2)}$ for the ^{56}Fe matter.

Figures 6-9. For comparison we have also included the case where we have neglected the effects of the Debye-Waller factor and set $e^{-2W} = 1$. We also show the results of Raikh and Yakovlev (5) which are

$$[I_\sigma]_{RY} = 2 - \beta^2, \tag{2.27}$$

$$[I_\kappa^{(2)}]_{RY} = \ln Z - \beta^2 + 1.583. \tag{2.28}$$

It is readily seen that the Debye-Waller factor reduces the resistivities (enhances the conductivities) by a factor of 2-4 near the melting temperature. This means that the results of Yakovlev and Urpin (4) and those of Raikh and Yakovlev (5) give too low conductivities by that factor. It is very interesting to observe that the present result is fortuitously rather close to the original Flowers-Itoh conductivity in the crystalline lattice phase near the melting temperature (1,3).

3. EVOLUTION OF PULSAR MAGNETIC FIELDS

The pulsar magnetic moment decay has been an unresolved problem for two decades since Ostriker and Gunn (18,19) and Gunn and Ostriker (20) introduced the hypothesis of the magnetic moment decay to account for the observed properties of the pulsars. The exponetial decay of the magnetic field with a time scale $(4-9) \times 10^6\, yr$ has been assumed to fit the observational data (20,21,22,23). Recently Kulkarni (24) has identified the companion of PSR 0655+64 to be a cool white dwarf whose cooling age is $\sim 2 \times 10^9\, yr$, and has noted that in any evolutionary scenario the white dwarf is formed after the primary, thus implying that the age of PSR 0655+64 is greater than $\sim 2 \times 10^9\, yr$. More recently Wright and Loh (25) have identified the companion of PSR 1855+09 to be a cool white dwarf whose cooling age is also $\sim 2 \times 10^9\, yr$, thereby showing that PSR 1855+09 is again a very old pulsar. Those two cases are to be viewed as strong counter-examples against the hypothesis that the pulsar magnetic fields decay exponentially with a time scale $\sim 10^7\, yr$.

Regarding the theoretical justification for the relatively short time scale of order $10^6 - 10^7\, yr$ for the decay of the magnetic field in the interior of the pulsars, no consensus has been reached. Baym, Pethick, and Pines (26) have calculated the electrical conductivity of the core of the pulsar, and have shown that the decay time of the magnetic field inside the core of the pulsar exceeds the age of the universe. Since the core region constitutes the most part of the pulsar, one expects that the pulsar magnetic field originates mostly in the core. Therefore it is extremely difficult to reconcile the idea of the pulsar magnetic field decay of order $\sim 10^7\, yr$ with the result of Baym, Pethick, and Pines (26). More recently the present author and his collaborators (6,7,8) have carried out detailed calculations of the electrical conductivity of the pulsar crust material.

3.1 Theory

Here we propose a new mechanism for the decay of the pulsar magnetic moment. Preliminary reports of the present work have been made elsewhere (27,28). The new mechanism due to radiation damping (29) comes into existence because of the magnetic dipole radiation by a pulsar. The equation for the energy loss of a pulsar due to magnetic dipole radiation is written as

$$\frac{d}{dt}\left(\frac{1}{2}I\Omega^2 + \frac{m^2}{R^3}\right) = -\frac{2}{3c^3}m^2\Omega^4, \tag{3.1}$$

where I is the moment of inertia of the pulsar, m is its magnetic moment, and Ω is the angular frequency of the rotation. Here we have neglected the energy loss due to gravitational radiation. The justification for that is given in Appendix. The second term on the left-hand side of equation (3.1) is the magnetic energy of the pulsar, R being on the order of the size of the pulsar. The exact value for R depends on the distribution of the electric current in the pulsar. In deriving equation (3.1) we have assumed that the magnetic moment is perpendicular to the rotation axis.

The conventional pulsar theory (18,19) assumes that the pulsar magnetic moment is infinitely rigid: It is assumed to be never affected by the magnetic dipole radiation. In quantum theory it is well known that atomic nuclei undergo change in the magnetic moment through magnetic dipole radiation (30). In classical theory the reaction of the radiation on the charges (29) causes the electric current inside the pulsar to decay, thereby reducing the pulsar magnetic moment. Since it is not feasible to carry out a detailed calculation of this process, we content ourselves with a phenomenological approach. Let us assume a simple relationship

$$m = m_0 \frac{\Omega}{\Omega_0}, \tag{3.2}$$

where m_0 and Ω_0 are the values at $t = 0$. Then we obtain the solution

$$\frac{m}{m_0} = \frac{\Omega}{\Omega_0} = [1 + \frac{4m_0^2\Omega_0^4}{3c^3\{(I\Omega_0^2/2) + (m_0^2/R^3)\}}t]^{-1/4}. \tag{3.3}$$

In terms of the pulsar period P this solution can be written as

$$t = \frac{1}{4}\frac{P}{\dot{P}}[1 - (\frac{P_0}{P})^4]. \tag{3.4}$$

Thus for $P \gg P_0$, the time t gives the age of the pulsar in the present model

$$\tau = \frac{1}{4}\frac{P}{\dot{P}}. \tag{3.5}$$

In other words the present model gives the pulsar braking index $n = 5$.

Here it is to be noted that the braking index $n = 5$ in the present theory must be considered as a long-term value defined for a very long time span comparable to the age of the pulsar. It is well known that the Crab pulsar has a braking index $n = 2.515 \pm 0.005$ (31). More recently Manchester, Durdin, and Newton (32) measured the braking index of PSR 1509-58 $n = 2.83 \pm 0.03$. However, these values should be considered as "short-term" values compared with the age of the pulsars. The Crab pulsar has a true age 935 years, whereas its value for $P/\dot{P}$ is about 2,480 years. Thus the "long-term" braking index for the Crab pulsar is $n = 3.66$, which is of course not very close to $n = 5$ but certainly greater than $n = 3$.

From equation (3,3) one obtains

$$\dot{P}P^3 = 4\pi^4 \frac{4m_0^2}{3c^3\left\{(I\Omega_0^2/2)+(m_0^2/R^3)\right\}}. \tag{3.6}$$

Thus a pulsar is predicted to evolve along a $\dot{P}P^3 = const.$ line in the $P - \dot{P}$ plane. Spread of the $\dot{P}P^3 = const.$ lines is caused by the spread in the initial magnetic moments of the pulsars. One can also show that in the present model the pulsar magnetic moment m is given by

$$m = \frac{(6c^3I)^{1/2}}{4\pi}(\dot{P}P)^{1/2}, \tag{3.7}$$

which is exactly the same as that in the conventional theory first derived by Ostriker and Gunn (19) in which the pulsar magnetic moment is assumed to be constant. The magnetic moment m is related to the strength of the magnetic field B through the relationship

$$m = BR^3. \tag{3.8}$$

3.2 Comparison with observations

Here we compare the predictions of the theory with the observations. The observational data are taken from Taylor and Manchester (33), Manchester and Taylor (34), Gullahorn and Rankin (35), Helfand et al. (36), Newton, Manchester, and Cooke (37), Ashworth and Lyne (38), Manchester and Taylor (39), Backus, Taylor, and Damashek (40), Seward, Harnden, and Helfand (41), Kulkarni et al. (42), Frucher et al. (43), Taylor (44), Lyne et al. (45), and Foster et al. (46).

3.2.1 *Kinetic Age - P/(4Ṗ) Relationship*   In order to examine the validity of the present theory we compare the predicted pulsar age $P/(4\dot{P})$ with the pulsar kinetic age deduced from the velocity of the pulsar and its position relative to the galactic plane (22). The comparison in Figure 10 shows that the kinetic age correlates reasonably well with $P/(4\dot{P})$, thereby proving that formula (3.5) is in reality a good indicator of the true age of the pulsar. It is to be noted that the deduced kinetic ages have significant error bars due to the lack of information concerning the z-distance from the galactic plane at the birth of the pulsar and also the radial velocity of the pulsar along the line-of-sight.

One the other hand the fit with $P/(2\dot{P})$ is evidently very much poorer. One might as well use an *ad hoc* assumption of the exponential decay of the magnetic moment with the time scale $\sim 10^7\,yr$ (47). But then one is confronted with the serious dilemmas of the ages $\sim 2 \times 10^9\,yr$ of PSR 0655+64 (24) and PSR 1855+09 (25).

3.2.2 *P – Ṗ Relationship*   In Figure 11 we show the $P - \dot{P}$ diagram of the observed pulsars. The pulsar death line is defined as (48)

$$B_{12}P^{-2} = 0.1, \tag{3.9}$$

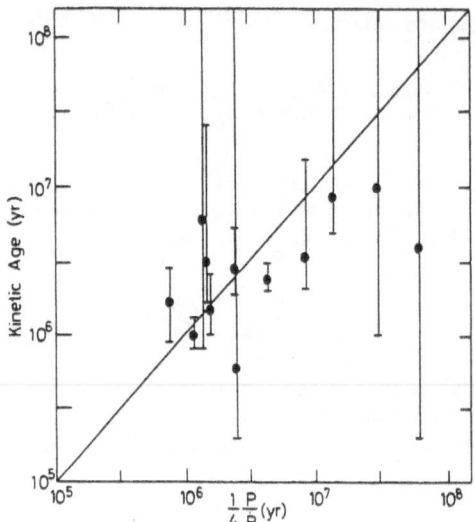

FIG. 10. The relationship between the kinetic age and $P/(4\dot{P})$. The data points
are taken from Lyne, Anderson, and Salter (22). The origin of the error
bars of the deduced kinetic age is explained in the text.

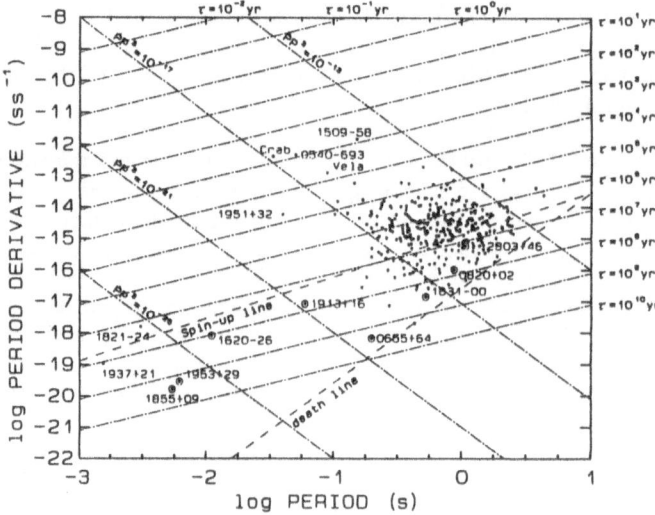

FIG. 11. The $P - \dot{P}$ diagram of the observed pulsars. The binary radio pulsars are
circled.

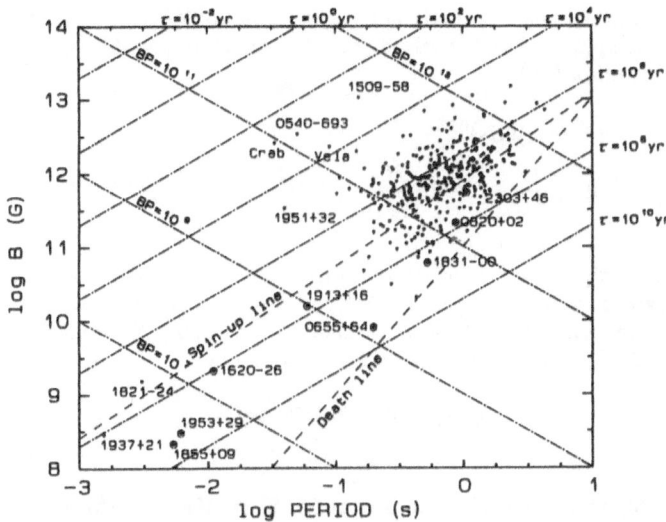

FIG. 12. The magnetic field-period diagram of the observed pulsars.

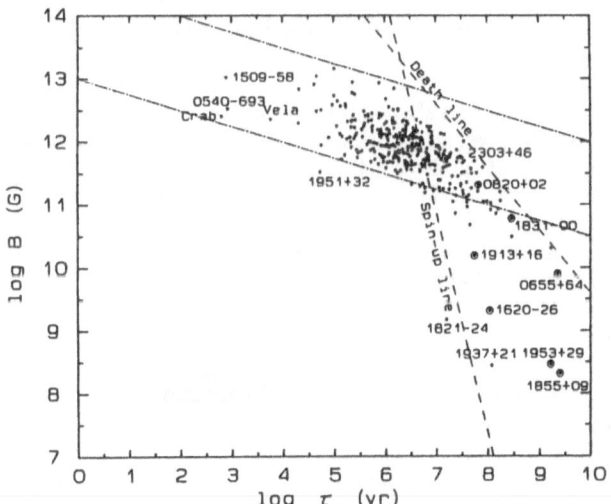

FIG. 13. The magnetic field-age diagram of the observed pulsars.

where B_{12} is the strength of the magnetic field in units of $10^{12}G$,and P is the period in units of s. The pulsar spin-up line is derived from the shortest possible spin period that can be reached by accretion (49)

$$P_{eq} = (3.0\,ms)(B_9)^{6/7}$$

$$\times \left(\frac{M}{1.4M_\odot}\right)^{-5/7}\left(\frac{\dot{M}}{\dot{M}_{Edd}}\right)^{-3/7}\left(\frac{R}{1.2} \times 10^6\,cm\right)^{18/7}, \qquad (3.10)$$

where B_9 is the surface dipole field strength in units of $10^9\,G$, M is the mass of the pulsar, $\dot{M}$ is the accretion rate, and $\dot{M}_{Edd} = 1.5 \times 10^{-8}M_\odot yr^{-1}$ is the maximum possible accretion rate. In order to obtain numerical values we assume $M = 1.4M_\odot$, $R = 1.2 \times 10^6\,cm$, and $I = 1.4 \times 10^{45}g\,cm^2$ (19). Then the magnetic field is given by

$$B = 2.2 \times 10^{19}(\dot{P}P)^{1/2} \quad [G]. \qquad (3.11)$$

The shortest attainable value of P_{eq}, which occurs for $\dot{M} = \dot{M}_{Edd}$, in this case is

$$P_{eq,min} = (3.0\,ms)\,B_9^{6/7}. \qquad (3.12)$$

It is interesting to observe that the observed pulsars other than the binary and millisecond pulsars are generally located between the two lines $\dot{P}P^3 = 10^{-13}$ and $\dot{P}P^3 = 10^{-17}$. One also notes that the binary radio pulsars PSR 0655+64, PSR 1953+29, and PSR 1855+09 are predicted to be older than $\sim 10^9\,yr$.

3.2.3 *B − P Relationship*   In Figure 12 we show the magnetic field-period relationship for the observed pulsars. In the present model pulsars are expected to evolve along the $BP = const.$ lines. The majority of the isolated pulsars are located between the two lines $BP = 10^{13}$ and $BP = 10^{11}$.

3.2.4 *B − τ Relationship*   In Figure 13 we show the magnetic field-age relationship for the observed pulsars. The pulsar magnetic field is found to decay $\propto \tau^{-1/4}$ in agreement with the present theory.

3.2.5 *P−τ Relationship*   In Figure 14 we show the period-age relationship for the observed pulsars. The pulsar period is found to increase $\propto \tau^{1/4}$ in agreement with the present theory.

3.2.6 *Rotational energy loss rate*   The rotational energy loss rate $|\dot{E}| = I\Omega|\dot{\Omega}|$ is an important quantity for the pulsar activity. In Figure 15 we show the evolution of the pulsar rotational energy loss rate as a function of the pulsar age τ. Except the binary radio pulsars and the millisecond pulsars, the majority of the pulsars are located within a band which has a slope $|\dot{E}| \propto \tau^{-3/2}$. This relationship agrees with the prediction of the radiation damping model. In fact the best-fit value for the slope excluding the binary radio pulsars and the millisecond pulsars with due consideration of the pulsar deathline (using the data for the

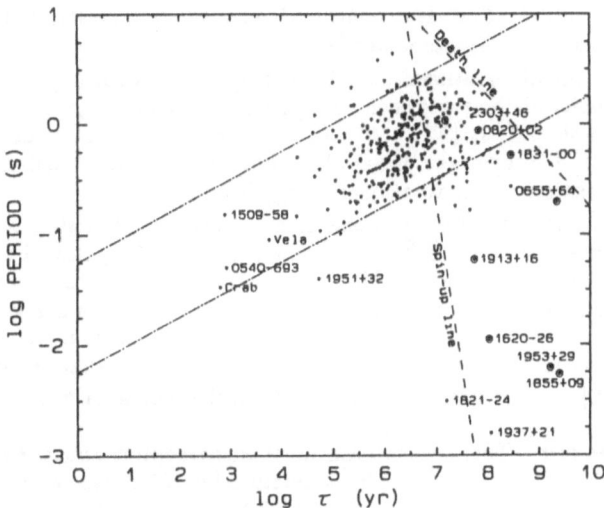

FIG. 14. The period-age diagram of the observed pulsars.

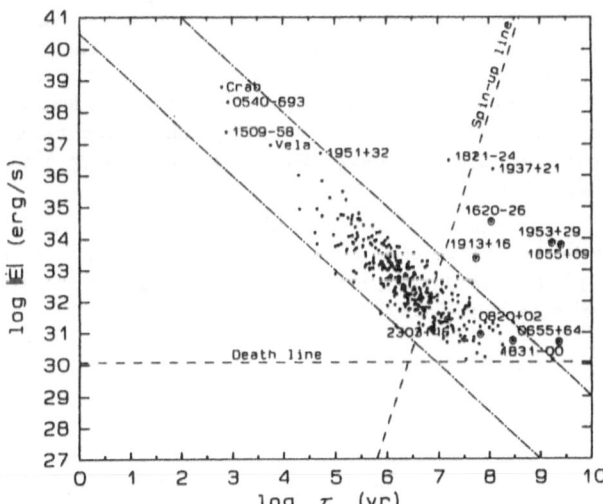

FIG. 15. Pulsar rotational energy loss rate $|\dot{E}|$ as a function of the age of the pulsar τ.

pulsars with $\tau \leq 10^7\,yr$) is -1.50. It is remarkable that the relationship $|\dot{E}| \propto \tau^{-3/2}$ holds for the range of 9 orders of magnitude of $|\dot{E}|$.

The scarcity of the observational data for young pulsars with $\tau \leq 10^5$ yr reflects the fact that the probability of observing such young pulsars is small. Figure 15 shows a natural limit for the observable pulsars $|\dot{E}| \geq 10^{30}\,erg\,s^{-1}$. This limit exactly corresponds to the pulsar deathline (48). Therefore the pulsar deathline can be reinterpreted as the line which corresponds to the critical pulsar rotational energy loss rate.

If the hypothesis that the pulsar magnetic fields decay with a time scale $\tau \sim 10^7\,yr$ (21,23) is correct, one should see a conspicuous change in the rotational energy loss corresponding to $\tau \sim 10^7\,yr$. However, this appears not to be the case in view of Figure 15. Rather, Figure 15 shows that the empirical law $|\dot{E}| \propto \tau^{-3/2}$ holds generally for the whole lifetime of the pulsar. If the pulsar magnetic field stays constant, the canonical magnetic dipole radiation model (34) predicts $|\dot{E}| \propto \tau^{-2}$. The fact that the observed result is better fitted by $|\dot{E}| \propto \tau^{-3/2}$ implies a deviation from the constancy of the magnetic field.

3.2.7 *Randomization test*

These exists traditional misconception about the genuineness of the result of the pulsar data analysis on $B - \tau$ relationship (21). Here we will point out the cause of this misconception, and demonstrate that the results of the $B - \tau$, $P - \tau$, $|\dot{E}| - \tau$ analysis are in fact statistically genuine.

The cause of the traditional misconception is easily understood from Figure 11. In this diagram it is rather difficult to find out pulsar evolutionary paths. We point out that this difficulty is caused by the existence of the pulsar deathline (48). The slope of the pulsar deathline has the same absolute value and the opposite sign as that of the line $\dot{P}P^3 = const.$ which we consider as the pulsar evolutionary path. This coincidence gives a special symmetry to the pulsar distribution in the $P - \dot{P}$ diagram. Thus it is important to recognize that the pulsar distribution possesses an inherent symmetry. In other words this symmetry is the very product of the pulsar evolution. We also note that random interchange of the combinations of the pulsar $P - \dot{P}$ data does not lead to a significantly different distribution because of this inherent symmetry. This is the reason why one often gets a similar distribution of the pulsar data after random interchange of the combinations of the pulsar $P - \dot{P}$ data (21).

Although the $P - \dot{P}$ diagram is most frequently used in order to analyze the pulsar evolution, the use of the $\Omega - |\dot{\Omega}|$ diagram is in some cases more suited for this purpose. In Figure 16 we show the $\Omega - |\dot{\Omega}|$ diagram of the observed pulsars. It is interesting to note that the observational data is best fitted by the relationship

$$|\dot{\Omega}| \propto \Omega^{5.0}, \tag{3.13}$$

except the binary radio pulsars and the millisecond pulsars. To be more precise, we remark that the actual best-fit has been obtained as

$$\Omega \propto |\dot{\Omega}|^{0.20}, \tag{3.14}$$

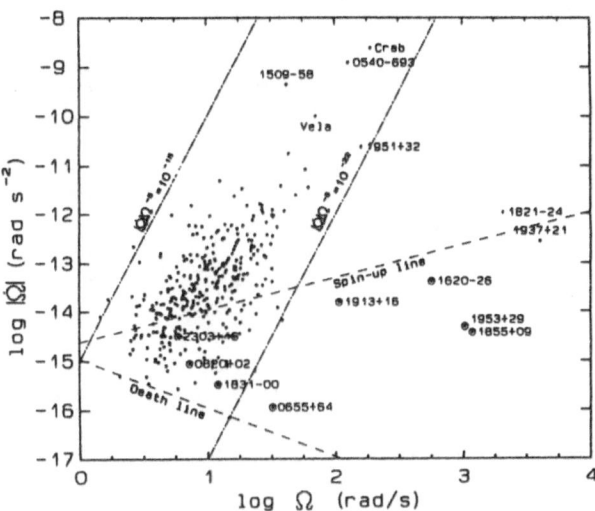

FIG. 16. Angular frequencies and their time derivatives of the observed pulsars.

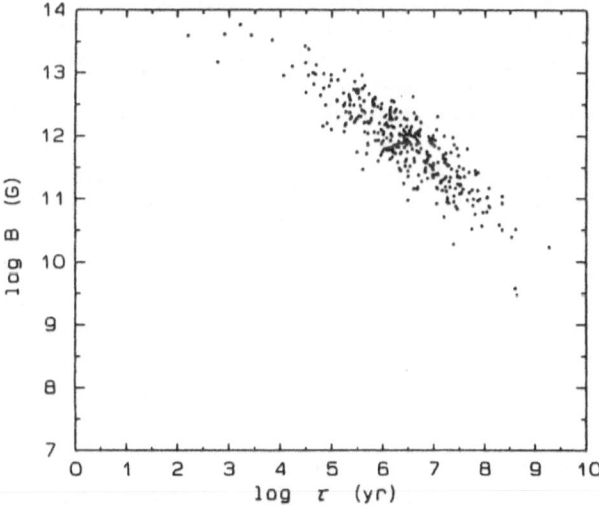

FIG. 17. The magnetic field-age diagram constructed from the randomized set $(\Omega_i, |\dot{\Omega}_j|)$ of the pulsars. Binary radio pulsars and millisecond pulsars are excluded.

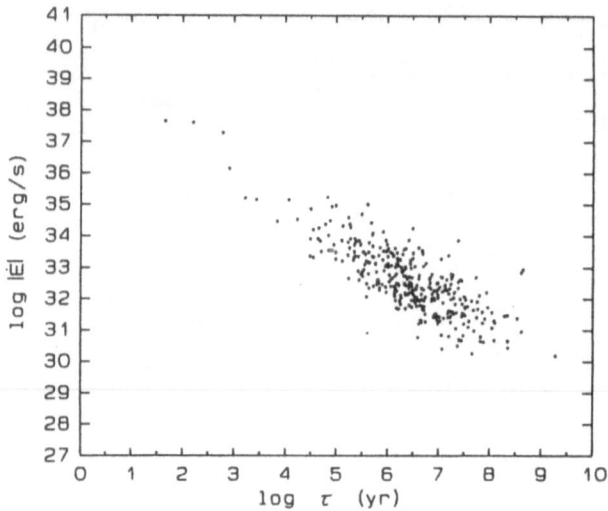

FIG. 18. The rotational energy loss rate-age diagram constructed from the randomized set $(\Omega_i, |\dot{\Omega}_j|)$ of the pulsars. Binary radio pulsars and millisecond pulsars are excluded.

thereby leading to the relationship (3.13). In order to make a least-squares analysis one should be cautious about dealing with the pulsar deathline. Otherwise the existence of the deathline produces a spurious correlation. One must choose the vertical axis for the least-squares analysis to be almost parallel to the deathline. It is worth noting that the observed global braking index $n = 5.0$ agrees with the prediction of the radiation damping model.

In the above we have explained the reason why one obtains similar $B - \tau$ and $|\dot{E}| - \tau$ distributions after the randomization of the combinations of the pulsar $(P, \dot{P})$ data. In order to demonstrate that the correlations observed in the $B - \tau$ and $|\dot{E}| - \tau$ data of the pulsars are statistically genuine, we take yet another way of proof. Let us now consider that B, $|\dot{E}|$, and τ are functions of Ω and $|\dot{\Omega}|$ rather than P and $\dot{P}$. The genuine $B - \tau$ and $|\dot{E}| - \tau$ distributions are of course the same as Figures 13, 15. Then we make randomization of the combinations of the $(\Omega, |\dot{\Omega}|)$ data except the binary radio pulsars and the millisecond pulsars, and obtain a randomized set of $(\Omega_i, |\dot{\Omega}_j|)$ data. In Figures 17, 18 we show the $B - \tau$ and $|\dot{E}| - \tau$ distributions constructed from this randomized set of $(\Omega_i, |\dot{\Omega}_j|)$ data.

As is readily seen the randomized $B - \tau$ and $|\dot{E}| - \tau$ distributions are totally different from the genuine $B - \tau$ and $|\dot{E}| - \tau$ distributions. The best-fit values for the slopes of the genuine $B - \tau$ and $|\dot{E}| - \tau$ distributions for the pulsars with $\tau \leq 10^7 \, yr$ are -0.25 and -1.50, respectively, whereas those for the randomized set are -0.496 and -1.01, respectively. The latter two values can be readily understood if one notes that the randomized $(\Omega, |\dot{\Omega}|)$ data with a much wider range of distribution of $|\dot{\Omega}|$ than that of Ω gives the slopes for $B - \tau$

and $|\dot{E}| - \tau$ distributions the values -0.5 and -1.0, respectively.

3.3 Binary radio pulsars

Kulkarni (24) has identified the companion of PSR 0655+64 to be a cool white dwarf whose cooling age is $\sim 2 \times 10^9\, yr$. He has noted that in any evolutionary scenario, the white dwarf is formed after the primary, thus implying that the age of PSR 0655+64 is greater than $\sim 2 \times 10^9\, yr$. He has further remarked that this fact is in conflict with the conventional scenario that the pulsar magnetic field decays exponentially with a time scale $\sim 10^7\, yr$.

More recently Wright and Loh (25) have identified the companion of PSR 1855+09 to be a $0.3\, M\odot$ white dwarf with a surface temperature of $\sim 5,900\, K$. They have given the age of the pulsar deduced from the characteristic cooling time for the companion white dwarf $2 \times 10^9\, yr$.

TABLE 1 $P, \dot{P}$, AND PREDICTED AGES OF BINARY AND MILLISECOND PULSARS

PSR	P(ms)	$log\dot{P}\,(ss^{-1})$	Predicted age $\tau = \frac{1}{4}\frac{P}{\dot{P}}(yr)$	Observed age(yr)
1937+21	1.6	-19.0	1.3×10^8	
1821 −24	3.05	-17.8	1.5×10^7	
1855+09	5.4	-19.7	2.1×10^9	2×10^9
1953+29	6.1	-19.5	1.6×10^9	
0655+64	195.6	-18.2	2.5×10^9	2×10^9
1913+16	59.0	-17.1	$6 \ \times 10^7$	
1831 −00	520.9	< -17.0	$> 4 \ \times 10^8$	
0820+02	864.9	-16.0	$7 \ \times 10^7$	
2303+46	1066.4	-15.4	2.1×10^7	

According to the present theory the age of the pulsar is $P/(4\dot{P})$, which are $2.5 \times 10^9\, yr$ and $2.1 \times 10^9\, yr$ for PSR 0655+64 and PSR 1855+09, respectively. Thus the theoretical ages are in excellent agreement with the ages deduced from the cooling time of the companion white dwarfs. In Table 1 we show the predicted ages of the binary and millisecond pulsars. It is of great interest to know if this prediction is substantiated by the observations of the companion stars or not. The case of PSR 1953+29 is particularly interesting since Taam and van den Heuvel (50) predict that its age is less than $\sim 10^8\, yr$.

3.4 Alignment of the pulsar magnetic axis

Recently Lyne and Manchester (51) have analyzed the slope of pulsar radio beams, and have concluded that for young pulsars the distribution of the inclination angle of the magnetic axis is uniform whereas old pulsars have preferentially aligned fields. Here we make independent tests of the alignment hypothesis. In Figure 19 we show the relationship between the perpendicular component of the pulsar magnetic field derived from the $(P, \dot{P})$ data and the inclination angle. If the decay of the magnetic field is due to the alignment of the magnetic axis with the rotation axis, then one should observe a relationship

$$B = B_0 \sin \alpha, \tag{3.15}$$

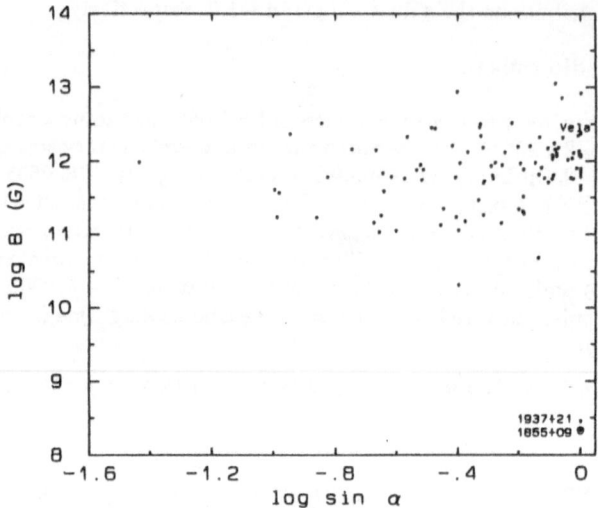

FIG. 19. The magnetic field-sin α relationship of the observed pulsars.

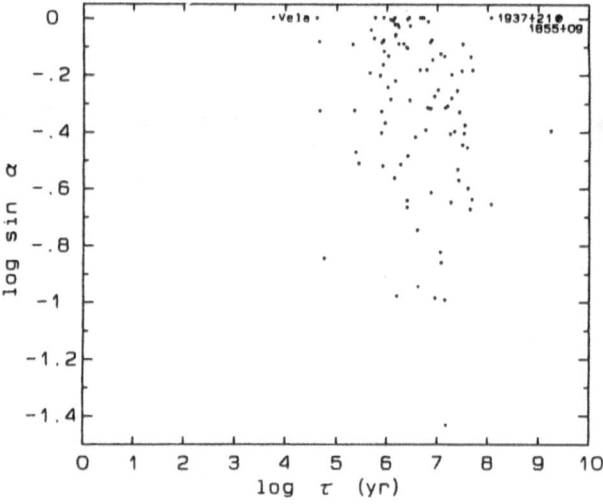

FIG. 20. The relationship between sin α and the age of the observed pulsars.

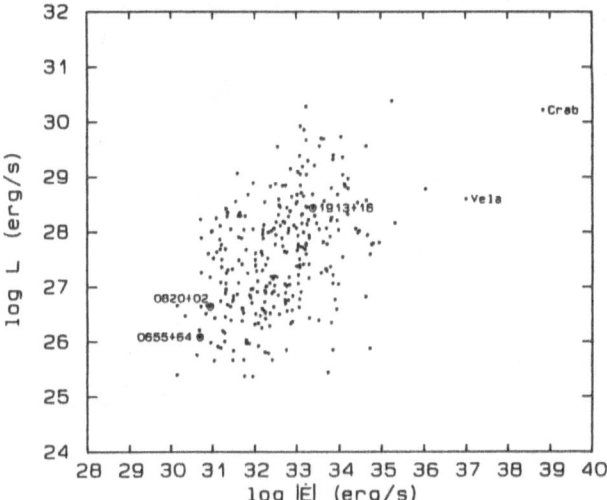

FIG. 21. The radio luminosity-rotational energy loss rate relationship of the observed pulsars.

where B is the absolute value of the magnetic field and considered to be more or less constant. However, one does not sees a strong correlation $B \propto \sin \alpha$ in Figure 19.

In Figure 20 we show the relationship between $\sin \alpha$ and the pulsar age τ. Again one does not see a strong correlation between $\sin \alpha$ and τ. This is in clear contrast with Figure 13 in which one sees a conspicuous correlation $B \propto \tau^{-1/4}$. Therefore, we conclude that it is still premature to state that the decay of the pulsar magnetic field is due to the alignment of the pulsar magnetic axis with the rotation axis.

3.5 Radio luminosity-rotational energy loss rate relationship

In Figure 21 we show the relationship between the radio luminosity at 400MHz and the rotational energy loss rate of the observed pulsars. Apart from the Crab pulsar and the Vela pulsar one does not see strong correlation between the two quantities.

4. CONCLUDING REMARKS

In this article we have reported on two subjects: transport processes in dense stars and the evolution of the pulsar magnetic fields. The two subjects are closely related to each other. The problem of the evolution of the pulsar magnetic fields has been one of the most challenging problems in stellar astrophysics. The magnetic field is the most important physical quantity of the pulsars. Yet its origin and evolution are not satisfactorily understood. The model outlined in this paper is expected to serve as a good working hypothesis.

ACKNOWLEDGMENT

I thank N.Sato for his invaluable assistance in the preparation of this paper.

APPENDIX

In the main body of the present paper we have neglected the effect of the gravitational radiation. Here we give the justification for that.

The energy loss due to gravitational radiation is given by (29)

$$\frac{dE}{dt} = -\frac{1}{45} GD_{\perp}^2 \frac{\Omega^6}{c^5}, \tag{A1}$$

where G is the gravitational constant, and D is the component of the mass-quadrupole moment perpendicular to the rotation axis. The perpendicular component of the mass-quadrupole of a ragid, slightly aspherical body is given by (19)

$$D_{\perp}^2 = 288 \, \epsilon_e^2 \, I^2, \tag{A2}$$

where ϵ_e is the ellipticity in the equatorial plane. Let us suppose that the pulsar magnetic moment is zero, and that the energy loss is due only to gravitational radiation. Then one can show that pulsars evolve along the line

$$\dot{P}P^3 = \frac{16\pi^4 G}{45c^5} 288 \, \epsilon_e^2 I. \tag{A3}$$

Taking $I = 1.4 \times 10^{45} \, gcm^2$, one obtains

$$\dot{P}P^3 = 4 \times 10^{-11} \epsilon_e^2, \tag{A4}$$

where $\dot{P}$ is in units of ss^{-1}, and P is in units of s. For the millisecond pulsar PSR 1937+21 one knows $P = 1.6 \, ms$, $\dot{P} = 1.0 \times 10^{-19} \, ss^{-1}$ from the observation. Thus in this case one has

$$\dot{P}P^3 = 4 \times 10^{-28}. \tag{A5}$$

Therefore the ellipticity of PSR 1937+21 is evaluated to be

$$\epsilon_e < 3 \times 10^{-9}. \tag{A6}$$

Let us next evaluate the ratio of the gravitational radiation loss to the magnetic dipole radiation loss. This is give by

$$\frac{L_g}{L_m} = \frac{48}{5} \frac{G}{c^2} \epsilon_e^2 (\frac{\Omega_0}{m_0})^2 I^2 . \tag{A7}$$

Thus for $\Omega_0 = 2\pi \times 10^3 \, s^{-1}$, $m_0 = 10^{32} G \, cm^3$, $I = 1.4 \times 10^{45} \, gcm^2$, one obtains

$$\frac{L_g}{L_m} = 6 \times 10^6 \, \epsilon_e^2 . \tag{A8}$$

Therefore if $\epsilon_e \ll 4 \times 10^{-4}$, the gravitational radiation loss is negligible compared with the magnetic dipole radiation loss. Although it is not easy to rigorously verify that this condition holds for all pulsars, the case of PSR 1937+21 can be considered as strong evidence in support of the idea that this condition is in reality valid.

REFERENCES

 (1) E.Flowers, and N.Itoh: Ap.J.,**206**,218,(1976).
 (2) E.Flowers, and N.Itoh: Ap.J.,**230**,847,(1979).
 (3) E.Flowers, and N.Itoh: Ap.J.,**250**,750,(1981).
 (4) D.G.Yakovlev, and V.A.Urpin: Soviet Astr.,**24**,303,(1980).
 (5) M.E.Raikh, and D.G.Yakovlev: Ap.Space Sci.,**87**,193,(1982).
 (6) N.Itoh, S.Mitake., H.Iyetomi, and S.Ichimaru: Ap.J.,**273**,774,(1983).
 (7) S.Mitake, S.Ichimaru, and N.Itoh: Ap.J.,**277**,375,(1984).
 (8) N.Itoh, Y.Kohyama, N.Matsumoto, and M.Seki: Ap.J.,**285**,758,(1984).
 (9) R.Nandkumar, and C.J.Pethick: M.N.R.A.S.,**209**,511,(1984).
(10) N.Itoh, Y.Kohyama, and H.Takeuchi: Ap.J.,**317**,733,(1987).
(11) S.Ichimaru: Rev.Mod.Phys.,**54**,1017,(1982).
(12) W.L.Slattery, G.D.Doolen, and H.E.DeWitt: 1982, Phys.Rev.,**A26**,2255,(1982).
(13) J.Ziman: Phil.Mag.,**6**,1013,(1961).
(14) B.Jancovici: Nuovo Cimento,**25**,428,(1962).
(15) H.Iyetomi, and S.Ichimaru: Phys.Rev.,**A25**,2434,(1982).
(16) E.L.Pollock, and J.P.Hansen: Phys.Rev.,**A8**,3110,(1973).
(17) R.A.Coldwell-Horsfall, and A.A.Maradudin: J.Math.Phys.,**1**,395,(1960).
(18) J.P.Ostriker, and J.E.Gunn: Nature,**223**,813,(1969).
(19) J.P.Ostriker, and J.E.Gunn: Ap.J.,**157**,1395,(1969).
(20) J.E.Gunn, and J.P.Ostriker: Ap.J.,**160**,979,(1970).
(21) A.G.Lyne, R.T.Ritchings, and F.G.Smith: M.N.R.A.S.,**171**,579,(1975).
(22) A.G.Lyne, B.Anderson, and M.J.Salter: M.N.R.A.S.,**201**,503,(1982).
(23) A.G.Lyne, R.N.Manchester, and J.H.Taylor: M.N.R.A.S.,**213**,613,(1985).
(24) S.R.Kulkarni: Ap.J.(Letters),**306**,L85,(1986).
(25) G.A.Wright, and E.D.Loh: Nature,**324**,127,(1986).
(26) G.Baym, C.Pethick, and D.Pines: Nature,**224**,674,(1969).
(27) N.Itoh: Proc. Conference "The Physics of Neutron Stars and Black Holes", ed. Y.Tanaka,(Tokyo, Universal Academy Press, 1988),p.487;

N.Itoh: Proc. 20th Yamada Conference "Big Bang,Active Galactic Nuclei and Supernovae", ed. S.Hayakawa, and K.Sato,(Tokyo, Universal Academy Press, 1989),p.587.

(28) N.Itoh: Nuovo Cimento, in press (1989).

(29) L.D.Landau, and E.M.Lifshitz: *The Classical Theory of Fields* (Pergamon Press: Oxford), (1983).

(30) A.Bohr, and B.R.Mottelson: *Nuclear Structure*, vol.I (Benjamin: New York), (1969).

(31) E.J.Groth: Ap.J.Suppl.,**29**,431,(1975).

(32) R.N.Manchester, J.M.Durdin, and L.M.Newton: Nature,**313**,374,(1985).

(33) J.H.Taylor, and R.N.Manchester: A.J.,**80**,794,(1975).

(34) R.N.Manchester, and J.H.Taylor: *Pulsars* (Freeman: San Francisco),(1977).

(35) G.E.Gullahorn, and J.M.Rankin: A.J.,**83**,1219,(1978).

(36) D.J.Helfand, J.H.Taylor, P.R.Backus, and J.M.Cordes: Ap.J.,**237**,206,(1980).

(37) L.M.Newton, R,N,Manchester, and D.J.Cooke: M.N.R.A.S.,**194**,841,(1981).

(38) M.Ashworth, and A.G.Lyne: M.N.R.A.S.,**195**,517,(1981).

(39) R.N.Manchester, and J.H.Taylor: A.J.,**86**,1953,(1981).

(40) P.R.Backus,J.H.Taylor, and M.Damashek: Ap.J.(Letters),**255**,L63,(1982).

(41) F.D.Seward, F.R.Harnden,Jr., and D.J.Helfand: Ap.J.(Letters),**287**,L19,(1984).

(42) S.R.Kulkarni, T.R.Clifton, D.C.Backer, R.S.Foster, A.S.Fruchter, and J.H.Taylor: Nature,**331**,50,(1988).

(43) A.S.Fruchter, J.H.Taylor, D.C.Backer, T.R.Clifton, R.S.Foster, and A.Wolszczan: Nature,**331**,53,(1987).

(44) J.H.Taylor: Proc. IAU Symposium 125, The Origin and Evolution of Neutron Stars, ed. D.J.Helfand, and J.-H.Huang (Reidel: Dordrecht), p.383, (1987).

(45) A.G.Lyne, A.Brinkow, J.Middleditch, S.R.Kulkarni, D.C.Backer, and T.R.Clifton: Nature,**328**,399,(1987).

(46) R.S.Foster, D.C.Backer, J.H.Taylor, and W.M.Goss: Ap.J.(Letters),**326**, L13,(1988).

(47) A.G.Lyne: Proc. IAU Symposium 125, The Origin and Evolution of Neutron Stars, ed. D.J.Helfand, and J.-H.Huang (Reidel: Dordrecht),p.23,(1987).

(48) M.A.Ruderman, and P.G.Sutherland: Ap.J.,**196**,51,(1976).

(49) E.P.J.van den Heuvel: Proc. IAU Symposium 125, The Origin and Evolution of Neutron Stars, ed. D.J.Helfand, and J.-H.Huang (Reidel: Dordrecht),p.393,(1987).

(50) R.E.Taam, and E.P.J.van den Heuvel: Ap.J.,**305**,235,(1986).

(51) A.G.Lyne, and R.N.Manchester: M.N.R.A.S.,**234**,477,(1988).

MATHEMATICAL PROBLEMS IN DYNAMO THEORY

D. Lortz

Max-Planck-Institut für Plasmaphysik, IPP-EURATOM Association,

D-8046 Garching bei München

Abstract

The kinematic, stationary dynamo problem is reformulated as the "inverse dynamo problem". The inverse problem is useful for constructing explicit solutions. This is done both for helical symmetry with isotropic diffusivity and for axisymmetry with anisotropic diffusivity.

Introduction

The simplest self-sustained dynamo configuration is the disc dynamo :

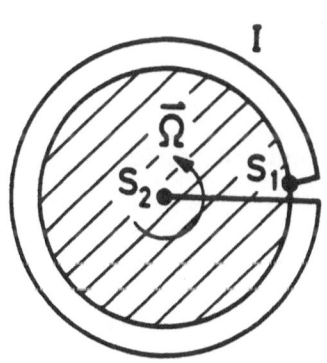

A circular metal disc (hatched) rotates around its centre with angular velocity $\vec{\Omega}$. At the point S_1 of the periphery there is a sliding contact with a wire which is turned around the disc in the same direction as $\vec{\Omega}$. After one revolution the wire is led to a second sliding contact S_2 at the centre of the disc. If $\vec{\Omega}$ is large enough, then the system acts as a self-generating dynamo and produces a dipol-like magnetic field.

The main difference to a fluid dynamo is that the conducting region of the disc dynamo is doubly connected, whereas that of a fluid dynamo is supposed to be singly connected. If one were to try to embed the doubly-connected configuration in a singly-connected conductor, the current density distribution would change because there is an electric field outside the disc dynamo conductor. As a result, the disc dynamo configuration would not be self-sustained.

221

W. Brinkmann et al. (eds.), Physical Processes in Hot Cosmic Plasmas, 221–234.
© 1990 *Kluwer Academic Publishers.*

The kinematic dynamo problem is described by the induction equation of resistive MHD:

$$(1) \qquad \frac{\partial \vec{B}}{\partial t} = \nabla \times (\vec{v} \times \vec{B} - \eta \nabla \times \vec{B}), \quad \eta = \frac{1}{\sigma \mu},$$

where $\vec{v}$ is the (non-relativistic) flow velocity and $\vec{B}$ the magnetic field. Usually, the kinematic dynamo problem is formulated as follows: What velocity field $\vec{v}$ satisfying only the conservation equation of mass leads to a solution of eq. (1) which is non-decaying in time and is continuous at the surface of the conductor with a magnetic vacuum field which vanishes at infinity? For such a velocity field the state without a magnetic field becomes unstable.

For the time-independent case eq. (1) reduces to

$$(2) \qquad \vec{J} + \nabla U - \vec{v} \times \vec{B} = 0, \quad \vec{J} = \eta \nabla \times \vec{B},$$

$$\nabla \cdot \vec{B} = 0$$

in the conductor, where the electric potential U should be single-valued in space in order that the configuration be self-sustained. Outside the conductor there is a vacuum governed by the equations

$$(3) \qquad \nabla \times \vec{B} = \nabla \cdot \vec{B} = \Delta U = 0.$$

If the vacuum and fluid are separated by a discontinuity surface S, then the fields have to satisfy

$$(4) \qquad \vec{B}, U \text{ continuous across } S,$$

while the magnetic and electric fields should tend to zero for large distances from the conductor:

$$(5) \qquad \vec{B}, \nabla U \to 0, \vec{r} \to \infty.$$

If the magnetic diffusivity η is a scalar, then Cowling's theorem [1] states that eqs. (2)-(5) do not admit axisymmetric poloidal solutions. This result was generalized to the poloidal and toroidal axisymmetric case in [2]. Further generalization was achieved in [3], where it was shown for a time-dependent solenoidal flow that all axisymmetric magnetic fields decay in time. The first non-existence proof for a non-solenoidal flow was given in [4] for the stationary case. The magnetic field of the earth and of Saturn are very close to axisymmetry, however. It was speculated in [5] that compressibility and time dependence together could allow axisymmetric fields to grow. In [6] − [11] it was shown that this is impossible. So there are no dynamo solutions in simple geometry.

The inverse dynamo problem

The usual procedure to solve eqs. (2)-(5) is to prescribe $\vec{v}$ in a normalized way and solve for $\vec{B}$ and U, considering the amplitude of the velocity field , say R_c, as an eigenvalue of a non-selfadjoint eigenvalue problem. The kinematic dynamo problem can then be formulated as follows: Does there exist a real eigenvalue R_c? Sometimes, however, it is easier to solve the inverse problem [12]. This consists in prescribing a sufficiently smooth solenoidal magnetic field $\vec{B}$ with the property that $\vec{B}$ is a spatially decaying vacuum field outside the conducting fluid and in asking what additional constraints on $\vec{B}$ are necessary so that eq. (2) admits meaningful single-valued solutions U and $\vec{v}$. This procedure has a practical and a formal advantage. In the case of the planets the magnetic field outside the planet, unlike the velocity field, can be measured direct. The formal advantage is that eq. (2) is algebraic in $\vec{v}$, but not in $\vec{B}$.

A consequence of the latter is that $\vec{v}$ can be eliminated:

(6) $$\vec{B} \cdot \nabla U = -\vec{B} \cdot \vec{J}.$$

Equation (6) is a so-called magnetic differential equation [13], [14] for U. The existence and types of solutions U depend on the topology of the magnetic lines of force. Let us assume that the lines of force lie on surfaces, the so-called magnetic surfaces, which are topologically nested toroids. This means that the case of chaotic field lines is excluded. Let V be the volume inside such a toroid. One then has

$$\vec{B} \cdot \nabla V = 0.$$

The quantity V can serve as coordinate. In addition, two angle-like coordinates θ and ς are introduced. Lines of constant θ close on themselves after one trip the long way around the toroid. If one goes around the toroid the short way, θ increases by unity. Lines of constant ς close on themselves after one trip the short way around the toroid and ς increases by unity on one trip the long way around. Then the following surface functions can be defined [13]:

$$V = \int_V d^3\tau$$

is the volume contained in the toroid;

$$\Phi(V) = \int_V \vec{B} \cdot \nabla\varsigma\, d^3\tau$$

is the longitudinal flux inside the surface $V = const$;

$$\chi(V) = \int_V \vec{B} \cdot \nabla\theta\, d^3\tau$$

is the flux through a ribbon whose one side is the magnetic axis (defined by $V = 0$) and whose other lies on the surface and does not encircle the axis the short way. The rotational transform ι is defined by

$$\iota(V) = \frac{\dot{\chi}}{\dot{\Phi}} = \frac{d\chi}{d\Phi},$$

where the dot denotes the derivative with respect to V. Let us further assume that the inverse Jacobian

$$D = (\nabla\theta \times \nabla\varsigma) \cdot \nabla V$$

is everywhere positive, so that θ, ς, V is a right-handed coordinate system. The coordinates θ, ς can be chosen such that [15]

$$\vec{B} = \dot{\chi}\nabla\varsigma \times \nabla V + \dot{\Phi}\nabla V \times \nabla\theta$$

$$= D(\dot{\chi}\frac{\partial\vec{r}}{\partial\theta} + \dot{\Phi}\frac{\partial\vec{r}}{\partial\varsigma}),$$

where $\vec{r} = (x, y, z)$ is the position vector in Cartesian coordinates. Thus, the contravariant components of the $\vec{B}$-field in the natural coordinate system are

$$B^\theta = \vec{B} \cdot \nabla\theta = D\dot{\chi},$$

$$B^\varsigma = \vec{B} \cdot \nabla\varsigma = D\dot{\Phi},$$

$$B^V = \vec{B} \cdot \nabla V = 0.$$

The l.h.s of eq. (6) yields

(7)
$$\vec{B} \cdot \nabla U = D(\frac{\partial U}{\partial\theta}\dot{\chi} + \frac{\partial U}{\partial\varsigma}\dot{\Phi}).$$

The volume element can be written as

$$d^3\tau = D^{-1}d\theta d\varsigma dV = d^2 S|\nabla V|^{-1}dV,$$

where

$$d^2 S = |d^{\vec{2}}S|$$

and $d^{\vec{2}}S$ is the surface element on a magnetic surface, which, expressed in terms of θ, ς, is then

$$d^{\vec{2}}S = D^{-1}d\theta d\varsigma \nabla V.$$

The form (7) of the l.h.s. of eq. (6) shows that

(8)
$$\int\int \vec{B} \cdot \vec{J}\frac{d^2 S}{|\nabla V|} = o$$

is a necessary condition for eq. (6) to have a single-valued solution U. Condition (8), however, is not sufficient. Newcomb [14] has given a more stringent condition to be satisfied by the right-hand side of eq. (6) which is necessary and sufficient for the existence of a single-valued solution U. This condition reads

(9)
$$\oint \frac{\vec{B}}{B} \cdot \vec{J} \, dl = 0$$

for every closed line of force. Condition (9) is, in general, complicated, because if the configuration has shear ($i \neq 0$), one has to distinguish between "rational "surfaces, on which all lines of force are closed (ι rational) , and "irrational "surfaces, on which the lines of force are ergodic (ι irrational).

Once eq. (6) is satisfied, eq. (2) for $\vec{v}$ can be solved:

$$(10) \qquad \vec{v} = \alpha \vec{B} + \vec{v}_\perp, \quad \vec{v}_\perp = \frac{\vec{B}}{B^2} \times (\vec{J} + \nabla U).$$

The free function α in expression (10) is determined by satisfying an equation of continuity. Let us consider a solenoidal flow

$$(11) \qquad \nabla \cdot \vec{v} = 0.$$

One then has

$$(12) \qquad \vec{B} \cdot \nabla \alpha = -\nabla \cdot \vec{v}_\perp.$$

Equation (12) is again a magnetic differential equation. The necessary and sufficient condition that α be single-valued is thus

$$(13) \qquad \oint \nabla \cdot \vec{v}_\perp \frac{dl}{B} = 0$$

for every closed line of force, whereas the surface condition

$$(14) \qquad \int \int \nabla \cdot \vec{v}_\perp \frac{d^2 S}{|\nabla V|} = 0$$

is only necessary.

Finally, it can be stated that a magnetic field which is a vacuum field outside the conductor, has nested magnetic surfaces everywhere, and satisfies conditions (9) and (13) is a dynamo field because the single-valued scalar field U and vector field $\vec{v}$ can be computed from it.

Condition (9) raises the question whether it is possible to find a field such that not only the integral (9) but even the integrand of eq. (9) vanishes:

$$(15) \qquad \vec{B} \cdot \vec{J} = 0$$

If η is a scalar then eq. (15) implies that the $\vec{B}$-field possesses orthogonal surfaces $\gamma = const$:

(16) $$\vec{B} = \psi \nabla \gamma.$$

If there is such a field, it would admit the solution $U \equiv 0$, which might be called a "short-circuited dynamo". In [16] it was claimed that this is impossible. Unfortunately, it was assumed that the function γ is single-valued, which is not true in general [17]. In [12] the impossibility proof was given for the case of helical symmetry. However, for the general geometry the problem of the existence of a solution with $U \equiv 0$ is still unsolved.

Helical symmetry

Let s, ϕ, z be cylindrical coordinates. Helical scalars only depend on s and u where

$$u = \phi + kz$$

and k is a constant describing the pitch of the helices. The helically symmetric field is of the form

$$\vec{B} = f\vec{w} + \vec{w} \times \nabla F = \begin{pmatrix} -F_u/s \\ q(F_s - ksf) \\ q(ksF_s + f) \end{pmatrix}, \quad f(u, s), \quad F(u, s),$$

$$\vec{w} = \nabla s \times \nabla squ = q \begin{pmatrix} 0 \\ -ks \\ 1 \end{pmatrix}, \quad q = (1 + k^2 s^2)^{-1}, \quad w^2 = q,$$

which yields

$$\nabla \times \vec{B} = h\vec{w} - \vec{w} \times \nabla f,$$

$$h = q^{-1} LF - 2kqf,$$

$$L = \frac{1}{s} \frac{\partial}{\partial s} sq \frac{\partial}{\partial s} + \frac{1}{s^2} \frac{\partial^2}{\partial u^2}.$$

If the continuity equation (11) is satisfied by the respective ansatz

$$\vec{v} = g\vec{w} + \vec{w} \times \nabla G,$$

then the vector equation (2) leads to the three component equations [18],

$$(17) \qquad \eta sh + F_s G_u - F_u G_s = 0,$$

$$(18) \qquad \eta sqf_s + q(fG_u - gF_u) - U_u = 0,$$

$$(19) \qquad \eta \frac{1}{s} f_u - q(fG_s - gF_s) + U_s = 0.$$

Combination of eqs. (17)-(19) gives

$$(20) \qquad \eta sq(fh - \nabla F \cdot \nabla f) + F_s U_u - F_u U_s = 0,$$

$$(21) \qquad \frac{\eta}{s}(F_s f_u - f_s F_u) + \nabla U \cdot \nabla F - qf \nabla G \cdot \nabla F + qg|\nabla F|^2 = 0.$$

Equations (17), (20), (21) in the fluid can be considered as three equations for the five unknown scalar functions F, f, G, g, U. Outside the conducting fluid the equations are

$$(22) \qquad LF = 2kq^2 f,$$

$$(23) \qquad f = const, \ G = g = \Delta U = 0,$$

and on the discontinuity surface the conditions

$$(24) \qquad F, \nabla F, f, G, U \quad \text{continuous on } S$$

have to be satisfied.

Rather than prescribe the scalars G, g of the velocity field and solving for F, f, U it is easier to solve the inverse problem [18] by prescribing F, f. This inverse problem can here be formulated as follows: Prescribe F such that it is zero for $s \to \infty$, solves equation (22) in the external region, and has exactly one minimum in the conducting region such that the curves $F = const$ in the plane

$z = const$ are non-intersecting. This can be done by, for instance, choosing for the external region the field of a straight wire $(F = C \ln s, \; f = -C \, k)$ at $s = 0$ and continuing F into the interior so that it has exactly one minimum. Because of

$$\frac{1}{s}(F_s G_u - F_u G_s) = (\nabla z \times \nabla F) \cdot \nabla G = |\nabla z \times \nabla F|\frac{dG}{dl'} = \vec{B} \cdot \nabla G = B\frac{dG}{dl},$$

where l' is the arc length along the curve $F = const, z = const$, eq. (17) can be considered as an inhomogeneous first-order ODE for G on the closed curve $F = const, z = const$. The necessary and sufficient condition that G be a single-valued solution of eq. (17) leads to

$$(25) \qquad \oint_{F=const,z=const} \eta h \frac{dl'}{|\nabla z \times \nabla F|} = 0.$$

Analogously, the condition that eq. (20) have a single-valued solution U is

$$(26) \qquad \oint_{F=const,z=const} \eta q(fh - \nabla F \cdot \nabla f)\frac{dl'}{|\nabla z \times \nabla F|} = 0.$$

Satisfying the integral relations (25), (26) is thus necessary and sufficient for the existence of single-valued solutions G, g, U of eqs. (17)-(19).

When condition (9) is compared with condition (26), the open helical geometry is considered as a topological torus where the z-period $2\pi/k$ corresponds to one trip the long way around. This means that the integral (9) is carried out along the field line till the x, y components of the position vector coincide with those of the starting point and z has increased by an integer number, say n, of periods. The line elements dl and dl' are related by

$$\frac{dl}{B} = \frac{dl'}{|\nabla z \times \nabla F|}.$$

Integrals of the form (9) and (26) are proportional:

$$(27) \qquad \oint \cdots \frac{dl'}{|\nabla z \times \nabla F|} = \frac{1}{n} \oint \cdots \frac{dl}{B}.$$

Because of

$$\vec{J} \cdot \vec{B} = \eta q (fh - \nabla F \cdot \nabla f)$$

conditions (9) and (26) are equivalent. The helical symmetry implies that integrals of the form (26) do not depend on z. From

$$\int\int \cdots \frac{d^2 S}{|\nabla V|} = \int\int \cdots \frac{d\varsigma d\mu}{D} = \int_0^{\frac{1}{2}} d\mu \int_0^n \cdots \frac{d\varsigma}{D} = \dot{\Phi} \int_0^{\frac{1}{2}} d\mu \oint \cdots \frac{dl}{B} = \frac{\dot{\Phi}}{n} \oint \cdots \frac{dl}{B},$$

where the variable $\mu = \theta - \iota\varsigma$ is constant on a field line, it thus also follows that condition (9) is equivalent to the surface condition (8).

The comparison of condition (25) with condition (13) is more complicated. Formula (25) suggests that for helical symmetry in condition (14) the electric potential can be eliminated. This is achieved by the following chain of equations:

$$(\vec{B} \times \nabla U) \cdot \nabla F = -[f(\vec{w} \times \nabla F) + (\vec{w} \times \nabla F) \times \nabla F] \cdot \nabla U =$$

$$= -f(\vec{w} \times \nabla F) \cdot \nabla U = -f\vec{B} \cdot \nabla U = \eta q f(fh - \nabla F \cdot \nabla f),$$

$$(\vec{B} \times \vec{J}) \cdot \nabla F = \eta q (f \nabla f \cdot \nabla F + h|\nabla F|^2).$$

Because of

$$B^2 = q(f^2 + |\nabla F|^2)$$

this gives

$$\vec{v}_\perp \cdot \nabla F = \eta h.$$

Then differentiating the relations

$$\int\int \eta h \frac{d^2 S}{|\nabla F|} = \int\int \vec{v}_\perp \cdot \nabla F \frac{d^2 S}{|\nabla F|} = \int\int \vec{v}_\perp \cdot d^2 S =$$

$$= \int\int\int div\, \vec{v}_\perp d^3\tau = \int\int\int div\, \vec{v}_\perp d^2 S \frac{dF}{|\nabla F|}$$

with respect to F finally yields

$$\frac{d}{dF} \int\int \eta h \frac{d^2 S}{|\nabla F|} = \int\int div\, \vec{v}_\perp \frac{d^2 S}{|\nabla F|},$$

which shows the connection between conditions (25) and (14).

Conditions (25) and (26) can be satisfied by [18]

(28) $F(s), \; f = f_0(s) + f_1(s)\cos u$

and then yield simple explicit algebraic solutions whose geophysical relevance has been discussed in [19] − [21]. Moreover, it is also possible with ansatz (29) to make the surface charge vanish on S, which is a necessary condition for embedding the torus in a singly-connected conductor without altering the fields [22]. This has the consequence that the fields $\vec{v}, \nabla U, \nabla \times \vec{B}$ vanish on S.

Axisymmetry and anisotropy

Let ρ, θ, ς be cylindrical coordinates. The fields $\vec{B}, \vec{v}$ are written as

(29) $\vec{B} = \nabla\theta \times \nabla F + f\nabla\theta, \; \vec{v} = \nabla\theta \times \nabla G + g\nabla\theta, \; F, f, G, g(\rho, \theta).$

One then has

$$curl \; \vec{B} = \nabla f \times \nabla\theta + (\Delta_* F)\nabla\theta,$$

where

$$\Delta_* = \frac{\partial^2}{\partial\rho^2} - \frac{1}{\rho}\frac{\partial}{\partial\rho} + \frac{\partial^2}{\partial\varsigma^2} = \rho^2 \nabla \cdot \frac{\nabla}{\rho^2} = \nabla \cdot (\nabla - 2\frac{\nabla\rho}{\rho})$$

is the Stokes operator.

Suppose that the scalar η is replaced by the tensor

$$\eta \begin{pmatrix} 1 & 0 & 0 \\ 0 & 1 & e \\ 0 & e & 1 \end{pmatrix}, \; \eta = const$$

in cylindrical components, then the vector $\eta \; curl \; \vec{B}$ is replaced by the vector

$$\frac{\eta}{\rho} \begin{pmatrix} -f_\varsigma \\ \Delta_* F + f_\rho e \\ e\Delta_* F + f_\rho \end{pmatrix}$$

and the vector equation (2) yields the three component equations

(30) $-\frac{\eta}{\rho}f_\varsigma + U_\rho - \frac{1}{\rho^2}(G_\rho f - F_\rho g) = 0,$

(31)
$$\frac{\eta}{\rho}(\Delta_*F + f_\rho e) - \frac{1}{\rho^2}(G_\varsigma F_\rho - G_\rho F_\varsigma) = 0,$$

(32)
$$\frac{\eta}{\rho}(e\Delta_*F + f_\rho) + U_\varsigma - \frac{1}{\rho^2}(G_\varsigma f - F_\varsigma g) = 0.$$

As in the case of helical symmetry with isotropic η, the existence conditions are [23]

(33)
$$\oint_{F=const} \eta\rho(\Delta_*F + f_\rho e)\frac{dl}{|\nabla F|} = 0,$$

(34)
$$\oint_{F=const} \frac{\eta}{\rho}(\nabla f \cdot \nabla F - f\Delta_*F + eF_\rho\Delta_*F - ef_\rho f)\frac{dl}{|\nabla F|} = 0.$$

Satisfying the integral relations (33), (34) is thus necessary and sufficient for the existence of single-valued solutions G, g, U of eqs. (30)-(32).

That conditions (33), (34) can be satisfied is demonstrated for the large-aspect-ratio case. Let f, Δ_*F, G, g be non-zero only inside a torus with large radius R and circular cross-section with radius r. Then toroidal coordinates s, ϕ, z are introduced by [24]

$$\rho = R + x, \quad x = s \cos\phi, \quad \theta = -R^{-1}z, \quad \varsigma = y, \quad y = s \sin\phi.$$

The interior of the torus is described by

$$0 \le s \le r < R.$$

In the limit of small inverse aspect ratio $\epsilon = r/R$ the axisymmetry reduces to plane symmetry. In order that the fields stay finite in this limit, it is useful to scale the scalars as

$$F = RH, \quad f = -Rh, \quad G = RK, \quad g = -Rk.$$

Let the function H be chosen such that it depends only on s (cylindrically symmetric magnetic surfaces). One then has

(35)
$$\eta(e\Delta H \sin \phi + \frac{1}{s}h_\phi) + U_s + hK_s - kH' = 0,$$

$$(36) \qquad \eta(e\Delta H \cos\phi - h_s) + \frac{1}{s}U_\phi + \frac{1}{s}hK_\phi = 0,$$

$$(37) \qquad \eta\left[\Delta H - e(h_s \cos\phi - \frac{1}{s}f_\phi \sin\phi)\right] - \frac{1}{s}H'K_\phi = 0,$$

where the prime denotes the derivative with respect to s and

$$\Delta H = H'' + \frac{1}{s}H'.$$

For the case that e does not depend on ϕ, the ansatz

$$(38) \qquad \begin{cases} h = h_1(s) \cos\phi, \quad K = K_2(s) \sin 2\phi, \\ \quad k = k_1(s) \sin\phi + k_3(s) \sin 3\phi, \\ \quad U = U_1(s) \sin\phi + U_3(s) \sin 3\phi. \end{cases}$$

reduces the system (35)-(37) of PDE's to a system of ODE's which is algebraic and explicit in the variables K_2, U_1, U_3, k_1, k_3.

These solutions satisfying eqs. (38) have the property that the toroidal flux vanishes, which means that all field lines are closed.

If e depends on s, then it can be shown [23] that the function H can be chosen such that all functions have the right behaviour at $s \to 0$ and $s \to r$.

If the anisotropy becomes small, $e \to 0$, then the externally visible field H stays finite and

$$h \to O(e^{-1}), \quad K \to O(e^0), \quad U \to O(e^{-1}), \quad k \to O(e^{-1}).$$

A question which has not been discussed here is what physical effects could produce such an anisotropy of the diffusivity η. One possibility is inhomogeneities in the temperature distribution [25].

References

[1] Cowling, T. G., Monthly Notices Roy. Astron. Soc. 94, 39 (1934)
[2] Backus, G. and Chandrasekhar, S. Nat. Acad. Sci 42, 105 (1956)
[3] Braginskii, S. I., Sov. Phys. JETP 20, 726 (1964)

[4]   Lortz, D., Phys. Fluids 11, 913 (1968)

[5]   Todoeschuck, J. P. and Rochester, M. G., Nature 284, 250 (1980)

[6]   Hide, R., Nature 293, 728 (1981)

[7]   Lortz, D., and Meyer-Spasche, R., Math. Meth. in the Appl. Sci. 4, 91 (1982)

[8]   Lortz, D. and Meyer-Spasche, R., Z. Naturforsch. 37a, 736 (1982)

[9]   Ivers, D. J. and James R. W., Phil. Trans. Roy. Soc. Lond. A312, 179 (1984)

[10]   Lortz, D., Meyer-Spasche, R., and Stredulinsky, E. W., Comm. Pure Appl. Math. 37, 677 (1984)

[11]   Stredulinsky, E. W., Meyer-Spasche, R., and Lortz, D., Comm. Pure Appl. Math. 39, 233 (1986)

[12]   Lortz, D., Workshop on Mathematical Aspects of Fluid and Plasma Dynamics, Trieste, May 30- June 2 (1984)

[13]   Kruskal, M. D. and Kulsrud, R. M., Phys. Fluids 1, 265 (1958)

[14]   Newcomb, W. A., Phys. Fluids 2, 362 (1959)

[15]   Greene,J. M. and Johnson, J. L., Phys. Fluids 5, 510 (1962)

[16]   Pichakhchi, L. D., Sov. Phys. JETP 23, 542 (1966)

[17]   Moffat, H. K., Magnetic Field Generation in Electrically Conducting Fluids, Cambridge University Press, Cambridge (1978)

[18]   Lortz, D., Plasma Physics 10, 967 (1968)

[19]   Benton, E. R., Geophys. J. R. astr. Soc. 42, 385 (1975)

[20]   Benton, E. R., Geophys. Astrophys. Fluid Dynamics 12, 313 (1979)

[21]   Benton, E. R., Geophys. Astrophys. Fluid Dynamics 12, 345 (1979)

[22]   Lowes, F., private communication

[23]   Lortz, D., Z. Naturforsch. (1989), in print

[24]   Lortz, D., Z. Naturforsch. 27a, 1350 (1972)

[25]   Pfirsch., D., private communication

Generation of Cosmic Magnetic Fields

A. A. RUZMAIKIN
Institute of Terrestrial Magnetism
Ionosphere and Radio Wave Propagation
Troitsk Moscow Region, 142092, USSR

ABSTRACT: The generation of magnetic fields by dynamo action of cosmic turbulent plasmas is considered. Three levels of the consideration are discussed: (1) the mean field behaviour, (2) deviations from the mean field, i.e. magnetic fluctuations, and (3) the generation and distribution of random magnetic field in the random flow of a conducting plasma. Examples of magnetic fields generated in the Uranus and Neptune interiors, in the Sun, and in the clusters of galaxies are presented.

1. Introduction

The idea of cosmic magnetic field generation has its source from the Thames. Soon after the discovery the induction law M. Faraday who knew that an electrical current can pass through a solution of salt in water decided to find the action of extended masses of moving water on the Earth's magnetic field. Having asked for his majesty's permission he realized the experiment on the Waterloo bridge. A copper wire of nine hundred and sixty feet in length was stretched along the railing of the bridge. The ends were attached to large metallic plates in order to have a good contact with water were sunk into the river. Thus the wire and the water made up a conducting circuit. The flow action on a perpendicular to the river surface component of the Earth magnetic field creates the induction e.m.f. (Figure 1). The experiment proceeded over three days and was unlucky due to an imperfection of the registering device used by Faraday. However his concept, used later in 1919 by J. Larmor to explain the origin of the Earth and solar magnetic fields, appeared to be true and profound (two years later the experiment was repeated by W. Wollaston with positive results).

Instead of water a natural component of most celestial bodies is a well conducting moving plasma. A relative action of the plasma hydrodynamical motions and magnetic diffusion on the magnetic field is characterized by the dimensionless magnetic Reynolds number

$$R_m = \frac{\ell v}{\nu_m} = \frac{\ell^2/\nu_m}{\ell/v}$$

where ℓ and v are the spatial and amplitude scale of the velocity field, ν_m is a magnetic diffusivity. The magnetic Reynolds number is equal to the ratio of the rate at which magnetic energy can be generated by Joule dissipation at the scale ℓ. Unfortunately, R_m is

W. Brinkmann et al. (eds.), Physical Processes in Hot Cosmic Plasmas, 235–245.

very small in the Thames (the Mediterranean sea gives more hopes!), however, it is larger by a factor of 10^4 in the fluid planetary core and 10^8 in the solar convective shell.

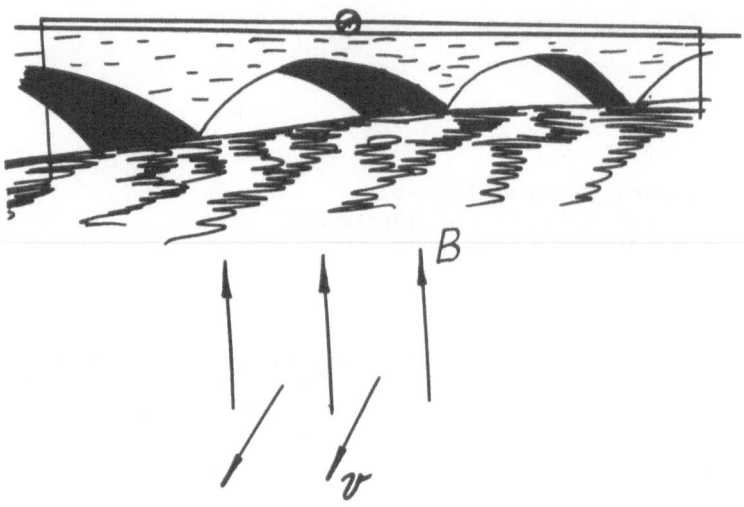

Fig.1: The flow motion, $v \approx 5$ km/sec, on the Earth magnetic field, $B \approx 0.3$ Gauss, induces in the circuit, $L \approx 2 \times 960$ ft, the e.m.f. $\approx vBL \approx 20$ mV.

Parallel with R_m the hydrodynamic Reynolds number is also very large in astrophysical conditions. It means that the plasma motions are turbulent.

The condition $R_m \gg 1$ is necessary however not sufficient for the magnetic field amplification because the small scale motions play the role of a turbulent diffusion for the large scale magnetic field.

Now it is known that the presence of a mean helicity $< v\ rotv >$ is sufficient to generate a mean, large scale magnetic field (Moffatt, 1978; Parker, 1979; Krause and Rädler, 1980). The mean helicity breaks any reflectional (mirror) symmetry of the velocity field. Note that a laboratory turbulence, e.g. of the Kolmogorov type, is typically mirror symmetric. In the celestial bodies this symmetry is broken due to the rotation and density or velocity stratification, the reflectionally non-invariant pseudoscalar $\Omega \nabla \rho$ characterizes this quality. The problem of the mean magnetic field generation is discussed in Section 2.

The subsequent development is to consider deviations from the mean magnetic field, i.e. the magnetic fluctuations. The best way to do it is the use of correlation function language (Section 3).

And the most principal step is an attempt to understand the behaviour of the real (non averaged) magnetic field in the random turbulent flow. It is discussed in Section 4.

The process of magnetic field self-excitation under consideration is called "the hydromagnetic dynamo". It can act also under some conditions in a laminar well ordered conducting fluid, this part of the problem is discussed in particular in the review by D. Lortz this volume. My talk is devoted to the turbulent dynamo, more exactly to the fast turbulent dynamo (the term was invented by Ya. B. Zeldovich and myself, for details see Zeldovich et al., 1983).

2. Magnetic thermodynamics; the Uranus, Neptune and solar cycle

The mean field approach is similar to thermodynamics. All essential features of the turbulence are parametrized by phenomenological coefficients. Since in the Maxwell equations the velocity enters only in the Ohm law it is sufficient to have an averaged form for this law. The usual form is

$$\mathbf{j} = (\nu_m^{-1} + \nu_T^{-1})\mathbf{E} + \mathbf{V} \times \mathbf{B} + \alpha \mathbf{B}, \tag{1}$$

where $\mathbf{j}, \mathbf{E}, \mathbf{B}$ are the mean (averaged over the turbulent pulsations) values of the current, electrical and magnetic fields, $\mathbf{V}$ is the mean (regular) velocity of the flow, ν_T is a turbulent viscosity and α is a phenomenological pseudoscalar function producing the electrical current parallel or antiparallel to the mean magnetic field. The form (1) proved to be correct at last in two cases: (1) the two scale situation when the basic scale of the turbulence is smaller than the scale of the mean magnetic field (Moffatt, 1978); (2) the short–correlated turbulence when its velocity field has a δ–type time correlation (see the review by Molchanov et al., 1985; the original idea is due to A. Kazantsev). In these cases the coefficients in (1) can be found in the exact form. It appears that ν_T is proportional to $< v^2 >$ and α is proportional to the mean helicity. In a more general case the averaged Ohm's law probably has an integral form as shown for a renovating flow with a finite correlation time by Dittrich et al., (1984).

Large–scale magnetic fields are observed around some planets, on stars and in the galaxies. The presence of plasma turbulent motions in these celestial bodies gives a good chance to explain the origin of the magnetic fields by dynamo action. The problem is to construct non–decaying (static, growing or oscillating) solutions of the mean field equations without any e.m.f. except the induction one in a proper geometry, e.g. a spherical shell in the case of the star or a disk in the galactic case. A number of such solutions (the dynamo models) are constructed, see the books cited above. I will try to give here a fresh example of magnetic field generation in Uranus and Neptune and to mention an old problem concerning the mechanism of the solar activity.

In January 1986 the magnetometer on the spacecraft "Voyager–2" registered a magnetic field in the close vicinity of the Uranus. The geometry of the planetary field reconstructed from these data differs essentially from the magnetic configurations of other planets with known fields. The main mode of the Uranian magnetic field can be represented by a dipole tilted at 60 degrees to its rotation axis, while for example the tilt of the Earth magnetic dipole is very small, 11 degrees. Thus, the non–symmmetric components of the Uranian magnetic field are profound.

A model of mean magnetic field generation to explain the origin of this large tilt was constructed by Ruzmaikin and Starchenko (1989). The magnetic field can be generated in the fluid external part of the planetary core similar to the Earth fluid core. The magnetic diffusivity ν_m is estimated as $10^4 cm^2/s$. The assumption that there is convection due to the gravitational differentiation of a matter between the fluid and a hard internal core gives an estimate for the convective velocity, $v \approx 0.1$ cm/s. By use of a characteristic size of the core, $R \approx 8 \cdot 10^8$ cm, we obtain for the magnetic Reynolds number the estimate, $R_m = vR/\nu_m \approx 10^4$.

In addition to the fluid core there is a thin $(0.06R_U)$ conducting layer in the lower part of the Uranian ocean consisting of water, ammonium hydrate and methane. Analogous estimates give the magnetic Reynolds number in this layer of the order of 10^5 .

The sources for the mean magnetic field generation are the mean helicity (α) and the non homogeneous (differential) rotation ($\nabla\Omega$). The direct measurements of the Uranian radiation belt rotation made by Voyager–2 gave for the angular velocity in the internal regions the value $10^{-4}s^{-1}$ or 17.3^h in period. A characteristic value of α can be estimated as $v/3$ from a simple mixing–length type theory.

The averaged induction equation in a spherical coordinate system has the form

$$\left(\frac{\partial}{\partial t} + q\epsilon^{-2}\Omega\frac{\partial}{\partial \phi} - \Delta\right)\mathbf{B} = q\epsilon^{-2}\mathbf{r}\sin\theta \; \nabla\left(\Omega\mathbf{B}\right)\mathbf{e}_\phi + q^{-1}\epsilon^{-1}\mathrm{rot}\alpha\mathbf{B}$$

where the dimensionless parameters $\epsilon = R\left(\alpha_0\Omega_0\nu_T^2\right)^{-1/3}, q = \left(\Omega_0\nu_T\alpha_0^{-2}\right)^{1/3}$ define the intensity of the sources and their ratio, (α_0, Ω_0 are characteristic values of α and Ω). The solutions growing in time are possible only for sufficiently intense sources, i.e. at small ϵ. Hence, they can be searched in the asymptotic WKB form

$$\mathbf{B} = \Psi(r, \theta, \epsilon)\exp\left[\gamma t + \imath m\phi + \imath\epsilon^{-1}S(r,\theta,q)\right],$$
$$\Psi = \Psi_0 + \epsilon\Psi_1 + \ldots ,$$
$$\gamma = \epsilon^{-2}\left(\gamma_0 + \epsilon\gamma_1 + \ldots\right),$$

By substitution of these expressions into the evolution equation and equalizing the terms with the same powers of ϵ one finds the equations for unknown functions S, Ψ, etc. The analysis made by Starchenko and Ruzmaikin shows that a preferable magnetic mode to be excited in the fluid core is the axisymmetric one, $m = 0$, while in the ocean layer (because it is thin) the non–axisymmetric mode $m = 1$ is preferably excited. The axisymmetric mode corresponds to a magnetic dipole aligned along the rotational axis. The mode $m = 1$ describes a dipole lying in the equatorial plane. The resulting field looks like a tilted magnetic dipole (Figure 2).

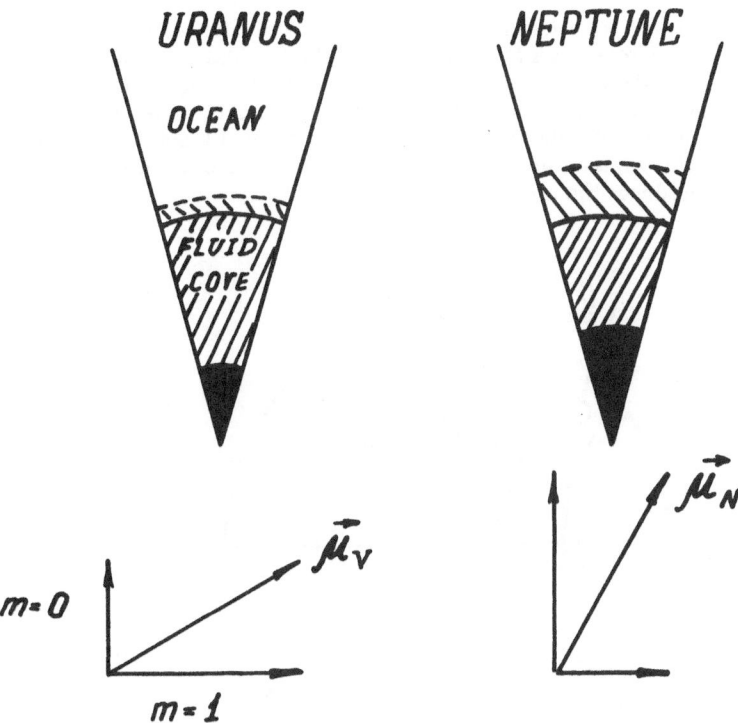

Fig.2: A schematic view of the interiors of Uranus and Neptune. The magnetic field can be generated in the fluid core and in the lower shell of the ocean. The shell is very thin in Uranus which results in generation of the non–symmetric m = 1 mode.

Neptune is similar to Uranus, however, the width of the conducting layer in its ocean is larger. It makes a preferable excitation of the non–axisymmetric mode there unprobable. So the expected tilt for the Neptunian magnetic dipole is smaller compared to the Uranian one. This statement may be checked after the approach of the spacecraft to Neptune in August 1989.

The mean field approach is used to solve the problem of solar activity mechanism. All manifestations of the solar activity, sunspots, prominences, flares etc., are connected with magnetic fields. Along with the local magnetic activity there is a global process which displays itself in the form of the 11–year cycle, modulated with secular variations and the Grand Minima of prolonged disappearence of the solar activity. This process is described as an evolution of dynamo waves of the mean magnetic field excited by the action of differential rotation and mean helicity in the solar convective zone (Parker, 1979; Zeldovich et.al., 1983; Stix, 1989). Let me explain in short the nature of the dynamo waves and mention some up to date problems.

In a demonstrative way the action of differential rotation reduces to conversion of a given poloidal magnetic field into an azimuthal (toroidal) one because different parts of the poloidal magnetic line are rotating with different angular velocities. The helical convective motion is able to lift a Greek letter Ω-type loop from the toroidal field and twist it around its axis to create a poloidal component of the field. The turbulent diffusion helps to extend the loop size. A nonzero mean helicity means that the number of left-screwed loops is different from the number of right-screwed ones so that there is a net contribution to the poloidal magnetic field. Figure 3 illustrates how a joint operation of the differential rotation and mean helicity produces the propagating magnetic wave.

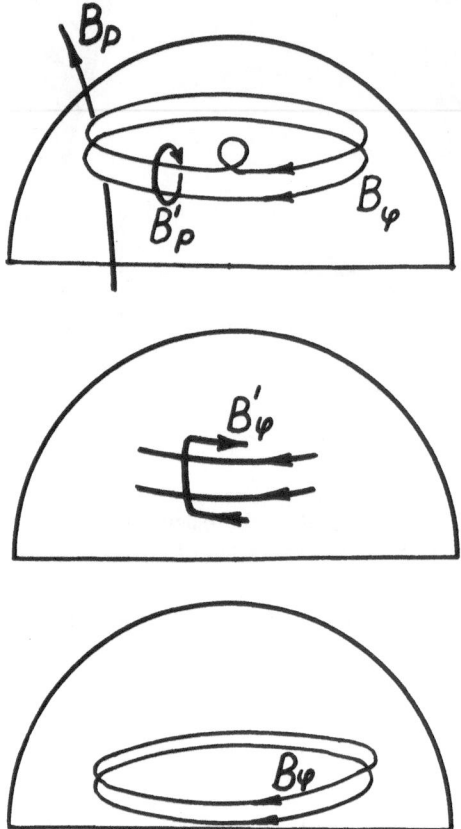

Fig.3 (a): The radial dependence of angular the angular velocity in the solar hemisphere creates a toroidal component of magnetic field B_ϕ from the poloidal B_p. The mean helicity produces a new poloidal field B_p from this toroidal component. (b) The action of the differential rotation on the B'_p produces a new toroidal contribution B'_ϕ. (c) The sum of B_ϕ and B'_ϕ gives a distribution shifted relative to the distribution of B_ϕ. The direction of the shift depends on the sign of the product $\alpha\nabla\Omega$. In order to reproduce the Maunder's butterfly diagram for the sunspots this product must be negative as it used in the Figure.

Modern helioseismology is able to reconstruct the distribution of the angular velocity in the solar convective zone. The problem is to adjust this information to the magnetic waves. A first attempt was made by Makarov et al. (1988) who showed that there may be waves propagating both from the poles to the equator i.e., the Maunder's butterfly type, and waves propagating from the equatorial regions to the poles.

The other interesting problem is the study of the non linear evolution of these magnetic waves to explain the origin of the Grand Minima. The strange attractor concept is fruitfully used for this purpose (Ruzmaikin, 1981,1989; Weiss et al., 1984).

3. Magnetic fluctuations

The magnetic fluctuations may be of two kinds; the deviations from the mean magnetic field and small-scale magnetic fields excited by motions having no mean helicity and which are therefore unable to generate the mean magnetic field

The spectrum of fluctuations that arises from the action of turbulent motions on the large-scale, mean magnetic field is discussed in Section VII.13 of the book by Ruzmaikin, Shukurov and Sokoloff (1988). At low and intermediate region wave numbers the energy coming from the large-scale magnetic field is balanced by energy cascades to small scales. At high wave numbers the energy income is balanced by Ohmic dissipation. In the intermediate part there is an equipartition between the magnetic and kinetic energies resulting in the Kraichnan spectrum $k^{-3/2}$. The expected form of the spectrum is

$$M(k) = \begin{cases} B^2 k^{-1}, & \text{if } k_0 < k < k_e; \\ (v_A \epsilon)^{1/2} k^{-3/2}, & \text{if } k_e < k < k_d; \\ \epsilon^{2/3} \nu_m^2 B^2 k^{-11/3}, & \text{if } k \le k_d. \end{cases}$$

Unfortunately, the spectrum does not take into account the phase correlations between the Fourier modes so we lose information about the structure of the small-scale magnetic field. It is better to use a correlation function, for example

$$w(r) = < \mathbf{H}(\mathbf{r_1}, t)) \mathbf{B} >, < \mathbf{H}(\mathbf{r_2}, t) - \mathbf{B} > \mathbf{z}$$

where $r - |r_1 - r_2|$. It is positive at small r and goes to zero at infinity. The correlation function which changes its sign once corresponds to a simple magnetic loop (check it!). Thus the correlation function approach gives some evidence about the structure of the magnetic field.

This approach is used to explain the origin of secular variations of the Earth magnetic field (Ruzmaikin et al., 1989a), the solar self-exciting magnetic fields (Kleeorin et al., 1986), and magnetic fluctuations in galaxies (Ruzmaikin, Shukurov and Sokoloff, 1988). Recently, we made an attempt to apply it to the galaxy clusters (Ruzmaikin et al., 1989b).

A few galaxy clusters have non thermal synchrotron radio emission which indicates the presence of magnetic fields in the intergalactic well conducting gas. The motion of galaxies through the intergalactic gas creates turbulence in the wakes behind them. Of course, we cannot speak about a general cluster differential rotation and mean helicity. However, the mirror symmetrical hydromagnetic dynamo can act producing the intermittent rope-like magnetic structure. This concept is supported by observations. In fact, the observed rotation measures do not exceed a few hundred rad per square meter (Dennison, 1979); this value implies that the magnetic field averaged along the line of sight does not exceed 0.2 Gauss, whereas equipartition arguments predict a characteristic strength of about 2 Gauss.

4. Real random fields

The third (and the last) step in understanding the magnetic field behaviour under the action of the well conducting turbulence is to consider not the mean or mean squared values but the real field which is a random function of time and space.

On a first glance the problem looks untractable. In fact, even in the kinematic approach when the velocity distribution is given one should solve the induction equation with coefficients wich are random functions. However, by using the principle "the more randomness the better" it is possible to make some important conclusions concerning the asymptotic behaviour of a random magnetic field in the random velocity field.

What can the velocity field do with the magnetic field? Locally it can turn the magnetic line and increase or decrease a number of the lines in a small volume under consideration, i.e. increase or decrease the intensity of the magnetic field. In a finite volume it results in some topological transformations of the magnetic lines, namely, the stretching, twisting and folding.

Consider first for the sake of simplicity a single magnetic loop.The motions can stretch this loop, twist it into a figure eight form and then fold thus increasing two times the intensity of the field and a magnetic flux through the crossection of the loop (the Zeldovich dynamo, see for details Zeldovich et al., 1983). However, it is more probable instead of converting the loop into an eight to make the opposite directed parts of the loop approach each other and stimulate a reconnection process which decomposes the loop into two or more small loops. These transformations can be desribed by a simple one dimensional map (see the paper by Molchanov et al., 1989 which develops the one dimensional maps suggested by Finn and Ott, 1988) graphically presented in Figure 4.

Imagine now an ensemble of such initial loops. Most of them will undergo reconnections and only a small number of the loops at some rare places will be intensified due to the twisting and folding. The resulting magnetic structure becomes very intermittent; the strong concentrations in the form of magnetic loops (not necessary simple!) will rise among the extended regions of a weak background magnetic field. The picture essentially differs from the distribution of the mean magnetic field. It is determined by some non typical realizations of the magnetic field. In fact, for a statistically independend sequence of the transformations

$$H_n = \xi_n \xi_{n-1} \ldots \xi_1 H_0$$

where ξ_i are the random values characterizing the velocity field (Molchanov et al., 1989). To have a growing field it is necessary that all these values are non-zero and exceed the unit value. One may think that such growing realizations are extremely rare. However, for a normally distributed velocity field for example the distribution of the magnetic field according to the induction equation appears to be similar to the log-normal distribution having a sufficiently high tail (Molchanov et al., 1984; Zeldovich et al., 1987).

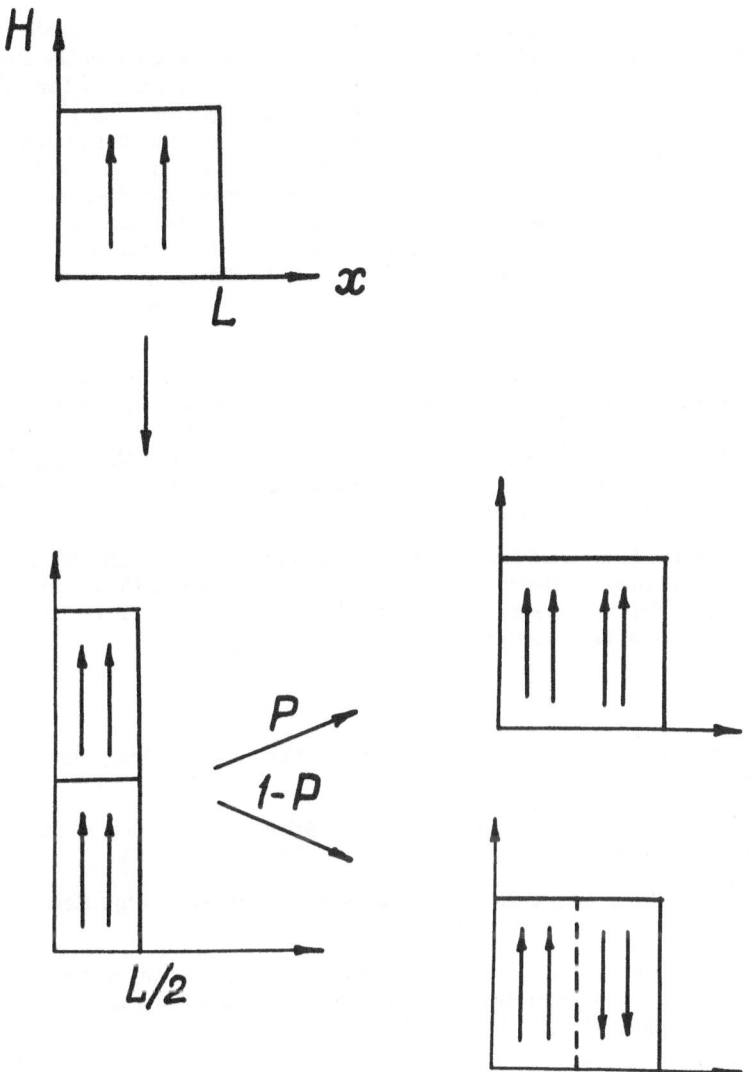

Fig.4: *The initial magnetic field y−directed and distributed in a region of lenght L over the x−axis is stretched twice then cut in the middle (this is the price for one−dimensional simplicity) to form the L−region of double field with a small probability p or with the probability* 1 − p *the L−region of the opposite directed magnetic lines subjected to the reconnections.*

According to the maximal principle the behaviour of a typical realization of a scalar like the temperature in a random velocity field is asymptotically (in time) similar to the evolution of its mean value. Unlike this, the realizations of (pseudo)vectorial magnetic fields in well conducting fluid evolve quite differently from the evolution of the mean and mean squared magnetic field. It is even possible that all realizations of the magnetic field will decay while the mean squared field will exponentially rise! This situation may be realized in the infinite space due to the existence of infinite number of temporarily growing magnetic realizationa so that at any given time there is a growing field. Possibly this situation takes place in cosmic conditions when the kinematic viscosity is much smaller compared with the magnetic diffusivity so that the velocity field transforming the magnetic field has almost fractal structure. For the case of a smooth velocity field each realization of the magnetic field grows exponentially (Molchanov et al., 1984). The rate of the growth is smaller than the growth rate of the averaged statistical moments that is the reason for the intermittency.

The generation of intermittent magnetic fields can give an answer to the question concerning the origin of the sunspots (Ruzmaikin, 1989). The origin of the sunspots is associated with the solar magnetic field. Pairs of sunspots are joined by a loop of a large-scale subphotoshere magnetic field emerging through the solar surface due to the magnetic buoyancy (Parker, 1979). However, why do the sunspots occupy a very small area (not more than one percent) of the solar surface?

The strength of the large-scale magnetic field itself is probably not sufficient to reach a critical value to float through the surface. Only a sum of the field with the intermittent, fluctuative magnetic field of the same sign can reach in some regions this critical value. It explains also the random character of appearence of the sunspots.

REFERENÇES:

Dennison, B. 1979, On the intercluster Faraday rotation. Observations, *Astron J.*, **84**, 725–729.

Dittrich,P., Molchanov, S.A., Ruzmaikin, A.A., and Sokoloff, D.D. 1984, Mean magnetic field in renovating random flow, *Astron. Nachr.*, **305**, 119–125.

Finn, J., and Ott, E. 1988, Chaotic flows and fast magnetic dynamo, *Phys.Rev. Lett.*, **60**, 760–763.

Kleeorin, N.I., Ruzmaikin, A.A., and Sokoloff, D.D. 1986, Correlation properties of self-exciting fluctuating magnetic fields, Proceedings Varenna–Abastumani Intern.School "Plasma Astrophysics", ESA SP–251, pp.557–561, Paris.

Krause, F. and Rädler, K.–H. 1980, Mean–field magneto-hydrodynamics and dynamo theory. Akademie Verlag, Berlin.

Makarov, V.I., Ruzmaikin, A.A., and Starchenko, S.V. 1987, Magnetic waves of solar activity, *Solar Physics*, 111,267–277.

Moffatt, H.K. 1978, Magnetic field generation in eletrically conducting fluids, Cambridge Univ.Press, Cambridge.

Molchanov, S.A., Ruzmaikin, A.A., and Sokoloff, D.D. 1984, A dynamo theorem, *Geophys. Astrophys. Fluid Dyn.*, 30,241–259.

Molchanov, S.A., Ruzmaikin, A.A., and Sokoloff, D.D. 1989, Self–excitation and reconnections of magnetic field in random flow, Proceedings of the workshop "Topological fluid mechanics", ed. by K. H. Moffatt, Cambridge Univ.Press, Cambridge.

Parker, E.N. 1979, Cosmic magnetic fields, Clarendon Press, Oxford.

Ruzmaikin, A.A. 1981, Solar cycle as strange attractor, *Comments on Astrophysics*, **9**, 85–93.

Ruzmaikin, A.A. 1989, Order and chaos in solar cycle, Proceedings of IAU Symposium N138 "Solar Photoshere", ed. I. Stenflo, Kluwer, Dordrecht.

Ruzmaikin, A.A., Shukurov, A.M., and Sokoloff, D.D. 1988, Magnetic fields of galaxies, Kluwer, Dordrecht.

Ruzmaikin, A.A., Shukurov, A.M., and Sokoloff, D.D. 1989a, On the origin of secular variations of the Earth magnetic field, *Geomagnetism and Aeronomy* (in press).

Ruzmaikin, A.A., Shukurov, A.M., and Sokoloff, D.D. 1989b, The dynamo origin of magntic fields in galaxy clusters, *Mon. Not. Roy. Astr. Soc.* (in press).

Ruzmaikin,A.A., and Starchenko, S.V. 1989, Generation of large-scale magnetic fields of the Uranus and Neptune by turbulent dynamo, *Kosmicheskie issledovaniya (in Russian)*, **27**, 297–303.

Stix, M. 1989, The Sun, Kluwer, Dordrecht.

Weiss, N.O., Cattaneo, F., and Jones, C.A. 1984, Periodic and aperiodic dynamo waves, *Geophys. Astrophys. Fluid. Dyn.*, **30**, 305–341.

Zeldovich,Y.B., Ruzmaikin,A.A., and Sokoloff, D.D. 1983, Magnetic fields in astrophysics, Gordon and Breach, New–York, London, Paris.

Zeldovich, Ya.B., Molchanov, S.A., Ruzmaikin, A.A., and Sokoloff, D. 1987, Intermittency in random media, *Soviet Phys.Uspehi*, **152**, 3–32.

THE ORIGIN OF GALACTIC MAGNETIC FIELDS

R. M. KULSRUD
Department of Astrophysical Sciences and
ThePlasma Physics Laboratory
Princeton University
Princeton N.J
U.S.A.

ABSTRACT.   There are two suggested origins for the observed galactic magnetic fields: the primordial origin and the dynamo origin. In this lecture the dynamo origin is discussed and criticized. It is pointed out that if the interstellar medium in which the dynamo operates is infinitely conducting the dynamo will not behave properly but will amplify the chaotic part the magnetic field more rapidly than the mean part. An approximate criteria for flux freezing is derived taking into account the fact that the interstellar medium is partially ionized. It is shown that flux freezing occurs for fields larger than approximately 10^{-10} gauss. For larger fields the dynamo amplification only results in chaotic fields, at variance with observed fields in galaxies.

1. Introduction

It is well known that our galaxy has a large scale magnetic field, whose field strength is roughly two to three microgauss. Evidence for this field follows from a number of observations, such as Faraday rotation measures from pulsars and extragalactic radio sources, Zeeman splitting in dense clouds, interstellar polarization of starlight by interstellar grains, and nonthermal synchrotron radiation by cosmic ray electrons in this interstellar field. Similar magnetic fields are observed in other spiral galaxies and even in the intracluster medium of clusters of galaxies.

An important question is, what is the origin of these large scale fields? The two most reasonable explanations are: (1) The fields were originally of primordial origin and were frozen in the plasma from which a galaxy formed. During the life of a galaxy its field would continue to be frozen in the interstellar medium, and the motions of the interstellar medium would modify the large scale structure of the field so that it would no longer resemble the original primordial field. (2) The second explanation of the existence of galactic fields is that they were generated by dynamo action from a primordial seed field of much smaller strength than the primordial field of explanation (1).

W. Brinkmann et al. (eds.), Physical Processes in Hot Cosmic Plasmas, 247–254.
© 1990 *Kluwer Academic Publishers.*

Both of these explanations suffer from certain difficulties that bear closer examination. The first explanation, that, the field is really of primordial origin, merely pushes the problem of the origin back to an earlier era. At the present time it is very difficult to see how a large scale field could be generated in the early universe. If the origin was at a very early epoch, the scale of coherence of the field could hardly be larger than the horizon size and extrapolating forward to the present epoch, this size would be much smaller than the scale of the protogalaxies from which galaxies are imagined to form. Such a field would not be sufficiently uniform to serve as a suitable origin for the galactic fields. If the origin of the field occurs at a late epoch, then considerably more is known about the properties of the cosmic plasma and on the basis of this knowledge, it is difficult to invent a mechanism which would generate a magnetic field of sufficient strength for a primordial origin.

Because of these difficulties with the primordial origin of galactic magnetic fields, the second explanation, of a dynamo origin for the field, has become very popular. The theory of the dynamo generation of magnetic fields has advanced considerably since its origin twenty years ago. It has been very successful in explaining the earth's field and has had some success with the solar field. However, in both of these cases the magnetic Reynolds number, $R_M = VL/\eta$, where V is a typical velocity L a typical scale size, and η the normal Spitzer resistivity, is not too large compared to one. For such values of R_M there is a well-founded theory. The real question is, can this theory reasonably be extended to the the case of the galactic field where the length scales are so large that the magnetic Reynolds number is enormous? The magnetic Reynolds number is a measure of the extent to which the magnetic field is frozen in the plasma, and to which flux is conserved by the plasma. In building up a field from a small value the constraint of flux conservation must be broken on the largest scales of coherence of the magnetic field. No rigorous theory has yet been developed to show how this will happen. (Dynamos operating when the Reynolds number is very large are referred to as fast dynamos)

In any theory of dynamo generation at large magnetic Reynolds numbers, it is usually assumed that there is a large effective turbulent resistivity, η_T. On the basis of this assumption it is maintained that the effective Reynolds number that applies is one in which, in its definition, the normal Spitzer resistivity is replaced by the turbulent resistivity. If this assumption were valid, it would justify the use of standard dynamo theory, and would provide the mechanism for the breaking of the flux constraint.

To decide whether the turbulent resistivity acts as claimed in the dynamo context one must examine the turbulence on very small scales. The behavior of the small scale turbulence depends on the nature of the plasma medium in which the dynamo is operating. For the interstellar medium one must take into account the fact that, during the early phases of the galaxies when the dynamo is postulated to act, the bulk of the interstellar medium is only partially ionized. (At the present time most of the interstellar medium is in the form of diffuse clouds) For a partially ionized medium the normal fluid viscosity is quite large. This leads to the fluid turbulence being cut off at waves lengths much larger than the appropriate resistive scale at which the flux constraint is destroyed. (This resistive scale, λ_R is defined to be the resistive skin depth for the time of dynamo amplification, T_D, that is $\lambda_R = \sqrt{\eta T_D/4\pi}$.) Further, although the larger scale fluid motions lead to smaller scale magnetic perturbations because of velocity shear, it turns out that these smaller scale magnetic fluctuations are cut off by ambipolar diffusion at wave lengths which are larger than the appropriate resistive scales if the mean magnetic field is sufficiently weak.. In effect, this cutoff stops the amplification of the mean magnetic field at strengths much smaller than the dynamic limit, the limit at which the total magnetic energy is equal to the total energy in fluid motions.

In summary, the principal criticism of the dynamo theories in the interstellar context is that, when one considers the question of turbulent resistivity in microscopic detail, one finds its operation depends on the physics of the medium in which the dynamo is operating. As a consequence it would appear that the dynamo will cease to operate smoothly above a certain critical magnetic field strength, and this field strength is much smaller than the dynamic limit.

In these lectures we attempt to estimate this critical field strength for the galactic case and to show that the value is quite small, of order 10^{-10} gauss. Above this field the dynamo will still act to increase the mean field at the usual rate. However, the magnetic field becomes frozen into the plasma, and the mean square field will increase much faster. Thus, the dynamo will yield a very tangled field, at variance with the rather regular field observed in the galaxy. It will finally stop amplifying when the energy in the mean square field equals the turbulent motion in the interstellar medium.

2. The Dynamo Theory for Galactic Fields

In this section we briefly review the ideas behind the dynamo origin of the galactic magnetic field. The theory comes in two parts: First the dynamo equation is derived from first principles. The derivation applicable to large magnetic Reynold's numbers (the case of interest for galactic fields) is usefully referred back to an early paper of Vainshtein's [1] . The dynamo equation can be written

$$\frac{\partial \bar{\mathbf{B}}}{\partial t} = \nabla \times (\mathbf{V} \times \bar{\mathbf{B}}) + \eta \nabla^2 \bar{\mathbf{B}} + \nabla \times (\alpha \bar{\mathbf{B}}). \tag{1}$$

This equation describes the evolution of the mean field $\bar{\mathbf{B}}$ in the presence of Gaussian turbulence. The term "mean" refers to the fact that an average has been taken over an ensemble of all the realizations of the turbulence. The velocity $\mathbf{V}$ is the large scale systematic velocity. For the galactic case it consists primarily of the differential rotation of the galactic disk. η is the sum of the turbulent resistivity η_T and the Spitzer resistivity η_S. η_T is related to the isotropic part of the turbulence, by $\eta_T \sim vl$ where v is the turbulent velocity and l is the largest scale of the turbulence. ($\eta_T \gg \eta_S$ if the magnetic Reynolds number is very large.) α is the famous alpha parameter of dynamo theory and relates to the net helical component of the turbulence. Such a helical component arises naturally for turbulence in a rotating system with a density scale height L. It can be shown to be of order $l^2\omega/L$, where ω is the angular velocity of the galaxy. [2].

The dynamo equation has been solved for galactic disks, and has been shown to lead to an exponential growth of $\bar{B}$ for certain mode structures of $\bar{B}$. The growth time is short, of order of several galactic rotation periods, while the modes resemble the observed magnetic structures in galaxies [2]. On the basis of these results, one could reasonably conclude that the present magnetic fields arose from the amplification of very small initial seed magnetic fields. It has been suggested that such seed fields of order 10^{-20} gauss could arise in the early universe[3]. Thus, the need for a large scale primordial field with a strength in the range of 10^{-9} gauss is avoided by the dynamo theory.

But, can we be certain that the dynamo theory is really correct? Let us examine the assumptions that underlie it.

A principal assumption, that seems to be made in nearly all derivations of the dynamo equation, is that the field is too small to affect the turbulence. The reasonable presumption is that this will be the case until the total energy density associated with the mean field is equal to the total kinetic energy density in the turbulent motions. The observed energy densities of galactic fields are comparable to the observed energy densities of the turbulent motions. Hence, this assumption is in accord with the observations.

A second assumption, which Vainshtein makes, [1] is that the correlation time for the turbulence is very small in some sense. This assumption allows him to develop a precise and rigorous derivation of the dynamo equation from the induction equation

$$\frac{\partial \mathbf{B}}{\partial t} = \nabla \times (\mathbf{v} \times \mathbf{B}) + \eta_S \nabla^2 \mathbf{B}. \tag{2}$$

which governs the field $\mathbf{B}$ in terms of $\mathbf{v}$ for a given realization in the ensemble. These two assumptions allow Eq.(2) to be solved by iteration in $\mathbf{v}$, and make it possible to keep all orders in the iteration.

An additional important assumption, which is tacitly made, but whose significance has not received much emphasis, is that the magnetic field $\mathbf{B}$ in any given realization does not differ by a large amount from the mean $\mathbf{B}$. This assumption is a principal weakness in the large magnetic Reynolds number dynamo theories. We shall attempt to show that this assumption is not valid in the galactic case. To put the matter simply, suppose that $\overline{B^2}$ increases much faster than $\bar{B}$. Then the result of the interaction of the magnetic field with the turbulence could be interpreted as leading to a very chaotic field, in each realization, but when the chaotic field is averaged over the ensemble, there is a much smaller mean field that would survive the averaging and not vanish. This small mean field $\bar{B}$ is the one described by Eq.(1). However, the dynamic limit on the field leads to much smaller values for $\bar{B}$ than that obtained by balancing its energy against the turbulent energy. Further, such fields are at variance with the observed galactic fields. which satisfy $\overline{B^2} \sim (\bar{\mathbf{B}})^2$

To properly justify the dynamo theory, one should derive a "dynamo equation" for $\overline{B^2}$ and compare the results with those of the standard dynamo equation for $\bar{\mathbf{B}}$. An attempt is being made to derive such an equation following the derivation of Vainshtein for $\bar{\mathbf{B}}$[4]. However, it is found that even the first iteration in $\overline{B^2}$ diverges at the small wave length end for the case of a Kolmogoroff spectrum, unless the turbulent spectrum is cut off. This indicates that the theory must involve the small scale behaviour of $\mathbf{B}$. This is plausible since the question of flux freezing also depends on the magnetic turbulence at small scales. It is clear that for incompressible turbulence with flux conservation, the length of a line of force must increase at the same rate as B increases. If the field amplifies by ten orders of magnitude, a piece of the line of force must increase its length by the same factor, and the field must become very chaotic.

At the moment there is no *rigorous* way to criticize the consequences of the dynamo theory (which is, of course, is not a rigorous theory). However, the dynamo theory operates differently if the plasma is infinitely conducting, in the sense that any amplification of $(\bar{\mathbf{B}})^2$ is accompanied by a much larger increase in $\overline{B^2}$. In this lecture the Ansatz is made that flux is frozen for a time T if the magnetic spectrum is cut off at wave lengths larger than $\lambda_R \approx \sqrt{\eta T}$. Then in the next section it is shown that flux freezing commences in the galactic case for a mean magnetic field of strength of order 10^{-10} gauss.

3. The Limit on the Field Produced by a Galactic Dynamo

In this section we estimate an upper limit to the magnetic field strength above which the magnetic field becomes frozen into the interstellar plasma. As discussed in section 2 this occurs when the turbulent magnetic spectrum is suppressed at wave lengths shorter than λ_R. Our estimate involves a careful consideration of the properties of the plasma in which the dynamo operates. In particular, we consider the consequences of taking into account the fact that the interstellar medium is a partially ionized plasma.

Let us assume that the Kolmogoroff spectrum can be written as

$$\tilde{v}_k^2 = \tilde{v}_0^2 (k_0/k)^{2/3} \quad k_0 < k < k_\nu \tag{3}$$

where $\tilde{v}_k^2$ is the mean square velocity at the wave number k, $k_0 = \pi/l_0$ is the smallest wave number, and $\tilde{v}_0^2$ is the mean square velocity at k_0. k_ν is the wave number at which the hydrodynamic spectrum is cut off by viscosity in the neutral gas. and is given by

$$k_\nu = k_0 R^{3/4} = k_0 \left(\frac{n_0 \sigma_0 v_0}{k_0 v_{th}} \right)^{3/4} . \tag{4}$$

R is the hydrodynamic Reynold's number, $R = v_0/k_0\nu_0 = v_0 n_0 \sigma_0/k_0 v_{th}$, n_0 is the neutral density, v_{th} is the thermal velocity, and ν_0 is the kinematic viscosity of the neutral component of the plasma, and σ_0 is the atomic cross section.

For $l_0 = 100$ pc, $v_0 = 10$ km/sec, $n_0 = 1/cm^3$, and a neutral temperature of 1 ev, $k_\nu \approx 1/(3 \times 10^{17}$ cm). This scale is large compared to the resistive scale, $\lambda_R = 10^{13}$ cm. for $T_e = 1$ eV and $T = 10^9$ years.

Thus, the hydrodynamic turbulence spectrum does not reach down to the resistive scale. However, as is well known, the magnetic spectrum will extend further down in scale than the hydrodynamic spectrum[5]. This is because the shear motions of the large scale turbulent velocities distort the magnetic field on ever smaller scales. The magnetic spectra does not continue indefinitely to shorter scales because the shear strains are in the neutral component while the distorted magnetic field acts only on the ionized component of the medium. This force is thus balanced only through a frictional force between the two components. These considerations lead to a cut-off in the magnetic spectrum. It is critical for flux conservation whether or not the cut-off is at wave lengths smaller than $k_R = \lambda_R^{-1}$.

Let us try to estimate k_c, the wave number at which the magnetic turbulence stops, by the following simplified model:

Imagine a magnetic perturbation of scale k^{-1}. It will propagate at the Alfven velocity $V_A = B/\sqrt{4\pi n_0 m}$. (For B we may take the value of the mean field $\bar{B}$, since we assume that $\overline{B^2}$ has not grown larger than $(\bar{B})^2$.) Now, because of ambipolar damping, this perturbation will decay at a rate equal to the minimum of

$$\Gamma(k) = \frac{(kV_A^*)^2}{2\nu_{in}} = \frac{k^2 B^2}{8\pi n_i m \langle \sigma_{in} v \rangle} \tag{5}$$

and

$$\Gamma(k) = n_i \langle \sigma_{in} v \rangle / 2 \tag{6}$$

where V_A^* is the Alfven velocity in the ionized component, n_i is the neutral density, and $\nu_{in} = n_i \langle \sigma_{in} v \rangle$ is the ion–neutral collision frequency[6].

During its life time, Γ^{-1}, the perturbation will be distorted by hydrodynamic shear velocity at k' at a rate $k' \bar{v}_{k'}$. For a Kolmogoroff spectrum of $\bar{v}_k$ this rate is a maximum for $k' = k_\nu$. If the distortion is small compared to unity during this time, then the magnetic energy will not be transported to smaller wave scales. Hence, k_c is determined by the critical condition

$$k_\nu v_\nu = \Gamma(k_c) \tag{7}$$

Combining Eqs.(3),(4),(5), and (7), we find

$$k_c^2 = \frac{8\pi n_0 n_i m}{B^2} \langle \sigma_{in} v \rangle \frac{k_0^{1/2} v_0^{3/2}}{v_{th}^{1/2} n_0^{1/2}} \sigma_0^{1/2} \tag{8}$$

(We have chosen Eq.(5) as being the minimum value for the ambipolar damping rate). k_c must be compared with the resistive $k_R = (\eta T/4\pi)^{1/2}$ where T is the dynamo time. If $k_c < k_R$, the flux is frozen into the medium and the dynamo is inhibited from working. Since k_c^2 is inversely proportional to B^2, we see that as, B increases, k_c decreases, and this continues as the dynamo amplifies the field until the condition that $k_c < k_R$ is violated. Thus from

$$k_c^2 < k_R^2 = \eta T/4\pi \tag{9}$$

and Eq.(8) we get for the upper limit on B

$$B^2 < B_c^2 = 2\eta T m n_0^{3/2} n_i \sigma_0^{1/2} \langle \sigma_{in} v \rangle k_0^{1/2} v_0^{3/2} / v_{th}^{1/2} \tag{10}$$

or

$$B < B_c = 1.5 \times 10^{-10} v_6^{3/4} T_9^{1/2} n_0^{3/4} n_i^{1/2} / l_{100}^{1/4} T_e \text{ gauss} \tag{11}$$

where we have chosen m to be the proton mass, $\langle \sigma_{in} v \rangle = 2 \times 10^{-9} cm^3/\text{sec}$ and $\sigma_0 = 10^{-16} cm^2$. Also, v_6 is v_0 in units of 10 km/sec, T_9 is T in units of 10^9 years. l_{100} is l_0 in units of 100 parsecs, and T_e is the electron temperature in units of eV. n_0 and n_i are in cgs units.

(The other value for the ambipolar damping, Eq.(6), is the correct rate if $B > 5 \times 10^{-9} n_i^{1/2} n_0$ gauss and for this limit flux freezing prevails if $n_0^{1/2} > 1.7 \times 10^{-3} cm^{-3}$. This limit applies only at very low values for either n_0 or n_i.)

In summary, according to our picture, the smooth behavior of the dynamo saturates when B reaches a field, B_c which for the above parameters is many orders of magnitude smaller than the observed galactic fields. If the seed field is much smaller than B_c, it would be amplified by the dynamo to the value B_c with no large random component. At this point the flux would become frozen into the hydromagnetic media. From this time on, the field would become very tangled until the energy in the random component of the field equals that in the hydrodynamic motions, at which point the chaotic amplification of the field would cease. One would thus be left with a very chaotic field, and this disagrees with the rather regular observed fields in galaxies.

Apparently the only way to avoid the production of such chaotic fields is to have the original seed field large enough to prevent dynamo amplification. This returns us to the primordial field as the most natural origin for galactic fields.

4. Conclusion

In this lecture we have discussed the two principal theories of origin for the presently observed galactic fields. These are a primordial origin and a dynamo origin. Both of these theories suffer from serious difficulties. We have concentrated on a discussion of the dynamo theories because more is known about the physics of this process than the rather problematical primordial theory. We point out a serious difficulty in connection with the dynamo theory. This is that flux freezing commences for magnetic field strengths that are very weak. This follows from the physics of the interstellar medium when proper account is taken of its partially ionized nature. Another important factor is also the extreme size of the magnetic Reynold's number. Flux freezing interferes with the smooth operation of the dynamo amplification of the field leading to a chaotic field whose random component is much larger than its mean component. It appears that the only way to avoid such chaotic fields is to start with initial seed fields large enough to suppress dynamo amplification altogether. In other words to start with a primordial field.

5. Acknowledgments

I would like to acknowledge useful discussions on dynamo theory with Catherine Cesarsky, Steve Cowley, and Keith Moffatt. I have benefitted from many helpful discussions concerning the galactic fields with Jerry Ostriker, Lyman Spitzer Jr.,David Spergel, and Bruce Draine. This work was supported by U.S.DOE Contract No. DE-AC02-76-CHO-3073.

References

[1] Vainshtein, S.I. (1970) 'The Generation of Large-Scale Magnetic Fields by a Turbulent Fluid', Sov. Phys. JETP **31**, 87 [Zh. E.T. F. (1970) **58** , 153]

[2] Vainshtein, S.I. and Ruzmaikin, A.A. (1972) 'Generation of the Large-Scale Galactic Magnetic Field', Soviet Astronomy **15**, 714 [Astronomicheskii Zhurnal (1970) **48**, 982]

[3] Harrison, E.R. (1970) 'Generation of Magnetic Fields in the Radiation Era' M.N.R.A.S. **147**, 279

[4] Anderson, S. and Kulsrud R. (1989) 'Limits on the Field Strength that a Dynamo Can Generate', Abstract, Plasma A.P.S. meeting Anaheim, California, November 1989

[5] Moffatt H. K. (1983) 'Transport Effects Associated with Turbulence with Particular Attention to the Influence of Helicity', Rep. Prog. Phys. **46**, 621

[6] Kulsrud, R.M. and Pearce, W.P. (1969) 'The Effect of Wave-Particle Interactions on the Propagation of Cosmic Rays', Ap.J. **156**, 445

Recent Developments in the Theory
of Magnetic Reconnection

Dieter Biskamp

Max-Planck-Institut für Plasmaphysik
8046 Garching bei München, Federal Republic of Germany

Abstract

The talk briefly reviews previous stationary models, mainly configurations of
the Petschek type, pointing out their shortcomings and basic failure in accounting
for fast magnetic reconnection in the limit of large magnetic Reynolds number.
It is shown that in this limit no relevant stationary states exist. Instead strong
small-scale MHD turbulence develops even in 2D geometry, giving rise to energy
dissipation and reconnection rates independent of the value of the collisional resistivity.

I Introduction

In the last decade it has been realized that the presence of magnetic fields is a
ubiquitous phenomenon in cosmic systems. On the one hand, magnetic fields serve
as a large energy reservoir which may be tapped in a fast dynamic process leading
to various kinds of explosive events such as flares. On the other hand, magnetic
fields tend to be compressed in processes such as protostar formation and are
computed to dominate the dynamics in the later phases in a nonrealistic way if
not dissipated sufficiently fast. To account for such processes of fast magnetic field
annihilation is the main objective of the theory of magnetic reconnection.

The term magnetic reconnection refers to the picture of magnetic field lines.
These have a well-defined meaning in a highly conducting fluid, viz. thin magnetic
flux tubes which are carried along with the fluid, maintaining their individuality,
though they may be wound in a very complex manner. Only owing to finite
electrical resistivity or some equivalent process may two field lines coming close
together lose their identities by being cut and reconnected in a different way.
Though this is a local process, it leads to a change of field topology permitting
new types of large-scale plasma motions that would otherwise be inhibited. The

W. Brinkmann et al. (eds.), Physical Processes in Hot Cosmic Plasmas, 255–269.
© 1990 *Kluwer Academic Publishers.*

change of the magnetic field is described by Faraday's law:

$$\frac{\partial \vec{B}}{\partial t} = \nabla \times \left(\vec{v} \times \vec{B}\right) + \eta \nabla^2 \vec{B} \ . \tag{1}$$

Here the ratio of the diffusion term and the convection term

$$\frac{\left|\eta \nabla^2 \vec{B}\right|}{\left|\nabla \times \left(\vec{v} \times \vec{B}\right)\right|} \sim \frac{\eta}{vL} = \frac{1}{R_m} \tag{2}$$

is a convenient dimensionless measure of the resistivity, R_m being the magnetic Reynolds number. In practically all astrophysical plasmas R_m is large, essentially because of the large scales L. Hence magnetic diffusion is in general a very weak process. Magnetic processes such as solar flares, however, seem to require fast reconnection with time scales practically independent of R_m. The main theoretical problem therefore is to find models allowing sufficiently high reconnection rates.

Fast reconnection is not a diffuse process, but is strongly localized in current sheets. Such current sheets may arise at any point with non-vanishing magnetic shear and a velocity gradient along the direction of the shear perpendicular to the field, i.e. virtually everywhere in the plasma, as visualized in Fig. 1. The simplest models are quasi-stationary configurations with one current sheet at a well defined location determined by the overall geometry, which have been investigated in the conventional theory of magnetic reconnection. The basic assumption in these theoretical approaches is the existence of a two-dimensional subsystem around an X-type magnetic neutral point which is small compared with the global magnetic configuration but large compared with the so-called diffusion region around the neutral point, where the diffusion term in (1) is important. In this subsystem conditions would rapidly adjust to changes in the global configuration, so that the evolution of the latter would correspond to a sequence of stationary states in the former which are steady-state solutions with the boundary conditions determined by the global system. This is the idea of stationary forced reconnection. The prototype of such configurations is Petschek's reconnection model[1], which is given schematically in Fig. 2.

In fact, much of the theoretical work on magnetic reconnection[2],[3] consists of modifications and refinements of this model. The theory is based on the effect that the motion of a plasma may be supersonic at arbitrarily low speed with respect to the slow mode. Hence, by analogy with a system of two supersonic gas jets colliding head-on, two pairs of slow shocks are generated back to back against the incoming plasma flow, diverting it into the outflow cone and accelerating it up to the Alfvén speed corresponding to the upstream magnetic field intensity. Petschek's configuration is characterized by a single parameter, the angle α of the outflow cone, which determines the ratio of the inflow and outflow velocities, the so-called reconnection rate $M = u/v_A$. The diffusion region is assumed to be small with dimensions $O(\eta)$ and to automatically adjust to the external configuration. Since α is a free parameter, this class of solutions, considered as solutions of a small section of the global magnetic configuration, seems to guarantee that reconnection and corresponding energy conversion rates depend only on the asymptotic plasma velocities, implying that M is essentially independent of η though the reconnection process of course requires magnetic diffusion.

II Failure of stationary models of fast reconnection

It has been taken for granted that such steady-state solutions actually exist and are stable in the limit of small η. In recent years, however, evidence has become convincingly strong, that this is not the case. The fundamental difference between a magnetized plasma and its gas dynamic analog is that the plasma motion is not truly supersonic. Since plasma velocities are usually small compared with the compressional Alfvén mode - in fact the theory usually assumes incompressible plasma motions - information about the plasma behavior in front of the diffusion region can easily propagate upstream and affect the asymptotic inflow velocity. Mathematically speaking, the inconsistency in the Petschek-type solutions consists in essentially ignoring the boundary layer problem, i.e. the matching of the solution in the diffusion layer to that in the external region. The failure of Petschek's solution to apply for large Reynolds number has become apparent by exact numerical solutions of the two-dimensional resistive MHD equations revealing a completely different behavior. Figure 3 shows a set of three numerical solutions (taken from Ref. 4) each computed from the same initial state in time until a stationary state had been reached, with the same boundary conditions but with different values of the resistivity η. Here ϕ is the stream function, $\vec{v} = \hat{z} \times \nabla\phi$, and ψ is the flux function, $\vec{B} = \hat{z} \times \nabla\psi$. The conspicuous feature is that by reducing η the size of the diffusion region, i.e. the length of the current sheet does not decrease as assumed for Petschek-like behavior, but increases finally reaching the global system size.

Detailed scaling laws obtained from a series of numerical runs are given in Ref. 4. The physical picture is that for the inflow velocity exceeding the natural magnetic diffusion rate in a current sheet $\sim R_m^{-1/2}$, reconnection becomes inefficient. Consequently, magnetic flux piles up in front of the diffusion region with corresponding slowing down of the upstream plasma flow compared with the prescribed boundary value of the inflow velocity. In addition to the increase in size the diffusion region develops an increasingly complex structure (for details see Refs. 4, 5) such that a rigorous analytical treatment of the diffusion region appears to be practically impossible, not to speak of the matching problem mentioned above.

III Turbulent reconnection

Stability investigations have shown[4] that current sheets arising during the process of magnetic reconnection (so-called Sweet-Parker current sheets), which carry a strong inhomogeneous flow, are substantially more stable with respect to tearing modes than static current sheets. Only if the length/thickness ratio L/d exceeds 10^2, compared with $L/d > 10$ in the static case, does tearing instability set in. Since this threshold is exceeded for sufficiently small η, steady-state reconnection does not exist in the limit $\eta \to 0$. Experimental observations seem to indicate that at sufficiently high Reynolds numbers all fluids, magnetic or not, become turbulent, if allowing fully three-dimensional motions. In the framework of two-dimensional reconnection theory the question arises whether current sheet configurations becoming unstable lead to turbulence even within 2D geometry. It is true that previous numerical studies (see, for example, Ref. 4) exhibit rather regular motions such as plasmoid formation and acceleration, which suggests that, as in 2D hydrodynamics, turbulent behavior characterized by, for instance, η-independent energy dissipation rates, does not occur in 2D MHD either. (Another argument frequently given is the conservation of ψ, corresponding to the persistence of exact flux surfaces, however convoluted, in contrast to 3D dynamics, which in general leads to rapid destruction of such surfaces.)

For sufficiently general, i.e. nonsymmetric, flows, however, we find that genuine turbulence also develops in 2D MHD. Let us describe some features of these investigations[6], which are relevant to the problem of fast magnetic reconnection. We study freely decaying turbulence solving the equations of 2D incompressible MHD written in the convenient scalar form

$$\frac{\partial \psi}{\partial t} + \vec{v} \cdot \nabla \psi = (-1)^{\nu-1} \eta_\nu \nabla^{2\nu} \psi , \tag{3}$$

$$\frac{\partial \omega}{\partial t} + \vec{v} \cdot \nabla \omega = \vec{B} \cdot \nabla j + (-1)^{\nu-1} \mu_\nu \nabla^{2\nu} \omega , \tag{4}$$

where j is the current density $j = \nabla^2 \psi$, and $\omega = \nabla^2 \phi$ is the vorticity. We include generalized dissipation operators, with $\nu = 1$ corresponding to normal diffusion, while $\nu \geq 2$ is called hyperdiffusion (η_2 has the physical meaning of an electron viscosity). Such higher-order dissipation terms are often used in numerical studies to concentrate dissipation at small scales. These equations are solved numerically by a pseudospectral de-aliased method with up to $1,024^2$ collocation points. The resolution thus achieved appears to be sufficient to reveal the basic properties of high-Reynolds-number systems. We consider the following large-scale initial conditions (a generalization of the Orszag-Tang vortex[7]):

$$\phi(x,y) = \cos(x + 1.4) + \cos(y + 0.5), \tag{5}$$

$$\psi(x,y) = \cos(2x + 2.3) + \cos(y + 4.1), \tag{6}$$

which turn out to be particularly suitable to study the processes of spontaneous generation of small-scale turbulence and its effect on the energy dissipation rate $\epsilon = -dE/dt$, $E = 1/2 \int \left(v^2 + B^2 \right) d^2 x$. Figures 4, 5 illustrate a high-Reynolds-number case $\nu = 2$, $\eta_2 = \mu_2 = 10^{-8}$. At $t = 1.6$, Fig. 4a, the configuration has developed extended current sheets, the characteristic feature observed in most previous 2D MHD computations. Subsequently, however, these sheets are disrupted, leading to the build-up of small-scale turbulence. Figure 4b shows the configuration at $t = 2.15$. Two different processes are clearly discernible:

i) A folding process, seen particularly in the upper-left corner. Considering the temporal evolution in greater detail, it becomes clear that this is not a simple kinematic folding of two separate current sheets, but that new sheets are generated in addition. In fact, any local velocity fluctuation $\delta \vec{v}$ with a component δv_n normal to the magnetic field may give rise to a current sheet, which will be the more pronounced the longer the time interval during which the local flow configuration persists. The process takes place primarily in regions where the magnetic field intensity is weak, $\delta v \gtrsim B$, i.e. around the X-points of the large-scale magnetic configuration (see Fig. 5), which gives the ψ-contours of the state shown in Fig. 4b.

ii) The tearing instability, occurring most prominently in the center-right current sheet of Fig. 4b. As mentioned above, dynamically generated and maintained current sheets associated with strong inhomogeneous parallel flows have a significantly higher tearing mode threshold than static sheets, which explains the obvious stability of the elongated current sheets observed in many MHD computations. Nevertheless, since sheets become thinner with decreasing η, instability will occur for sufficiently high Reynolds number. Similar to process i), the tearing mode transforms a single current sheet into a layer of multiple sheets.

The generation of small-scale turbulence has the ultimate effect of making the turbulent energy dissipation rate ϵ independent of the Reynolds number. Figure 6 gives $\epsilon(t)$ for five simulation runs with identical initial conditions, but different values of $\eta(=\mu)$ and different orders ν of the dissipation operators, $\eta_1 = 2.5 \times 10^{-3}$, 1.25×10^{-3} , 6.25×10^{-4} , 3.125×10^{-4} and $\eta_2 = 10^{-8}$ (from above). With decreasing value of the dissipation coefficients a two-stage saturation process in the time development of ϵ becomes more and more pronounced. The first bend-over of $\epsilon(t)$ from exponential growth occurs at $t \simeq 1.5$ and corresponds to the formation of elongated current sheets, as shown in Fig. 4a. Comparing the values of ϵ at this time for the four $\nu = 1$ cases, we find the Sweet-Parker (see Ref. 4) scaling $\epsilon \propto \eta_1^{1/2}$. The further increase of ϵ is due to the excitation of turbulent small scales resulting from processes i) and ii), covering increasingly broad areas until ϵ becomes independent of η. It is noteworthy that the $\nu = 2$ case (broken line in Fig. 6) also reaches the same energy dissipation level, which hence depends neither on the Reynolds number nor on a particular type of dissipation operator.

Let us now discuss the properties of the fully developed turbulent state, as illustrated in Fig. 7, corresponding to $\eta_2 = 5 \times 10^{-9}$ (and initial conditions somewhat different from eqs. (5), (6)). Apart from a few dissipation-free regions around the O-points of the large magnetic eddies small-scale turbulence is rather uniformly distributed. Dissipative structures consist of current micro-sheets. (Each current sheet is also associated with a vorticity sheet, but for $\mu \lesssim \eta$ viscous dissipation is found to be weak compared with resistive dissipation.) Figure 8 shows these structures more clearly, indicating each major current sheet by a single contour line. It gives essentially the same macro-state for different values of η, decreasing from above to below. It is easily seen that current sheets become thinner, shorter and more numerous. These highly intermittent structures can be described in terms of an ensemble of micro-sheets with thickness l, and width and spacing λ, where l is the smallest possible turbulent scale, called the Kolmogorov length, $l = (\epsilon^{-1}\eta^3)^{1/4}$ for $\nu = 1$, depending only on the micro-state quantities ϵ and η, and λ is an intermediate scale, called the Taylor micro-scale defined by $\lambda^2 = E\eta/\epsilon$, which also depends on the macro-state quantity E.

IV Conclusions

The focus of magnetic reconnection theory in the past has been to find sufficiently simple models of fast reconnection in the limit of almost vanishing dissipation coefficients. For reasons of analytical feasibility these models were restricted to 2D steady-state configurations. Unfortunately, however, simple as they look, these models could only be treated with crude approximations, concerning in particular the diffusion region (where the actual process of field line cutting occurs) and its coupling to the outside MHD region. As a result it has not been noticed that these configurations of the Petschek type, on which most of the previous reconnection theory was centered, do not exist, not even approximately. The true behavior of a plasma flow across an X-type neutral point could only be revealed by high-resolution numerical solutions of the resistive MHD equations, which also show a complexity which seems to escape any purely analytical approach. These numerically obtained configurations are dominated by extended quasi-stationary current sheets. Direct extrapolation to almost vanishing η yields very small reconnection rates $\propto \eta^{1/2}$, as with the Sweet-Parker model, which was the first selfconsistent reconnection model proposed. Since, however, such elongated current sheets become unstable for sufficiently small η, no stationary configuration appears to exist for large enough Reynolds number. We have shown that in general the resulting dynamics leads to strong small-scale turbulence even if one remains within 2D geometry, with energy dissipation rates and corresponding magnetic reconnection rates independent of the value of η. It therefore appears that the view that in high-Reynolds-number plasmas the value of the weak collisional dissipation coefficients becomes irrelevant owing to turbulence is essentially correct. We find that 2D resistive MHD gives the probably simplest selfconsistent model of such behavior. The only change compared with the previous erroneous reconnection models is to relax the artificial requirement of steady state. Hence fast reconnection rates are automatically established and need not be of particular theoretical concern. Time scales of explosive magnetic events, such as a flare, are hence not determined by reconnection processes but mainly by the amount of free energy available; loosely speaking, a big flare occurs on a faster time scale than a weak one.

References

1) Petschek, H.E. (1964) "Magnetic Field Annihilation", AAS/NASA Symposium on the Physics of Solar Flares, edited by W.N. Hess (NASA, Washington, DC), p. 425

2) Vasyliunas, V.M. (1975) "Theoretical Models of Magnetic Field Line Merging", Rev. Geophys. 13, 303

3) Forbes, T.G. and Priest, E.R. (1987) "A Comparison of Analytical and Numerical Models for Steadily Driven Magnetic Reconnection", Rev. Geophys. 25, 1583

4) Biskamp, D. (1986) "Magnetic Reconnection via Current Sheets", Phys. Fluids 29, 1520

5) Biskamp, D. (1987) "Magnetic Reconnection in Astrophysical Plasmas", Proc. Cargèse Workshop on Magnetic Fields and Extragalactic Objects, E. Asseo and D. Grésillon editors, p. 37

6) Biskamp, D. and Welter, H. (1989) "Dynamics of Decaying Two-dimensional Magnetohydrodynamic Turbulence", Phys. Fluids, to appear

7) Orszag, S.A. and Tang, C.M. (1979) "Small-scale Structure of Two-dimensional Magnetohydrodynamic Turbulence", J. Fluid Mech. 90, 129

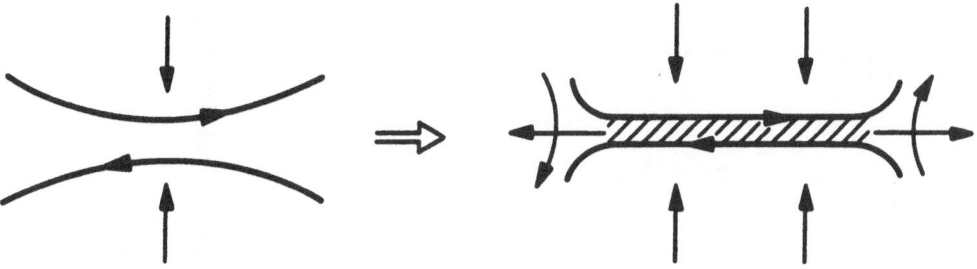

Fig. 1 Current sheet formation. Magnetic field and velocity components in the plane perpendicular to and in the frame moving with the central field line.

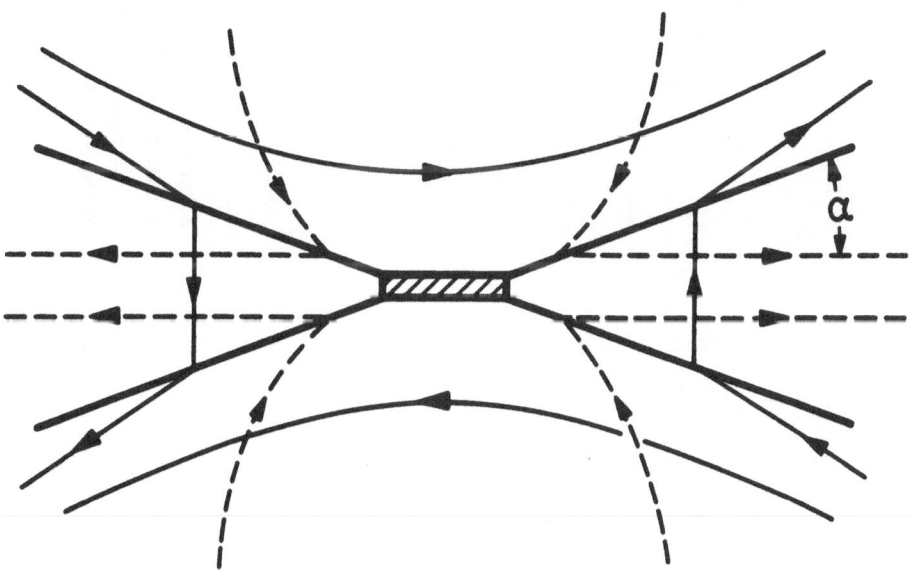

Fig. 2 Schematic representation of Petschek's reconnection configuration.

264

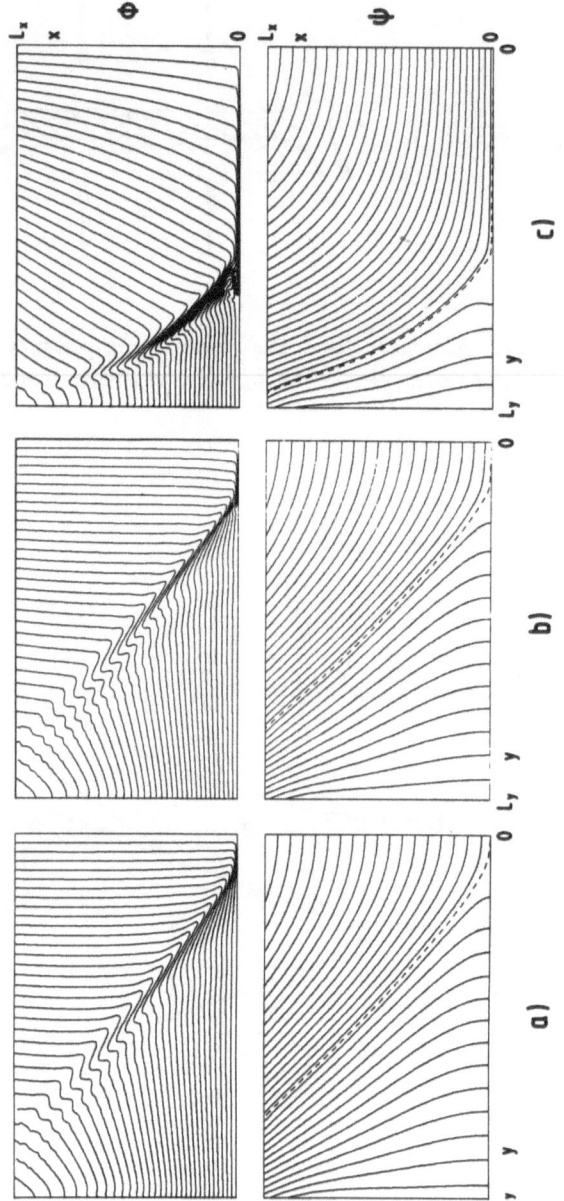

Fig. 3 Steady state forced reconnection configurations with identical boundary conditions, differing only in the value of η: a) $\eta = \eta_0$, b) $\eta = \eta_0/2$, c) $\eta = \eta_0/4$. From Ref. 5.

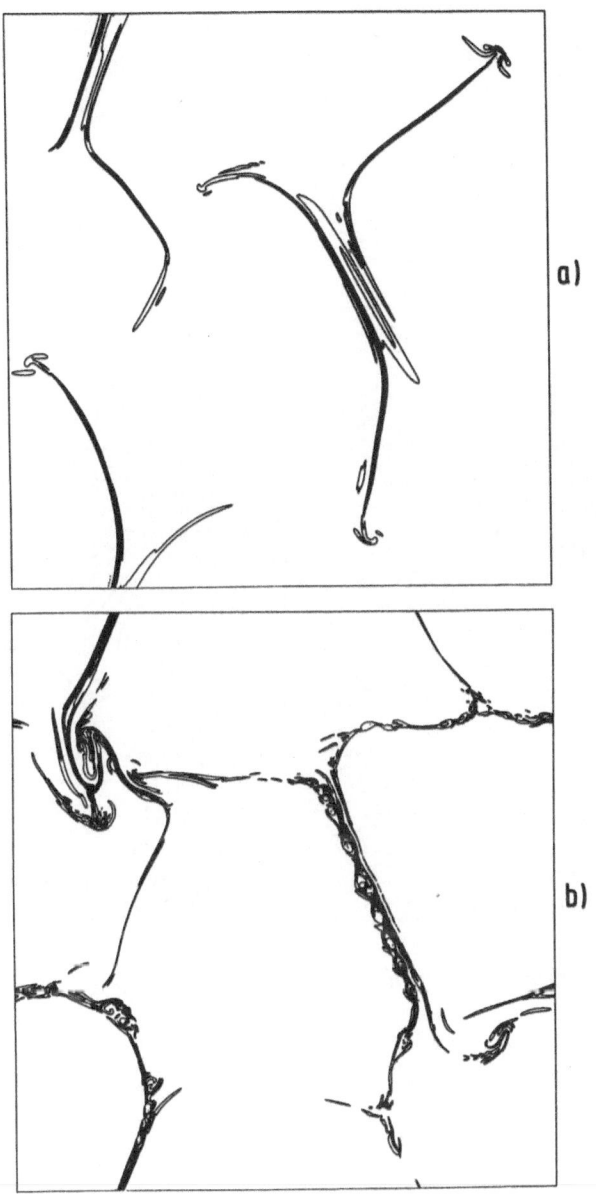

Fig. 4 *Transition to turbulence, j contours.*
a) t = 1.6; b) t = 2.15.

266

Fig. 5 ψ contour of the state shown in Fig. 4 b).

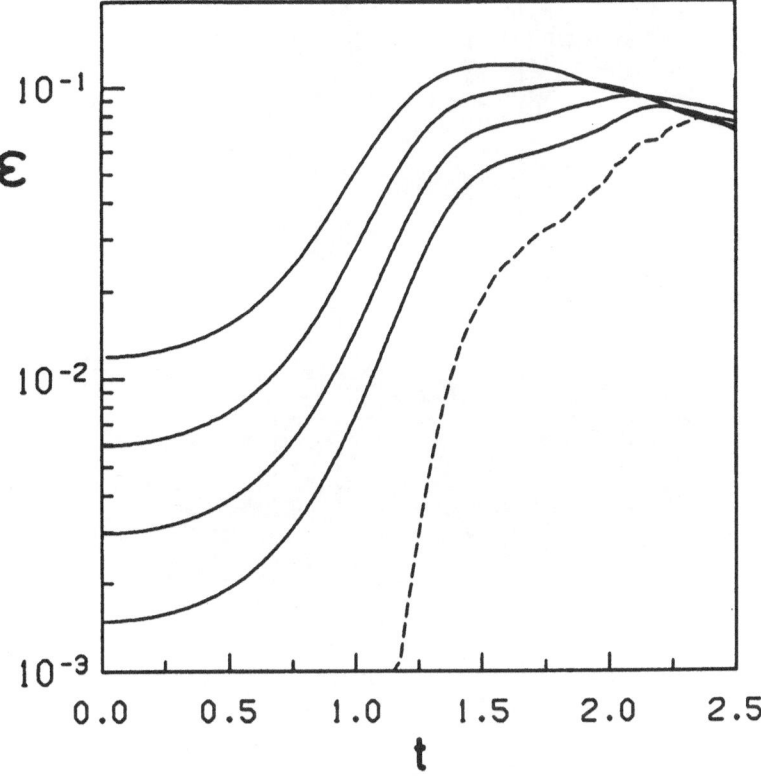

Fig. 6 Energy dissipation rates $\epsilon(t)$ *during the phase of turbulence generation for five simulation runs with identical initial conditions but different values* η.

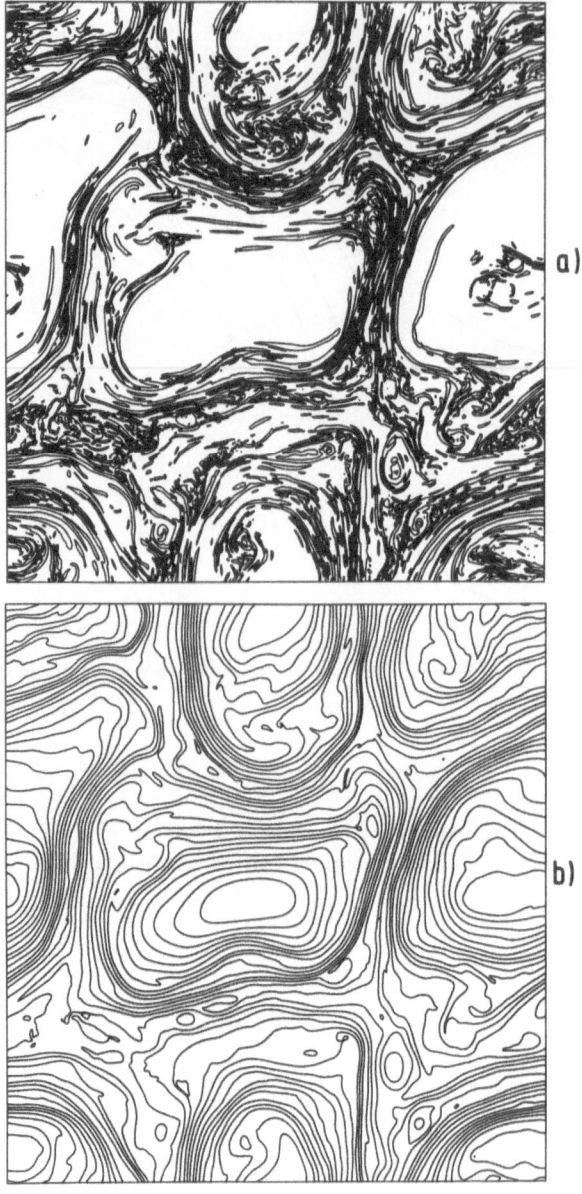

Fig. 7 Fully developed turbulent state
a) j contours, b) ψ contours.

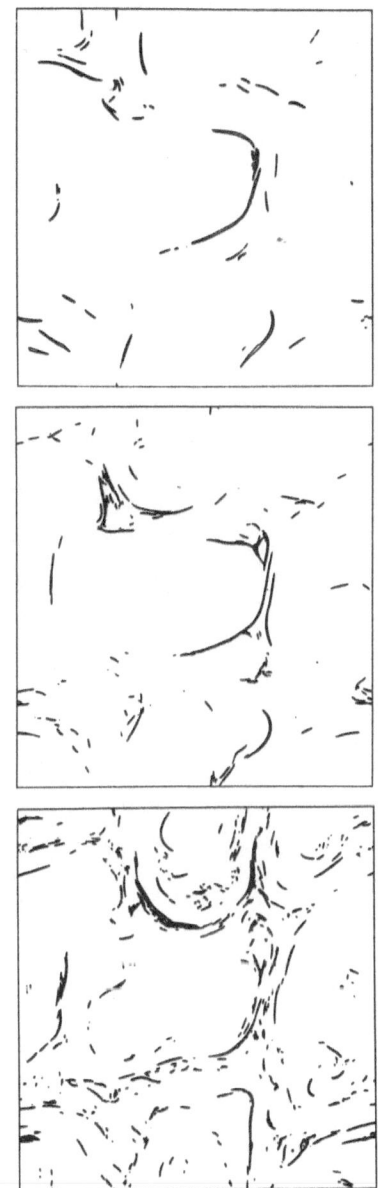

Fig. 8 Distribution of current sheets for turbulent states with
a) $\eta_1 = 3.125 \times 10^{-4}$, b) $\eta_1 = 1.56 \times 10^{-4}$, c) $\eta_2 = 5 \times 10^{-9}$.

CONDUCTION AND TURBULENCE IN INTRACLUSTER GAS

A.C. Fabian
Institute of Astronomy
Madingley Road
Cambridge CB3 OHA
U.K.

ABSTRACT.

Clusters of galaxies typically contain about 10^{14} M$_\odot$ of metal-enriched gas which is at temperatures of between 10^7 to 10^8 K. After briefly reviewing the general properties of this gas, a detailed account is given of cooling flows, which occur in the cores of most clusters. The intracluster gas is shown to be inhomogeneous, posing strict constraints on thermal conduction. Optical line-emitting clouds and filaments are observed in the centres of many clusters. Studies of these suggest that the hot intracluster gas is turbulent.

1. INTRODUCTION

Enormous quantities of diffuse hot gas are observed in clusters and groups of galaxies. The thermal pressure of the gas supports it against the large gravitational field of the cluster requiring that the sound speed of the gas is similar to the typical velocity of a cluster galaxy (the velocity dispersion of the cluster). This is generally in the range 500 to 1200 $km\,s^{-1}$, implying that the gas temperature must be $10^7 - 10^8$ K. Such high temperatures mean that the main energy loss of the gas is bremsstrahlung. The radiation is the origin of the diffuse X-radiation from clusters of galaxies and is our principal source of information on their intracluster medium (ICM). There is further indirect evidence for the gas in 'head-tail' radio sources and from theories of the propagation of double-lobe radio sources. Reviews of the properties of the ICM are given by Sarazin (1986, 1988) and Fabian (1988b, on which parts of this paper is based).

Most of the *observed* intracluster gas has an electron density, n_e, in the range of $10^{-4} - 10^{-2}\,cm^{-3}$ and a temperature $T \sim 10^7 - 10^8$ K, and is contained within a radius of 1 to 2 Mpc. The total bremsstrahlung luminosity of a cluster is $\sim 10^{42} - 10^{46}\,erg\,s^{-1}$. An emission line at 6.7 keV is observed in all clusters that are bright enough for detection to be possible (see e.g. Rothenflug & Arnaud 1986). The line is due to highly-ionized iron and shows that the gas has ~ 0.3 times solar abundance in iron. The work of Canizares *et al.* (1979; 1982) and Mushotzky *et al.* (1981) on cooling regions in clusters shows O, Ne, Si and S to be also present at abundances close to solar (although O may be super-solar).

Some cluster properties show correlations which are fairly good if just X-ray quantities are used (e.g. X-ray luminosity L_X versus gas temperature T_X). The gas temperature also correlates with the optically-determined velocity dispersion (a somewhat uncertain quantity) and with the number density of galaxies in the cluster. Generally, the deeper the potential well, the more gas and galaxies it contains and the more luminous it

271

W. Brinkmann et al. (eds.), Physical Processes in Hot Cosmic Plasmas, 271–297.

is in X-rays. The gas is at a temperature, T_X, close to that given by the Virial Theorem for a cluster of total mass M and scale size R, i.e.

$$kT_X \approx \frac{GMm_p}{R}. \tag{1}$$

The observations suggest (Edge 1989),

$$kT_X \approx 7 \times L_{45}^{0.4} \text{ keV}, \tag{2}$$

where $L_X = 10^{45}L_{45} \text{ erg s}^{-1}$.

The origin of the gas is uncertain. Being metal-enriched, it cannot all be primordial. It is generally assumed that much of the gas was processed through an early-population of stars which then blasted it into intracluster space in supernova explosions. It could also have been stripped from the young galaxies during the formation of the cluster. Whichever of these is correct, gas shares the kinetic energy of the galaxies, which is ultimately gravitational in origin, and has a sound speed similar to the galaxy motions. The presence of so much gas, comparable to all the the mass in observable stars ($\sim 10^{14} \text{ M}_\odot$ in a rich cluster) and about 10 per cent of the virial mass of the cluster, suggests that galaxy formation is an inefficient business as far as gas is concerned.

Many theories of large-scale structure in the Universe propose that hierarchical clustering takes place so that richer clusters are built up by the merger of smaller ones. This offers a way of understanding the temperature, luminosity correlation if the characteristic size, R, of the units (subclusters) is unchanged by the merger. Loose groups and rich clusters both seem to have a scale size of about a Mpc, so that can be taken for R. Then $T_X \propto M/R$ and $L_X \propto M_g^2 R^3 T_X^{0.5}$ from the properties of bremsstrahlung, which is a 2-body process and so proportional to density squared. If R is unchanged on merging 2 subclusters, then $T_X \propto L_X^{0.4}$, as found. Conversely, the X-ray results show that R must be relatively unchanged. Some evidence that this should occur is found in the cluster-collision simulation of McGlynn & Fabian (1984). The half-mass radius of the galaxies, which best probes the core of the cluster where most of the observed gas resides, changes little during a violent collision and merger.

2. THE GAS DISTRIBUTION

The intracluster gas acts as a fluid on galactic scales. The electron-electron coupling time, $t_{e-e} \approx 2.10^5 T_8^{\frac{3}{2}} n_{-3}^{-1} \text{ yr}$, where $T = 10^8 T_8 \text{ K}$ and $n = 10^{-3} n_{-3} \text{ cm}^{-3}$, and the electron-ion coupling time is 1840 times larger. The gas is therefore expected to behave locally as a Maxwellian distribution. The crossing time of the cluster by member galaxies moving at velocity $v = 10^3 v_3 \text{ km s}^{-1}$ exceeds these 2-body timescales,

$$t_{cross} = \frac{R}{\langle v^2 \rangle^{1/2}}$$

$$\approx 10^9 R_{\text{Mpc}} v_3^{-1} \text{ yr.} \tag{2}$$

This timescale is also roughly equal to the free-fall timescale in the cluster potential. The radiative cooling time of the gas, due to bremsstrahlung, is

$$t_c \approx 7.10^{10} T_8^{\frac{1}{2}} n_{-3}^{-1} \text{ yr.} \tag{3}$$

Cooling is therefore important ($t_c < t_a$, the cluster age) in high density regions (see §3) and where the temperature is low. The mean-free-path of an electron in the intracluster gas,

$$\lambda_e \approx \lambda_i \approx 23 T_8^2 n_{-3}^{-1} \text{ kpc} \tag{4}$$

$$\sim R_{galaxy},$$

whereas the gyroradius,

$$r_g \approx 3.10^8 Z^{-1} T_8^{\frac{1}{2}} (\frac{m}{m_e})^{\frac{1}{2}} B_{\mu G}^{-1} \text{ cm.} \tag{5}$$

Although $\lambda_e \sim R$ in the outskirts of a cluster, tangled weak magnetic fields will keep the gas behaviour fluid-like.

The intracluster gas is optically thin to electron scattering since its Thomson depth is ~ 0.1 per cent (but see Sunyaev 1982 on how it may be a very weak source of diffuse emission if the cluster contains a quasar, and Fabian 1989 for how the Thomson depth may be much higher in a cooling flow and give rise to an extended optical continuum around powerful cluster radio galaxies). Gilfanov et al. (1987) have noted that certain emission lines of the highly-ionized hot gas may be resonantly scattered. This is relevant to abundance gradient determinations if the gas is static, particularly if low-energy lines are used. Turbulent motions of $> 100 \text{ km s}^{-1}$ will minimize this effect. Cooled gas may provide a distributed source of optical absorption (Crawford et al. 1987). Krolik & Raymond (1988) and Sarazin (1989) have pointed out that the hot gas is a (weak) source of X-ray absorption for a background object, such as a quasar.

Non-hydrostatic pressure variations are eliminated on the sound crossing time of the gas,

$$t_s = \frac{R}{c_s} \approx t_{cross} = 10^9 R_{\text{Mpc}} T_8^{-\frac{1}{2}} \text{ yr,} \tag{6}$$

where the size of the region is R_{Mpc} Mpc. In the inner Mpc of the cluster, which is the region well-studied by X-ray detectors, t_s is much less than the age of the cluster. Any radial flows of the gas must be subsonic otherwise impossibly large mass flow rates are implied ($> (10^{14} \text{ M}_\odot)/(10^{10} \text{ yr}) = 10^4 \text{ M}_\odot \text{ yr}^{-1}$). Intracluster gas is then close to hydrostatic support so that

$$\frac{dP_{gas}}{dr} = -\rho_{gas} \frac{d\phi}{dr} = -\rho_{gas} g, \tag{7}$$

where $g = GM(<r)/r^2$. This means that measurements of P_{gas} and ρ_{gas}, the gas pressure and density as a function of radius r (i.e. $n_e(r)$ and $T_e(r)$), allow $\phi(r)$ and so $M(r)$ to be determined (see later). The gas will arrange itself (convect) so that isobaric surfaces are on equipotentials which are roughly spherical, even for quite a lumpy or flattened mass distribution. Analysis therefore assumes spherical equipotentials, or at most two sets of spherical equipotentials (e.g. A754; Fabricant et al. 1986). Hydrostatic equlibrium will break down somewhere at the edge of a cluster where $t_s \sim t_a$, i.e.

$$R > 10 T_8^{\frac{1}{2}} \left(\frac{t_a}{10^{10} \text{ yr}}\right) \text{ Mpc.} \tag{8}$$

The relevant age for the cluster, t_a, is about 10^{10} yr, although as clusters are probably forming now by the accretion of subclusters (Geller 1984), a more appropriate time may be half that value. If the outer regions of the cluster gas have an adiabatic profile with temperature decreasing outwards then hydrostatic equilibrium may be lost at 3 to 5 Mpc.

If the gas is isothermal (which it probably is not) then hydrostatic equilibrium implies

$$kT\frac{dn_e}{dr} = -n_e \mu m \frac{GM(r)}{r^2},$$ (9)

and

$$\rho_{gas} \propto \exp\left(-\frac{\phi(r)}{(kT/\mu m)}\right).$$ (10)

Also, if the galaxies have an isothermal velocity distribution (also unlikely) then

$$\rho_{gal} \propto \exp\left(-\frac{\phi(r)}{\sigma_{los}^2}\right).$$ (11)

where σ_{los} is the line-of-sight velocity dispersion of the cluster. Then

$$\rho_{gas} \propto (\rho_{gal})^\beta,$$ (12)

where

$$\beta = \frac{\mu m \sigma_{los}^2}{kT}$$ (13)

(introduced by Cavaliere & Fusco-Fermiano 1976 as τ). We might expect $\beta \sim 1$ so $c_s^2 \approx \sigma_{los}^2$. The above isothermal-isothermal model suggests the use of King's (1966) approximation of an isothermal distribution where

$$\rho_{gal}(r) = \rho_{gal}(0)(1 + (\frac{r}{a})^2)^{-\frac{3}{2}},$$ (14)

so

$$\rho_{gas}(r) = \rho_{gas}(0)(1 + (\frac{r}{a})^2)^{-\frac{3\beta}{2}},$$ (15)

where a is the core radius. This profile can then be fitted to X-ray images of clusters (emissivity$\propto \rho_{gas}^2$). Jones & Forman (1984) obtain

$$\langle \beta_{image} \rangle = 0.65,$$ (16)

so

$$\rho_{gas} \approx \rho_{gas}(0)(1 + (\frac{r}{a})^2)^{-1}.$$ (17)

In two-thirds of clusters studied by Edge (1989), X-ray spectral and optical velocity-dispersion measurements of T and σ_{los} give

$$\langle \beta_{spec} \rangle \approx \langle \beta_{image} \rangle.$$ (18)

However, in the remaining one-third of clusters, $\beta_{spec} > 1$, and in the well-known case of the Perseus cluster, $\beta_{spec} \sim 1.5$, which is significantly greater than β_{image}. The reasons for this discrepancy are not yet clear, although it is unlikely that the gas and galaxies are isothermal. β_{spec} is therefore not defined for each cluster, although some mean for a cluster involving $\bar{T}$ and $\bar{\sigma}_{los}$ could be constructed. This need not correspond to β_{image} as they are weighted differently. There may also be considerable velocity anisotropy in the galaxy distribution. If clusters are lumpy, as seems to be the case, and some significant component of σ_{los} is due to individually bound subclumps, then the relevant σ_{los} for estimating the potential may be overestimated. There may be no discrepancy in the Coma cluster. The high quality data obtained with the Einstein Observatory allow more sophisticated fits to be made so that a single parameter like β loses its usefulness.

Earlier attempts to overcome a lack of knowledge of the equation of state of the cluster gas assumed that it is polytropic i.e $P \propto \rho^\gamma$. This does not necessarily mean that γ is the ratio of specific heats * and is little more than a mathematical expediency. Using it in the equation of hydrostatic support yields

$$\frac{\gamma}{(\gamma - 1)} \frac{k}{\mu m} \frac{dT}{dr} = -\frac{d\phi}{dr}, \tag{19}$$

so that

$$T = T_c + \frac{(\gamma - 1)}{\gamma} \frac{\mu m}{k} (\phi_c - \phi) \tag{20}$$

and

$$\frac{\rho}{\rho_c} = \left(1 + \frac{(\gamma - 1)}{\gamma} \frac{(\phi_c - \phi)}{(kT_c/m)}\right)^{1/(\gamma-1)}. \tag{21}$$

The subscripts refer to values at the centre. The density equation limits to the exponential isothermal form as $\gamma \to 1$.

The polytropic approach is still in common use, especially when needing to extrapolate to large radii. There is no particular reason, however, why the polytropic γ has to have a single fixed value throughout a cluster. It just measures how the cluster was set up and the conditions when the core was formed may have been quite different from those when the outermost atmosphere arrived. Furthermore, the clumpiness of clusters (Geller 1984, Forman et al. 1981) means that the cluster potential is time-dependent on large-scales (see also work by Cavaliere and colleagues). The core radius is also poorly defined. Most estimates of the total mass of gas in a cluster rely on some large extrapolation assuming γ is constant (but see Hughes 1989). This is the main reason why there are conflicting results in the literature (see e.g. Cowie, Henriksen & Mushotzky 1987 and The & White 1988). Note that the expected breakdown of hydrostatic equilibrium discussed earlier means that the extrapolation should not be extended much beyond $\sim 3\,\mathrm{Mpc}$.

It is possible to obtain gas density and temperature profiles without assuming an equation of state (Fabian et al. 1981). The X-ray surface brightness profile can be deprojected, assuming some geometry (e.g. spherical) and a distance to the cluster, to yield count emissivities as a function of radius. The emissivity depends upon n_e and T_e as well as the detector response and the effects of intervening photoelectric absorption, the last two of which are assumed known. A further relationship between n_e and T_e is obtained from the pressure via the equation of hydrostatic equilibrium. The densities obtained in this way are usually determined to better than 10 per cent, whereas the temperatures are somewhat dependent upon the assumed value of g (usually estimated from σ_{los}) that is used in the hydrostatic equilibrium equation. One pressure or temperature, typically the outer value is required to start the solution and this is usually chosen so that most of the cluster gas has

---

* The polytropic γ, γ_p, parametrizes the entropy profile in the cluster atmosphere which is presumably due to initial (or early) conditions. It does not mean that $P \propto \rho^\gamma$ if work is done on the gas. The ratio of specific heats γ, γ_{sh}, is relevant when work is done on the gas or it is displaced. γ_{sh} is probably always 5/3 for rapid changes. Note that gas with an isothermal profile, $\gamma_p = 1$, can have $\gamma_{sh} = 5/3$.

a temperature consistent with X-ray spectral measurements. These spectra have usually
been obtained with wide field-of-view experiments. When spatially resolved spectra become
available with future X-ray satellites such as ROSAT, AXAF and XMM, we shall be able
to solve directly for n_e and T_e without requiring g.

X-ray measurements of the hot gas in elliptical galaxies and clusters will eventually provide a powerful means for determining their gravitational mass profiles (see e.g.
Mushotzky 1987a). The equation of hydrostatic equilibrium may be written as

$$\frac{d\phi}{dr} = -\left(\frac{kT_{gas}}{\mu m}\right)\left(\frac{d\ln\rho_{gas}}{dr} + \frac{d\ln T_{gas}}{dr}\right). \tag{22}$$

$$\phi = \int \frac{GM}{r^2} dr \tag{23}$$

is thus obtained from the gas pressure and density profiles. Although the density profiles
measured so far are robust with respect to potential changes, the temperatures carry large
uncertainties. Some progress has been made using Einstein Observatory IPC spectra of
the cluster emission around M87 (Fabricant & Gorenstein 1983) and Focal Plane Crystal
Spectrometer spectra of the inner regions (Stewart et al. 1984). These measurements all
indicate that M87 is surrounded by an extensive dark halo. Confirming evidence is obtained
from velocity measurements of the globular clusters surrounding that galaxy (Mould, Oke
& Nemec 1987; Huchra & Brodie 1987). An upper limit to the total gravitational mass of a
cluster can be obtained simply from knowledge of the temperature at some radius, provided
that the gas is convectively stable (Fabian et al. 1986).

3. COOLING FLOWS

The intracluster gas is, of course, densest in the core of a cluster and its cooling time due
to the emission of X-rays such as those observed, t_{cool}, is shortest there. A cooling flow is
formed when t_{cool} is less than the age of the system, $t_a(\sim H^{-1})$.

In the cases considered here, t_{cool} exceeds the gravitational free-fall time, t_{grav},
within the cluster (except perhaps in some very small region at the centre), so, for a cooling
flow,

$$t_a > t_{cool} > t_{grav}. \tag{24}$$

The flow takes place because the gas density has to rise to support the weight of the overlying
gas.

If that is not immediately clear, consider the gaseous atmosphere trapped in
the gravitational potential well of the cluster or galaxy to be divided into two parts at the
radius, r_{cool}, where $t_{cool} = t_a$. The gas pressure at r_{cool} is determined by the weight of the
overlying gas, in which cooling is not important. Within r_{cool}, cooling is tending to reduce
the gas temperature and so the gas density must rise in order to maintain the pressure at
r_{cool}. The only way for the density to rise (ignoring matter sources within r_{cool}, which is a
safe assumption in a cluster of galaxies) is for the gas to flow inward. This is the cooling
flow.

If the initial gas temperature exceeds the virial temperature of the central galaxy
(which is generally the case for rich clusters but not for poor ones or individual galaxies)
then the gas continues to cool as it flows in. However, when the temperature has dropped
to the virial temperature of the central galaxy, the gas heats up as it flows further in due to
the release of gravitational energy. The gas temperature eventually drops catastrophically

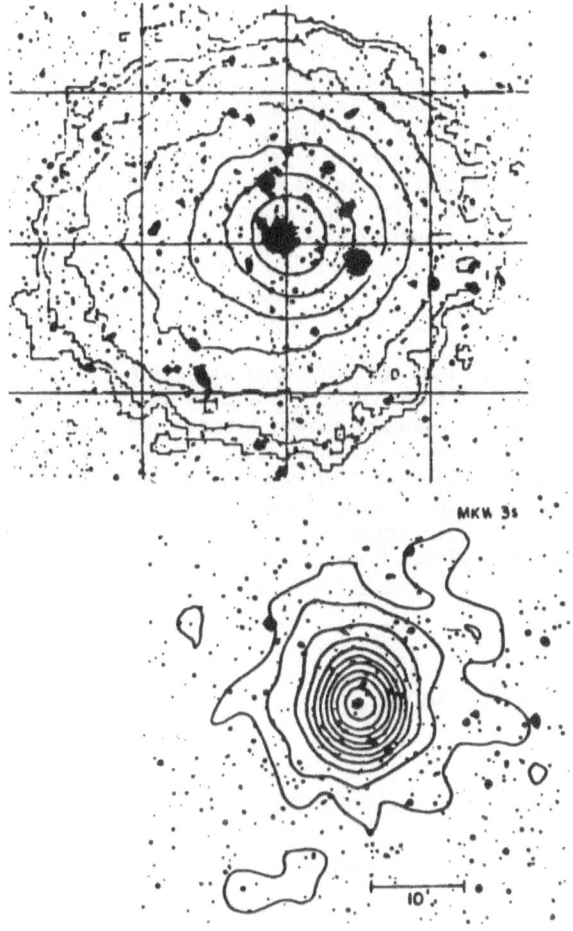

Figure 1. IPC X-ray surface brightness contours of the Perseus cluster (upper) and MKW3s (lower, from Kriss, Cioffi & Canizares 1983) superimposed on an optical image of the cluster. Note that the contours peak onto the central cluster galaxy. The mass deposition rates are about 200 and $100\,M_\odot\,\mathrm{yr}^{-1}$ respectively.

in the core of the galaxy if its gravitational potential flattens there. The net result is that the gas within r_{cool} radiates its thermal energy plus the PdV work and gravitational energy released in the flow.

This is how an idealized, homogeneous cooling flow, in which the gas has a unique temperature and density at each radius, will behave. Observations of real cooling flows shows that they are inhomogeneous and must consist of a mixture of temperatures and densities at each radius. The homogeneous flow is, however, still a fair approximation of the mean flow.

General reviews of cooling flows have been made by Fabian, Nulsen & Canizares (1984) and Sarazin (1986, 1988) and some other points of view may be found in the Proceedings of a NATO Workshop (Fabian 1988a). As explained above, the cooling flow mechanism is very simple, although the details of its operation are not. The primary evidence for them is in the X-ray observations. There is no evidence at other wavelengths for the large mass deposition inferred from the X-ray data. I discuss this point more fully later, but it should be stressed that large amounts of distributed low-mass star formation at other wavelengths need not be detectable if the gas is initially at X-ray emitting temperatures. This is, perhaps, the crux of the controversial aspect of cooling flows. They are difficult to prove or disprove in wavebands other than the X-ray. The X-ray evidence is, for me, sufficiently compelling that the existence of large cooling flows is a reasonable and straightforward conclusion.

It was Uhuru observations of clusters that first showed the mean cooling time of the gas in the cores of clusters to be close to a Hubble time (Lea et al. 1973). X-ray measurements from the Copernicus satellite showed that the core emission in the Perseus and Centaurus clusters was highly peaked (Fabian et al. 1974; Mitchell et al. 1975). These, and theoretical considerations, led Cowie & Binney (1977), Fabian & Nulsen (1977) and Mathews & Bregman (1978) to independently consider the effects of significant cooling of the central gas, i.e. cooling flows. The process was noted by Silk (1976) as a mechanism for the formation of central cluster galaxies from intracluster gas at early epochs. It has been pointed out to me by T. Gold that Gold & Hoyle (1959) proposed the process for galaxy formation thirty years ago.

3.1 X-ray evidence for cooling flows

3.1.1 X-ray Images

A sharply-peaked X-ray surface brightness distribution is indicative of a cooling flow. It shows that the gas density is rising steeply towards the centre of the cluster or group since the emissivity depends upon the square of the gas density and only weakly on the temperature*.

Most of the images have been obtained with the *Einstein Observatory* and with EXOSAT, although the peaks were anticipated with data from the Copernicus satellite (Fabian et al. 1974; Mitchell et al. 1975), from rocket-borne telescopes (Gorenstein et al. 1977) and with the modulation collimators on SAS 3 (Helmken et al. 1978).

Deprojection, or modelling, of the X-ray images shows that $t_{cool} < H_0^{-1}$ within the central 100kpc or so of more than 30 to 50 per cent of the clusters well-detected with the Einstein Observatory (Stewart et al. 1984b; Arnaud 1988) †. Whether H_0^{-1} should be used for t_a is debatable (but see §2), and it is not obvious how to extrapolate from 'well-detected'

---

* The spectroscopic data discussed in Section 3.1.2 rules out models in which the increased X-ray emission is due to populations of point sources around the central galaxy or to inverse Compton emission. Other points against such interpretations are the lack of peaks around the dominant ellipticals in the Coma cluster and the lack of any detailed spatial correlation with radio emission.

† More than two-thirds of the 50 X-ray brightest clusters in the Sky (see list in Lahav et al. 1989) have cooling flows (Pesce et al. 1989). Since the luminosity associated with the flow does not dominate the total X-ray emission, this high fraction is not a simple

clusters to all clusters. Inspection of the results shows that reducing t_a by 2, say, does not much change the fraction of clusters which contain cooling flows. The spatial resolution of the commonly-used IPC was not sufficient to resolve the central regions of the fainter or more distant clusters. The measured t_{cool} is then an upper limit. The overall picture is that the prime criterion for a cooling flow, $t_{cool} < 10^{10}$ yr, is satisfied in a large fraction of clusters. It is also satisfied in a number of poor clusters and groups (Schwartz, Schwarz & Tucker 1980; Canizares, Stewart & Fabian 1983; Singh, Westergaard & Schnopper 1986). Cooling flows must be both common and long-lived, in order that such a high fraction of peaked clusters is observed.

The mass deposition rate, $\dot{M}$, due to cooling (i.e. the accretion rate, although this is a poor term since most of the gas does not much change its radius) can be estimated from the X-ray images by using the luminosity associated with the cooling region (i.e. L_{cool} within r_{cool}) and assuming that it is all due to the radiation of the thermal energy of the gas, plus the PdV work done.

$$L_{cool} = \frac{5}{2}\frac{\dot{M}}{\mu m}kT, \tag{25}$$

where T is the temperature of the gas at r_{cool}. Values of $\dot{M} = 50 - 100\,M_\odot\,\text{yr}^{-1}$ are fairly typical for cluster cooling flows. (L_{cool} is similar to the excess luminosity measured by Jones & Forman 1984.) Some clusters show $\dot{M} \sim 500\,M_\odot\,\text{yr}^{-1}$ (e.g. PKS0745, A1795, A2597 and Hydra A). The main uncertainties in the determination of $\dot{M}$ are the gravitational potential within the cluster core and t_a. Assuming $t_a \sim 10^{10}$ yr, the estimates of $\dot{M}$ are probably accurate to within a factor of 2 (Arnaud 1988).

Since we often measure a surface brightness profile for the cluster core (where The X-ray emission is well-resolved), we have $L_{cool}(r)$ which can be turned into $\dot{M}(r)$, the mass deposition rate within radius r. Generally,

$$\dot{M}(r) \propto r. \tag{26}$$

This means that the surface brightness profiles are less peaked than they would be if all the gas were to flow to the centre. This means that the gas must be inhomogeneous, so that some of the gas cools out of the flow at large radii and some continues to flow in. The actual computation of $\dot{M}(r)$ is in detail complicated, since we need to take into account how the gas cools and any gravitational work done, but since plain cooling dominates in clusters, a simple analysis gives a fair approximation to the profile (see Fabian, Arnaud & Thomas 1986; Thomas, Fabian & Nulsen 1987; White & Sarazin 1987abc).

3.1.2 X-ray Spectra

Key evidence that the gas does actually cool is given by moderate to high resolution spectra of the cluster cores. Canizares et al. (1979, 1982), Canizares (1981); Mushotzky et al. (1981) and Lea et al. (1982) used the Focal Plane Crystal Spectrometer (FPCS) and the Solid State Spectrometer (SSS) on the Einstein Observatory to show that there are low temperature components in the Perseus and Virgo clusters, consistent with the existence of cooling flows. Detailed examination of the line fluxes and of the emission measures of the cooler gas by Canizares, Markert & Donahue (1988) and Mushotzky & Szymkowiak (1988) shows that,

---

consequence of the clusters being X-ray bright. It is due to the data on them generally being of the best quality (i.e. many X-ray counts detected and the core well-resolved).

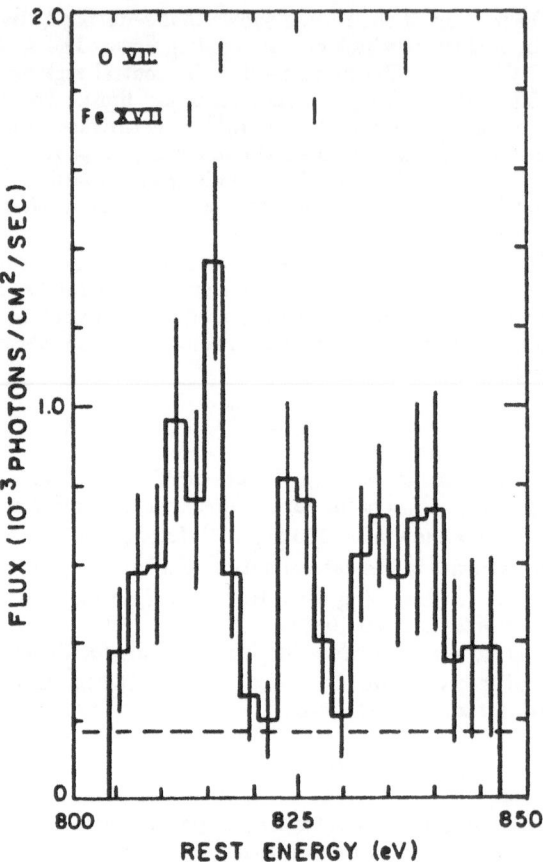

Figure 2. Part of the FPCS spectrum of the Perseus cluster (from Canizares, Markert & Donahue 1988). Note the prominent emission lines of OVIII and Fe XVII. The emission measure of the gas producing the Fe XVII lines is consistent with about $200\,M_\odot\,yr^{-1}$ of gas cooling through $\sim 5.10^6 - 10^6$ K. This is the $\dot{M}$ found from imaging and SSS studies, which are most sensitive to higher temperature gas (typically $5.10^6 - 3.10^7$ K).

in the case of the Perseus cluster, the gas loses at least 90 per cent of its thermal energy and that the mass deposition rates are in agreement with those obtained from the images. Good agreement is obtained also in several other clusters. The SSS results show that the emission measures vary with temperature in the manner expected from a cooling gas. *The importance of these data cannot be overemphasized since they show that the gas does cool.* Any 'alternative interpretation' of the images must confront this spectroscopic evidence successfully.

The cooling time of the gas in the Perseus cluster which emits the FeXVII line ($T < 5 \times 10^6$ K) is less than 3×10^7 yr. Since the emission measure of this gas agrees with that inferred from the gas cooling at the higher temperatures which dominate the images and the SSS result, we must conclude that the flow is steady (Nulsen 1988). The shape of

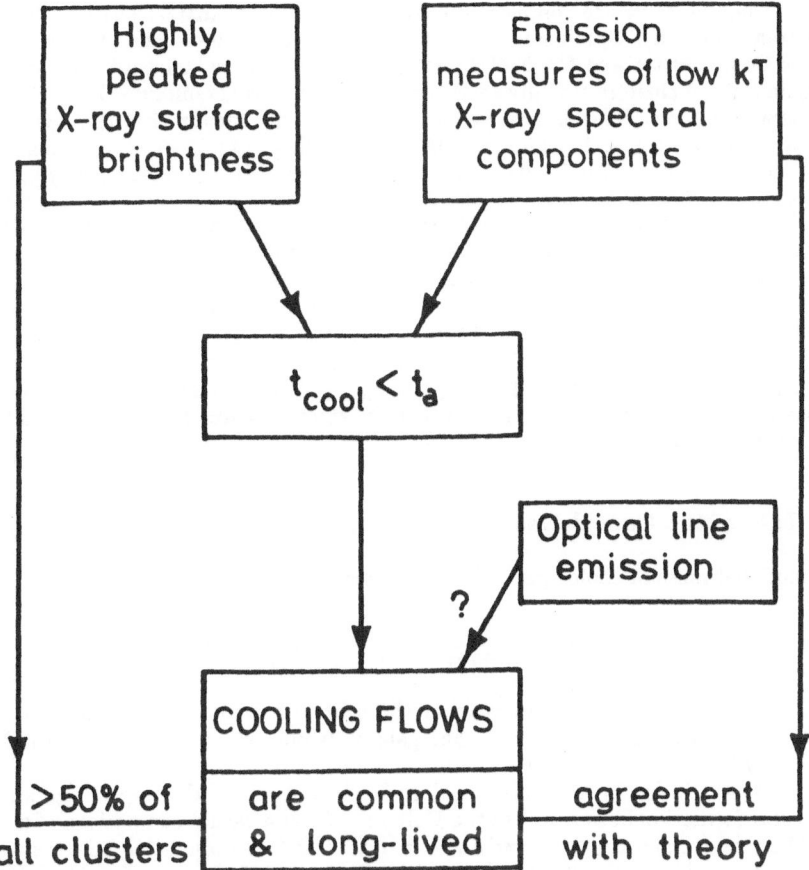

Figure 3. The evidence for Cooling Flows.

the continuum and line spectrum observed with the SSS is consistent with the same mass deposition rate at all X-ray temperatures (as expected) so we must again conclude that the flow is long-lived. It cannot be some intermittent or transient phenomenon only a billion years old.

3.1.3 Summary

The overwhelming evidence of the images and spectra shows that cooling does occur at a steady rate over long times (at least several billion years). Since mass is then cooling out of the hot phase at rates of hundreds of solar masses per year an inflow must occur. We do not expect yet to have direct evidence of any inward flow since the velocity is highly subsonic at $\sim 10\,\mathrm{km\,s^{-1}}$.

Cooling flows are common and all of the nearest clusters (Virgo, Centaurus, Hydra, Fornax, Perseus, Ophiuchus) contain one apart from the Coma cluster. Clusters such as the Coma cluster and CA 0340-54 with 2 large central galaxies orbiting each other are the main class of clusters that does not show strongly peaked emission. Even they could contain disrupted flows. The motion of the central galaxies means that there is no focus for the flow (Fabian, Nulsen & Canizares 1984). Many flows are observed out to a redshift of 0.1. A more distant one is in 3C295 at $z \sim 0.5$ (Henry & Henriksen 1987) and there is evidence from optical spectra for cooling flows being common around radio-loud quasars (Crawford & Fabian 1989). Most of the clusters detected by the Medium Sensitivity Survey of the *Einstein Observatory* must have the peaked, surface-brightness profile characteristic of cooling flows in order to have been detected (Pesce *et al.* 1989). Probably *all* clusters of galaxies will be found to have gas cooling out at rates exceeding a few solar masses per year when we have the improved spectral and spatial response of future missions such as AXAF.

The current values of $\dot{M}$ are probably good to a factor of 2 (Arnaud 1988) and could be higher if there are denser blobs beyond r_{cool} (Thomas *et al.* 1987). I am not aware of any alternative interpretation of the X-ray data which explains both the peaked images and the X-ray spectra.

3.2 The Fate of the Cooled Gas

The accumulated mass of cooled gas can be considerable;

$$\dot{M} t_a = 10^{12} \left(\frac{\dot{M}}{100\,\mathrm{M_\odot\,yr^{-1}}} \right) \left(\frac{t_a}{10^{10}\,\mathrm{yr}} \right) \mathrm{M_\odot}. \qquad (27)$$

This is a significant fraction of the mass of the central galaxy. It suggests that we are witnessing the continued formation of that galaxy, which is typically one of the largest galaxies known.

If the gas forms stars, then cooling flows are some of the largest and strongest regions of star formation in our part of the Universe. Even a casual comparison of a central cluster galaxy and a spiral galaxy such as our own, which is thought to be forming stars at a rate of $3 - 10\,\mathrm{M_\odot\,yr^{-1}}$, shows that cooling flows must form low-mass stars (Fabian, Nulsen & Canizares 1982; Sarazin & O'Connell 1983). Massive stars would make central cluster galaxies much bluer than they are. The absence of massive blue stars means that star formation with a spiral galaxy initial-mass-function (imf) is almost non-existent in the central galaxies of cooling flows.

It should be stressed that the cooling gas is not directly detected once it has cooled below about 3×10^6 K. If it recombines and forms low-mass stars ($\langle M_* \rangle \ll 0.5\,\mathrm{M_\odot}$) in a distributed manner ($M(r) \propto r$) then there is no reason for it to have been seen.

There is, however, plenty of evidence for dark matter in clusters and low-mass stars are one plausible form of dark matter. The manner of the mass deposition with radius, $\dot{M}(r) \propto r$, leads to an isothermal halo which is consistent with the dark matter distribution around large galaxies (*e.g.* M87, Stewart *et al.* 1984a; Mould *et al.* 1987). Cooling flows are a source of baryonic dark matter.

3.3 Heating

Since the implied star formation rates are so large and there is little sign of it optically, there have been a number of studies suggesting that the rates have been grossly over-estimated.

Some heat source that balances the cooling is the obvious solution. Cosmic rays (Tucker & Rosner 1982), conduction (Bertschinger & Meiksin 1986), supernovae (Silk *et al.* 1986) and galaxy motions (Miller 1986) have all been invoked as heat sources. Unfortunately for these models, the X-ray spectra indicate cooling without heating. None of the models proposed so far is able (or even attempts!) to account for the X-ray line emission. There are other problems with these heat sources as well (see Fabian 1988b; Bregman & David 1988).

To consider heat sources in more detail, we note that it is generally difficult to keep the gas stable whilst heating it (Stewart *et al.* 1984a). Most heat sources, such as cosmic rays, heat at a rate proportional to the gas density, whereas the cooling varies as the density squared. Heating may then cause the gas to become more unstable by tending to increase the temperature and pressure of the lowest density phases but allowing the denser gas to carry on cooling. If conduction occurs unimpeded then cooling of the core gas can set up a temperature gradient that is offset by a conductive heat flux. Whilst such a situation can occur, it is restricted to only a small part of parameter (initial density and temperature) space. The energy equation (at constant pressure) is

$$\frac{\rho v}{\mu m}\frac{5}{2}k\frac{dT}{dr} = n^2\Lambda - \frac{1}{r^2}\frac{d}{dr}\left(r^2\kappa\frac{dT}{dr}\right). \tag{28}$$

The first term on the r.h.s. is radiative cooling, which at constant pressure and for bremsstrahlung, varies as $T^{-\frac{3}{2}}$. The second term is the conductive heating flux which varies as $T^{\frac{1}{2}}$ (it does not depend on density). The widely different temperature dependences of these terms makes it difficult to allow a balance where one term does not dominate. Conduction dominates where the temperature is high and cooling where it is low. Generally, cooling dominates at the centre of a flow and conduction can be important further out (Nulsen *et al.* 1982). If conduction is dominant, then it tends to make the gas almost isothermal, in disagreement with observations. The X-ray spectroscopic observations (Canizares *et al.* 1979, 1982, 1988; Mushotzky *et al.* 1981, 1988) then demonstrate that conduction is inhibited, probably by tangled magnetic fields which greatly decrease the effective electron mean-free-path. We shall return to this later in §4.4.

Supernova heating from stars formed from cooled gas (Canizares *et al.* 1982; Lea *et al.* 1982; Silk *et al.* 1986) can at most change the mass deposition estimates by a factor of two. This is because the energy from a supernova can only heat a mass of gas sufficient to form another supernova progenitor (and an IMF's worth of lower-mass stars) to 8.10^6 K. There is no evidence for supernovae around the central galaxies in cooling flow clusters (see e.g. Caldwell & Oemler 1981).

Finally on the topic of heating, it is worth remembering that cooling flows occur in a wide variety of clusters, both with and without strong radio sources (e.g. Cyg A vs. AWM4) and in deep and shallow potential wells (e.g. Perseus vs. Hydra). Any heat source necessary to counteract the radiative cooling would represent a major heat flow, of $\gtrsim 10^{62}$ erg per cluster. In my view, the current lack of understanding of star formation in our own Galaxy means that it is not a simple business to extrapolate to other situations. Star formation can proceed at hundreds of solar masses per year without necessarily being evident optically, provided that only low mass stars are formed. There is no problem with the total mass deposited, which is distributed out to 100 - 300 kpc from the central galaxy and does not pile up within its centre. Clusters are full of dark matter and there can be no problem with some (or even all) of it being baryonic.

It has been suggested by Hu (1988) that cooling flows began only recently. This reduces that total accumulated mass but does not explain the lack of blue stars. The imf must be different from our local imf even if the flow is recent.

The total level of heating necessary to balance the cooling is very large, $\sim$ 10^{62} erg for a large flow over t_a and so if some heat source is found that can accommodate the X-ray spectral measurements successfully, it must be one of the major (unseen!) energy flows in the Universe! Whilst the luminosity of a cooling flow may be only 10 per cent of the total cluster X-ray luminosity and the mass lost through cooling a negligible drain on the enormous outer atmosphere, the cooling luminosity is a major loss of energy from the cluster core. Whatever is eventually decided about cooling flows, they cannot be an insignificant process.

3.4 Star Formation in Cooling Flows

As already mentioned, the average mass of a star formed in a cooling flow must be considerably smaller than in our Galaxy. In particular, the fraction of the mass turned into massive OB stars must be very small, since there is little ultraviolet light seen with the IUE (*e.g.* Fabian, Nulsen & Arnaud 1984). (The shortest cooling times for the X-ray emitting gas ($\sim 3 \times 10^7$ yr) are comparable to the lifetime of B stars, so intermittency of the flow cannot be important here.) In understanding why there are these differences, it would be helpful to have a predictive theory of star formation for our Galaxy. As we do not, we must look for differences. Some are a) the lack of dust* in gas that has cooled from $T > 10^7$ K (Draine & Salpeter 1979), which presumably means that there are no molecular clouds such as give birth to massive stars in our Galaxy; b) the thermal pressure of the gas is 100 – 1000 times higher than in our interstellar medium; c) differential motions, cloud masses, angular momentum and magnetic field strengths may be different.

The general statement about the necessity for low mass stars applies to the bulk of the cooled gas. In the centre, there are often seen optical emission line blobs or filaments, which may be atypical of most of the flow. These blobs may give rise to higher mass stars. There is some excess blue light observed at the centres of many cooling flows (see *e.g.* Hintzen & Romanishin 1988) and it does correlate in strength with the mass deposition rate (Johnstone, Fabian & Nulsen 1987). Spectral fits of the blue light together with upper limits from IUE spectra show that the upper mass limit for stars must be around $1.5 - 2 \, M_\odot$ there (Crawford 1988; Crawford *et al.* 1989). Some F and early G stars are seen. Of course, the best place to look for the bulk of the cooled gas is at large radii where the underlying stellar light of the galaxy is least, so the contrast is highest.

Small cooling flows occur continuously in most elliptical galaxies which are not in rich clusters. Ram-pressure stripping removes the gas from galaxies within clusters. This is observed in the Virgo cluster where M86 has a plume of X-ray emission to the NW (Forman *et al.* 1979). A faint diffuse patch of optical light is observed coincident with this plume and may be due to stars formed from the cooling gas (Nulsen & Carter 1986).

An exciting possibility is that the Giant, Red Envelope Galaxy (GREG; Maccagni *et al.* 1988) found from the Einstein Medium Sensitivity Survey is showing the low-mass stars formed in a cooling flow (Johnstone & Fabian 1989). This galaxy lies in the centre of an X-ray luminous poor group of galaxies and has a normal de Vaucouleurs profile in the V-band. A large r^{-1} envelope appears in the i-band, which can plausibly be explained as due to $0.5 \, M_\odot$ stars. If due to starlight, then the envelope requires an exceptional imf of the

---

* Hintzen & Romanishin (1988) find that the central parts of the cooling flow galaxy 3A0335+096 are red. Whether this is due to red stars or to dust is not clear. NGC 4696 in the Centaurus cluster, which has a cooling flow, also has a dust lane, so some dust is found in the middle of cooling flows. It may just be due to stellar mass loss from the galaxy itself.

kind required by cooling flows. Its profile is consistent with mass deposition in a cooling flow, which must have been more massive in the past.

4. INHOMOGENEOUS INTRACLUSTER GAS

The distributed manner of the mass deposition shows that the cooling flow is inhomogeneous. This means that it contains blobs of gas that are denser than the surrounding gas. The evidence for distributed mass deposition is obtained principally from the X-ray surface brightness profiles, which are less peaked than expected if all the gas flowed into the very centre of the cluster. The gas emitting the Fe XVII line must also be distributed with a small filling factor or its surface brightness would exceed that observed (Canizares et al. 1988).

4.1 Distributed Mass Deposition

The mean conditions of an inhomogeneous flow are represented by the equations of a homogeneous flow. We have the equation of continuity;

$$\dot{M} = 4\pi r^2 \rho v, \tag{29}$$

the pressure equation (ignoring highly subsonic flow terms)

$$\frac{dP}{dr} = -\rho \frac{d\phi}{dr}, \tag{30}$$

and an energy equation,

$$\rho v \frac{d}{dr}\left(\frac{5}{2}\frac{kT}{\mu m} + \phi\right) = n^2 \Lambda, \tag{31}$$

where Λ is the cooling function. If the cooling region (where $t_{cool} < H_0^{-1}$) is at constant pressure ($d\phi/dr = 0$), then $n \propto T^{-1}$ and

$$\rho v \frac{d}{dr}\left(\frac{5}{2}\frac{kT}{\mu m}\right) = n^2 \Lambda \tag{32}$$

and if

$$\Lambda \propto T^\alpha, \tag{33}$$

then

$$n v \frac{dn^{-1}}{dr} \propto n^{2-\alpha}, \tag{34}$$

so

$$v \frac{dn}{dr} \propto n^{3-\alpha}. \tag{35}$$

From continuity, if $\dot{M}$ is constant,

$$v \propto n^{-1}r^{-2}, \tag{36}$$

so

$$\int_\infty^n \frac{dn}{n^{4-\alpha}} \propto \int_0^R r^2 dr \tag{37}$$

and

$$n \propto R^{-3/(3-\alpha)}. \tag{38}$$

This is proportional to $R^{-\frac{2}{3}}$ for bremsstrahlung. The density rises inward as the temperature falls. Constant pressure is a fair approximation to the core region of a cluster. Gravity is not particularly important, except perhaps for focussing the flow, until the gas has cooled to about the virial temperature of the central galaxy. Then the gas heats up as it flows in further and the pressure rises (Fabian & Nulsen 1977). The flow velocity $v \simeq r/t_{cool}$, which is highly subsonic.

When the flow is inhomogeneous we can estimate $\dot{M}(r)$ by assuming that the gas is composed of a number of phases, the densest of which cools out of the flow at the radius under consideration (Thomas et al. 1987), or by model fitting (White & Sarazin 1987). In the first approach, the cooling region is divided into a number of concentric shells of size compatible with the instrumental resolution. The luminosity, δL_i of the ith shell can then be considered to be the sum of the cooling luminosity of the gas cooling out at that radius from the mean temperature T_i at rate $\delta \dot{M}$ and the luminosity of gas flowing across the shell experiencing temperature and potential changes ΔT_i and $\Delta \phi_i$;

$$L_i = \delta \dot{M}_i \frac{5}{2} \frac{kT_i}{\mu m} + \left(\frac{5}{2} \frac{k \Delta T_i}{\mu m} + \Delta \phi_i \right). \tag{39}$$

In our most detailed approach (Thomas, Fabian & Nulsen 1987), we have allowed for as many phases at a radius as there are shells within that radius and have integrated the cooling function and spectrum carefully. A typical mass deposition profile is shown in Fig. 2. It agrees fairly well with that obtained by assuming that the gas is homogeneous, principally because most of the energy is lost on cooling from the average cluster temperature T_X at temperatures close to T_X. This new approach does allow us to measure the spread of densities in the gas at any radius. It is this which determines the manner in which mass is deposited (Nulsen 1986). We infer that the intracluster gas must contain a density spread of at least a factor of two. This may not be surprising when it is recalled that it has been enriched in metals which must have mixed different gases together.

The result that $\dot{M}(r) \propto r$ means that the deposited matter has $\rho \propto r^{-2}$ which is essentially an isothermal halo such as inferred for the dark matter around galaxies. It is assumed that whatever condenses out of the cooled gas orbits about, or through, the central galaxy such that its mean radius is similar to that where it was formed.

4.2 The Behaviour of Gas Blobs

How the cooling gas blobs behave is ill-understood. Malagoli et al. (1987) and Balbus (1988) have shown that cooling flows are not expected to be thermally unstable and cannot generate sizable blobs from initially infinitesimal perturbations. A region that is slightly overdense with respect to its surroundings will fall ahead of the flow under gravity and join a region of similar properties to itself. Computations of the oscillations of overdense blobs are discussed by Loewenstein (1989); further computations have been made by Hatori et al. (1989) and by Brinkmann & Massaglia (1989).

As already mentioned, in the limit of zero viscosity the gas in a cooling flow should be homogeneous and cool into a central singularity. This is because gravity causes any denser gas to fall ahead and join gas with similar properties. Only near the centre of the flow, where $v \sim v_{ff}$ the free-fall velocity, would the flow become inhomogeneous. However a real flow does have viscosity, is turbulent and contains magnetic fields. Consequently,

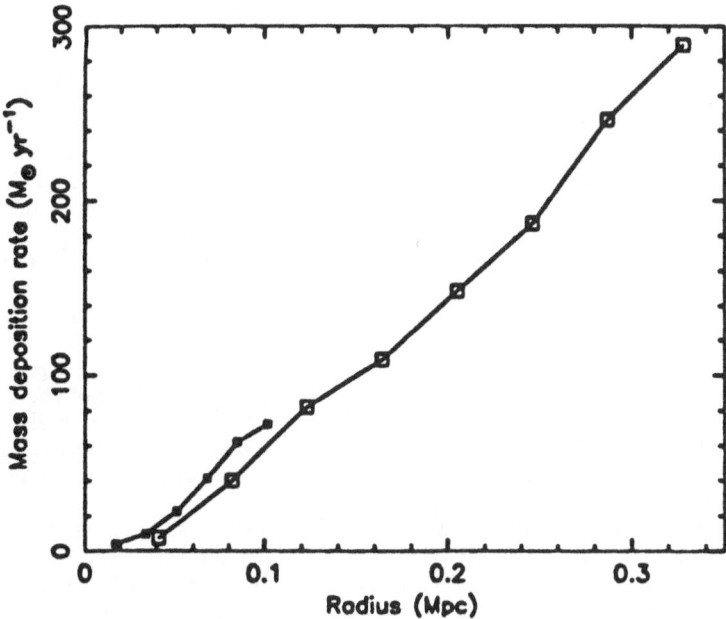

Figure 4. Mass deposited within radius r by the cooling flow in A2199, from Thomas *et al.* (1987).

the linear perturbation analyses are not particularly relevant to a real cooling flow. The non-linear behaviour of gas blobs in a flow has been explored by Nulsen (1986). A large gas blob of size r and overdensity $\delta\rho$ will try to move ahead of the mean flow and reaches a terminal velocity

$$v_T \simeq v_{Kepler}\sqrt{\left(\frac{\delta\rho}{\rho_0}\frac{r}{R}\right)}. \tag{40}$$

This relative motion will then cause the blob to spread out and fragment (Nittman, Falle & Gaskell 1982). r/R is reduced and the relative velocity of the overdense gas, v_T, is reduced. Magnetic fields can help to pin the gas to the mean flow so that it comoves (see also Loewenstein 1989). The net result is that large, slightly overdense blobs at large radii from the centre of the flow are turned into an emulsion of smaller and very overdense blobs at smaller radii.The densest gas will cool out of the flow (i.e. $T \rightarrow 0\,\mathrm{K}$) at intermediate radii. The density distribution of gas at a given radius will tend to evolve a 'cooling tail' (volume filling fraction $f \propto \rho^{-(4-\alpha)}$, where α is the exponent of the cooling function; Nulsen 1986; Thomas, Fabian & Nulsen 1987). This allows mass to be deposited in a distributed manner. If a spread of densities exists throughout the cluster, then gas may be deposited by cooling well beyond the radius where the mean cooling time is H_0^{-1}.

Density inhomogeneities will have been introduced early into the cluster gas by the production of metals, by any quasars in the cluster and from the stripping of gassy galaxies and winds. Later it will be further mixed and inhomogenized by the infall of subclusters. A further spread of density will occur because of the motion of the central cluster galaxy and turbulence. Convection, with magnetic fields and viscosity, will then create a limited range of densities throughout the gas.

4.3 Direct Observations of Gas Clouds in Cooling Flows

Optical observations of the central galaxies in many cooling flows show patches of line-emitting nebulosity (Hu, Cowie & Wang 1985; Johnstone, Fabian & Nulsen 1987; Heckman *et al.* 1989). The spectra indicate that the gas has a relatively low ionization ([OII]>[OIII]). The Hβ luminosity of the gas ranges from 10^{39} to 10^{42} erg s^{-1} which means, at the pressure of intracluster gas, that the mass of warm, line-emitting gas ($10^3 < T < 10^4$ K) is in the range of $\sim 10^5 - 10^8$ M$_\odot$. To be spread over the observed regions, the gas must be in the form of small clouds or filaments with a volume filling factor of $10^{-5} - 10^{-7}$. The emission lines show doppler widths indicating both random and systematic (rotation?) velocities of up to several hundred km s^{-1}.

The origin of this gas and its source of ionization is unclear. It probably originates from the cooling flow, but why only such a small fraction is observed is puzzling. When observed (there are flows in which no optical emission lines are observed, *e.g.* A2029), the emission is strongly peaked to the centre (Heckman *et al.* 1989) and is not consistent with, say, $L_{H\beta}(<r) \propto r$. It is far too luminous to be plain recombination of the cooling gas. Each proton must recombine 100 to 1000 times to produce the observed luminosity. Most simple sources of ionization, such as photoionization by massive stars or by an active nucleus, have been ruled out (Johnstone & Fabian 1988).

Whether there is more colder (recombined) gas also associated with it is unknown. Limits on the amount of HI by its emission are not very restrictive ($\sim 10^9$ M$_\odot$; Burns, White & Haynes 1981; Valentijn & Giovanelli 1982), especially when it is realized that any HI is probably optically thick (Loewenstein & Fabian 1989). HI has been seen in absorption against NGC1275 in the Perseus cluster by Crane *et al.* (1982) and by Jaffe *et al.* (1988). Lazareff *et al.* (1989) and Mirabel *et al.* (1989) have also observed CO in NGC1275 indication that there is a very large quantity of gas in the centre, if a Galactic CO to H_2 conversion factor is used. (This factor is probably a gross overestimate, since it assumes that line widths are due to self-gravitation, which is unlikely in a turbulent cooling flow where the velocity widths just reflect the motions of the hotter gas which carries the cold gas around, as I discuss later.)

Whatever the origin or amount of the warm gas, it can be used to probe the proerties of the hotter gas, such as its pressure and its motions. The gas pressure, for example, can be obtained from the [SII] lines, which are collisionally de-excited at densities above a few 100 cm^{-3}. In the Perseus cluster, these lines change their ratio indicating high density and thus high pressure within 5kpc of the nucleus (Johnstone & Fabian 1988). Since the pressure has risen above the X-ray inferred pressure (from the X-ray surface brightness) at 20 kpc, the mean gas temperature must be down to the virial temperature of the central galaxy, NGC1275, of about 10^7 K. (Hydrostatic equilibrium requires that the pressure increases inward once the gas has cooled to the local virial temperature.) This is further confirmation that the gas has cooled there below the outer temperature of $\sim 6 \times 10^7$ K. Since the gas pressure is so high there, the magnetic pressure cannot be more than about twice the gas pressure (and is probably less than that).

4.4 The Survival of Clouds

The properties of the warm clouds in the cores of cooling flows can be used as a diagnostic of the properties of the hot gas. The mere existence of the clouds argues that thermal conductivity is suppressed by a large factor below the 'Spitzer value', κ_S. The motions of the clouds implied by the optical line widths pose enormous survival problems for the clouds if they are moving relative to a static intracluster medium. Cloud break-up is a very significant problem, as discussed in §4.2.

Turning first to the problem of conduction, consider a small sheet of thickness λ at temperature T_s embedded in the hot intracluster gas at temperature, T_X. Then, from the definition of conductivity, the heat flow per unit area

$$\frac{dq}{dt} = f\kappa_S T_X^{5/2} \frac{dT}{dx}, \tag{41}$$

where a factor f is included to account for the suppression of conductivity by a tangled magnetic field. If we assume that the temperature gradient can be approximated by T_X/λ, then the timescale for the cloud to be evaporated

$$t_{cond} \approx \frac{nkT_s\lambda^2}{f\kappa_S T_X^{\frac{1}{2}}}. \tag{42}$$

The cloud will not be evaporated if the conducted heat can be radiated away on a timescale, t_{rad}, less than t_{cond}. Instead it will grow (Bohringer & Fabian 1989). Now,

$$t_{rad} = \frac{3kT_s}{n\Lambda}, \tag{43}$$

where Λ is the appropriate cooling function at T_s. The cloud will then survive against evaporation if its size

$$\lambda > 8f^{\frac{1}{2}}T_7^{\frac{7}{4}}n^{-1}\Lambda_{-22} \text{ kpc}, \tag{44}$$

where $T_7 = 10^7 T_X$ K and $\Lambda = 10^{-22}\Lambda_{-22}$ erg cm^{-3} s^{-1} ; $\Lambda_{-22} \sim 1$ is appropriate if $T_s \sim 10^4$ K. If we also assume that the region is at constant pressure, we obtain a steeper temperature dependence by eliminating the density. For the whole cooling flow, $T_7 \sim 5$ and we see that f must be less than 1 for cooling to dominate. For small parsec-size clouds embedded in the flow, $f < 10^{-4}$.

The thermal conductivity must be highly suppressed. Tangled magnetic field lines are the obvious reason since it is the electrons that dominate the conductivity and they spiral around the field lines. It is possible that magnetic mirrors trap the electrons and cause f to be very small. It should be remembered that the magnetic pressure in the intracluster gas is less than one per cent of the thermal energy so that the field is very tangled if the gas is turbulent. The cooled blobs may have their own separate magnetic field structure that is little related to their surroundings. A cloud would not have cooled out if it was thermally couple to the hotter gas. Consequently, it may be reasonably thermally isolated from the hot gas in which it is embedded.

Tribble (1989) has produced an interesting model of thermal conductivity in a plasma with very tangled magnetic field. The electrons random walk rather than flow. The resulting conductivity coefficient involves the length scale of the temperature gradient, meaning that the conventional approach has totally broken down.

Now to turn to cloud break-up by motion through the hotter gas. A cooled cloud is destroyed by the gravitational (Jeans) instability and by the Rayleigh-Taylor and Kelvin-Helmholtz instabilities. The first will turn a cloud into stars and limit the population of large, cool clouds in the flow. The second pair of instabilities act on a cloud moving relative to the mean flow in such a way as to shred it and make it smaller (see also §4.2 and Nulsen 1986). Loewenstein & Fabian (1989) have studied the temperature – mass parameter space available to clouds with the velocities implied by the observed large line widths. The relative motion eventually destroys all clouds unless they have sufficient self-gravity (or magnetic tension) to keep intact. Sufficient self-gravity means Jeans unstable, or almost so, so there is very little parameter space available to the observed clouds. Magnetic fields are of little help since the [SII] pressure measurements show that the field cannot be large enough. The parameter space is opened up further if we require only that the clouds accelerate to some observed velocity before being destroyed. Even then the parameter space available to the observed clouds is very restricted.

4.5 Turbulence in Intracluster Gas

Loewenstein & Fabian (1989) resolve the cloud destruction problem posed in the last section by concluding that the velocity linewidths are the result of emission from small clouds formed in cluster gas with large-scale chaotic motions, *i.e.* the hotter gas is turbulent. Provided that these motions exceed $\sim 100\,\mathrm{km\,s^{-1}}$, clouds formed by cooling from the hot gas will have the required velocity width. Clouds slowly pick up a velocity relative to the hot gas, if its motions are chaotic, and are destroyed or broken into much smaller clouds. The timescale for destruction is $\sim l/u$, where l is the lengthscale of the motions (or the radius of curvature of their streamlines) and is 10^8 yr when l is 10 kpc and u is 100 km s^{-1}. This is considerably longer than the radiative cooling time of the warm clouds. It is likely that large clouds are slowly broken into smaller clouds which have smaller relative velocities with the flow and so last longer. The line emission from the whole population of clouds will have a velocity width appropriate to the velocity spread of the hot gas.

We can obtain limits on the amount of turbulence throughout the cooling flow region from the following arguments. The rate of energy dissipated by the smallest eddies will be on the order of

$$\varepsilon_d \sim \frac{u^3}{l}. \tag{45}$$

If the turbulent velocity scale, u, and the length scale, l, are 100 km s^{-1} (the order of observed linewidths) and 10 kpc (the order of the observed maximum scale of magnetic field tangling from Faraday depolarization studies; Garrington *et al.* 1988, Dreher *et al.* 1987; Laing 1988), respectively, then ε_d exceeds the inferred cooling rate $\rho\Lambda$ at radii greater than 20 – 50 kpc. If such a strong constant heating source were present it would give a much flatter surface brightness profile than observed. Similarly, turbulent diffusion would also affect the distribution of cooling material since the diffusion time

$$t_d = \left(\frac{l}{u}\right)\left(\frac{r}{l}\right)^2 \tag{46}$$

is $\sim 10^{10}$ yr out to 100 kpc which is on the order of the cooling time there. If turbulence and mixing are important out to the cooling radius, the resulting mass deposition profile ($\dot{M} \propto r^3$) would be steeper than that observed ($\dot{M} \propto r$). These arguments suggest that the turbulent velocity field within the flow is such that either the length scale increases, or the velocity decreases, outward.

These considerations do not rule out the possibility of turbulence being an important physical process in the *central* regions (the inner few kpc) of cooling flows where the *observed* optical emission-line gas lies and where the X-ray emission is unresolved. The gas will be more susceptible to large-scale (and therefore longer-lived) turbulence where the gravitational potential flattens. This can occur at cluster radii where the overall cluster gravitation declines (inside the core radius around ~ 250 kpc) and the gravity of the central galaxy (possibly including an extended non-luminous component) has yet to become important, and also well inside the central galaxy itself. The former region may be a fertile ground for the growth of density perturbations; the latter can be effectively extended by the motion of the central galaxy.

Pringle (1989) has pointed out that the steep negative density gradient in a cooling flow can focus sound waves (noise) in the intracluster gas into the core of the cooling flow. This can help to heat the cooling gas there and may provide a source of energy for the warm clouds and could even 'shut-off' the flow if the noise energy can be dissipated in the hot gas at the necessary (large) rate. Balbus & Soker (1989) have considered the propagation of gravity waves in the core of a cluster. These can transport energy outward, meaning that the mass deposition rates can be higher than originally estimated.

There are several sources of energy that could drive turbulent motions, and noise, in the inner 10-50 kpc in cooling flows. In addition to galaxy motions (including motion of the central galaxy), there is likely to be dissipation of magnetic energy from fields frozen in and advected inwards with the cooling flow as well as dissipation of rotational energy and angular momentum. As discussed by Nulsen, Stewart & Fabian (1984), turbulent viscosity is the most likely transport process of angular momentum in a cooling flow. Even small amounts of angular momentum carried in with the flow would become dynamically important at radii of a few kpc (Cowie, Fabian & Nulsen 1980). The tangential motions provoked by the angular momentum should pump turbulence in the hot gas.

Turbulence may be important on the scale of the entire ICM. It can be driven by the infall of subclusters and by the motion of the cluster potential well if it is highly structured, as seems likely for many clusters. Turbulence is damped by the subadiabatic gas distribution which is stably stratified in the cluster gravitational field. (It requires the expenditure of energy to exchange denser gas deeper in the potential well with less dense gas further out.) If not continually excited, turbulence dissipates into smaller scales on a turnover time ($\sim l/u$). Consequently, we can expect that motions persist on scale l with a speed $u = l/t_{sub}$ for a time t_{sub} after the last subcluster joined the cluster (assuming that it was massive enough for its kinetic energy to be sufficent to power turbulence throughout the cluster). Such motions can be large (hundreds of $km\,s^{-1}$ on scales of 0.5 Mpc or so) for a few billion years after a large subcluster joins a cluster. The resulting substructure in the gravitational potential of the cluster will also drive more turbulent motions for a longer time.

The chaotic motions of the intracluster gas will dissipate eventually as heat, and will also mix the gas up, making it inhomogeneous. Consider two clusters colliding, both of which contain dense gas in their cores (and probably cooling flows). During the collision, some of the outer gas will be shocked to high temperatures ($> 10^8$ K), whilst the bulk of the gas is compressed and heated. The denser cores of the subclusters attempt to sink together to the core of the final cluster, partly breaking up into many blobs as described in §4.2. The resulting intracluster gas should be inhomogeneous on all scales smaller than that on which the breakup time exceeds the time since the collision. A cooling flow will be detectable if one central galaxy dominates in the core and acts as a gravitational focus for the cooling gas. If clusters are assembled in this hierarchical manner, then an inhomogeneous, mildly turbulent, intracluster medium should be the norm.

Returning now to the inner parts of a cooling flow, we note that the presence of turbulent hot gas with entrained cool clouds in this region can simultaneously explain both the observed velocity widths and the unaccounted for heating mechanism of these clouds if the bulk kinetic energy can be successfully converted into heat. The optical (and UV) emission-line energy can be a significant fraction of the total energy radiated inside $\sim$ 30 kpc in some of the optically spectacular flows (Heckman et al. 1989).

4.5.1 Turbulence and the Wave Heating of Clouds

As already discussed, the motion of the dense clumps relative to intercloud gas, the transport of angular momentum, the motion of the central cluster galaxy and of the other galaxies, all contribute to making the intercloud gas turbulent. This means that there is chaotic kinetic energy that is feeding into smaller and smaller scales at a rate given by equation (45). Eventually it must dissipate, possibly on scales of the order of an ion mean-free-path. The intracluster medium is thus expected to be very noisy, especially in its innermost regions. Mechanical heating of embedded clouds can then occur, much as in our ISM (Cox 1979). There is a difficulty presented by the enormous density contrast between the intercloud gas and the cloud gas which is an impediment to high transmittance of power. It can be overcome if the power is transported via magnetic fields which are a significant fraction of the total pressure. Magnetic fields passing from the warm to the hot gas (the lines may be closed) can be forced to oscillate by the noise and so cause the ionized particles within the cloud to oscillate and collide with the neutral particles. This can efficiently transport (primarily by Alfven waves), and dissipate, wave energy (Kulsrud & Pearce 1969)

This process of 'plasma slip' (Spitzer 1982) has been suggested as a heat source for clouds in the Galactic ISM by Arons & Max (1975); Silk (1975); Elmegreen, Dickinson & Lada (1978) and by Spitzer (1982). It can balance the cooling rate around 10^4 K and can penetrate well into a large cloud. A lot of power will be available on the appropriate length scales if the clouds are a major source of dissipation of this energy. The turbulent power then cascades down to the scales where it can be efficiently dissipated in the clouds.

This magnetic viscous heating transports the energy without requiring a shock. Direct mechanical heating, at a rate $\rho v^3/l$, where v is the velocity of the motion and l is the dissipation length scale, is only competitive if $v \sim c_s$, the speed of sound in the warm gas and so leading to shocks if large. Of course, the real situation is likely to be very complicated and involve both of the above processes to some extent. Magnetic fields are expected to be amplified in the flow, however, and so magnetic processes can be important.

The optical emission-line spectra are consistent with the low ionization of $\sim 100 \, \mathrm{km \, s^{-1}}$ shocks. It is not clear that plasma slip can do more than 'hang up' the cooling of some clouds around a few 10^3 K. However, this makes the clouds susceptible to cloud-cloud collisions, which occur at appropriate relative velocities, and also to continual shredding and mixing of their outer layers. This last process means that a cloud is surrounded by small cool pieces mixing in with the hotter gas. A temperature gradient is thereby established, depending upon the details of the mixing and evaporation (which in turn rely on the magnetic field configuration). This *may* lead to a radiative interface at 10^5 K above the cloud where fragments radiate energy, conducted from the hotter gas, at the peak of the cooling curve. Such an interface then irradiates the surface of the cloud and produces most of the observed emission lines (the result would be similar to the shock spectrum) at the expense of the thermal energy in the hot gas.

Turbulence may be greatest in the core of a cooling flow where the optical emission is observed. Further out any rapid motions must be on larger scales leading to less noise on small scales so that clouds can collapse below 10^4K without being halted there. The gravitational potential of the dark halo of the central galaxy is probably responsible for the damping of turbulence beyond $\sim 20 \, \mathrm{kpc}$ radius. The cooling flows that have little, or no, detectable emission-line gas presumably have a lower level of turbulence.

Note that turbulence cannot significantly affect the total mass deposition rates since the X-ray emission lines give such good agreement with the X-ray imaging results. It may however mean that $\dot{M}$ does not vary linearly with r within $\sim 20 \, \mathrm{kpc}$.

5. DISCUSSION

Studies of intracluster gas provide a splendid opportunity to investigate the properties of large bodies of hot, optically thin plasma. The major observing band is, of course, the X-ray one. At the present time we only have relatively poor data on many aspects of the gas, but future prospects with the new generation of space-borne telescopes to be launched in the 1990s are very good.

So far, we know that the bulk of the gas has a sound speed close to the velocity dispersion of the galaxies, *i.e.* $\beta \sim 1$. This relates to the past history of heating and cooling of the gas. Looked at simply, it suggests that most of the heating is gravitational (*i.e.* similar for both the gas and galaxies) and little cooling has taken place in the gas we observe (this does not mean that a lot of cooling has not occurred in the past – that gas would not now be detectable). It also suggests that most of the pressure supporting the gas is thermal. Looked at in more detail though, differences in β between clusters does indicate those in which bulk, kinetic energy (turbulence) may be important. For example, the Perseus cluster has a higher velocity dispersion and lower gas temperature than the Coma cluster, which means that they have different values of β. I suspect that this is due to gas in the Perseus cluster having large scale motions due, perhaps, to the infall of a subcluster in that past few billion years. These motions should dissipate (or pass to smaller scales) on a turnover time – the gas can be 'oscillating' at up to $\sim 500 \, \mathrm{km \, s^{-1}}$ on a scale of a Mpc for up to $\sim 2.10^9 \, \mathrm{yr}$. This could in part explain the offset of the outer contours of the X-ray emission in the Perseus cluster (Branduardi-Raymont *et al.* 1981). β should then reduce with time (t_{sub} measured in billions of years) as the kinetic energy is thermalized. High observed values of β then indicate small t_{sub}, *i.e.* a recent subcluster interaction.

As discussed in §4.4, the gravitational potential of the cluster will prevent much turbulent energy being fed directly into the core. A recent subcluster merger will, however, mean a fesh injection of denser gas, some of which will accumulate in the core as blobs. The infall of the blobs carries energy inward and their higher density will promote a more massive cooling flow. Depending upon the cluster and subcluster masses, and upon the impact parameter, a merger could either disrupt a cooling flow for some time, or enhance an existing one.

Future work in which the spatial distribution of the gas temperature(s) is determined and compared with the spatial distribution of the density will measure the extent of these effects. A more direct determination can be made from high-resolution X-ray spectroscopy, in which the velocity distribution of the hot gas is obtained from the line width distribution. The extent to which kinetic energy contributes to the pressure of the intracluster gas must be found before accurate measure of the gravitational potential can be made through equation (22).

On the smaller scale of the inner few 100 kpc in most clusters there is a cooling flow. The X-ray surface brightness distribution shows that the mass flow rate is not constant and therefore that the gas must be inhomogeneous. A key observation here is of the FeXVII line in the Perseus cluster, and of other lines, that show the gas to be definitely cooling. *Ad hoc* models in which some distributed heat source (including the dissipation of turbulence) stems widespread cooling are most unlikely to work in practice since the spectra show the same mass rate cooling at all observable X-ray temperatures. The fact that the gas is inhomogeneous indicates that thermal conductivity is highly suppressed. This in turn means that the weak magnetic field in the intracluster gas does have a role to play – tangling up and reducing the mean free path of the electrons (by a very large factor).

The observations of relatively small amounts of warm gas in the centres of many cooling flows poses new problems, of the origin of the gas and its large linewidths and of its heat source. We suspect that some of the answers are provided by turbulence of the intracluster gas. This is relatively more important in the innermost regions of the flow. Some of the gas is heated by 'plasma-slip' and other processes when it has cooled to $\sim 10^3$ K and it then hangs around for some time before cooling further. The covering fraction of such gas may be relatively high and collisions common in the centre of the flow. Optically-thick HI may also be present. The observed clouds and filaments may then be the result of the coalescence of many smaller clouds. The magnetic fields of the clouds are important in the heating process and in increasing the cross sections and coherence of the clouds. Just how the magnetic field (which ought to have been amplified in the cooling collapse) is finally eliminated so that stars can form is unclear. Clouds further out in the flow are presumably able to cool and form low mass stars without too much bother. It is interesting that the mass of clouds likely to survive intact at large radii is about $1 \, M_\odot$, which, with some fragmentation, should collapse into low mass stars if they continue to cool to 10 K or less.

In conclusion, the intracluster medium is complex, at least as complex as the interstellar medium in our own Galaxy. There are large mass gains and losses, with gas being supplied sporadically from infalling subclusters and more continuously from stellar mass-loss in member galaxies and gas being lost from the medium by cooling in the cluster core. Subclusters also bring in gravitational energy, much of which is converted into chaotic kinetic energy before dispersing into heat. The study of this medium is worthwhile in its own right and as a model of gravitational collapse elsewhere, particularly during galaxy formation.

Acknowledgements
I thank Mike Loewenstein, Paul Nulsen and Hans Böhringer for many discussions and collaboration and the Royal Society for supporting my work.

References

Arnaud, K.A., 1988. In *Cooling Flows in Clusters and Galaxies*, ed. A.C.Fabian, Reidel, 31.
Arons, J. & Max, C.E., 1975. *Astrophys. J.*, **196**, L77.
Balbus, S., 1988. *Astrophys. J.*, **328**, 395.
Balbus, S. & Soker, N., 1989. Preprint.
Bertschinger, E. & Meiksin, A., 1986. *Astrophys.J*, **306**, L1.
Bohringer, H. & Fabian, A.C., 1989. *Mon. Not. R. astr. Soc.*, **247**, 1147.
Branduardi-Raymont, G. *et al.* 1981. *Astrophys. J.*, **248**, 55.
Burns, J. O., White R. A. & Haynes, M. P., 1981.*Astr. J.*, **86**, 1.
Bregman, J.D. & David L.P., 1988. *Astrophys. J.*, **326**, 639.
Brinkmann, W.P. & Massaglia, 1989. Preprint.
Caldwell, C.N. & Oemler, A., 1981. *Astr. J.*, **86**, 1424.
Canizares, C.R., Clark, G.W., Markert, T.H., Berg, C., Smedira, M., Bardas, D., Schnopper, H. & Kalata, K., 1979, *Astrophys. J.*, **234**, L33.
Canizares, C.R., 1981. In *X-ray Astronomy with the Einstein Satellite* ed. R. Giacconi, Reidel, 215.
Canizares, C.R., Clark, G.W., Jernigan, J,G. & Markert, T.H., 1982. *Astrophys. J.*, **262**, L33.
Canizares, C.R., Stewart, G.C. & Fabian A.C., 1983. *Astrophys. J.*, **272**, 449.
Canizares, C.R., Markert, T.H. & Donahue, M.E., 1988. In *Cooling Flows in Clusters and Galaxies*, ed. A.C.Fabian, Reidel, 63.
Cavaliere, A. & Fusco-Femiano, R., 1976., *Astr. Astrophys.*, **49**, 137.

Cowie, L.L. & Binney, J., 1977. *Astrophys. J.*, **215**, 723.

Cowie, L.L., Fabian, A.C. & Nulsen, P.E.J., 1980. *Mon. Not. R. astr. Soc.*, **191**, 399.

Cowie, L.L., Henriksen, M.J. & Mushotzky, R.F., 1987. *Astrophys. J.*, **312**, 593.

Cox, D.P., 1979. *Astrophys. J.*, **234**, 863.

Crane, P., van der Hulst, J. & Haschick, A., 1982. In: *Proc. IAU Symp. No. 97, Extragalactic Radio Sources*, eds. Heeschen, D.S. & Wade, C.M., Reidel, Dordrecht, Holland, p307.

Crawford, C.S., Crehan, D.A., Fabian, A.C. & Johnstone, R.M., 1987. *Mon. Not. R. astr. Soc.*, **224**, 1007.

Crawford, C.S., 1988. PhD Thesis, University of Cambridge.

Crawford, C.S., Arnaud, K.A., Fabian, A.C. & Johnstone, R.M., 1989. *Mon. Not. R. astr. Soc.*, **236**, 277.

Crawford, C.S. & Fabian, A.C., 1989. *Mon. Not. R. astr. Soc.*, **239**, 219.

Draine, B.T. & Salpeter, E.E., 1979. *Astrophys. J.*, **231**, 77.

Dreher, J.W., Carilli, C.L. & Perley, R.A., 1987. *Astrophys. J.*, **316**, 611.

Edge. A.C., 1989. Ph.D. Dissertation, Univ. of Leicester.

Elmegreen, B.G., Dickinson, D.F. & Lada, C.J., 1978. *Astrophys. J.*, **220**, 853.

Fabian, A.C. *et al.*, 1974. *Astrophys. J.*, **189**, L59.

Fabian, A.C. & Nulsen, P.E.J., 1977. *Mon. Not. R. astr. Soc.*, **180**, 479.

Fabian, A.C., Hu, E.M., Cowie, L.L & Grindlay, J.,1981. *Astrophys. J.*, **248**, 47.

Fabian, A.C., Nulsen, P.E.J. & Canizares, C.R., 1982. *Mon. Not. R. astr. Soc.*, **201**, 933.

Fabian, A.C., Nulsen, P.E.J. & Canizares, C.R., 1984. *Nature*, **311**, 733.

Fabian, A.C., Nulsen, P.E.J. & Arnaud, K.A., 1984. *Mon. Not. R. astr. Soc.*, **208**, 179.

Fabian, A.C., Arnaud, K.A. & Thomas, P.A., In Dark Matter in the Universe, eds. J. Kormendy & G.R. Knapp, Reidel, 201.

Fabian, A.C., Thomas, P.A., Fall, S.M. & White, R.A., 1986. *Mon. Not. R. astr. Soc.*, **221**, 1049.

Fabian, A.C., Crawford, C.S., Johnstone, R.M. & Thomas, P.A., 1987. *Mon. Not. R. astr. Soc.*, **228**, 963.

Fabian, A.C., 1988a. In *Cooling Flows in Clusters and Galaxies*, ed. A.C.Fabian, Reidel, 315.

Fabian, A.C., 1988b. In *Hot Thin Plasmas in Astrophysics*, ed. R. Pallavicini, Reidel, 293.

Fabian, A.C., 1989. *Mon. Not. R. astr. Soc.*, **238**, 41P.

Fabricant, D. & Gorenstein, P., 1983. **267**, 535.

Fabricant, D., Beers, T.C., Geller, M.J., Gorenstein, P., Huchra, J.P. & Kurtz, M.J., 1986. **308**, 580.

Forman, W., Bechtold, J., Blair, W.,Giacconi, R., Van Speybroeck, L. & Jones, C., 1981, *Astrophys. J.*, **243**, L133.

Forman, W., Schwarz, J., Jones, C., Liller, W. & Fabian, A.C., 1979. *Astrophys. J.*, **234**, L27.

Garrington, S. T., Leahy, J. P., Conway, R. G., Laing, R. A., 1988. *Nature*, **331**, 147..

Geller, M.J., 1984. *Comments on Astrophys. Space. Sci.*, **10**,47.

Gilfanov, M.R., Syunyaev, R.A. & Churazov, E.M., 1987. *Sov. Astr. Lett.*, **13**, 17.

Gold, T. & Hoyle, F., 1958. In *Paris Symposium on Radio Astronomy*, ed. RN Bracewell, Stanford Univ. Press, 574.

Gorenstein, P., Fabricant, D., Topka, K., Tucker, W. & Harnden, F.R., 1977. *Astrophys. J.*, **216**, L95.

Hattori, M., & Habe, A., 1989. *Mon. Not. R. astr. Soc.*, in press.

Haynes, M. P., Brown, R. L. & Roberts, M. S., 1978. *Astrophys J.*, **221**, 414.

Heckman, T.M., Baum, S.A., van Breugel, W.J.M. & McCarthy, P.,1989. *Astrophys. J.*, **338**, 48.

Helmken, H., Delvaille, J.P., Epstein, A., Geller, M.J., Schnopper, H.W. & Jernigan, J.G., 1978. *Astrophys. J.*, **221**, L43.

Hintzen, P. & Romanishin, W., 1988. *Astrophys. J.*, **327**, L17.

Henry, J.P & Henriksen, M.J., 1986. *Astrophys. J.*, **301**, 689.

Hu, E.M., 1988. In *Cooling Flows in Clusters and Galaxies*, ed. A.C.Fabian, Reidel, 73.

Hu, E.M., Cowie, L.L. & Wang, 1985. *Astrophys. J. Suppl.*, **59**, 447.

Huchra, J & Brodie, J. 1987. *Astr. J.*

Hughes, J.P., 1989. *Astrophys. J.*, **337**, 21.

Jaffe, W., de Bruyn, A.G. & Sijbreng, D., 1988. In: *Cooling Flows in Clusters and Galaxies*, ed. Fabian, A. C., Reidel, Dordrecht, Holland, p145.

Jones, C. & Forman, W., 1984. *Astrophys. J.*, **276**, 38.

Johnstone, R.M., Fabian, A.C. & Nulsen, P.E.J., 1987. *Mon. Not. R. astr. Soc.*, **224**, 75.

Johnstone, R.M. & Fabian, A.C., 1988. *Mon. Not. R. astr. Soc.*, **233**, 581.

Johnstone, R.M. & Fabian, A.C., 1989. *Mon. Not. R. astr. Soc.*, **237**, 27P.

Jones, C. & Forman, W., 1984. *Astrophys. J.*, **276**, 38.

King, I., 1966. *Astr. J.*, **71**, 64.

Kriss, G.A., Cioffi, D.F. & Canizares, C.R., 1983. *Astrophys. J.*, **272**, 439.

Krolik, J.H. & Raymond, J.C., 1988. *Astrophys. J.*, **335**, L39.

Kulsrud, R.M. & Pearce, W.P., 1969. *Astrophys. J.*, **156**, 445.

Lahav, O., Edge, A.C., Fabian, A.C. & Putney, A., 1989. *Mon. Not. R. astr. Soc.*, **238**, 881.

Laing, R.A., *Mon. Not. R. astr. Soc.*, **331**, 149.

Lazareff, B., Castets, A., Kim, D-W. & Jura, M., 1989, **336**, L13.

Lea, S.M., Silk, J., Kellogg, E. & Murray, S., 1973. *Astrophys. J.*, **184**, L105.

Lea, S.M., Mushotzky, R.F. & Holt, S.S., 1982. *Astrophys. J.*, **262**, 24.

Loewenstein, M., 1989. *Mon. Not. R. astr. Soc.*, **238**, 15.

Loewenstein, M., 1989. Preprint.

Loewenstein, M. & Fabian, A.C., 1988. *Mon. Not. R. astr. Soc.*, in press.

Maccagni, D., Garilli, B., Gioia, I.M., Maccacaro, T., Vettolani, G. & Wolter, A., 1988. *Astrophys. J.*, **334**, L1.

Mathews, W., G. & Bregman, J.N., 1978. *Astrophys. J.*, **244**, 308.

Malagoli, A., Rosner, R. & Bodo, G., 1987. *Astrophys. J.*, **319**, 632.

McGlynn, T.A. & Fabian, A.C., 1984. *Mon. Not. R. astr. Soc.*, **208**, 709.

Miller, L., 1986. *Mon. Not. R. astr. Soc.*, **220**, 713.

Mirabel, F. *et al.* 1989. Preprint.

Mitchell, R.J., Charles, P.A., Culhane, J.L., Davison, P.J.N. & Fabian, A.C., 1975. *Astrophys. J.*, **200**, L5.

Mould, J.R., Oke, J.B. & Nemec, J.M., 1987. *Astr. J.*, **92**, 53.

Mushotzky, R.F., Holt, S.S, Smith, B.W., Boldt, E.A. & Serlemitsos, P.J., 1981. *Astrophys. J.*, **244**, L47.

Mushotzky, R.F., 1987a. *Astrophys. Lett.*, **26**, 43.

Mushotzky, R.F. & Szymkowiak, A.E. , 1987b. In *Cooling Flows in Clusters and Galaxies*, ed. A.C.Fabian, Reidel, 47.

Nittmann, J., Falle, S.A.E.G. & Gaskell, P.H., 1982. *Mon. Not. R. astr. Soc.*, **201**, 833.

Nulsen, P.E.J., Stewart, G.C., Fabian, A.C., Mushotzky, R.F., Holt, S.S, Ku, W.H.M. & Malin, D.F., 1982. *Mon. Not. R. astr. Soc.*, **199**, 1089.

Nulsen, P.E.J., Stewart, G.C. & Fabian, A.C., 1984. *Mon. Not. R. astr. Soc.*, **208**, 185.

Nulsen, P.E.J., 1986. *Mon. Not. R. astr. Soc.*, **221**, 377.

Nulsen, P.E.J. & Carter, D., 1987. *Mon. Not. R. astr. Soc.*, **225**, 935.

Nulsen, P.E.J., 1988.In *Cooling Flows in Clusters and Galaxies*, ed. A.C.Fabian, Reidel, 378.

Pesce, J.E., Edge, A.C., Fabian, A.C. & Johnstone, R.M., 1989. *Mon. Not. R. astr. Soc.*, in press.

Pringle, J.E., 1989. *Mon. Not. R. astr. Soc.*, **239**, 479.

Rothenflug, R. & Arnaud, M., 1986. *Astr. Astrophys.*, **147**, 337.

Sarazin, C.L., 1986. *Rev. Mod. Phys.*, **58**, 1.

Sarazin, C.L., 1988. *X-ray Emission from Clusters of Galaxies*, C.U.P.

Sarazin, C.L. & O'Connell, R.W., 1983. *Astrophys. J.*, **258**, 552.

Sarazin, C.L., 1989. *Astrophys. J.*, **345**, 12.

Schwartz, D.A., Schwarz, J. & Tucker, W.H., 1980. *Astrophys. J.*, **238**, L59.

Silk, J., 1975. *Astrophys. J.*, **198**, L77.

Silk, J., 1976. *Astrophys. J.*, **208**, 646.

Silk, J., Djorgovski, G., Wyse, R.F.G. & Bruzual, G.A., 1986. *Astrophys. J.*, **307**, 415.

Singh, K.P., Westergaard, N.J. & Schnopper, H.W., 1986. *Astrophys. J.*, **308**, L51.

Spinrad, H. & Djorgovski, G., 1984. *Astrophys. J.*, **280**, L9.

Spitzer, L. 1982. *Astrophys. J.*, **262**, 315.

Stewart, G.C., Canizares, C.R., Fabian, A.C. & Nulsen, P.E.J., 1984a. *Astrophys. J.*, **278**, 536.

Stewart, G.C., Fabian, A.C., Jones, C. & Forman, W., 1984b. *Astrophys. J.*, **285**, 1.

Sunyaev, R.A., 1982. *Sov. Astr. Lett.*, **8**, 175.

The, L.S. & White, S.D.M., 1988. *Astrophys. J.*, **95**, 15.

Thomas, P.A., Fabian, A.C. & Nulsen, P.E.J., 1987. *Mon. Not. R. astr. Soc.*, **228**, 973.

Tribble, P. 1989. *Mon. Not. R. astr. Soc.*, **238**, 1247.

Tucker, W.H. & Rosner, R., 1982. *Astrophys. J.*, **267**, 547.

Valentijn, E. A. & Giovanelli, R., 1982. *Astr. Astrophys.*, **114**, 208.

White, R.E. & Sarazin, C.L., 1987. *Astrophys. J.*, **318**, 612.

White, R.E. & Sarazin, C.L., 1987. *Astrophys. J.*, **318**, 621.

White, R.E. & Sarazin, C.L., 1987. *Astrophys. J.*, **318**, 629.

A HOT INTERGALACTIC MEDIUM

XAVIER BARCONS
Departamento de Física Moderna
Universidad de Cantabria
39005 Santander
Spain

ABSTRACT. A critical review of the possible existence of a hot $(T \geq 10^8 K)$ diffuse intergalactic medium (IGM) is given. Attention is paid to three main issues: its contribution to the X-ray background (XRB) by bremsstrahlung radiation, the Compton upscattering of the microwave background radiation (MBR) and the degree of clumpiness of the IGM. It is concluded that a hot IGM could produce a substantial fraction of the XRB (if enough energy to heat it up can be found), that the Compton distortion of the MBR is compatible with (not the source of) recently detected spectral distortions and that significant clumping is unlikely.

1. Is There any IGM?

Before trying to answer this question, it must be pointed out that what is meant here by IGM is any significant amount of diffuse matter outside galaxies, clusters and other structures. This is not to be confused with the intracluster gas (Fabian, these proceedings) nor with the Lyman α clouds which are pressure or gravity bound systems intercepting our line of sight to distant QSOs.

The above question is not a trivial one, since there is no clear observational evidence for the existence of such a medium. I shall then outline several facts *suggesting* rather than proving its existence. The first and perhaps strongest argument is a theoretical one: galaxy formation (and cluster formation) is unlikely to be 100 per cent efficient therefore leaving significant amounts of baryonic gas outside these structures. Currently fashionable ideas ('biased' galaxy formation) imply that most of the gas would reside outside galaxies and clusters. In addition, during galactic evolution, some gas outflow is expected due to supernovae and winds.

Another suggestive fact is that the contribution of the gas found on several scales to the density of the Universe (as measured by $\Omega_{gas} = \rho_{gas}/\rho_{cr}$ where $\rho_{cr} = 1.9\,10^{-29}\,h^{-2}\,\mathrm{gr\,cm^{-3}}$, h being the Hubble constant in units of $100\,\mathrm{km\,s^{-1}\,Mpc^{-1}}$) grows with the size of the structure. Thus, the interstellar gas of galaxies only contributes $\Omega_{gas} \sim 0.01$, but intracluster gas gives $\Omega_{gas} \sim 0.1$. It could then be that $\Omega_{\mathrm{IGM}} \sim 1$, although there is no *a priori* physical basis for this extrapolation. Incidentally, it also happens that T_{gas} grows with size, from $10^4 - 10^6 \mathrm{K}$ in galaxies to $10^7 - 10^8 \mathrm{K}$ in clusters. It seems then that the bigger the scale the more the gas is found and the hotter it is. As already mentioned, however, there is no reason to believe that an extrapolation to scales larger than clusters must be true.

There are also indirect observational arguments pointing towards the existence of an IGM. For instance, Lyman α clouds could be pressure confined by a diffuse IGM (Sargent *et al.* 1980), although they could also be gravitationally self-bound. Also, the rapid increase of the size of radio sources with cosmic time (Rosen & Wiita 1988) has a natural explanation if there is a high-pressure gas filling the intergalactic space.

As a conclusion, it could be fair to say that some sort of IGM does exist, although we have no direct observational evidence for it.

W. Brinkmann et al. (eds.), Physical Processes in Hot Cosmic Plasmas, 299–306.

2. What is its Physical State?

If the existence of an IGM is not clearly established, its composition, density and temperature are even more controversial. There are, however, a few observational limits that are worth discussing.

The first one is the so-called Gunn–Peterson test. Any signiicant amount of diffuse neutral hydrogen (HI) in the Universe should produce a continuous depression in the spectrum of QSOs shortward of their Lyman α emission line. No such effect is seen and current upper limits to the density of cosmologically distributed HI are $\Omega_{IGM}(\text{HI}) \leq 10^{-7}$ (Steidel & Sargent 1987). This shows that if there is an IGM it should be very highly ionized. A lower bound to its temperature is $T_{IGM} \geq 2\,10^5 \text{K}$ (Sherman 1982), because otherwise some Gunn–Peterson effect would be seen.

An obvious upper limit to the density of any IGM is $\Omega_{IGM} \leq 1$, because of the age of the Universe. A more interesting limit comes from the standard theory of primordial nucleosynthesis (Boesgaard & Steigmann 1985) which is $\Omega_{IGM} < \Omega_{baryon} < 0.19$. This limit comes from the fact that if the baryonic density in the Universe is too high, less Deuterium than it is observed today was produced in the early Universe, and stars very easily destroy rather than produce this element. However, recent work (Malaney & Fowler 1988) shows that inhomogeneities in the nucleosynthesis epoch (coming from the quark–hadron phase transition) can give rise to the observed abundances of light elements with $\Omega_{bayon} \sim 1$. As a consequence, limits on the baryonic content of the Universe coming from primordial nucleosynthesis have to be taken with some care.

Upper limits to the temperature of the IGM cannot be easily stated. A maximum $T_{IGM} \leq 2\,10^6 \text{K}$ is usually invoked in order not to overproduce the soft X–ray background, but if $T_{IGM} \geq 10^8 \text{K}$ this constraint does no longer apply, since the gas will mainly radiate in the hard X–ray band. It has also been said that an IGM temperature exceeding a few million degrees will destroy the Lyman α clouds. However, if these objects are considerably smaller and flattened than it is usually assumed, much higher temperatures will be possible (Barcons & Fabian 1987). Finally, a firm upper limit for T_{IGM} seems to appear from the extragalactic background flux at $100\mu\text{m}$, because temeperatures much greater than 10^9K would comptonize the MBR spectrum exceeding this flux (see Section 4).

One major problem with the temperature of the IGM is how has it been reheated. If T_{IGM} is only a few times 10^5K, QSOs can just photoionize it at this temperature (Donhaue & Shull 1987, although there has been some controversy, Shapiro & Giroux 1987). Therefore if T_{IGM} is higher, some extra heat source has to be present in the Universe.

The purpose of this contribution is to show what are the effects of a hot ($10^8 - 10^9 \text{K}$) IGM on the cosmic background radiation (CBR) and also to consider the possibility that the gas could be clumpy. The main influence on the CBR comes from bremsstrahlung of the gas, which will contribute to the XRB, and from Compton scattering of MBR photons, which will distort the spectrum in the $100 - 1000\mu\text{m}$ band. Inhomogeneities will show up as fluctuations and anisotropies in the CBR.

3. Can a Hot IGM Give a Substantial Contribution to the XRB?

Although the XRB was the first component of the CBR discovered (Giacconi et al.1962) it would not be surprising that will be one of the last to be understood. After the HEAO–1 missions, its spectrum is well known from 3 to 300 keV and it fits remarkably well thermal bremsstrahlung at 40 keV (see Fig. 1; Marshall et al.1980, Gruber et al.1984; see Boldt 1987 for a review). There is probably a bump at energies ~ 1 MeV and from 1 to 3 keV it has not been properly measured,

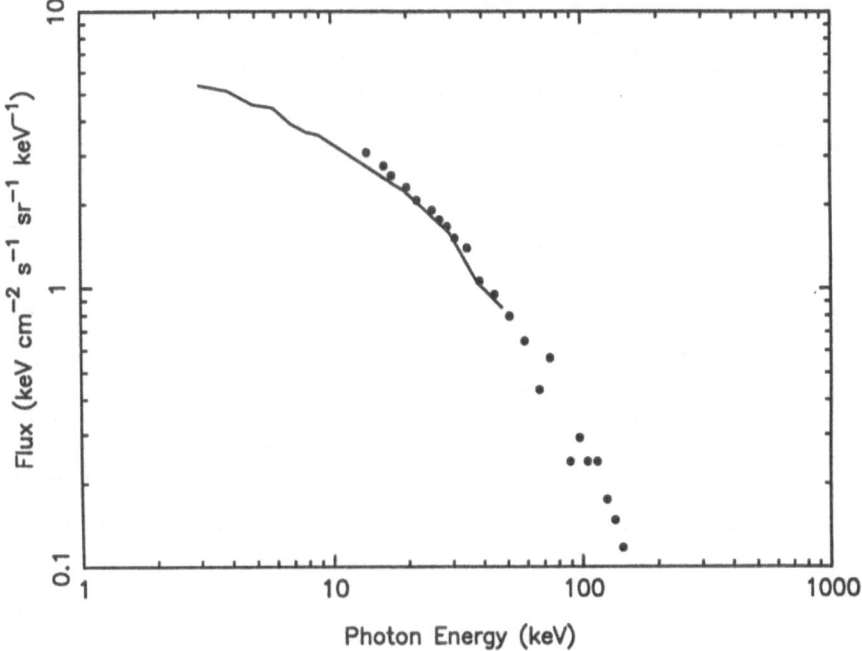

Figure 1. The X–ray background. The solid curve corresponds to the Marshall *et al.* (1980) measurements and the filled dots to the Gruber *et al.* (1984) measurements.

although it has been argued that there could be an excess with respect to the extrapolated spectrum from higher energies (Garmire & Nousek 1981).

The XRB is also remarkably isotropic. In the 2–10 keV band, its intensity fluctuations are of the order of 2 per cent on $5° \times 5°$ beams (Shafer & Fabian 1983).

It is also known that cosmic X–ray sources contribute to the XRB. However, as it will be argued below, known classes of X-ray sources are unlikely to produce the bulk of the XRB. The first problem is that over the 2–10 keV energy band (where the XRB spectrum can be approximated by a power law $\propto \epsilon^{-\alpha}$, with $\alpha = 0.4$), Active Galactic Nuclei have significantly steeper spectra ($\alpha \sim 0.7$). Clusters of Galaxies, whose X-ray continuum has also a bremsstrahlung spectral shape, have temperatures of the order of ~ 2 keV and therefore their contribution to the 2–10 keV band is small.

Apart from this spectral mismatch, there are other arguments against known classes of sources producing the XRB. Radio–loud QSOs, which have about the right spectral shape, at least in the 0.3–3 keV band ($\alpha \sim 0.5$, Wilkes & Elvis 1987, Canizares & White 1989) produce less than 10 per cent of the 2–10 keV XRB. The most numerous radio–quiet QSOs have much steeper spectra ($\alpha \sim 1$ to 1.4) in the 0.3–3 keV band. This fact, together with limits on the extragalactic soft XRB (0.1–0.2 keV) sets an upper limit of 10 to 40 per cent of the 2–10 keV being produced by these objects (Fabian, Canizares & Barcons 1989).

Furthermore, and using fluctuation analyses of the XRB it has been recently shown (Hamilton & Helfand 1987, Barcons & Fabian 1989) that an *euclidean* population of sources (i.e., the number of sources with flux greater than S is proportional to $S^{-3/2}$) cannot produce more than ~ 40 per cent of the 1–3 keV XRB (and presumably less in the 2–10 keV band). In the above works it has also been found that a population of at least a few thousand sources per square degree is needed to produce such an isotropic background.

The above facts indicate that a substantial fraction (from 60 to 75 per cent) of the 2–10 keV XRB has yet to be explained, the remainder being produced by steep spectrum sources.

The idea that a hot IGM could produce, via bremsstrahlung radiation, a substantial fraction of the XRB was first suggested by Hoyle (1953) whithin the steady–state cosmological framework. Important changes and improvements to this model have been made since then by many authors. The point now is, can the *residual* XRB (i.e., after removing the contribution from steep spectrum sources) arise in a hot IGM?. From the point of view of plasma physics, this reduces to the following problem: take a reheating redshift z_H, assume a termal distribution for the electrons (the Coulomb e–e coupling time is less than the age of the Universe), let the gas cool by relativistic adiabatic expansion (Barcons & Lapiedra 1985) as well as Compton cooling by the MBR (Guilbert & Fabian 1986; radiative cooling is not important) and then compute the integrated relativistic bremsstrahlung spectrum (including e–p and e–e bremsstrahlung). The results show that (Guilbert & Fabian 1986, Barcons 1987)

[i] $z_H \leq 5$, because otherwise there is too much Compton scattering with the MBR and the integrated X–ray spectrum is too steep.

[ii] $T_{IGM}(z=0) \approx 3\,10^8 \mathrm{K}$ gives a good fit.

[iii] The fractional contribution of sources to the 2–10 keV XRB must not exceed ~ 30 per cent.

[iv] The density of the IGM needed to fit the XRB goes from $\Omega_{IGM} = 0.24$ if $z_H = 4$ to $\Omega_{IGM} = 0.41$ if $z_H = 2$.

This model for the XRB is indeed very attractive because it provides a physical explanation for the XRB spectrum, but it has a very severe problem: the IGM must be reheated at a temperature of a few hundreds of keV at $z \sim 4$. Actually, the kinetic energy contained in the IGM is ~ 50 per cent of the energy in the MBR, but we only see a tiny fraction of it in the XRB (because bremsstrahlung is a very inefficient process here). The main point against such a hot IGM is therefore the large energy input required: about 10^{64} ergs/galaxy are needed which exceeds, by far, their gravitational energy.

As a consequence, some very energetic events are required at high redshift in this framework. Galactic explosions (Bookbinder *et al.*1980) do not seem to provide enough energy. The observed dependence of the Initial Mass Function on metallicity (in the sense that primeval galaxies formed a lot of high mass stars and hence supernovae) could eventually help (R. Terlevich, private communication). More exotic phenomena, such as superconducting cosmic strings (Ostriker, Thompson & Witten 1986) or heating by a zero–point field (Rueda 1989) have also been proposed. The general conclusion is that our ignorance of the Universe at high redshift does not allow a clear conclusion about the existence of such a hot IGM, although it seems rather extravagant.

4. What is the Influence of a Hot IGM on the MBR?

At wavelenghts $\lambda \geq 1$ mm (i.e., in the Rayleigh–Jeans zone) the spectrum of the MBR has a blackbody shape with a temperature $T \approx 2.75$ K. However, the Nagoya–Berkeley group (Matsumoto *et al.*1987) has reported several excesses at shorter wavelengths (see Fig. 2). The corresponding temperatures are 2.799 ± 0.018 K at $\lambda = 1160\mu$m, 2.955 ± 0.017 K at $\lambda = 709\mu$m, 3.175 ± 0.027 K

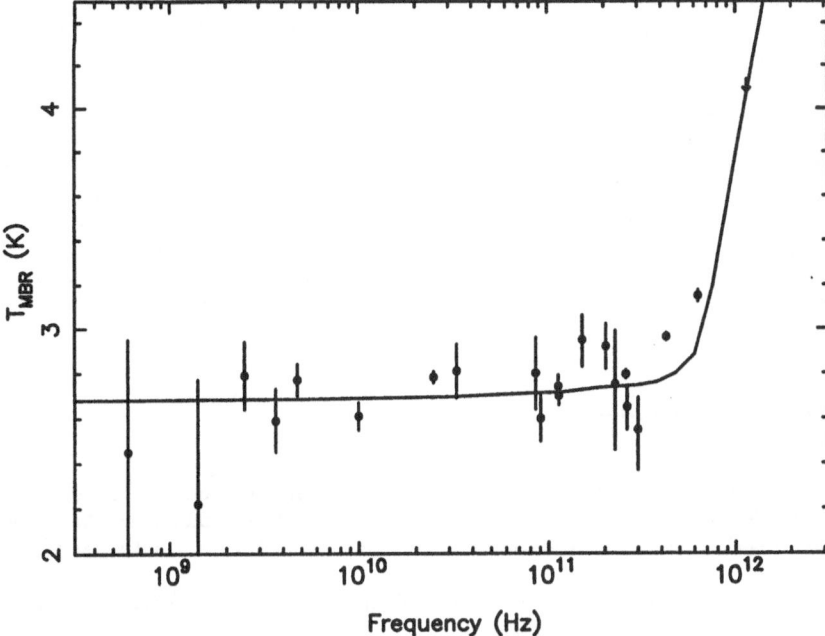

Figure 2. Temperatures of the MBR at several frequencies. The solid curve corresponds to the Compton–scattered spectrum by a hot IGM as explained in the text.

at $\lambda = 481\mu$m and an upper limit of 4.105 K at $\lambda = 262\mu$m. It must be stressed that these data still await for confirmation and therefore have to be taken with some caution.

The energy contained in this excess is ~ 10 per cent of the energy in the MBR. Several explanations (none of them satisfactory) have been proposed. These include high–redshift dust, decaying particles and Comptoniztion (see De Zotti et al.1989 for a recent review).

Compton distortion by a hot IGM upscatters photons from the Rayleigh–Jeans part of the spectrum with mean frequency shift per scattering $\frac{\Delta\nu}{\nu} \sim 4\theta_{IGM}(1+4\theta_{IGM})$ where $\theta_{IGM} = kT_{IGM}/m_ec^2$. The result is a decrease in the MBR temperature at frequencies below the peak and an increase above the peak. The hotter the IGM, the higher the frequency where the distortion will be more important. The mean number of collisions per photon is given by the Thomson scattering depth $\tau_{IGM} = \sigma_T N_e$ where σ_T is the Thomson cross-section and N_e the electron column density.

As the excesses reported by Matsumoto et al.(1987) are relatively near the peak, values of $\theta_{IGM} \ll 1$ and $\tau_{IGM} > 1$ are needed to obtain a reasonable fit at all frequencies. In this case, the non–relativistic diffusion (Kompane'ets) equation can be used to find the comptonized spectrum. The best fit is obtained for $y = 4\tau_{IGM}\theta_{IGM} = 0.018 \pm 0.001$ (see, e.g., Wilkinson 1988). However, the χ^2 is bad because the decrease of the temperature below the peak conflicts with the most accurate measurements at those frequencies.

A hot IGM producing the residual XRB, as discussed in last Section, does not match the observed distortion, since photons are scattered towards much higher frequencies (in this case $\tau_{IGM} \ll 1$ and diffusion approximations are no longer valid). The predicted curve is shown in Fig. 2. The decrease in the MBR temperature below the peak cannot be completely corrected without conflicting with the upper limit at $\lambda = 262\mu$m (De Zotti *et al.*1989). As a consequence, this hot IGM cannot give rise to the observed distortion, although it does not exceed the observed limits.

As already mentioned, most of the photons Compton scattered by a hot IGM will go to shorter wavelengths. At $\lambda = 100\mu$m there is an upper limit to the extragalactic flux (from the IRAS satellite) which was first set to 5–6 MJy sr^{-1} (Rowan–Robinson 1986). More recent work by Boulanger & Pérault (1988) shows that the high galactic latitude flux at 100μm is well correlated with the HI column density, and extrapolation to zero HI gives an extragalactic flux of 1.8 ± 0.3 MJy sr^{-1}. This flux is similar to what is predicted by the hot IGM proposed in last Section (Guilbert & Fabian 1986). Higher temperatures of the IGM would conflict with this datum.

5. Can a Hot IGM be Clumpy?

Since the bremsstrahlung emissivity of the IGM is proportional to the square of its density, less gas would be required to produce the XRB if the IGM is clumpy. This could remove the conflict with the standard nucleosynthesis models. However, clumping such a hot gas is not an easy task as we shall see.

Two ways have been proposed to confine the clumps (as in the case of the Lyman α clouds): pressure confinement and gravity confinement.

5.1 PRESSURE CONFINEMENT

This was first suggested by Guilbert & Fabian (1986) and further explored by Barcons & Fabian (1988). As the sound speed of a hot IGM is so high, pressure fluctuations will be washed out (in less than a Hubble time) on scales ≥ 20 Mpc. As a consequence, in the absence of gravity, the IGM has to be isobaric on very large scales.

Guilbert & Fabian (1986) proposed a two–phase model for the IGM: dense 'cold' blobs ($\sim 3\,10^8$K) in pressure equilibrium with a yet hotter tenuous gas. In order not to distort the XRB spectrum, the temperature of the hotter phase had to be at least $\sim 10^9$K, which could, in turn, give rise to the 1 MeV bump (all this with $\Omega_{IGM} \approx 0.1$).

This model, however (and pressure confinement models in general), can be rejected in many ways. First, the Compton distortion of the MBR produces a flux of ~ 4 MJy sr^{-1} at 100μm, exceeding the values found by Boulanger & Pérault (1988). Secondly, this Compton scattering will be anisotropic, and enormous fluctuations and anisotropies would be imprinted in the MBR unless the size of the 'cold' blobs is very small ($\leq 10\,h^{-1}$kpc, Barcons & Fabian 1988). In addition the 'hot' phase cools very efficiently via Compton scatterring with the MBR and would disappear in less than a Hubble time (De Zotti, private communication). All of these facts make pressure confinement of the IGM very unlikely.

5.2 GRAVITY CONFINEMENT

A general problem with confining the IGM in potential wells is that the gas will not have a single temperature. If hydrostatic equilibrium works (in a similar way as in clusters of galaxies) the deepest parts of the potential well will have the densest and coolest gas and therefore it will be difficult to get good agreement with the XRB spectrum, since a soft excess (forbidden by the observations) would be present.

The other problem is that in order to confine a gas with such a high thermal energy density, either a very deep or a very extended potential well is needed. Too deep potential wells make the gas too dense in such a way that radiative cooling becomes important and soft excesses appear again. For a $\frac{\delta \varrho}{\varrho} \sim 1$ potential well, its required size to have a virial temperature of $\sim 3\,10^8$K is about $\sim 20h^{-1}$Mpc (Daly 1987, Subrahmnyan & Cowsik 1989). This usually produces more fluctuations in the XRB than it is actually observed, unless these potential wells are anticlustered or perhaps at a very high redshift (Barcons & Fabian 1988).

The conclusion is that, although gravitationally confined clumps of hot gas are not completely ruled out, very special potential wells are needed, with very small dispersion in the parameters (sizes, depths, redshifts, etc.) to fit the XRB spectrum. This makes this way of clumping also unlikely.

6. Outlook

A brief summary of our present knowledge (or rather ignorance) about the possible existence and properties of a hot IGM could be as follows:

[i] A hot IGM could produce the residual XRB. The energy required to heat the IGM is rather high, but so is our level of ignorance of the high redshift Universe.

[ii] The distortions in the MBR, if real, cannot be explained with a hot IGM, although there is no conflict among them.

[iii] It is unlikely that the IGM is significantly clumpy.

The above facts seem to tell us that we have a lot to learn about the high redshift Universe. Current works on Cosmology, seem to be only worried about what happened at the recombination era ($z \sim 1000$) and at the present epoch ($z \sim 0$). However all of the theories of galaxy formation predict a rather strong evolution of the perturbations in the Universe at intermediate redshifts. These epochs have so far escaped our observations, but it seems that the answer to the question 'How is the Universe we live in?' will come with the improvement of our knowledge of the $z \sim 5$ Universe. As an example, the XRB has to be explained (see Section 3) either by a numerours faint (high redshift) population of sources or by a general reheating of the IGM. None of these alternatives is predicted by currently fashionable theories of galaxy formation. My impression is that singnificant progress in this field will only come when the next generation of space telescopes (optical, infrared, and X–ray) have done their work.

Acknowledgements

I am very grateful to the organizers of this meeting for asking us to write an 'understandable' version of the lectures we presented in Vulcano. I am not sure I succeded, but writting these notes has been indeed very useful to me, since I had to go to the most deep fundamentals of this issue. I am grateful to Andy Fabian for teaching me almost everything I know about the IGM. Very valuable comments about the IGM by Gianfranco De Zotti and Roberto Terlevich are also acknowledged. Partial financial support for this work was provided by the Comisión Interministerial de Ciencia y Tecnología.

References

Barcons, X. & Lapiedra, R., 1985. *Astrophys. J.*, **289**, 33

Barcons, X., 1987. *Astrophys. J.*, **313**, 54

Barcons, X. & Fabian, A.C., 1987. *Mon. Not. R. astr. Soc.*, **224**, 675

Barcons, X. & Fabian, A.C., 1988. *Mon. Not. R. astr. Soc.*, **230**, 189

Barcons, X. & Fabian, A.C., 1989. *Mon. Not. R. astr. Soc.*(submitted)

Boesgaard, A.M. & Steigmann, G., 1985. *Ann. Rev. astr. astrophys.*, **23**, 319

Boldt, E., 1987. *Phys. Rep.*, **146**, 215

Bookbinder, J., Cowie, L.L., Krolik, J.H., Ostriker, J.P. & Rees, M.J., 1980. *Astrophys. J.*, **237**, 647

Boulanger, F. & Pérault, M., 1988. *Astrophys. J.*, **330**, 964

Canizares, C.R. & White, J., 1989. *Astrophys. J.*, **339**, 27

Daly, R.A., 1987. *Astrophys. J.*, **322**, 20

De Zotti, G., Danese, L., Toffolatti, L. & Franceschini, A., 1989. In: *IAU Symp. 139: Galactic and Extragalctic Background Radiations*, ed. Bowyer, S. & Leinert, C.. Kluwer (in the press)

Donhaue, C.M. & Shull, J.M., 1987. *Astrophys. J.*, **323**, L13

Fabian, A.C., Canizares, C.R. & Barcons, X., 1989. *Mon. Not. R. astr. Soc.*(in the press)

Garmire, G. & Nousek, J., 1981. *B.A.A.S.*, **12**, 853

Giacconi, R., Gursky, H., Paolini, F. & Rossi, B., 1962. *Phys. Rev. Lett.*, **9**, 439

Gruber, D.E., Rothschild, R.E., Matteson, J.L. & Kinzer, R.L., 1984. *MPE Report*, **184**, 129

Guilbert, P.W. & Fabian, A.C., 1986. *Mon. Not. R. astr. Soc.*, **220**, 439

Hamilton, T.T. & Helfand, D.J., 1987. *Astrophys. J.*, **318**, 93

Hoyle, F., 1953. *Astrophys. J.*, **118**, 513

Malaney, R.A. & Fowler, W.A., 1988. *Astrophys. J.*, **333**, 14

Marshall, F.E. *et al.* ,1980. *Astrophys. J.*, **235**, 4

Matsumoto, T. *et al.* , 1987. *Astrophys. J.*, **329**, 567

Ostriker, J.P., Thompson, C. & Witten, E., 1986. *Phys. Lett.*, **B280**, 231

Rosen, A. & Wiita, P.J., 1988. *Astrophys. J.*, **330**, 16

Rowan–Robinson, M., 1986. *Mon. Not. R. astr. Soc.*, **219**, 737

Rueda, A., 1989. In: *IAU Symp. 139: Galactic and Extragalactic Background Radiations*, eds. Bowyer, S. & Leinert, C. (in the press)

Sargent, W.L.W., Young, P.J., Boksenberg, A. & Tytler, D., 1980. *Astrophys. J. Suppl.*, **42**, 41

Shapiro, P.R. & Giroux, M.L., 1987. *Astrophys. J.*, **321**, L107

Shafer, R.A. & Fabian, A.C., 1983. In: *IAU Symp. 104: Early Evolution of the Universe and its present Structure*, eds. Abell, G.O. & Chincarini, G., Reidel

Sherman, R.D., 1982. *Astrophys. J.*, **256**, 370

Steidel, C. & Sargent, W.L.W., 1987. *Astrophys. J.*, **318**, L11

Subrahmnyan, R. & Cowsik, R., 1989. *Astrophys. J.*(in the press)

Wilkes, B.J. & Elvis, M., 1987. *Astrophys. J.*, **323**, 243

Wilkinson, D.T., 1988. In: *IAU Symposium 130: Large Scale Structures of the Universe*, eds. Audouze, J., Pelletan, M.–C. & Szalay, A.S., Kluwer

THE EVOLUTION OF NON–SPHERICAL THERMAL INSTABILITIES IN COOLING FLOWS

S. MASSAGLIA
Istituto di Fisica Generale dell'Università
Via Pietro Giuria 1
I-10125 Torino, Italy

ABSTRACT. We have followed numerically the evolution of a thermally unstable blob in two dimensions. This analysis is related to the problem of the excess of accreting mass in cooling flows. We have introduced in a homogeneous medium a non–spherical entropy perturbation, and we have found that it evolves to form a flat, pancake–like blob with density ∼ 100 times the initial one.

1. Introduction

The thermal instability mechanism (Field 1965) is considered to be responsible of removing the accreting mass excess from the cooling flows. In fact, the presence of cooling flows in the central regions of many X–ray clusters of galaxies raises the question for a possible sink of the accreting mass, since any observational evidence of where this matter eventually goes is actually lacking (see, e.g., the reviews by Fabian, Nulsen and Canizares, 1984 and by Sarazin, 1986). Linear analyses (see e.g. Schwarz, McCray and Stein 1972) have shown that a uniform, cooling gas is unstable to thermal instabilities whenever the logarithmic temperature derivative of the cooling function $\delta\,(\equiv d\ln P(T)/d\ln T)$ is < 2 for isobaric, and $\delta < 1$ for isochoric perturbations (see also Balbus, 1986). These conditions are usually fulfilled for astrophysical plasmas with temperatures $\gtrsim 10^5$ K, as is typical for the intracluster gas. Mathews and Bregman (1978) discussed how the thermal instability can lead to the formation of isobaric condensations in the inflowing matter (see also Nulsen, 1986), which eventually drop out of the flow to form low-mass stars (White and Sarazin, 1987). However there is a wide gap between the expectations of a linear instability theory, as discussed in Schwarz et al. (1972) and Field (1965), and the actual star formation in cooling flows, in which the fluid behavior is scarcely explored.

David, Bregman and Seab (1988) (hereinafter DBS) have discussed the time dependent evolution of an isobaric perturbation in plane–parallel symmetry for initial density $n_0 = 10^{-2}$ cm^{-3} and temperature $T_0 = 3 \times 10^7$ K. They have found that condensation of density contrast $\sim 10^3$ can actually form. Brinkmann, Massaglia and Müller (1989) (hereinafter BMM) have then examined a similar case in spherical symmetry by means of explicit Eulerian codes, a PPM version (Fryxell et al., 1987) and SADIE (Arnold 1985, Mair et al., 1988), modified to accomodate the effects of cooling and thermal conduction. The main

307

difference between the two physical scenarios is the much higher gas density in the cooling medium which results, in turn, in a much more violent evolution of the condensing blobs in the nonlinear phase. In fact, the density contrasts between the condensation and the unperturbed gas BMM reached were $\sim 10^7$, in the case of initial density $n_0 = 10^{-2}$ cm^{-2}, and $\sim 10^3$ for $n_0 = 0.1$ cm^{-3}. In both cases, the physical conditions in the "final" blob fulfilled the conditions for gravitational instability.

We examine here the temporal evolution of an initially non–spherical perturbation by means of the SADIE code, using initial density of $n_0 = 0.1$ cm^{-3} and temperature $T_0 = 10^7$ K.

2. The evolution of a non–spherical perturbation

We have followed the temporal evolution of a blob of matter from an initial entropy perturbation; the calculations have been carried out in spherical geometry with a coordinate system centered on the blob. The standard hydrodynamical equations describing this system are reported in BMM.

In order to study the two dimensional evolution of a non–spherical condensation, we will start with an ellipsoidal density perturbation of the form

$$\rho = \rho_0\{1 + 0.1 \times \exp[-(\frac{r}{s})^2(\cos^2\theta + 0.2\sin^2\theta)]\} \tag{1}$$

where the scalelength of the perturbation, s, is 10 kpc and, correspondingly, $P =$ constant everywhere initially. We have used only one quadrant of a polar coordinate system, divided into 15 zones ($\Delta\theta = 6°$), taking advantage of the apparent symmetries at $\theta = 0$, $\pi/2$. In the radial direction we employed a logarithmically spaced Eulerian grid of 200 cells, with a spatial resolution of 6×10^{-2} pc at the origin.

BMM have argued that small deviations from an ideal spherical geometry for the blob will be eventually overcome, in a physically realistic situation, by self–gravity effects of the forming condensation and that, therefore, a 1–D spherical treatment gives a good description of the actual process. However, if the initial perturbation is "largely" non spherical (due to, e.g., some residual angular momentum of the condensation), it is not clear at all, whether similar high density contrasts can be reached as in the case of a purely spherical inflow. We will see later that the term "non–spherical" has not an absolute meaning, but largely depends on the choice of the initial parameters.

Figure 1 shows, in polar coordinates, the full computational domain of 200 radial grid points and 15 angular directions; the properties of the flow are reported when entering the non–linear phase of the instability. Arrows indicate the velocity vectors and contours represent the lines of constant density. The elapsed time is $\sim 2 \times 10^7$ yr and the density contrast is still the initial one. Figure 2(a,b,c) shows a later phase of the evolution ($\sim 3 \times 10^7$ yr) of relevant flow quantities along three radial cuts through the computational grid: 1) on the X–axis ($\theta = 0°$), dots; 2) a cut under $\theta = 45°$, dash–dotted line; and 3) along the Z–axis ($\theta = 90°$), full line. The number density (Figure 2a) has only increased slightly and the three temperature cuts (Figure 2b) clearly show the effect of the different radial density distribution onto the cooling times. The radial components of the matter flow (Figure 2c), however, already show marked differences: along the Z–axis (the minor axis of the perturbation) the inflow velocity is by a factor of ~ 2 higher than along the X–axis. This anisotropy in the inflow can be physically understood as follows: the ratio

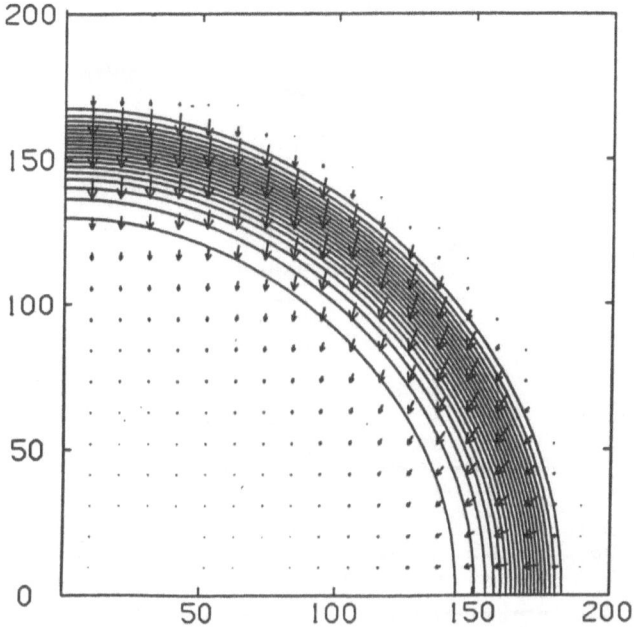

Figure 1. Two dimensional density contours in polar coordinates, overlayed with the appropriate velocity field. The full computational grid of 200 grid points is shown, corresponding to 200 kpc. The perturbation has evolved for $\sim 10^{14}$ sec; the Figure shows the end of the linear phase.

between the cooling time and the sound crossing time is initially about 2 on the X-axis and about 5 on the Z-axis, this means that the linear instability time scale is lower along the minor axis since the perturbation can still be considered isobaric, not so for the X-axis along which the perturbation is not isobaric any longer (see Schwarz et al. 1972, Field 1965). As a consequence the nonlinear phase is reached earlier along the minor axis of the perturbation leading to a higher temperature gradient along the Z-axis (Figure 1b); that drives a higher velocity inflow in the same direction. From this reasoning it is clear that the term "non-spherical perturbation" has a relative meaning since, with a different choice of the parameters, a perturbation isobaric in both X and Z directions would lead to a more isotropic evolution.

In Figure 3a we show the situation of the flow as it has entered the non-linear phase of its evolution. The plot, now in cartesian coordinates, extends up to 32 kpc. We see that the matter inflow remains asymmetric along the Z direction. Figure 3b depicts the flow pattern after $\sim 10^8$ yr; the density contrast in the grid has reached a factor ~ 3 and the maximum inflow velocity is $\sim 3.5 \times 10^6$ cm/sec which is supersonic, since the whole flow has cooled to a nearly constant temperature of $\lesssim 10^4$ K.

Figure 4 shows the central region of the condensation, extending up to 8 kpc, at a later phase. The velocity vectors are almost aligned with the Z axis and a flat, pancake-like dense structure is forming. Figure 5a shows the very inner part of the blob; in this phase the flow has reached its maximum density contrast of slightly above 100 over the full computational grid. The temperature over the whole flow is nearly constant, apart from some small, numerically introduced jitter. The inflow is highly supersonic, giving rise

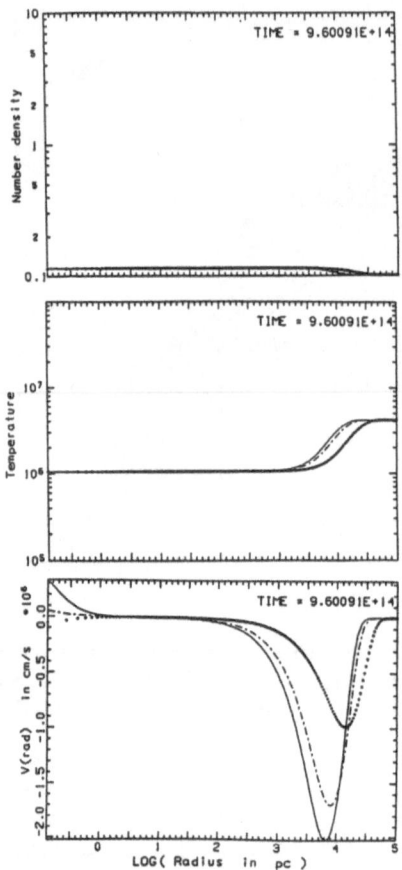

Figure 2. Radial cuts through the flow in one quadrant of a 2-dimensional polar coordinate system. Dots: profiles along the X–axis, ($\theta = 0°$), dash-dotted line: along $\theta = 45°$, and full line: along the Z–axis ($\theta = 90°$). Panel (a) plots the number density, (b) the temperature and (c) the radial inflow velocity.

to strong accretion shocks and the innermost parts of the condensation started already to reexpand (as can be seen in Figure 5b). The density plot demonstrates the highly flattened structure, the half length of the minor axis is only about 200 pc, that of the major axis $\gtrsim$10 kpc. In Figure 5b we see the reexpansion of the condensation; however, the magnitude of the velocities of this complex flow pattern is still very small compared with the inflow. Note in these physical conditions the Jean's critical length for a spherical blob in ~ 1 kpc, and applying, somewhat arbitrarily, the criterion to our pancake–like structure we find the condition for gravitational instability is matched on the X–axis only. Therefore, since again physical effects not included in out treatment may become important at this stage, we have not followed the further evolution of the blob.

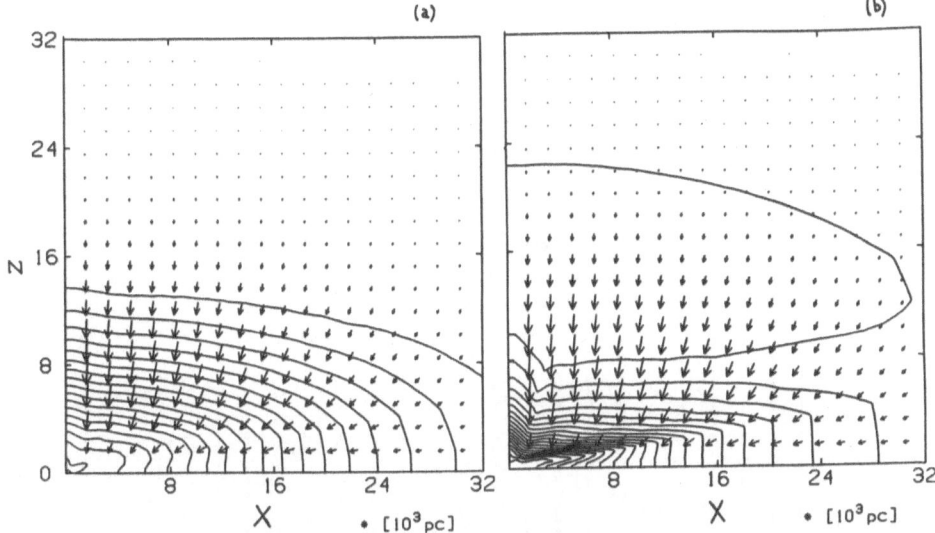

Figure 3. The same as in Figure 1 but in cartesian coordinates showing the inner 32 kpc. Panel (a): maximum density 0.115 cm^{-3}, maximum velocity 2.37×10^6 cm/sec; time of model 9.8×10^{14} sec. Panel (b): maximum density 0.313 cm^{-3}, maximum velocity 3.45×10^6 cm/sec; time of the model 3.47×10^{15} sec.

3. Discussion and Conclusions

The evolution of a grossly non-spherical perturbation proceeds in a very complex way. Due to the shorter cooling time along its minor axis (Z–axis), the perturbation evolves faster in this direction and a non–radial inflow pattern with higher velocity in the Z direction is set up. Because of this non isotropic inflow, the central parts of the blob assumes a flatter density distribution, and the evolution of the blob is more similar to a 1–D plane–parallel case of DBS than a quasi–spherical condensation. Consistently, the density contrast reached is lower with respect to the 1–D spherical case with the same initial parameters treated before. At later times in the reexpansion phase, the flow pattern seems to become very complex and non uniform. As discussed before, a main open question is whether self gravity of the formed blob will enforce a much higher radially directed inflow.

The same 2–D perturbation, in an initially lower density medium, would evolve in a different way: since fully isobaric perturbations, as defined above, have linear growth rate that is independent on scale length, in this case the anisotropy in the matter inflow along X and Z directions will be very small, therefore the evolution will be more spherical leading to higher density contrasts and more isotropic blobs.

We are aware that the results obtained cannot be directly applicable to the problem of cooling flows since neglecting the cluster's central gravitational pull and restricting the problem to a one or two dimensional perturbation on an otherwise infinite, homogeneous background, we fail to take into account the effects of buoyancy which are shown to be

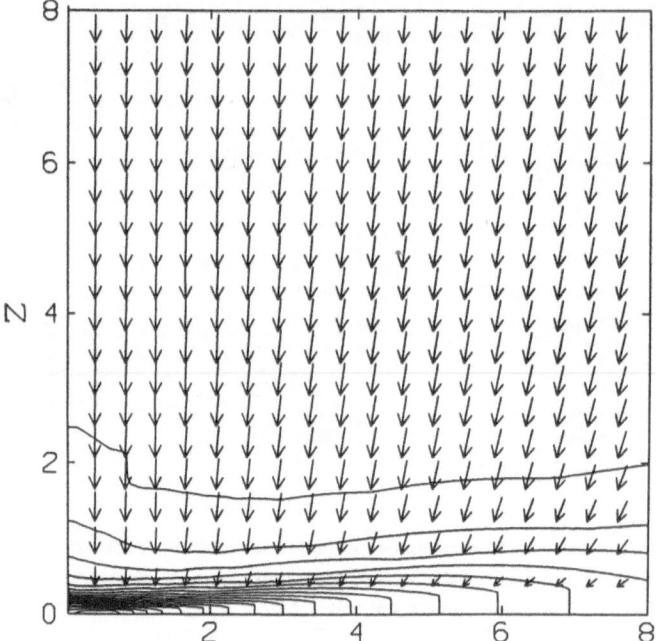

Figure 4. The same as in Figure 3 for the inner 8 kpc. Maximum density 2.69 cm^{-3}, maximum velocity 3.38×10^6 cm/sec; time of model 5.39×10^{15} sec.

important, at least in the linear regime (see Malagoli et al. 1987 and Balbus 1988). It is however very important, in our opinion, to tackle the problem step by step in order to separate and understand the effects of the various and complex physical mechanisms that compete in the instability evolution. Therefore, as a next step, a study of the evolution of the thermal instability in cooling flows in two dimensions including the central cluster's gravitational potential seems necessary. First attempts in that direction, performed in plane–parallel geometry (see Hattory and Habe, 1989) seemed not too promising as (initially) small perturbations, falling into the cluster from far outside, showed the tendency to get dispersed before reaching the phase of a non-linear instability, perhaps due to the onset of Rayleigh–Taylor instability.

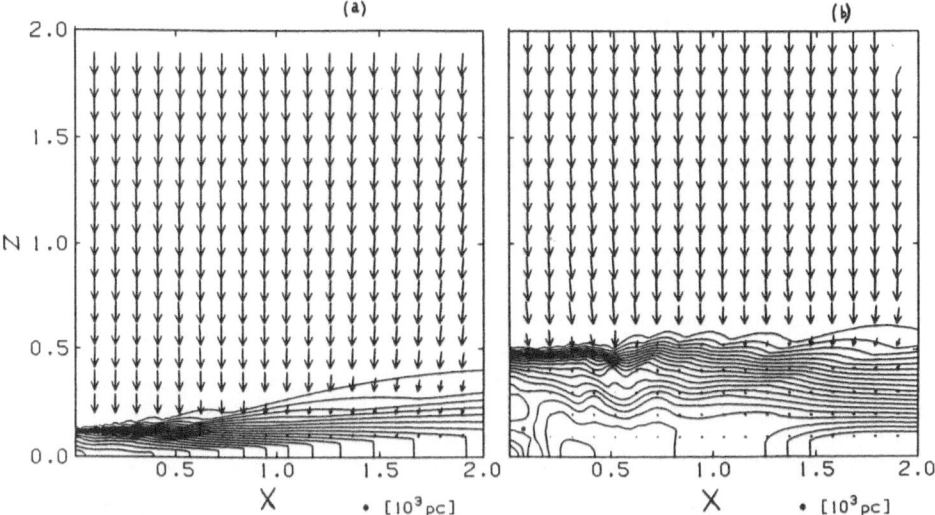

Figure 5. The same as in Figure 3 for the innermost 2 kpc. Panel (a): maximum density 10.5 cm^{-3}, maximum velocity 3.36×10^6 cm/sec; time of model 6.39×10^{15} sec. Panel (b): maximum density 5.85 cm^{-3}, maximum velocity 3.34×10^6 cm/sec; time of the model 8.48×10^{15} sec.

References

Arnold, C. N., 1985, 'Computational hydrodynamics in astrophysical studies: Highly shocked flows in three dimensions', Ph.D. Thesis, Univ. of Michigan.

Balbus, S. A. (1986), Ap. J. Lett. 303, L79.

Balbus, S. A. (1988), Ap. J. 328, 395.

Brinkmann, W., Massaglia, S., and Müller, E. (1989), Astron. Astrophys., submitted (BMM).

David, L. P., Bregman, J. N., and Seab, C. G. (1988), Ap. J. 329, 66 (DBS).

Fabian, A. C., Nulsen, P. E. J., and Canizares, C. R. (1984), M.N.R.A.S. 201, 933.

Field, G. B. (1965), Ap. J. 142, 531.

Fryxell, B. A., Taam, R. E., and McMillan, S. L. W. (1987), Ap. J. 315, 536.

Hattory, M. and Habe, A. (1989), M.N.R.A.S., submitted.

Malagoli, A., Rosner, R., and Bodo, G. (1987), Ap. J. 319, 632.

Mair, G., Müller, E., Hillebrandt, W., and Arnold, C. N. (1988), Astron. Astrophys. 199, 114.

Mathews, W. G. and Bregman, J. N. (1978), Ap. J. 224, 308.

Nulsen, P. E. J. (1986), M. N. R. A. S. 221, 337.

Sarazin, C. L. (1986), Rev. Mod. Phys. 58, 1.

Schwarz, J., McCray, R., and Stein, R. F. (1972), Ap. J. 175, p.673.

White, R. E. and Sarazin, C. L. (1987), Ap. J. 318, 612.

THE MASS SPECTRUM OF ULTRA HIGH ENERGY COSMIC RAYS

G. Auriemma
Physics Department and Sezione I.N.F.N.
University of Rome "La Sapienza"
Piazzale A. Moro, 2- 00185 Rome, Italy.

ABSTRACT. In this paper it is discussed a prediction upon the mass composition of Cosmic Rays in the PeV region, deduced from a self consistent model of acceleratio in early SN remnants. In this model the knee in the calorimetric "all particle" spectrum is due to a transition of the acceleration site, from the outer region of the remnants, to the inner plerionic region created by a powerfull pulsar, which quite naturally, takes place around 1 PeV. We observe that the mass composition versus energy distribution of the CR captured at the Earth, in this model, should be drastically affected by this transition, showing an increase of the heavy component while approching the knee and a rather sharp transition to an almost complete A=1 composition above it, due to the photodissociation of the accelerated nuclei with synchrotron photons in the plerion.

1 Introduction

The very probable connection existing among the origin of Cosmic Rays and SN explosions, postulated since 1934 by Baade and Zwicky [1], is strongly supported by an energetic argument, originally due to ter Haar [2]. The inference chain could be delineated as follows: we start considering that the energy density of the Cosmic Rays in the neighborood of the Earth is

$$w_{CR} \approx 1 \, \text{eV/cm}^3 \quad ,$$

then, if we assume this to be typical of the whole Galactic disk, the volume occupied by the C.R. ultrarelativistic gas should be:

$$V_{disk} \approx 5 \times 10^{66} \, \text{cm}^3 \quad .$$

A continous refilling of this volume is required, in order to mantain stable this energy density, because the particles leaks into the external space. Quantitatively this corresponds to a continous power supply of

$$Q_{CR} = \frac{w_{CR} \, V_{disk}}{\tau_{loss}} \approx 5 \times 10^{40} \quad \text{erg s}^{-1} \tag{1}$$

315

W. Brinkmann et al. (eds.), Physical Processes in Hot Cosmic Plasmas, 315–323.
© 1990 *Kluwer Academic Publishers.*

316

since it was shown by observations[3] that the time of confinement for the particles in the galactic disk is

$$\tau_{loss} \approx 3 \times 10^6 \, yrs \quad .$$

On the other side, the total energy released in a SN explosion ($\approx E_{51} \times 10^{51}$ ergs), multiplied by the frequency of SN in the Galaxy ($\approx \nu_{0.03} \times 0.03 \, yrs^{-1}$), gives an energy input of

$$Q_{SN} = 10^{42} \, E_{51} \, \nu_{0.03} \, \text{ergs sec}^{-1} \quad .$$

Therefore, if a mechanism could convert a small fraction (of the order of $\approx 5\%$) of the kinetic energy of the expanding SN envelope, into relativistic particles, we could expect to produce the observed energy density, since most of the energy released during the explosion goes into kinetic energy of the expanding shell [4], while only a small fraction ($\approx 10^{49-50}$ ergs) is radiated.

Nevertheless the proof that this is not a mere numerical coincidence should come from a detailed description of the scenario which realizes this conversion. One obvious possibility is that the acceleration of CR is due to the conversion of a tiny fraction of the kinetic energy of the expanding envelope, into kinetic energy of the particles, scattered by the magnetic field of the expanding plasma, as suggested long time ago by Fermi [5, 6].

2 Acceleration by the outer shock wave

The Fermi mechanism of particle acceleration was reviewed by Achterberg[7] during this Conference, and we will not repeat here. We recall that a particle cycling through the shock front increases, on the average, its momentum in each crossing by

$$\Delta p = \frac{4}{3} \frac{\Delta u}{\beta c} p \tag{2}$$

where p is the momentum before crossing the front, β the velocity and $\Delta u = u_1 - u_2$ is the differential velocity of the upstream and downstream plasma in the reference frame of the shock wave. The same particle will circulate through the shock front with a period

$$T_{cycle} = \frac{4}{\beta c} \left(\frac{D_1}{u_1} + \frac{D_2}{u_2} \right) \tag{3}$$

where D_1 and D_2 are the upstream and downstream diffusion coefficient. Approximating

$$\frac{dp}{dt} \approx \frac{\Delta p}{T_{cycle}} = \frac{u_1 - u_2}{3} \frac{p}{\left(\frac{D_1}{u_1} + \frac{D_2}{u_2} \right)} \tag{4}$$

we find the time required to accelerate a CR particle

$$t = \int_{p_0}^{p} \frac{3}{u_1 - u_2} \left(\frac{D_1}{u_1} + \frac{D_2}{u_2} \right) \frac{dp'}{p'} \tag{5}$$

For a strong shock we have $u_2 = u_1/4$ and for the diffusion coefficient we have $D \geq D_{min} = (1/3) r_L \beta c$, where $r_L = pc/ZeB$ is the Larmor radius of the particle. Integrating Eq. (5) we obtain

$$E_{max} \simeq \frac{3}{20} \frac{Z e B}{c} u_1^2 t_a \tag{6}$$

the maximum energy at which a particle can be accelerated by the Fermi mechanism. It is worth noticing that the limit of Eq.(6) corresponds to the situation in which $D/u_1 \approx R_{SN} = u_1 t$, or in other words to the energy of the particle whose diffusion lenght is comparable to the dimensions of the shock front. It is clear that in this situation the particle can hardly pass again through the shock front, for further acceleration.

We can estimate the velocity of the expanding shell from

$$E_{SN} \approx \frac{1}{2} M_{ej} u_1^2$$

the total kinetic energy. Then Eq. (6) becomes, for the shock wave generated by the expansion of the SN shell into the interstellar medium

$$E_{max} \approx 10^2 \, Z \, B_{\mu G} \, \frac{E_{51}}{M_{ej}} \, t_{yrs} \quad \text{GeV}, \tag{7}$$

where Z is the charge of the particle and $B_{\mu G}$ the average value of the interstellar magnetic field, in μGauss, M_{ej} is the typical mass of the SN ejecta in solar masses with kinetic energy $10^{51} \times E_{51}$ ergs.

The Eq. (7) indicates that the acceleration by the outer shock wave generated in a SN explosion cannot explain the entire cosmic rays spectrum, due to the limited lifetime of the shock wave itself, as also concluded by Lagage & Cesarsky[8, 9] in recent papers. The crucial point is how long can a SN shell mantain a shock front. This point was discussed also by Cioffi[10] at this meeting, and the conclusion is an estimate in the range of $\approx 10^4$ yrs. Then weare forced to conclude that the outer shock produced by a typical type II SN ($E_{SN} \approx 10^{51}$ ergs and $M_{ej} \approx 10 \, M_\odot$) could accelerate particles up to an energy $E_{max} \approx 10^6 \, Z$ GeV, almost exactly the energy corresponding to the knee in the CR spectrum.

However the spectrum of CR as measured at the Earth, is originated by the superimposition of a large number of different explosions, whose number is estimated to be $\approx 10^5 \, \nu_{0.03}$. The maximum energy of the particles accelerated by each explosion will be slightly different, being $\propto E_{51} M_{ej}^{-1}$, but the predictable overall effect will be an increase of the $\langle A \rangle$ with the energy, as it was, in fact, observed in recent measurements [11, 12].

3 Acceleration by the inner shock wave

Another scenario for particle acceleration is suggested from the observation of the Crab Nebula. The important fact, now clear after twenty years of extensive observation of this remnant, and of the pulsar located in its center, is that the pulsar's spin down energy is converted into nebular synchrotron luminosity with very high efficiency (10–20 %). This implies that a comparable efficiency of particle acceleration should be assumed.

It was noted [13] that the rotation energy of a fast spinning pulsar (momentum of inertia $I = I_{45} \times 10^{45}$ g cm^2) multiplied by the frequency of SN in the Galaxy,

$$Q_{pulsar} = \frac{1}{2} I \Omega^2 \nu_{SN} \approx 3 \times 10^{41} I_{45} \nu_{0.03} P_{10}^{-2} \quad \text{erg s}^{-1}, \tag{8}$$

where $P = p_{10} \times 10$ msec is its period, is a sufficient supply of energy for the CR, if the initial period of the pulsar is of the order of $\stackrel{\sim}{<} 10$ ms.

Recently it was discussed by Gaisser *et al.* [14] if a similar model could be applied to the very early phase of the evolution of the SN remnants, to the point of producing detectable effects in the SN 1987a (Shelton) exploded in the Large Magelanic Cloud. The basic virtue of this model is that it could explain the extension of the spectrum of the cosmic rays to much larger energies. The acceleration of particle is produced in this case at the inner shock front, produced by the strong MHD winds of the pulsar inside the expanding shell, by the Fermi acceleration process, as considered above for the outer shock, but in this case the magnetic field is much stronger, because it is provided by the pulsar itself.

Rees and Gunn [15] in a classic paper have outlined the interaction between the pulsar and the nebula, which was further developped into a self–consistent magnetohydrodinamical model by Kennel and Coroniti [16, 17]. The basic idea is that the dipole electromagnetic radiation (mostly electron–positron pairs formed in the pulsar magnetosphere) of the spinning pulsar cannot propagate inside the external shell of the SN, hence this waves open a cavity inside the shell, called a "plerion"[18, 19] in the literature, with radius R_{pl} which is filled with a thin plasma, magnetic field and relativistic particles. The electrically conductive outer boundary of this cavity expands with uniform expansion velocity $u_{pl} \approx u_8 \times 10^8$ cm s^{-1}. At a certain radius R_s a standing reverse shock is formed, due to the balance of the cumulated energy density in the plerion and the wind pressure. Following ref. [15] the shock radius R_s is given by

$$\frac{L}{4\pi R_s^2 c} \approx \frac{\int L \, dt}{(4\pi/3) \, R_{pl}^3} \tag{9}$$

where the luminosity L is the total dipole luminosity of the pulsar[20], given by

$$L = \frac{2}{3} \frac{\mu^2 \sin^2 \theta \, \Omega^4}{c^3} \approx 3 \times 10^{39} \, B_{12}^2 \, P_{10}^{-4} \quad \text{erg s}^{-1}, \tag{10}$$

where $B_{12} \times 10^{12}$ Gauss is the surface magnetic field of the pulsar, and $P_{10} \times 10$ *ms* its rotation period.

The caracteristic spin down time of the pulsar, is

$$\tau_{puls} = \frac{1}{2} \frac{I\Omega^2}{L} \approx 3000 \, B_{12}^{-2} \, P_{10}^2 \quad \text{yrs}, \tag{11}$$

which is also presumably the scale of the duration of the acceleration of particles. From Eq.(9) we can estimate

$$R_s = R_{pl} \sqrt{u_{pl}/3 \, c} \approx 10^{14} \, u_8^{3/2} \, t_{yrs} \quad \text{cm} \tag{12}$$

where t_{yrs} is the time from explosion (in yrs).

Assuming that the magnetic field is originated by the leakage into the cavity of the pulsar magnetic field itself [1], we have

$$B(r) = B_0 \left(\frac{r_0}{r_{LC}}\right)^3 \frac{r_{LC}}{r} \tag{13}$$

---

[1]In other papers (see *e. g.*Bandiera *et al.*[18], it is assumed that the field in the plerion is due to the motion of relativistic electron plasma. It was shown [21] that this assumption gives exactly the same estimate for the intensity of average the magnetic field.

where the field is taken to be dipolar inside the light cylinder ($r \leq r_{LC}$) and toroidal outside. In particular we have

$$B_s \approx 10\, B_{12}\, P_{10}^{-2}\, u_8^{-3/2}\, t_{yr_s}^{-1}\, \text{Gauss} \quad , \tag{14}$$

for the field at the shock front.

We remark here that the wind is relativistic, therefore $u_1 \approx c$ in Eq.(5), but the shock front moves with a small velocity. In this situation the limit to the maximum attainable energy is not limited the slowness of the cycling through the shock front, which is fast due to the comparable high magnetic field, but from the limited size of the shock front, in fact the acceleration ceases when $D/c \approx R_s$ or

$$E_{max} = 3\, Z\, e\, B_s\, R_s \approx 3 \times 10^9\, Z\, B_{12}\, P_{10}^{-2} \quad \text{GeV}. \tag{15}$$

It is remarkable that in this equation the dependance from the time was cancelled by the fact that the "bending power" of the plerion $\int B\, dl \propto B_s\, R_s$ is constant in time.

4 The composite scenario

In the two previous sections we have summarized the arguments for a composite scenario of the origin of CR, while discussing the maximum attainable energy for the acceleration processes. The Fermi mechanism is the basic process in both cases, but it is clear that the slowness of this process, when applied to the outer shock front, makes acceleration to very large energies impossible. Therefore we could assume that the whole spectrum of CR (perhaps up to a maximum energy of the order of $10^8 - 10^9$ GeV) is supplied by SN explosions, but in a composite way, namely below the knee the acceleration is produced in the outer shock front, while above the knee particles comes from the inner shock front. In this case the observed steepening in the *"all particle"* spectrum could be in effect the mark–point of a transition in the acceleration regime, around this energy, which is roughly in agreement with the maximum energy for the outer shock according to Eq.(7).

This is not the most popular interpretation of the steepening of the cosmic ray spectrum around 3000 TeV, which is rather thought to be an effect of the confinement [22, 23]. It is very intriguing the fact that the composite scenario, illustrated above, allows to make some definite predictions over the mass composition of CR above the knee, which hopefully could be compared with experimental data. The rest of the paper will be dedicated to the discussion of these predictions. If the acceleration take place in a limited region with high electron density and large magnetic field, such as the inner shock environment, described in Sec. III, we expect a large density of synchrotron photons. As noted by Hillas [24] in a head-on collision between a nucleus of total energy E and mass $A m_N$, with a photon with energy $\hbar\omega$, nuclear photodisintegration occurs when $2\,\hbar\omega\,(E/A m_N) \approx 10 - 20$ MeV. This process does not occur for protons but considerable energy losses set in above the threshold for pions photoproduction, when $2\,\hbar\omega\,(E/m_N) \approx 200$ MeV. In the following this basic idea will be discussed quantitatively.

5 Syncrotron radiation in the inner region

Acceleration of electron at the shock front will also take place, but the final spectrum will be strongly modified by the fast energy loss of the electrons in the strong magnetic field. The maximum energy ϵ_{max} to which electrons can be accelerated can be estimated by equating acceleration time scales to the loss time scale. From Eq. (4) we have

$$\dot{\epsilon}_{acc} \approx \frac{1}{20} \frac{u_1^2}{D} \epsilon \tag{16}$$

The power radiated by a single electron with energy ϵ is:

$$\dot{\epsilon}_{rad} = \frac{4\alpha}{\hbar m_e^2} (\mu_B B)^2 \epsilon^2 \quad , \tag{17}$$

where $\alpha \simeq 1/137$ is the fine structure constant, $\hbar \simeq 6.6 \times 10^{-22}$ MeV s the Planck constant, $\mu_B \simeq 5 \times 10^{-15}$ MeV Gauss^{-1} the Bohr magneton, B the average field in Gauss. Assuming $D = D_{min} \equiv (1/3) \epsilon \beta^2 c/e B$ we have

$$\epsilon_{max} \approx 10^4 B_{12}^{-1/2} P_{10} u_8^{3/4} t_{yrs}^{1/2} \quad \text{GeV} \tag{18}$$

The spectral distribution of the accelerated electrons, calculated from the momentum transport equation by Schlickeser[25], is also in this case a power law $J_e(\epsilon) \propto \epsilon^{-\beta}$ where the spectral index is close to 2. But if one assumes a longer escape time the electrom spectrum shows a marked Wien bump, due to the pile up of relativistic electrons at ϵ_{max}. Therefore we expect that the synchrotron spectrum will be a power law, with a spectral index $\alpha = (\beta + 1)/2$ or flatter up to a photon energy

$$\hbar\omega_{max} \simeq \frac{1}{3}\mu_B B \left(\frac{\epsilon_{max}}{m_e}\right)^2 \quad , \text{MeV} \quad , \tag{19}$$

dropping exponentially above this limit. We remark that this energy is constant in time, because from Eq. (18) we have that $\epsilon_{max} \propto t_{yrs}^{1/2}$ while from Eq. (14) we have $B \propto t_{yrs}^{-1}$ therefore we conclude that the synchrotron radiation spectrum is stationary the maximum photon energy being $\hbar\omega_{max} = 2.6$ MeV.

Thus we can estimate the energy density of the synchrotron photons ρ_γ in the plerion, at a distance from the pulsar $R \gg R_s$, by the equation:

$$\rho_\gamma \approx \frac{\eta_{acc} L}{c R^2} \quad , \tag{20}$$

in stationary conditions, where η_{acc} is an efficiency for the acceleration of the electrons by the inner shock front. We can use this equation here for two reason, first because the electrons radiate very fast, in fact their lifetime is very short compared to the expansion time and second because the total power radiated by a single electron drifting in the magnetic field, is $\propto \int B \, dl$, the bending power of the plerion, which, as we remarked before, is approximately constant in our model. Taking into account the power law distribution of the synchrotron spectrum we have for the photon density at a distance from the pulsar $R \gg R_s$

$$\frac{dn_\gamma}{d\hbar\omega} \approx \frac{(2-\alpha)\,\eta_{acc} L}{c\,(\hbar\omega_{max})^2 R^2} \left(\frac{\omega_{max}}{\omega}\right)^\alpha \tag{21}$$

6 Photodissociation of heavy nuclei

In the radiation field, a nucleus has a mean free path against photodisintegration given by

$$(\lambda_{ph})^{-1} = \int \sigma(\hbar\omega') \frac{dn'_\gamma}{d\hbar\omega'} \, d\hbar\omega' \quad . \tag{22}$$

where ω' is the frequency of the photon in the rest frame, of the nucleus, and n'_γ the apparent photon density in the same frame.

The photonuclear cross section integrated over the giant resonance [26] is given by good approximation by the dipole sum rule

$$\int \sigma \, d\hbar\omega' \simeq \frac{2\pi^2\alpha(\hbar c)^2}{m_N} \frac{(A-Z)\,Z}{A} = 0.06 \frac{(A-Z)\,Z}{A} \quad \text{MeV barns.} \tag{23}$$

The energy of the resonance in the rest frame of the nucleus is given by

$$E_{res} \approx 22 \times \left(\frac{A}{16}\right)^{-1/3} \quad \text{MeV,} \tag{24}$$

The photodisintegration mean free path can be approximated by the formula

$$(\lambda_{ph})^{-1} \approx \left.\frac{dn'_\gamma}{\hbar\omega'}\right|_{\hbar\omega'=E_{res}} \int \sigma \, d\hbar\omega' \tag{25}$$

The probability for a nucleus of diffusing out from a sphere of radius $R \gg R_s$, avoiding the photodisintegration, is $\exp[-\xi]$ where in the diffusion limit

$$\xi = (\lambda_{ph})^{-1} \frac{R^2}{r_L} \tag{26}$$

where r_L is the Larmor radius. Hence the apparent strong dependance of the photon energy density from the expansion of the SN ejecta is compensated. The condition for a nucleus to be ejected is finally

$$\frac{2-\alpha}{\alpha} \frac{\eta_{acc} L}{c(\hbar\omega_{max})^2} \left(\frac{2\,\hbar\omega_{max}}{E_{res}}\right)^\alpha \frac{Z\,e\,B}{A\,m_N\,c} \left(\frac{E}{A\,m_N}\right)^{\alpha-1} \int \sigma\,\hbar\omega' \stackrel{\sim}{<} 1 \tag{27}$$

For a given energy the left hand side of this equation increases with time. The extreme condition will be reached for a time $\approx \tau_{puls}$, when the radiation emitted by the pulsars strarts to fall down. If we calculate the maximum energy which a nucleus could have, at this time, without being disintegrated, we obtain:

$$E/A \stackrel{\sim}{<} 350\, B_{12}^{-2}\, L_{39}^{-4}\, u_8^{-3} \quad \text{GeV/nucl,} \tag{28}$$

which is well below the knee, but the strong dependence from the pulsar (initial) luminosity makes all the prediction rather fuzzy. In fact, an underluminos pulsar, of only a factor 10, lets escape nuclei with total energy above the knee, even if this pulsar would give a comparatively smaller contribution to the CR population of the galaxy.

7 Conclusions

Summarizing we have shown, that, in the general framework of the Fermi first order acceleration in SN remnants, it is quite reasonable to expect:

1. The spectrum of cosmic rays up to 10^9 GeV could be generated in the firsts 10^{3--4} years of the expansion of a plerionic SN remnant, if the pulsar inside the remnant has a period of the order of 10 ms.

2. CR with energy below 10^6 GeV are accelerated in the outer shock wave, due to the expansion of the remnant in the interstellar medium, while particles with energies above this limit can only be accelerated by the inner shock wave, due to MHD waves in the plerion.

3. The latter transition in the acceleration site is probably marked by a steepening in the total energy spectrum due to photopion interactions with the synchrotron radiation produced by electrons also accelerated in the plerion

4. The mass spectrum predicted by this model shows an enhancement of the heavier components approching the knee, but the photodisintegration of nuclei in the synchrotron radiation field leads to an almost purely protonic spectrum above the knee.

The prediction of this model appears to be in agreement with the observed phenomenology of CR.

References

[1] W. Baade and F. Zwicky. *Phys. Rev.*, 45:138, 1934.

[2] D. ter Haar. *Science*, 110:285–286, 1949.

[3] J.F. Ormes and R.J. Protheroe. *Astrophys. J.*, 272, 1983.

[4] V. Trimble. *Rev. Mod. Phys.*, 55:511–563, 1983.

[5] E. Fermi. *Phys. Rev.*, 75:1169–1174, 1949.

[6] E. Fermi. *Astrophys. J.*, 119:1–6, 1954.

[7] B. Achterberg. *"Particle acceleration in shocks"*. In these proceedings.

[8] P.O. Lagage and C.J. Cesarsky. *Astron. Astrophys.*, 118:223–228, 1983.

[9] P.O. Lagage and C.J. Cesarsky. *Astron. Astrophys.*, 125:249–257, 1983.

[10] D.F. Ciofi. *"Using SNR for testing the ISM"*. In these proceedings.

[11] J.M. Grunsfeld et al. *Astrophys. J.*, 327:L31–L34.

[12] T.H. Burnett and others (The JACEE Collaboration). Technical Report VTL–PUB–125, University of Washington, Seattle, WA 98195 U.S.A., Decenber 9 1988.

[13] T.K. Gaisser. In F. Giovannelli and G. Mannocchi, editors, *Vulcano Workshop 1989:Frontier Objects in Astrophysics and Particle Physics*, pages 137–144, Bologna, Italy, 1989. Società Italiana di Fisica.

[14] A. Harding T.K. Gaisser and T. Stanev. *Nature*, 329:314, 1987.

[15] M.J. Rees and J.E. Gunn. *Mont. Not. R. A. S.*, 167:1, 1974.

[16] C.F. Kennel and F.V. Coroniti. *Astrophys. J.*, 283:694–709, 1984.

[17] C.F. Kennel and F.V. Coroniti. *Astrophys. J.*, 283:710–730, 1984.

[18] F. Pacini R. Bandiera and M. Salvati. *Astrophys. J.*, 285:134–140, 1984.

[19] R. Bandiera. *"Plerionic supernova remnants"*. In *these procedings*.

[20] S.L. Shapiro and S.A. Teukolsky. *Black Holes, White Dwarfs and Neutron Stars: The Physics of compact objects*. John Wiley & Sons, New York, 1983.

[21] T.K. Gaisser G. Auriemma and P. Lipari. *Nuovo Cimento*, 102 B:583–592, 1988.

[22] B. Peters. *Nuovo Cimento Suppl.*, 14:436, 1959.

[23] C.E. Fichtel. *Phys. Rev. Lett.*, 11:172, 1963.

[24] A.M. Hillas. In *Proc. 16th Intern. Cosmic Ray Conf.*, volume 8, page 7, Kyoto, 1979.

[25] R. Schlickeiser. *Astron. Astrophys.*, 137:227–236, 1984.

[26] E. Hayward. *Rev. Mod. Phys.*, 35:324, 1963.

PLERIONIC SUPERNOVA REMNANTS

R. BANDIERA
Arcetri Astrophysical Observatory
Largo E. Fermi 5
50125 Firenze
Italy

ABSTRACT. The nature of the non-thermal emission from the Crab Nebula and other Crab-like supernova remnants (plerions) is investigated. It is commonly accepted that the dominant radiation mechanism is synchrotron emission, and that the energy required for the emission is provided by an internal spinning neutron star; but the details of the physics involved are not known yet. A review is presented of the main distinctive features of plerions, as well as of models devised to explain them. Models can be divided in two classes: the scope of some of them is to induce physical constraints directly from observations, limiting the number of a *priori* assumptions; others instead aim at describing the physical link between pulsar and nebula, in order to reproduce in a deductive way the observed emission. The aspect of very young plerions is finally investigated by a study of possible candidates among remnants of recent supernovae: plerionic models are put forward as a possible explanation of phenomena like radio supernovae and soft X-ray emission from SN 1987A, in the Large Magellanic Cloud.

1. Introduction

The story of our understanding of supernova remnants can be divided by year 1970 in two eras. Before that date it was believed that among all remnants there was one, the Crab Nebula, so atypical to constitute a class by itself; while only after 1970 it became apparent that "the Crab Nebula is not alone!" (by citing the title of a review by Weiler (1983a)).

Although the Crab Nebula has long been recognized to represent a peculiar object, its nature was not questioned, since it has been proved that this object is located in the same region of the sky where chinese astronomers reported the discovery of a "new star" (a supernova, according to modern terminology) in 1054 AD (see Clark and Stephenson (1977)). Its relation with the historical supernova SN 1054 was confirmed when the expansion time scale of the remnant was found to be (almost) consistent with an explosion dated 1054 AD (Trimble (1971) and references therein).

Some peculiarities of the Crab Nebula, known already before 1970, are:
1) a filled center morphology, both in optical and in radio, instead of the typical shell-like morphology of classical remnants;
2) a radio spectrum much flatter than usual;
3) an optical spectrum composed not only of emission lines, but also of a non-thermal continuum.

The Crab Nebula is in fact peculiar under many other aspects, as the low total energetics and mass, the high helium abundance, the presence of a jet/spur. These aspects will not

W. Brinkmann et al. (eds.), Physical Processes in Hot Cosmic Plasmas, 325–340.
© *1990 Kluwer Academic Publishers.*

be discussed here, because we are mainly concerned with non-thermal processes; anyway, they are already treated in depth in other reviews (see e.g. Davidson and Fesen (1985)).

The search for objects somehow similar to Crab succeeded shortly after 1970. Weiler and Seielsted (1971) showed that the supernova remnant 3C 58 (related to the historical supernova SN 1181) is similar to Crab, at least as far as the features listed above are concerned, and put forward a handful of remnants that should be considered "Crab-like" objects.

The number of candidates became soon large enough to require the official introduction of a new class. Weiler and Panagia (1978), directly referring to the filled centre morphology typical of these objects, suggested the name of "plerions", from the greek word $\pi\lambda\acute{\eta}\varrho\eta\varsigma$, that means "filled". According to the present status of observations of plerions the main distinctive characteristics of this class are:

1) in radio: a filled center structure, with a flat non-thermal spectrum (a power-law index flatter than -0.3) and a rather high linear polarization;
2) in optical: a non-thermal continuum component, with a filled center structure and some linear polarization;
3) in X rays: a non-thermal (a pure power law) emission, with a filled centre structure.

As usual in classifications, there are strange objects that are not clearly belonging to the class of plerions; among the most relevant are CTB 80 and MSH 15-52 (see Weiler (1985) for a discussion). Moreover there are objects in which a filled center component coexists with a shell; they are the so-called "composite" supernova remnants (Helfand and Becker (1987)). Some recent catalogs of pure plerionic or composite supernova remnants are in Weiler (1983a, 1983b), Weiler and Sramek (1988), and Seward (1985, 1989). At present only one object can be considered not just a "Crab-like" remnant, but rather an actual "Crab twin": this is SNR 0540-693, in the Large Magellanic Cloud (see Reynolds (1985) for a review).

The non-thermal emission from plerions is mostly due to synchrotron radiation (Shklovskii (1960)). This emission mechanism is not very exotic in astrophysics, at least in the radio range: for instance synchrotron emission originates also in shell supernova remnants, from the interaction between relativistic particles and turbulent magnetic fields produced by the shock. More unusual is the presence of this kind of emission in the optical and even the X-ray range, indicating that some emitting particles are highly relativistic.

The engine is rather exotic indeed: the presence of a neutron star harboured in the Crab Nebula has been predicted (Pacini (1967)), and subsequently observationally confirmed (Staelin and Reifenstein (1968)). A fast spinning, highly magnetized neutron star, typically revealed as a "pulsar", provides both magnetic field and relativistic particles. A direct relationship between the total radiation output of the nebula and the pulsar's rotational energy losses is evident in the Crab Nebula, where the radiated energy is 10–20% of that lost by the neutron star; therefore the efficiency by which field and relativistic particles are generated must be very high.

In analogy with the case of Crab, a search for pulsars has been performed also inside other plerions. This search has been partially successful, in the sense that radio pulsars have been detected in some plerions (Vela, SNR 0540-69.3), but also in objects of difficult classification (CTB 80, MSH 15-52), and even in remnants without any plerionic component (W 44). In other remnants, even though there is no evidence of radio pulsation, the detection of a central compact X-ray source seems to indicate the presence of a neutron star: while one of these remnants is clearly a plerion (3C 58), some others look as normal shell-like remnants (Kes 73, RCW 103) (see Seward (1985, 1989)).

2. Inductive Models

There are two general classes of models which have been used to reproduce the properties of plerions: they could be characterized as "inductive" and "deductive" models, respectively.

We shall qualify as "inductive" those models that are mostly phenomenological. Since they contain just little physics inside, they obviously suffer of some disadvantages for being limited to an approximate description; however they have the big advantage of not being strongly model-dependent. Therefore they are the most suitable models to infer some general physical properties directly from observations, being their content of a priori assumptions intentionally limited.

For instance the original model for the emission from the Crab Nebula (Pacini and Salvati (1973)), as well as further revisions of it (Bandiera, Pacini and Salvati (1984), Reynolds and Chevalier (1984)), can be classified as inductive models. Their basic assumption is that the non-thermal emission is due to synchrotron radiation; the rotational energy lost by the star is converted (by a mechanism which is not investigated at this stage) into magnetic fields and relativistic electrons (or positrons). While the magnetic bubble is expanding, both components are subject to adiabatic losses; the electrons are also subject to synchrotron losses, due to their interaction with the magnetic field.

A common feature of inductive models is that they try to avoid any geometrical complication: whenever it is possible, the field is assumed to be homogeneous throughout the bubble, while the velocity distribution of particles is always assumed to be isotropic. In the following we shall review the equations on which these models are based.

Neutron star spin-down. The decrease of the neutron star angular velocity Ω is usually parametrized by $-\dot{\Omega} \propto \Omega^n$. There are at least two cases in which reliable estimates of the braking index n have been obtained by direct measurements, namely the Crab pulsar (Grooth (1975)) and PSR 1509-58 (Manchester, Durdin and Newton (1985)): in these objects n is 2.51 and 2.83, respectively. However, indirect estimates of n done by statistical methods give considerably larger values (see N. Itoh, in this meeting). Also the braking index of PSR 0540-69, in the Large Magellanic Cloud, has been recently measured; however there are still large discrepancies between different estimates ($n = 3.6$ by Middleditch, Pennypacker and Burns (1987), $n = 2.0$ by Manchester and Peterson (1989)).

Direct measurements are in fair agreement with $n = 3$, the favourite value of standard theories for pulsar magnetospheres. For instance, magnetic dipole radiation from a spinning neutron star gives an energy loss $L = -I\Omega\dot{\Omega} = 2\mu^2\Omega^4/3c^3 \propto B_*^2\Omega^4$, where I is the momentum of inertia of the neutron star, μ its magnetic dipole and B_* a typical field at its surface) (Pacini (1968), Gunn and Ostriker (1969)). A similar formula is obtained also for a unipolar inductor (Goldreich and Julian (1969)), as well as by estimating the Poynting flux at the light cylinder (see Michel (1982)): in fact, due to purely dimensional arguments, such a dependence tends to be satisfied in the majority of models based on a magnetized rotator. Models predicting braking indexes different from 3 are however feasible, for instance by assuming a multipolar field, or a magnetic moment varying with time (see Blandford and Romani (1988), and references therein).

With $n = 3$ the time evolution for the rotational losses is $L = L_o/(1 + t/\tau)^2$, where $\tau = I\Omega_o^2/2L_o \propto 1/B_*^2\Omega_o^2$ is a characteristic time; for $t < \tau$, L is approximately constant while, for $t > \tau$, $L \propto t^{-2}$. In the following we shall assume that a constant fraction p of rotational losses is converted into magnetic field, while the remaining $(1 - p)$ goes into accelerating particles.

Relativistic particles injection. The standard assumption is that electrons are ejected according to a power-law energy distribution $J(E) \propto E^{-\gamma}$. If $\gamma < 2$ (the case of interest for flat spectrum sources like plerions) the spectrum is energetically dominated by the hardest particles, and therefore an upper limit to particle energies (E_{max}) must be settled, in order to avoid divergencies. The injected spectrum turns out to be:

$$J = (1 - p)L\frac{(2 - \gamma)}{E_{max}^{2-\gamma}}E^{-\gamma}. \tag{1}$$

One can see from this formula that the value of E_{max} has an indirect influence on the injection at all energies, because it determines the level of injection by acting as a "squeezing factor". As a consequence, models are also characterized by the evolution of E_{max}: the standard assumption of a constant E_{max} (Pacini and Salvati (1973), Reynolds and Chevalier (1984)) leads to $J \propto L$; while, as an alternative probably more related to the actual acceleration mechanism, an assumed proportionality with the maximum potential drop in the magnetosphere, $E_{max} \propto B_* \Omega^2 \propto L^{1/2}$ (Bandiera, Pacini and Salvati (1984)), leads to $J \propto L^{\gamma/2}$: these different assumptions imply different evolutions when L is not constant ($t > \tau$).

Magnetic field evolution. If the magnetic field is homogeneously distributed into a bubble of radius R, the total magnetic energy is $W_B = B^2 R^3/6$. This energy changes depending on the energy input from the neutron star and on the adiabatic losses due to the bubble expansion, namely $\dot{W}_B = pL - W_B \dot{R}/R$; this equation can be integrated giving:

$$B^2 = \frac{6p}{R^4} \int LR \, dt. \tag{2}$$

The evolution of B is therefore strongly coupled to the expansion law of the bubble. The simplest case is linear expansion (Pacini and Salvati (1973)), which is approximately consistent with the expansion measured in Crab. Another possibility is power-law expansion, examples of which are the case of interaction with expanding ejecta ($R \propto t^{6/5}$), at very early stages, or with a circumstellar ($R \propto t^{2/3}$) or interstellar ($R \propto t^{2/5}$) environment, at later stages. A more detailed study of the bubble dynamics, including the effects of a reverse shock, can be found in Reynolds and Chevalier (1984).

Single particle evolution. The energies of the injected electrons will evolve in time depending on the amount of losses, due both to synchrotron radiation and to adiabatic expansion of the magnetic bubble in which they are confined. The energy evolution is described by the equation $-\dot{E} = c_1 B^2 E^2 + E\dot{R}/R$, where the coefficient c_1 is derived from the standard theory of synchrotron emission. By integrating this equation one gets:

$$\frac{1}{ER} - c_1 \int \frac{B^2}{R} \, dt = \text{const.} \tag{3}$$

Particles spectrum evolution. In order to describe the evolution of the distribution of particles, $N(E, t)$, we need a continuity equation in the energy space. The distribution at the present time t is obtained by summing all particles injected at previous times t_i, with energies E_i such as to evolve into E at t. By integrating the equation $dN \, dE = J(E_i, t_i) \, dt_i \, dE_i$, and making use of Eq. 3, one obtains the following relation:

$$N = \int J(E_i, t_i)\frac{E_i^2 R_i}{E^2 R} \, dt_i, \tag{4}$$

where the integral is performed following a single particle evolution, namely using explicitly the dependence $E_i(E, t, t_i)$.

The evaluation of N is rather difficult in the general case, when a particle is evolving through many different regimes, corresponding to changes in the behaviour of $L(t)$, $E_{max}(t)$ and $R(t)$. For the sake of simplicity, let us assume that L and E_{max} are constant, and $R \propto t^\alpha$; in this case $B^2 \propto Lt/R^3 \propto t^{1-3\alpha}$. The E–t plane can be divided in two regions, depending on which type of losses, either synchrotron or adiabatic, dominates the single particle evolution. The intermediate region corresponds to an energy $E_b \simeq 1/(c_1 B^2 t) \propto t^{3\alpha-2}$. In the adiabatic-dominated regime the typical time scale for particles evolution is the dynamical time t; while in the synchrotron-dominated one the time scale is $t_s = 1/(c_1 B^2 E)$, related to radiative losses and shorter than t. Therefore, in the adiabatic case the particles distribution can be approximated as $N \propto Jt \propto E^{-\gamma}t$, while in the synchrotron case as $N \propto Jt_s \propto E^{-\gamma-1}t^{3\alpha-1}$: two different power laws are therefore obtained at lower and higher energies, respectively, and a break in the distribution appears near E_b. In the more general case there are various mixed regimes, each of them with a specific spectral slope and time evolution.

Synchrotron spectrum. The transformation from the particles distribution to the emitted spectrum is straightforward, provided that the so-called monochromatic approximation is applied. The synchrotron losses from each particle, $c_1 B^2 E^2$, are assumed to give monochromatic radiation at a frequency $\nu = c_2 B E^2$ (also c_2 comes from the synchrotron theory): such approximation gives good results in a rather wide range of indexes for power-law particles distributions. With this approximation the emitted spectrum turns out to be:

$$S(\nu, t) = \frac{c_1}{2c_2} BEN. \tag{5}$$

By this formula one can compute spectra or light curves, depending whether t or ν is taken as a constant. For instance, in the simple case presented above, the spectrum is divided into two parts: at frequencies lower than $\nu_b = c_2/(c_1^2 B^3 t^2) \propto t^{(9\alpha-7)/2}$ the emitting particles are in adiabatic regime and the emitted spectrum is proportional to $\nu^{-(\gamma-1)/2} t^{1-(3\alpha-1)(\gamma+1)/4}$; while at frequencies higher than ν_b the emitting particles are in synchrotron regime and the spectrum is proportional to $\nu^{-\gamma/2} t^{(3\alpha-1)(2-\gamma)/4}$.

We can use the Crab itself as a workbench, since many relevant parameters have been well determined in this object. Input parameters, both referring to the nebula and to its central pulsar, are listed in Table 1, as well as some derived quantities. They are evaluated using a numerical model based on the equations presented in this Section, and assuming also linear expansion, $E_{max} \propto L^{1/2}$, and equipartition in the energy input between field and particles ($p = 0.5$); we have evaluated E_{max} by fitting the emission in the radio band. The resulting spectrum is represented by a dashed line in Fig. 1, where it is compared with the overall spectrum of the Crab Nebula (from Woltjer (1987)); this is in turn represented by a solid line complemented by some individual data points.

The model is rather successful in reproducing the actual spectrum also at higher frequencies; for instance the break in the infrared range is fitted by the model (the breaks are actually two, because we are presently at a time $t > \tau$). However the model fails in the X-ray region in two respects: it cannot reproduce power-law spectra as steep as those observed, and it cannot explain the presence of emission in the hard X-ray range; these problems will be discussed in more detail in Section 4.

TABLE 1. Parameters for Crab

	Measured or assumed:	
P	$= 0.0331\,\mathrm{s}$	Present pulsar period
$\dot{P}$	$= 4.23 \cdot 10^{-13}$	Period derivative
t	$= 935\,\mathrm{yr}$	Present remnant age
V_{exp}	$= 1500\,\mathrm{km\,s^{-1}}$	Expansion velocity
d	$= 2\,\mathrm{kpc}$	Remnant distance
F_{rad}	$= 1000(\nu/1\,\mathrm{GHz})^{-0.26}\,\mathrm{Jy}$	Radio flux
M_*	$= 1.4\,M_\odot$	Neutron star mass
R_*	$= 10\,\mathrm{km}$	Neutron star radius

	Derived:	
P_o	$= 0.0168\,\mathrm{s}$	Original pulsar period
τ	$= 317\,\mathrm{yr}$	Pulsar spin-down time
B_*	$= 4 \cdot 10^{12}\,\mathrm{Gauss}$	Pulsar magnetic field
L	$= 5 \cdot 10^{38}\,\mathrm{erg\,s^{-1}}$	Present pulsar energy loss
R	$= 1.4\,\mathrm{pc}$	Present nebular size
B	$= 7.6 \cdot 10^{-4}\,\mathrm{Gauss}$	Present nebular field
γ	$= 1.5$	Injection power-law index
E_{max}	$= 1.6 \cdot 10^8\,mc^2$	Present maximum particle energy

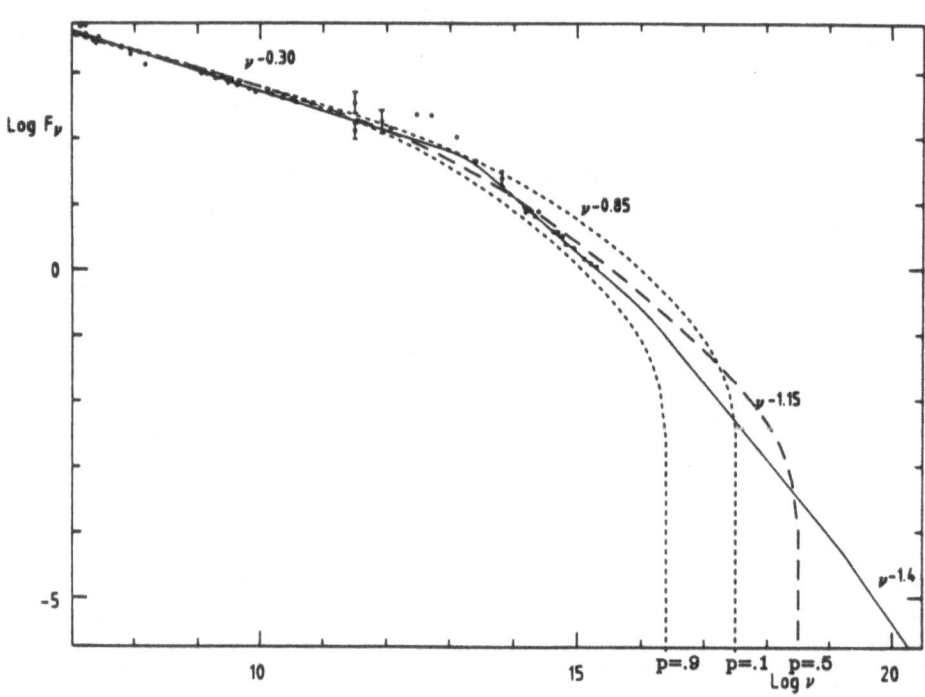

Figure 1. Comparison of theoretical spectra with the overall spectrum of the Crab Nebula.

3. Deductive models

A different approach consists in describing explicitly the physical link existing between the pulsar magnetosphere and the surrounding nebula. A model of this kind has been presented by Rees and Gunn (1974), and an improved version of it has been devised by Kennel and Coroniti (1984a, 1984b). This model is basically a spherically symmetric steady solution for a relativistic magnetohydrodynamic wind, matching the conditions of a pulsar magnetosphere at the inner boundary, and those observed at the edge of the Crab Nebula at the outer boundary. The energy output from the pulsar is shared in the following forms:
1) a wind of relativistic particles;
2) a toroidal magnetic field;
3) an electromagnetic wave at the pulsar spinning frequency.
While the low frequency electromagnetic wave is likely to be damped near the neutron star, both the high energy particles and the magnetic field will contribute to the global content present in the nebula.

An important constraint on the wind solution follows from the observed outer expansion of the nebula, $V_{\mathrm{exp}} \simeq 1500 \, \mathrm{km \, s^{-1}}$, which is clearly non-relativistic. Therefore the originally relativistic wind must heavily decelerate. The interaction of the wind with the surrounding nebula will take place on a stationary shock, at a radius R_s, where the internal ram pressure, $(L/c)/4\pi R_s^2$, is balanced by the nebular pressure (magnetic pressure has been neglected at this stage). The nebular energy increase per unit time is L, while its increase in volume is $4\pi R^2 V_{\mathrm{exp}}$: then the nebular energy density (and pressure) at equilibrium is approximately $(L/V_{\mathrm{exp}})/4\pi R^2$. This fixes the shock size to $R_s \simeq R(V_{\mathrm{exp}}/c)^{1/2}$: this result is appealing, because with the Crab Nebula expansion velocity the estimated size of the shock, $\simeq 0.07 \cdot R$, closely reproduces the actual size of optical wisps (e.g. Scargle (1969), van den Bergh and Pritchet (1989)) in Crab. Furthermore, the lack of emission in the region surrounded by wisps (Scargle (1969)) can be understood by thinking that, before reaching the shock, the particles in the wind are streaming along the field lines; while beyond the shock their directions are randomized, and therefore they emit synchrotron radiation more efficiently. A scenario that can include an explanation for wisps is even more appealing now, since wisps have been probably detected also in other plerions (see A. Ögelman, in this meeting).

Wind solutions for the shocked flow have been investigated by Kennel and Coroniti (1984a). A basic feature of these solutions is the conservation of the magnetic flux; as a consequence, for an almost toroidal field the quantity $R \cdot V \cdot B$ keeps constant with radius (in non-relativistic approximation). If the particles pressure is much higher than the field pressure, by an argument similar to that used for the standing shock one derives the relation $V \propto R^{-2}$ for the velocity profile, and then a magnetic field $B \propto R$. A field-dominated wind, instead, cannot decrease considerably its speed, and remains almost relativistic: this imply $B \propto R^{-1}$. In both cases the ratio between magnetic and particle pressure is increasing with radius: in the former case as R^2, while in the latter as $R^{2/3}$. As a result a particle-dominated wind tends to approach equipartition at larger radii, while the opposite is true for a field-dominated wind.

Kennel and Coroniti (1984a) showed that the wind flowing from the pulsar must be particle-dominated, because only in this case it fulfills the following requirements:
1) the wind velocity must fall considerably at outer radii, in order to match the expansion velocity of the nebula;
2) the magnetic field cannot be peaked at the centre, otherwise there would be a rapid synchrotron burn-off of particles, in contrast with the presence of an extended nebula;
3) the position of the standing shock must correspond to wisps positions.
The fact that the pulsar output is mostly in relativistic particles represents an important outcome of this model; Kennel and Coroniti (1984a) can fit the Crab Nebula expansion

by assuming that only 0.3% of the energy is injected in magnetic fields. This is however in contrast with models of pulsar magnetosphere (Cheng and Ruderman (1980)), inferring that the magnetic fields are produced more efficiently than non-thermal particles. A further problem concerns the spectrum of the emission; the spectrum at frequencies higher than optical can be fitted (Kennel and Coroniti (1984b)) fairly well, but the model fails because it does not predict any radio emission.

Two different kinds of approach can be followed in order to solve these drawbacks of theory: either to increase the complexity of a deductive model, or to use first inductive models in order to get more insight on the physical conditions.

4. Inductive models at work

There are many open problems regarding the physical conditions in a synchrotron nebula; in this Section we shall consider some of them, for which a simple inductive approach can be of use.

An important parameter in models is the relative efficiency of particles vs. field production. The model presented in Section 2 was assuming similar efficiencies ($p = 0.5$), while Kennel and Coroniti (1984a) argued that particles production is considerably favoured with respect to fields. In fact, inductive models have no way of deriving the "original" production rates, because they refer instead to "effective" production rates, which include also further energy exchanges between particle and field components.

Models have been computed with different values for p. The case $p = 0.5$ is that providing the best fit for high frequency emission. In fact models with a small p (energy input mostly in relativistic particles) or with $p \simeq 1$ (energy input mostly in magnetic field) give both spectra with a cutoff at much lower frequencies than observed (see dotted lines in Fig. 1, referring to models with $p = 0.1$ and $p = 0.9$, fitted to radio). Since the cutoff frequency can be expressed as $c_2 B E_{max}^2$, in the former case its decrease is due to the presence of a smaller magnetic field; while in the latter case to the decrease of E_{max}, in order to fit the radio emission with a smaller total number of particles. This result is not inconsistent with that of Kennel and Coroniti (1984a), because in their model for the wind the equipartition between particles and field is approached in the outer part of the nebula. On the other side, a field-dominated wind cannot be easily reconciled with observations: the field pressure will keep dominating also at larger radii, and the effective p will be therefore close to unity.

Another topic that deserves investigation is the degree of homogeneity of the nebular magnetic field. Although inductive models do not include geometry, a degree of inhomogeneity can be simulated by defining a filling factor f. If the particles are confined to regions containing magnetic field, the situation is equivalent to that of a homogeneous nebula expanding with a velocity $\dot{R} = V_{exp} f^{1/3}$; this implies $B \propto \dot{R}^{-3/2} \propto f^{-1/2}$, $E_b \propto \dot{R}^3 \propto f$ and finally $\nu_b \propto B E_b^2 \propto f^{3/2}$. Therefore f cannot be much smaller than unity, otherwise the spectral break will appear at much lower frequencies, in contradiction with observations.

Two interesting problems, somehow related each other, are the steepening of the X-ray spectrum and the decrease of the nebular size in X rays. Fig. 1 shows that theoretical spectra can give only an approximate description of the actual spectrum of the Crab Nebula. This problem arises because models can generally account for a change in the spectral index not larger than 1/2, from $-(\gamma - 1)/2$ to $-\gamma/2$, apart from a further steepening near the high-frequency cutoff. The actual spectral indexes range instead from -0.3 in radio to -1.15 in X rays, and the spectrum becomes even steeper in γ rays (Jung (1989)). Such a behaviour is a common feature of plerions, that typically present a smaller ratio between X-ray and radio luminosities than predicted by models (Reynolds and Chanan (1984)).

Reynolds and Chanan suggest that the distribution of injected particles is bimodal, and that X-ray emitting electrons are produced, with a steeper spectral slope, at the wind shock. This scenario relies however on a fine tuning: since the distribution is steeper at higher energies, the requirement of continuity in the distribution implies a high energy cutoff of the softer component at the same energy at which the harder tail is starting. Furthermore the relative intensities of the two spectral components must be such as to preserve continuity in the transition region. Another possible solution is that the bimodal distribution can be ascribed to only one acceleration process. However, at least for Fermi mechanisms, it is more natural to obtain pure power-law spectra (see B. Achtenberg, in this meeting).

An alternative possibility originates from the fact that particles with higher energies, which emit higher-energy photons, decay more quickly than the others. If all particles are produced near the centre, the most energetic ones will decay before reaching the boundary of the nebula: for this reason they are confined to a smaller central region. If the magnetic field is a function of radius these particles will be affected by an effective field different from that averaged out on the whole nebula.

Let us assume that the confining radius is $R \propto t_s^{1/\alpha}$, where t_s is the synchrotron time scale: α is equal to 1 for linear expansion, equal to 3 for a particle-dominated wind, or equal to 2 in the case of particle diffusion (Gratton (1972), Wilson (1972)). By expressing the nebular spectrum as a function of only frequency and magnetic field, one gets $S(\nu) \propto \nu^{-\gamma/2} B^{-(2-\gamma)/2}$. If the magnetic field is proportional to R^β, one can find the relation $B \propto \nu^{-\beta/(2\alpha+3\beta)}$; by direct substitution one gets then:

$$S(\nu) \propto \nu^{-(\gamma/2)+\beta(2-\gamma)/2(2\alpha+3\beta)}. \tag{6}$$

In the case of Crab ($\gamma = 1.5$), an X-ray spectral index of -1.15 can be obtained if $\beta = -0.56\,\alpha$, requiring a higher field near the centre.

On the other hand, the effect of a confining radius for higher-energy particles has been actually observed; the Crab size is smaller in optical than in radio, and even smaller in X rays (see Seward (1989)); a similar effect has been discovered also in other plerions (Becker, Helfand and Szymkowiak (1982)). By a procedure like that used for $S(\nu)$ one gets:

$$R(\nu) \propto \nu^{-1/(2\alpha+3\beta)}. \tag{7}$$

Since there is observational evidence (Ku *et al.* (1976)) that the size of the X-ray nebula is proportional to $\nu^{-0.148}$, the derived relation is $2\alpha + 3\beta = 6.76$. For reasonable values of α and β this relation is inconsistent with that obtained from the X-ray spectrum. In our opinion we can take the latter result as a likely constraint on the radial structure of the nebula, while the former result is only suggesting that the steep high-energy spectrum is due to a break, of unknown origin, in the injected particles distribution.

The presence of a break, instead of a cutoff, in the distribution of injected particles is also appropriate to explain the presence of some particles emitting γ-ray photons. The need of very energetic particles is even more stringent if the nebular magnetic field increases with radius, as suggested theoretically by deductive models, and observationally by the X-ray nebular size. The existence of particles with a Lorentz factor $\gamma_{\max} \simeq 1.6 \cdot 10^8$ (as derived from the homogeneous model) represents a strong constraint on the mechanisms for particles acceleration. If such acceleration is produced by coulombian forces acting in the pulsar magnetosphere, an upper limit to the energies of injected particles can be estimated as the maximum potential drop near the neutron star, giving $E_{\max} \simeq eB_* R_* (\Omega R_*/c)^2/2$ (Goldreich and Julian (1969)), namely $\gamma_{\max} \simeq 5 \cdot 10^{10}$ for the present day Crab pulsar;

dissipative mechanisms must however be small enough to allow large final energies.

5. Plerionic Nature of Very Young SNRs

An important and arduous task for modern research on supernova remnants is to establish a relationship between them and the various types of supernova events (e.g. Weiler and Sramek (1988)). A way to accomplish that task is to discover and study objects which represent an intermediate stage of evolution between supernovae and the already known remnants. The search for very young supernova remnants is twice as important in the case of plerions, because beyond the characteristics of very young nebulae we need to learn how newly born pulsars actually behave. It is unclear how many very young plerions are expected in our Galaxy. Plerions represent a minority among all supernova remnants, but they have typically shorter lifetimes (Weiler and Panagia (1978)): therefore their birth rate could be as high as that of shell-like remnants. Anyway, since the overall supernova rate in our Galaxy is rather low, it is convenient to extend the search to nearby galaxies, by taking advantage of the fact that very young remnants are expected to be very bright.

The earliest searches for radio emission in the positions of recent supernovae (de Bruyn (1973), Brown and Marscher (1978)) did not succeed due to lack of sensitivity; the only exception was the detection of a faint flux from SN 1970G (Gottesman et al. (1972)). Only with the beginning of the Very Large Array era it became possible to detect radio emission associated with various supernovae: a new class of objects, named "radio supernovae", was discovered in this way (e.g. Weiler et al. (1986)). The main characteristics of these objects are the following:
1) a non-thermal radio emission (high brightness temperature, negative spectral index);
2) a radio flux detected within 1 year from the supernova event;
3) a rapid rise of the emission, earlier at higher frequencies;
4) a constant spectral index at late times;
5) a steady decline in the late evolution;
6) radio detections only of Type II and Type Ib, but not of classical Type I supernovae.

A classification of radio supernovae has been attempted, limited to those objects for which enough data are available. Some convenient parameters are:
1) the time delay from the optical maximum to the appearance of radio (t_{del});
2) the maximum radio luminosity (L_{max});
3) the asymptotic spectral index (α);
4) the slope of the final decline (β).
In the following $\lambda = 6$ cm will be taken as a reference wavelength.

Most of radio supernovae can be divided into two classes: one is associated with Type II supernovae (SN 1950B, SN 1957D, SN 1970G, SN 1980K, SN 1981K) and characterized by $t_{del} \sim 1$ month, $L_{max} \sim 10^{26}$ erg s^{-1}Hz^{-1}, $\alpha \simeq -0.6$ (flatter spectra) and $\beta \simeq -0.7$ (slower decline); while the other is associated with Type Ib supernovae (SN 1983N, SN 1984L) and characterized by $t_{del} \sim -10$ days (even before the optical maximum), $L_{max} \sim 7 \cdot 10^{26}$ erg s^{-1}Hz^{-1}, $\alpha \simeq -1.0$ (steeper spectra) and $\beta \simeq -1.6$ (faster decline). Two objects, SN 1979C and SN 1986J, are atypically powerful in radio ($L_{max} \sim 2 \cdot 10^{27}$ erg s^{-1}Hz^{-1} and $L_{max} > 8 \cdot 10^{27}$ erg s^{-1}Hz^{-1}, respectively) and therefore do not belong to any of the classes previously defined.

A plerionic model has been suggested in order to explain the phenomenon of radio supernovae (Pacini and Salvati (1981), Shklovskii (1981), Bandiera, Pacini and Salvati (1984)). In fact, if scaled with Crab, it gives at an age of a few years radio luminosities similar to those measured in various radio supernovae. In Fig. 2 three models which assume a Crab-like pulsar, and expansion velocities of 1500, 2500 and 4000 km s^{-1} respectively, are superimposed on the available data for radio supernovae and young supernova remnants (plerions are represented by filled circles) (from Weiler *et al.* (1986)). In the data there is no apparent gap or discontinuity between radio supernovae and supernova remnants: this could indicate that there exists an evolutive connection between these two phases.

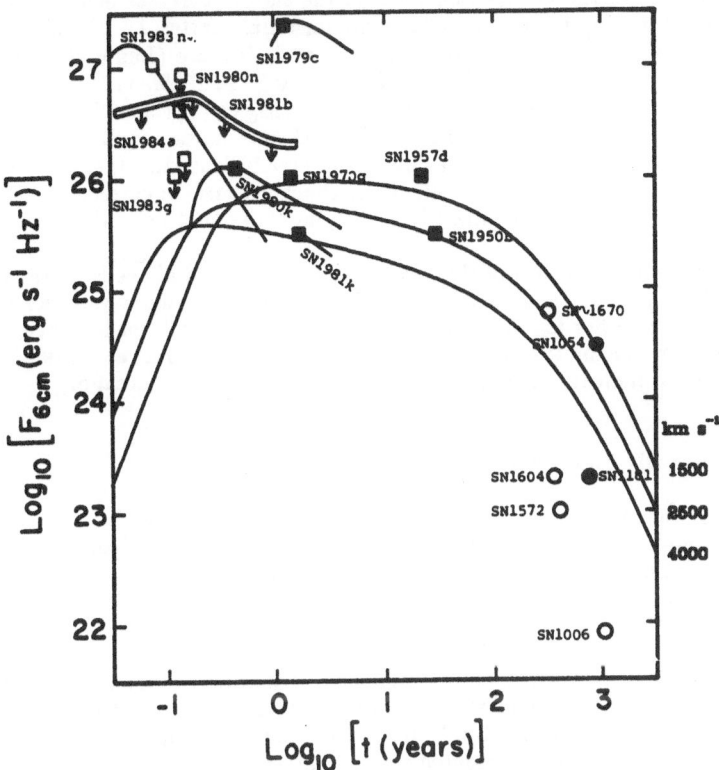

Figure 2. Radio light curves of plerionic models superimposed on data from radio supernovae and young supernova remnants.

A major problem intrinsic in plerionic models is that they assume no absorption from intervening material (the rapid rise at early times shown in Fig. 2 is due to synchrotron self-absorption). However, a large amount of material is ejected by the supernova: this is expected to be ionized and therefore to remain opaque at radio wavelengths for a very long time. For this reason a central source, like a newly born plerion, could not be seen unless a strong fragmentation in the ejecta (Pacini and Salvati (1981), Bandiera, Pacini and Salvati (1984)), or a protrusion of magnetic field beyond the ejecta (Shklovskii (1981)) is invoked. In the Crab Nebula we have in fact evidence that most of the mass is collected in filaments. Filaments are probably a common feature in plerions; for instance, some filaments have been seen in the remnant 3C 58 (van den Bergh (1978)). The origin of filaments is not clear, even though Rayleigh-Taylor instabilities, caused by the push of the relativistic bubble on the ejecta, could trigger their formation. It is also unclear when filaments, or generally clumps, start to form: if they form early enough, and holes appear in the ionized envelope, the radio emission of the inner plerion can be detected.

An alternative scenario (Chevalier (1982), Lundqvist and Fransson (1988)) places the synchrotron emission on the supernova shock which is moving through the circumstellar medium. Rayleigh-Taylor instabilities should be responsible for both magnetic field amplification and particles acceleration. The supernova will become visible in radio when the free-free opacity of the material external to the shock decreases below unity. By this scenario one can estimate the density of the circumstellar matter, i.e. the quantity $\dot{M}/V_w$ in the wind of the presupernova star. Estimates of $\dot{M}/V_w$ are (Lundqvist and Fransson (1988)) $1.2 \cdot 10^{-5}\,(M_\odot\mathrm{yr}^{-1})/(\mathrm{km\,s}^{-1})$ for SN 1979C, $0.3 \cdot 10^{-5}\,(M_\odot\mathrm{yr}^{-1})/(\mathrm{km\,s}^{-1})$ for SN 1980K (which are Type II supernovae), and $3 \cdot 10^{-7}\,(M_\odot\mathrm{yr}^{-1})/(\mathrm{km\,s}^{-1})$ for SN 1983N (a Type Ib supernova). The values obtained for Type II supernovae require a very strong and slow wind. It is unclear how long this superwind should last, and therefore how far the thick circumstellar matter would extend: if the shock scenario is correct, the existence of radio emission from older objects, like SN 1950B, indicates that more than one solar mass is lost in a strong wind.

It is difficult to decide which of the two scenarios presented above is more appropriate for explaining the phenomenon of radio supernovae. Probably one is valid for some radio supernovae, and the other for the rest of them; the shock model, for instance, looks very appropriate in reproducing the radio emission from Type Ib supernovae, while the plerionic one looks the most indicated to explain late emission (after a few decades).

6. X rays from SN 1987A: a very young plerion?

In the last year the attention of most people working on supernovae has been focussed on SN 1987A, in the Large Magellanic Cloud. In radio, a faint emission, compared with other supernovae, appeared with a delay of only 2 days (Turtle et al. 1987); its evolution is nicely fitted by the shock model, giving $1.6 \cdot 10^{-8}\,(M_\odot\mathrm{yr}^{-1})/(\mathrm{km\,s}^{-1})$ (Chevalier and Fransson (1987)).

An unexpected behaviour has been instead shown by SN 1987A in the X-ray band. Some X rays were expected (McCray, Shull and Sutherland (1987), Xu et al. (1988)), due to Compton downgrading of the γ rays released during ^{56}Co decay. However, the observed features deviated considerably from predictions. Compared to expectations, X rays appeared much earlier and presented a much flatter light curve: this has been taken as an indication of mixing of the radioactive debris with the outer layers of the stellar mantle (Pinto and Woosley (1988), Ebisuzaki and Shibazaki (1988)). One can show that the hard X-ray evolution is consistent with that in γ rays, and deduce a fading in hard X rays by the end of 1988 (Bandiera, Pacini and Salvati (1989)), in agreement with observations. One can even play a sort of deconvolution, and use the X-ray light curve to constrain the radial profile of ^{56}Co.

The above model however fails to explain the soft X-ray emission. In the case of Compton downgrading there should be no significant emission below photon energies of ~ 20 keV. A sort of bimodal spectrum has instead been detected by Ginga satellite (Dotani et al. (1987)), with one component at energies lower than the expected cutoff. X-ray spectra (Dotani et al. (1987)) show that the soft component has a spectral slope similar to the hard one, but is typically a factor 10 fainter than the extrapolation of the latter. There is also evidence for short time variability of the soft component; apart from a strong brightening during January 1988, secondary fluctuations are clearly seen in the light curve.

There is still a debate on the origin of this soft X-ray component. It has been proposed that it is due to the interaction of ejecta with circumstellar matter (Masai et al. (1988)); however a very high circumstellar density is required, $8 \cdot 10^{-7}\,(M_\odot\mathrm{yr}^{-1})/(\mathrm{km\,s}^{-1})$, namely 50 times higher than that inferred from the radio light curve. Other possible models are

based on the presence of a mini-plerion (Bandiera, Pacini and Salvati (1988)), or on that
of a central X-ray binary source (Fabian and Rees (1988)). It is still too early to give a
definite answer on which of these two models must be preferred: an interesting thing is
that both require an internal source, and therefore also for SN 1987A (as already shown
for radio supernovae) a strong fragmentation in the ejecta is necessary in order the central
source to be visible.

Let us review here the main results of plerionic models for the soft X-ray emission from
SN 1987A (Bandiera, Pacini and Salvati (1989)). In these models a Crab-like neutron star
has been assumed, except for the spinning period which is taken as a variable parameter;
two different regimes of nebular expansion have been considered:

1) linear expansion, $R = V_{exp}t$, with expansion velocity $V_{exp} = 5000\,\mathrm{km\,s^{-1}}$;
2) interaction with homogeneous ejecta in homologous expansion (Reynolds and Chevalier
   (1984)), $R = (50V_{exp}^3 L/77M_{ej})^{1/5}t^{6/5}$, with total mass of the ejecta $M_{ej} = 10\,M_{\odot}$ and
   maximum ejecta velocity $V_{exp} = 5000\,\mathrm{km\,s^{-1}}$.

A further assumption, largely justified in the X-ray range, is that the emitting particles
are in synchrotron-dominated evolution. By fitting the highest points in the soft X-ray
light curve (excluding the January 1988 event) the resulting pulsar period is, in the case
of linear expansion, $P = 55\,\mathrm{ms}$, corresponding to an energy output $L = 5.9 \cdot 10^{37}\,\mathrm{erg\,s^{-1}}$;
in the case of interaction with the ejecta, instead, $P = 36\,\mathrm{ms}$ and $L = 3.2 \cdot 10^{38}\,\mathrm{erg\,s^{-1}}$.
The energy output from the pulsar is expected to contribute to the bolometric light curve;
present observational upper limits on this contribution give nearly $10^{38}\,\mathrm{erg\,s^{-1}}$. In this
model the time variability can be justified as due to partial obscuration from intervening
clouds of which ejecta are constituted; however, the nature of the January 1988 flare is still
unclear.

The discovery of an optical pulsar has been recently announced (Kristian et al. (1989));
however the pulsar period, $0.5\,\mathrm{ms}$, is much shorter than expected. This discovery, if confirmed, will represent a severe constraint on models for the interior of neutron stars. Anyway it does not contradict the results presented above. In fact they are based on the
assumption of a stellar magnetic field like that in Crab ($B \simeq 4 \cdot 10^{12}$ Gauss), but there is
no physical reason why this should be the case. The quantity which is directly determined
in these models is L, rather than P; since $L \propto B_*^2 \Omega^4$, a direct measurement of the pulsar
period would translate into a limit on the stellar magnetic field. Further observations are
required to confirm, directly or indirectly, the presence of an active neutron star. In this
context, the main predictions of plerionic models are:

1) the energy output from the pulsar will eventually affect the bolometric light curve;
2) the soft X-ray component, apart from short time fluctuations, will show a slow rise,
   proportional to $t^{(3\alpha-1)(2-\gamma)/4}$ (see Section 2);
3) the hard X-ray component will continue fading until it will reach the spectral continu-
   ation of the soft component.

7. Conclusions

We showed in which ways Crab-like supernova remnants can be modelled, what physical
parameters can be inferred from observations, and what predictions can be derived from
theory. However many physical questions still remain unanswered. Just to mention a few

338

of these questions, we still do not know:
1) which mechanism is responsible for particles acceleration;
2) what is the exact shape of the injected particles spectrum;
3) how the nebular magnetic field is produced;
4) how pulsar losses are shared between field and particles;
5) what is the magnetic field distribution in the nebula;
6) how the interaction with field/waves affects particles directions;
7) how the interaction with field/waves affects particles energies;
8) how the relativistic bubble interacts with the ejecta;
9) why the spectra are steepening in the X-ray range;
10) what is the role of the nebular field in the formation of thermal filaments.
In our opinion, two are the necessary conditions for these and other points to become clearer in a near future. On a theoretical ground, there must be a closer interaction between inductive and deductive models. On an observational ground, in turn, a considerably large number of plerions should be studied in the same detail as the Crab Nebula presently is. This is necessary in order to discriminate those aspects which are specific of Crab from those which are instead common in plerions, and that therefore need a comprehensive theoretical explanation.

8. References

Bandiera, R., Pacini, F. and Salvati, M. (1984) "The Evolution of Nonthermal Supernova Remnants. II. Can Radio Supernovae Become Plerions?", *Ap. J.* **285**, 134–140.

Bandiera, R., Pacini, F. and Salvati, M. (1988) "A Synchrotron Model for the X-ray Emission from Supernova 1987A", *Nature* **332**, 418–419.

Bandiera, R., Pacini, F. and Salvati, M. (1989) "X Rays from Supernova 1987A: Beneath the Radioactive Layers", *Ap. J.* (in press).

Becker, R. H., Helfand, D. J. and Szymkowiak, A. E. (1982) "An X-ray Study of Crablike Supernova Remnants: 3C 58 and CTB 80", *Ap. J.* **255**, 557–563.

Bergh, S. van den (1978) "Observations of the Optical Remnant of SN 1181 = 3C 58", *Ap. J. (Letters)* **220**, L9–L10.

Bergh, S. van den and Pritchet, C. J. (1989) "The Crab Synchrotron Nebula at 0.5" resolution", *Ap. J. (Letters)* **338**, L69–L70.

Blandford, R. D. and Romani, R. W. (1988) "On the Interpretation of Pulsar Braking Indices", *M. N. R. A. S.* **234**, 57P–60P.

Brown, R. L. and Marscher, A. P. (1978) "Are Supernovae Radio Sources? A Search for Radio Emission from Young Supernova Remnants", *Ap. J.* **220**, 467–473.

Bruyn, A. G. de (1973) "A High-sensitivity Search for Radio Emission from Young Extragalactic Supernova Remnants at 1415 MHz", *Astr. Ap.* **26**, 105–112.

Cheng, A. and Ruderman, M. S. (1980) "Particle Acceleration and Radio Emission Above Pulsar Polar Caps", *Ap. J.* **235**, 576–586.

Chevalier, R. A. (1982) "The Radio and X-ray Emission from Type II Supernovae", *Ap. J.* **259**, 302–310.

Chevalier, R. A. and Fransson, C. (1987) "Circumstellar Matter and the Nature of the SN 1987A Progenitor Star", *Nature* **328**, 44–45.

Clark, D. H. and Stephenson, F. R. (1977) *The Historical Supernovae*, Pergamon Press, Oxford.

Davidson, K. and Fesen, R. A. (1985) "Recent Developments Concerning the Crab Nebula", *Ann. Rev. Astr. Ap.* **23**, 119–146.

Dotani, T. *et al.* (1987) "Discovery of an Unusual Hard X-ray Source in the Region of

Supernova 1987A", *Nature* **330**, 230–231.

Ebisuzaki, T. and Shibazaki, N. (1988) "The Effects of Mixing of the Ejecta on the Hard X-ray Emission from Supernova 1987A", *Ap. J.* **327**, L5–L8.

Fabian, A. C. and Rees, M. J. (1988) "On the Variability of the X-ray Emission from Supernova 1987A", *Nature* **335**, 50–51.

Goldreich, P. and Julian, W. H. (1969) "Pulsar Electrodynamics", *Ap. J.* **157**, 869–880.

Gottesman, S. T., Broderick, J. J., Brown, R. L., Balick , B. and Palmer, P. (1972) "First-epoch Radio Observations of Supernova 1987A", *Ap. J.* **174**, 383–388.

Gratton, L. (1972) "Source Models with Electron Diffusion", *Ap. Sp. Sci.* **16**, 81–100.

Grooth, E. J. (1975) "Timing the Crab Pulsar I. Arrival Times", *Ap. J. Suppl.* **29**, 431–441.

Gunn, J. E. and Ostriker, J. P. (1969) "Magnetic Dipole Radiation from Pulsars", *Nature* **221**, 454–456.

Helfand, D. J. and Becker, R. H. (1987) "G 0.9+0.1 and the Emerging Class of Composite Supernova Remnants", *Ap. J.* **314**, 203–214.

Jung, G. V. (1989) "The Hard X-ray to Low-energy Gamma-ray Spectrum of the Crab Nebula", *Ap. J.* **338**, 972–982.

Kennel, C. F. and Coroniti, F. V. (1984a) "Confinement of the Crab Pulsar's Wind by its Supernova Remnant", *Ap. J.* **283**, 694–709.

Kennel, C. F. and Coroniti, F. V. (1984b) "Magnetohydrodynamic Model of the Crab Nebula Radiation", *Ap. J.* **283**, 710–730.

Kristian, J. *et al.* (1989) "Submillisecond Optical Pulsar in Supernova 1987A", *Nature* **338**, 234–236.

Ku, W., Kestenbaum, H. L., Novick, R. and Wolff, R. S. (1976) "Energy Dependence of the Size of the X-ray Source in the Crab Nebula", *Ap. J. (Letters)* **204**, L77–L81.

Lundqvist, P. and Fransson C. (1988) "Circumstellar Absorption of UV and Radio Emission from Supernovae", *Astr. Ap.* **192**, 221–233.

Manchester, R. N., Durdin, J. M. and Newton, L. M. (1985) "A Second Measurement of a Pulsar Braking Index", *Nature* **313**, 374–376.

Manchester, R. N. and Peterson, B. A. (1989) "A Braking Index for PSR 0540-69", *Ap. J. (Letters)* **342**, L23–L25.

Masai, K., Hayakawa, S., Inoue, H., Itoh, H. and Nomoto, K. (1988) "Circumstellar Matter of SN 1987A and Soft X-ray Emission", *Nature* **335**, 804–806.

McCray, R., Shull, J. M. and Sutherland, P. (1987) "Inside Supernova 1987A", *Ap. J. (Letters)* **317**, L73–L77.

Michel, F. C. (1982) "Theory of Pulsar Magnetospheres", *Rev. of Mod. Phys.* **54**, 1–66.

Middleditch, J., Pennypacker, C. R. and Burns, M. S. (1987) "Optical Color, Polarimetric, and Timing Measurements of the 50 ms Large Magellanic Cloud Pulsar, PSR 0540-69", *Ap. J.* **315**, 142–148.

Pacini, F. (1967) "Energy Emission from a Neutron Star", *Nature* **216**, 567–568.

Pacini, F. (1968) "Rotating Neutron Stars, Pulsars, and Supernova Remnants", *Nature* **219**, 145–146.

Pacini, F. and Salvati, M. (1973) "On the Evolution of Supernova Remnants. I. Evolution of the Magnetic Field, Particles, Content, and Luminosity", *Ap. J.* **186**, 249–265.

Pacini, F. and Salvati, M. (1981) "Radio Emission from Very Young Supernova Remnants: the Case of Sn 1979C", *Ap. J. (Letters)* **245**, L107–L108.

Pinto, P. A. and Woosley, S. E. (1988) "X-ray and Gamma-ray Emission from Supernova 1987A", *Ap. J.* **329**, 820–830.

Rees, M. J. and Gunn, J. E. (1974) "The Origin of the Magnetic Field and Relativistic Particles in the Crab Nebula", *M. N. R. A. S.* **167**, 1–12.

Reynolds, S. P. (1985) "An Introduction to SNR 0540-693", in M. C. Kafatos and R.

B. C. Henry (eds.), *The Crab Nebula and Related Supernova Remnants*, Cambridge University Press, Cambridge, pp. 159–163.

Reynolds, S. P. and Chanan, G. A. (1984) "On the X-ray Emission from Crab-like Supernova Remnants", *Ap. J.* **281**, 673–681.

Reynolds, S. P. and Chevalier, R. A. (1984) "Evolution of Pulsar-driven Supernova Remnants", *Ap. J.* **278**, 630–648.

Scargle, J. D. (1969) "Activity in the Crab Nebula", *Ap. J.* **156**, 401–426.

Seward, F. D. (1985) "Supernova Remnants Containing Neutron Stars", *Comments Ap.* **11**, 15–51.

Seward, F. D. (1989) "The Crab-like Supernova Remnants", *Sp. Sci. Rev.* **49**, 385–423.

Shklovskii, I. S. (1960) *Soviet Astron.—AJ* **4**, 355.

Shklovskii, I. S. (1981) "Radio-emitting Supernovae as Young Plerions", *Soviet Astron. Lett.* **7**, 263–264.

Staelin, D. H. and Reifenstein, E. C. (1968) "Pulsating Radio Sources Near the Crab Nebula", *Science* **162**, 1481–1483.

Trimble, V. (1971) "Dynamics of the Crab Nebula", in R. D. Davies and F. G. Smith (eds.), *The Crab Nebula*, IAU Symp. No. 46, Reidel, Dordrecht, pp. 12–21.

Turtle, A. J. *et al.* (1987) "A Prompt Radio Burst from Supernova 1987A in the Large Magellanic Cloud", *Nature* **327**, 38–40.

Weiler, K. W. (1983a) "The Crab Nebula is not alone!", *Observatory* **103**,85-106.

Weiler, K. W. (1983b) "Supernova Remnants Resembling the Crab Nebula", in J. Danziger and P. Gorenstein (eds.), *Supernova Remnants and their X-ray Emission*, Reidel, Dordrecht, pp. 299–320.

Weiler, K. W. (1985) "A Catalogue of Plerionic and Composite Supernova Remnants", in M. C. Kafatos and R. B. C. Henry (eds.), *The Crab Nebula and Related Supernova Remnants*, Cambridge University Press, Cambridge, pp. 265–285.

Weiler, K. W. and Panagia, N. (1978) "Are Crab-type Supernova Remnants (Plerions) Short-lived?", *Astr. Ap.* **70**, 419–422.

Weiler, K. W. and Seielstad, G. A. (1971) "Synthesis of the Polarization Properties of 3C 10 and 3C 58 at 1420 and 2880 MHz", *Ap. J.* **163**, 455–478.

Weiler, K. W. and Sramek, R. A. (1988) "Supernovae and Supernova Remnants", *Ann. Rev. Astr. Ap.* **26**, 295–341.

Weiler, K. W. , Sramek, R. A., Panagia, N, Hulst, J. M. van der and Salvati, M. (1986) "Radio Supernovae", *Ap. J.* **301**, 790–812.

Wilson, A. S. (1972) "The Structure of the Crab Nebula — II. The Spatial Distribution of The Relativistic Electrons", *M. N. R. A. S.* **160**, 355–371.

Woltjer, L, (1987) "Recent Developments on the Crab Nebula", in F. Pacini (ed.), *High Energy Phenomena Around Collapsed Stars*, Reidel, Dordrecht, pp. 209–221.

Xu, Y., Sutherland, P., McCray R. and Ross, R. R. (1988) "X-rays from Supernova 1987A" *Ap. J.* **327**, 197–202.

ENERGY LOSS MECHANISMS FOR FAST PARTICLES

J. H. Beall

E. O. Hulburt Center for Space Research
Naval Research Laboratory
Washington, DC   20375-5000,

St. John's College, Annapolis, Maryland, and

SFA, Landover, Maryland

ABSTRACT.   Beams or jets of rapidly moving particles appear to be present in several classes of objects in astrophysics, including active galactic nuclei and quasars, compact binary systems, and supernovae. There is evidence from some radio bright active galaxies of a direct connection between the central, relativistically moving beams and the large, subrelativistic radio jets.   In this paper we review the various energy loss mechanisms for a beam of relativistic particles propagating through an ambient medium, and compare the energy loss rate of those processes with that of plasma (collective) processes. Plasma processes are typically found to be the dominant energy loss mechanism. Inverse Compton X-ray emission and proton-proton related γ-ray production yield photons at roughly equal rates, and X-ray and γ-ray emission represent less than 10% of the total energy loss rate of the beams for parameters likely to be associated with the nuclei of active galaxies and quasars.

1.   INTRODUCTION

Linear structures are a common occurence in observations of many astronomical objects, including active galaxies and quasars. In some radio bright active galaxies these structures maintain a coherence over scales from parsecs to kiloparsecs (see, for example, Feigelson et al. 1986, Wilson and Willis 1980, Ulvestad, Wilson, and Sramek 1981, Brown, Johnston, and Lo 1981, and Walker 1984). The persistence of such alignments argues for a highly ordered acceleration process. On the smallest scales the structures are one-sided, suggesting that the emitting material is moving relativistically.   Such a conclusion is also supported by observations of apparent superluminal expansion of these small scale objects.   The transition from bulk relativistic motion on small scales to supersonic flows occurs over relatively short distances compared to typical sizes of these linear structures.   It is an interesting observational point

W. Brinkmann et al. (eds.), Physical Processes in Hot Cosmic Plasmas, 341–355.

that the linear structures in the core region are not very luminous with respect to the central object or the more extended structures. Taken together, these observations suggest that a rather efficient energy loss mechanism operates on the moving material, and that the energy loss mechanism is not radiative.

There is also evidence for assymetries in the evolution of ejecta from supernova explosions (see, e.g. Geldzahler and Shaffer 1982, and especially Chevalier and Soker 1989 for a review of the literature on SN1987a) which is difficult to produce in simulations of the supernova blast wave (Chevalier and Soker 1989), but which can in principle be caused by relativistic particle beams generated within the central region of the exploding star (Beall 1989).

In this paper we consider the various energy loss mechanisms for a beam of relativistic particles propagating through an ambient medium, and compare the energy loss rate from collective (plasma) processes to the energy loss rate for collisional, non-relativistic and relativistic bremsstrahlung, synchrotron, inverse Compton scattering, and proton-proton interactions for parameters likely to be associated with the nuclei of active galaxies and quasars. In order to compare the various energy loss rates we will first outline the energy loss rates for the aforementioned particle loss mechanisms and then discuss in some detail the mechanisms for the collective (plasma) energy loss.

For the purposes of discussion and in the hope of helping to establish some uniformity in the language used in the literature, we distinguish between jets and beams in galactic and extragalactic sources in that jets are subrelativistic, while beams are relativistic. In active galaxies and quasars, beams propagate out of the cores, slow, and become the jets which eventually can form giant radio lobes (see e.g. Rose, et al. 1984, 1986, Beall 1986).

2. PARTICLE ENERGY LOSS MECHANISMS

There are a number of well-known energy loss mechanisms which extract energy from moving particles. In this section we will discuss collisional losses, non-relativistic bremsstrahlung, relativistic bremsstrahlung, synchrotron, inverse Compton, and hadronic collisions as energy loss mechanisms.

Derivations of collisional loss rates can be found in numerous sources (see e.g. Jackson 1960). For a particle of charge z traversing a gas of density N with Z electrons/atom, the energy loss is given by

$$dE_{Coll}/dx = -4\pi NZ(z^2e^4/mv^2)\ln B, \quad ergs/cm$$

$$where \ B = \gamma^2 mv^3/ze^2\omega,$$

(1)

and γ is the ratio of the particle energy over the rest mass energy, e is the charge of the electron, v is the electron velocity ($\ll$ c), ω is a characteristic atomic frequency of motion, M is the mass of the incident particle, and m is the mass of the electron.

It is interesting to compare this loss rate with that associated with non-relativistic bremsstrahlung. This can be expressed as

$$dE(\text{non-rel})_{rad}/dx = -(16/3)NZ^2(e^2/\hbar c)(z^4e^4/Mc^2), \tag{2}$$

where $(e^2/\hbar c)$ is the fine structure constant. For non-relativistic velocities, collisional energy losses are the dominant energy loss mechanism. When particle velocities become relativistic, however, bremsstrahlung energy losses dominate over collisional. We may express the energy losses associated with relativistic bremsstrahlung as

$$dE(\text{rel})_{rad}/dx = -(16/3)NZ^2(e^2/\hbar c) \, E \, \ln\beta,$$

$$\text{where } \beta = (\lambda 192M/mz^{1/2}), \tag{3}$$

and λ is a quantum-mechanical fudge factor of order unity.

It is interesting to note that the equation above has the form

$$dE/dx = -E/X_o \tag{4}$$

where X_o is appropriately defined. The solution to this equation has the form $E = E_o \exp(-x/X_o)$, where X_o is the characteristic radiation length. The form of the equation is identical to that of radiation propagating through an absorbing medium. Thus, the propagation length of a beam of particles can be considerably extended beyond some small multiple of the mean free path.

There are two more energy loss mechanisms which are relevant to our discussion: synchrotron and inverse Compton losses. Schott (1912) expressed the energy loss of a particle due to synchrotron radiation as

$$dE_{sync}/dx = -(2e^4/3m^2c^4)\gamma^2H^2 \tag{5}$$

where H is the magnetic field intensity. This equation can be rewritten as

$$dE_{sync}/dx = -(8\pi/3)\sigma_T\gamma^2u_H \tag{6}$$

with u_H expressing the energy density of the magnetic field. Written in this form, the expression for the energy loss associated with synchrotron emission is similar the the energy loss formula for inverse Compton scattering, which is (Felten and Morrison 1963,66)

$$dE_{IC}/dx = -\sigma_T\gamma^2u_{ph}, \tag{7}$$

where u_{ph} is the photon energy density.

In order to compare the energy loss rates associated with these mechanisms with those of plasma processes, we must involve ourselves in some detail with a discussion of the physics of the interaction of a relativistic particle beam with an ambient medium.

3. ENERGY LOSS MECHANISMS ASSOCIATED WITH PLASMA EFFECTS

3.1 The Interaction of a Beam of Relativistic Particles with an ambient medium

A relativistic charged particle beam heats and ionizes a background gas. The beam may be composed of electrons, electrons and ions, or electrons and positrons. The interaction between the beam and the background gas may be primarily the result of collisionless effects if the background gas is sufficiently ionized. In this section we will discuss collisionless effects.

A low-density relativistic electron beam entering a plasma produces perturbations that are localized to the region of the beam and induces currents that tend to cancel out its magnetic field. In this case steady state injection into the plasma occurs (Rukhadze and Rukhlin 1972).

A high-density electron beam produces hydrodynamic motion as it enters a plasma. In this paper we shall assume that the beam has entered the background plasma and that the diameter of the beam is sufficiently large that the time scales for subsequent hydrodynamic motions are much longer than the time scales for the growth of plasma instabilities. If magnetic fields produced by the beam-background plasma interaction can be neglected, then as the beam heats the background plasma, it will expand and thereby produce a plasma of lower density and higher temperature (channel plasma), which will continue to expand until there is an approximate pressure equilibrium between the channel plasma and the background plasma still unaffected by the beam. The electric self-field of the beam will be neutralized by the background plasma on a time scale that is approximately $1/\omega_p$, where ω_p is the plasma frequency of the background plasma. The interaction of the beam with the background plasma occurs through a number of instabilities. We shall discuss those instabilities below and then calculate their development in the next section.

3.2 Relevant Beam-Plasma Instabilities

The initial interaction of a charge neutralized beam traversing a collisionless plasma is through the two-stream instability. This instability transfers energy from the beam in the form of nearly longitudinal (electrostatic) waves and simultaneously acts through the wave-beam interaction to broaden the beam in momentum space. The propagation distance of a relativistic beam traversing an ambient medium depends on the time scales for beam broadening in momentum parallel to the beam and for scattering transverse to the direction of beam propagation.

If the electrom beam and background plasma are cold (i.e., the momentum spread is small) then the electrostatic dispersion relation for the case of a uniform medium and negligible field is (Godfrey, Shanahan, and Thode 1976)

$$1 - \frac{\omega_p^2}{\omega^2} - \frac{\omega_{pb}^2}{(\omega - k \cdot v_b)^2} \ (\sin \ \theta + \gamma^{-2} \cos \ \theta) = 0, \tag{8a}$$

where the wave vector k is at an angle θ with respect to the beam direction,

$$\gamma = \left[1 - \frac{v_b^2}{c^2} \right]^{-1/2} , \quad \omega_p^2 = \frac{4\pi n_p e^2}{m} \tag{8b}$$

is the background plasma frequency, and $\omega_{pb}^2 = 4\pi n_b e^2 / \gamma m$ is the relativistic beam plasma frequency. If the plasma is sufficiently magnetized, additional and shifted resonances appear due to gyromotion.

The nonrelativistic dispersion relation that applies even when there is significant momentum spread in the beam and plasma is

$$k^2 = \sum_j \frac{4\pi n_j e^2}{m_j} \int \frac{k \cdot (\partial f_j / \partial v) d^3 v}{k \cdot v - \omega} , \tag{9}$$

where the summation is over particle streams (beam and plasma) and f_j is the velocity distribution function for the jth stream. The corresponding relativistic dispersion relation is more complicated. In this paper we shall assume that initially the beam has some specified value of γ. If the beam has a power-law energy spectrum with a low-energy cutoff, then the value of γ used in our calculations corresponds to an energy close to the low-energy cutoff where the power-law energy distribution is N(E) $\propto$ E^{-a} $a >$ 2-3. When the plasma is collisionless and unmagnetized, the above dispersion relation can be solved for the growth rates of perturbations which vary as exp[-iωt + ikr] (with ω complex). When maximized over unconstrained k, the initial growth rate is (Godfrey, Shanahan, and Thode 1976, Bludman, Watson, and Rosenbluth 1960),

$$\gamma_1 = \text{Im}\omega \simeq \frac{3^{1/2}}{2\gamma} \left[\frac{n_b}{2n_p} \right]^{1/3} \omega_p \tag{10}$$

for a cold beam (hydrodynamic regime of instability). If the relativistic electron beam is warm or scattered (kinetic regime of instability), the growth is (Breizman and Ryutov 1971),

$$\gamma_1 = \text{Im}\omega \approx \frac{1}{\gamma} \frac{n_b}{n_p} \frac{1}{(\Delta\theta)^2} \omega_p \; , \tag{11}$$

where $\Delta\theta$ is the angular width of the beam momenta. The growth rates for the two-stream instability are approximately the same for electron, electron-proton, and electron-positron beams with similar n_b and γ, and consequently our calculations can be applied to all three cases.

After a sufficient number of e-folding times, localized electric field fluctuations caused by the two-stream instability can become large enough to trap beam electrons if no other nonlinear process intervenes first. Thode and Sudan (1973) found that when electron trapping determines the saturation, resonant waves saturate at an average energy density,

$$\epsilon_w = 2n_b \alpha mc^2 \frac{S}{2} (1 + S)^{-5/2} \; , \; \alpha \sim 1, \; S < 1/2 \quad \text{(cold beam)} \tag{12}$$

where

$$S = \beta^2 \gamma (n_b/2n_p)^{1/2}.$$

However, the growth of resonant waves can be halted or even reversed by nonlinear effects other than trapping of beam electrons. On the time scale of ion motion across a few Debye lengths, high electric field regions (spikes) are formed which tend to exclude plasma ions and electrons by ponderomotive pressure. As these spikes sharpen and collapse (Zakharov 1972), their Fourier spectrum is broadened. This leads to the production of localized density reductions (cavitons). These nonlinear effects imply that either before or after the saturation of resonant waves by beam electron trapping, the wave spectrum can change nonlinearly from waves with nearly a single wave vector to a spectrum that includes oscillation of zero phase velocity and waves in resonance with the tails of the plasma electron distribution. The oscillating two-stream (OTS) instability, which is formally the same as a modulational instability or AC instability (Papadopoulos 1975, Schamel, Lee, and Morales 1975 and Freund et al. 1980), develops as the ponderomotive force of high-frequency resonance waves produces cavitons. It causes the transfer of wave energy from waves resonant with the beam to waves of shorter wavelength.

There is a threshold energy for the onset of the OTS instability (Papadopoulos 1975). This threshold is reached when ϵ_1, the wave energy density in the resonant waves generated by the two-stream instability, becomes sufficiently large that the condition $\epsilon_1/n\kappa T > (k_1\lambda_D)^2$ is satisfied, where $\lambda_D = (\kappa T/4\pi ne^2)^{1/2}$ is the Debye length and k_1 is the most unstable wave vector associated with the two-stream instability. For the case of a cold beam, Freund et al. (1980) find the growth rate to be

$$G^{OTS} = \omega_p \epsilon_1 / (4 n \kappa T) \tag{13}$$

when $3(k_1 \lambda_D)^2 < \epsilon_1 / n \kappa T < m/m_i$, and

$$\Gamma^{OTS} = \omega_p \left[\frac{1}{3} \frac{m}{m_i} \frac{\epsilon_1}{n \kappa T} \right]^{1/2} \tag{14}$$

when $\epsilon_1 / n \kappa T > m/m_i$ and $3(k_1 \lambda_D)^2$.
   The conditions

$$k_1 = k_2 + k_s \tag{15}$$

$$\omega_1 = \omega_2 + \omega_s, \tag{16}$$

are approximately satisfied when a resonant plasma wave (k_1) is
transformed to a nonresonant plasma wave (k_2) and a low-frequency ion
acoustic wave (k_s) by means of the OTS instability. As the energy
density in the form of low-frequency fluctuations (i.e., ion acoustic waves)
increases, they interact nonlinearly with the resonant waves and increase
the energy density (ϵ_2 of nonresonant wave. The growth rate for this
interaction is (Dawson and Oberman 1962)

$$\Gamma^{D-O} = (k_s \lambda_D)^{-2} W_s \omega_p, \tag{17}$$

where k_s is the wave number of the ion density fluctuations and $W_s = \epsilon_s / n \kappa T$.
   The ion acoustic (sound) waves are damped at a rate (Smith,
Goldstein, and Papadopoulos 1976)

$$\Gamma^{D-W} = (m/m_i)^{1/2} (k_s \lambda_D)^{1/2} W_s^{1/4} \omega_p , \tag{18}$$

where $W_s = \epsilon_s / n \kappa T$ with ϵ_s the energy density of ion acoustic waves.
Ion acoustic waves are also produced when the threshold for a secondary
OTS is reached. Following Scott et al.(1980), we take this threshold to
be

$$W_2 = \frac{\epsilon_2}{n \kappa T} > 4 \frac{\Gamma_L}{\omega_p} , \tag{19}$$

where ϵ_2 is the energy density in high frequency nonresonant waves (i.e.,
k_2). We assume that initially the background plasma has a Maxwell-
Boltzmann distribution and therefore the Landau damping rate
is (Boyd and Sanderson 1969)

$$\Gamma_L = \left(\frac{1}{32\pi}\right)^{1/2} \omega_p \left(\frac{2\pi}{\lambda_D k_2}\right)^3 \exp\left[-\frac{2\pi^2}{\lambda_D^2 k_2^2} - \frac{3}{2}\right],$$

$$\left[\frac{\lambda_D^2 k_2^2}{4\pi^2} \ll 1\right]. \tag{20}$$

Jackson (1960) has shown that as $k_2\lambda_D \to 1$, $\Gamma_1 \to \omega_p$ and thus λ_D is the minimum wavelength for longitudinal oscillations that propagate through the plasma.

The rate equations that describe the growth of the wave energies are

$$\frac{\partial W_1}{\partial t} = 2[\Gamma_1 - \Gamma^{D-0}(W_s)]W_1 - 2\Gamma^{OTS}(W_1)\ W_2\theta(W_1 - k_1^2\lambda_D^2), \tag{21}$$

$$\frac{\partial W_2}{\partial t} = 2\Gamma^{D-0}(W_s)W_1 + 2\Gamma^{OTS}(W_1)W_2\theta(W_1 - k_1^2\lambda_D^2)$$

$$- 2\Gamma_L W_2 - 2\Gamma^{OTS}(W_2)\theta\left[W_2 - \frac{4\Gamma_L}{\omega_p}\right]W_s - \frac{W_2}{\tau_2} \tag{22}$$

$$\frac{\partial W_s}{\partial t} = \left[2\Gamma^{OTS}(W_1)\theta(W_1 - k_1^2\lambda_D^2) + 2\Gamma^{OTS}(W_2)\theta\left[W - \frac{4\Gamma_L}{\omega_p}\right]\right.$$

$$\left. - 2\Gamma^{D-W}\right]W_s - \frac{W_s}{\tau_s}, \tag{23}$$

where $W_1 = \epsilon_1/n\kappa T$, $W_2 = \epsilon_2/n\kappa T$, $W_s = \epsilon_s/n\kappa T$, and the θ's are step functions. The terms $-W_2/\tau_2$ and $-W_s/\tau_s$ take into account the circumstance that waves tend to move outside the path of the beam. We estimate that $\tau_2 \simeq (r_0/\sin 45°)/V_g$ and $\tau_s = (r_0/\sin 45°)(2\kappa T/m_i)^{-1/2}$, where r_0 is the beam radius, V_g is the approximate group velocity of the high-frequency non-resonant waves, and $(\kappa T/m_i)^{1/2}$ is the ion sound velocity. The group velocity $= V_g = d\omega/dk$ for electron plasma waves is obtained from the dispersion formula $\omega^2 = \omega_p^2 + (\kappa T/m)k^2$, which yields $V_g = (\kappa T/m)/(\omega/k)$. For the ion acoustic wave we have $V_g \cong [\kappa T_e/m_i + \kappa T_i/m_i]^{1/2}$.

Scott et al.(1980) have discussed steady state solutions for rate equations similar to the above equations.

3.3 Solutions to the Rate Equations

In this section we will discuss solutions to equations (21)-(23) subject to the boundary conditions that W_1, W_2 and W_s are small at t = 0. In obtaining these solutions we have replaced the θ functions with analytical functions of the form $\theta(x - y) = 1$ for $y > x$; $\theta(x - y) = \exp[z(x - y)]$, $z = -K/x$ (K = constant) for $x > y$. Following Freund et al.[18], we assume that $k_2 \lambda_D = 0.2$ (i.e., we let the phase velocity $v_{k2} = 5\omega_p \lambda_D = 5v_e$). Since $k_1 \cdot v_b \simeq k_1 c \simeq \omega_p$, we have $|k_1| << |k_2|$; $|k_s|$ and $|k_2| \simeq |k_s|$ from equation (15)

As a beam excites waves in traversing a background plasma, it loses energy and γ decreases. For an electron-proton beam the principal collisionless interaction is between the beam electrons and background plasma, and consequently the beam electrons will tend to slow down with respect to the beam protons. If the beam-plasma interaction is not very strong, the beam protons will drag the electrons along with them For an ultrarelativistic beam $v_b = c$, and the energy loss through a distance Δl can be found from the equation:

$$cn_b m'c^2 \frac{d\gamma}{dl} \Delta l \simeq - \frac{a\epsilon_1}{dt} \Delta l \ (\text{ergs cm}^{-2} \text{ s}^{-1}), \tag{24}$$

where m' = m for an electron beam, m' = m for an electron-positron beam, $m' = m_p + m$ for an electron-proton beam, ϵ_1 is the energy density of the resonant waves, and a is a factor (≥ 1) that corrects for the simultaneous transfer of resonant wave energy into nonresonant wave energy ϵ_2 and ion-acoustic wave energy ϵ_s. From equation (24) we find

$$\int_1^2 d\gamma = -\int_1^2 (d(a\epsilon_1)/dt)dl \ / \ (cn_b m \ c^2). \tag{25}$$

The propagation length L_p [i.e., the distance over which $\gamma = (1 - v^2/c^2)^{-1/2}$ decreases by a factor of ~ 2] becomes

$$L_p \simeq \frac{\gamma}{2} \frac{cn_b mc^2}{\langle da\epsilon_1/dt \rangle}, \tag{26}$$

where $\langle da\epsilon_1/dt \rangle$, the time average rate of excitation of wave energy density, can be obtained from the solutions described below.

Figures 1 and 2 show some solutions to equations (21)-(23) for a relativistic charged particle beam with large beam radius (i.e., the terms $-W_2/\tau_2$ in eq. (21) and $-W_s/\tau_s$ in eq. (23) are neglected). These solutions are parameterized in terms of the growth rate for the two-stream instability Γ_1, the plasma frequency ω_p, and the background plasma temperature T. The solutions demonstrate that, as the wave levels W_2 and W_s grow, the wave energy W_1 is stabilized. The wave energy levels

W_1, W_2, and W_s undergo periodic relaxation oscillations for a wide choice of parameters. However, if either the temperature T is increased sufficiently or Γ_1 decreased sufficiently, steady state solutions are obtained.

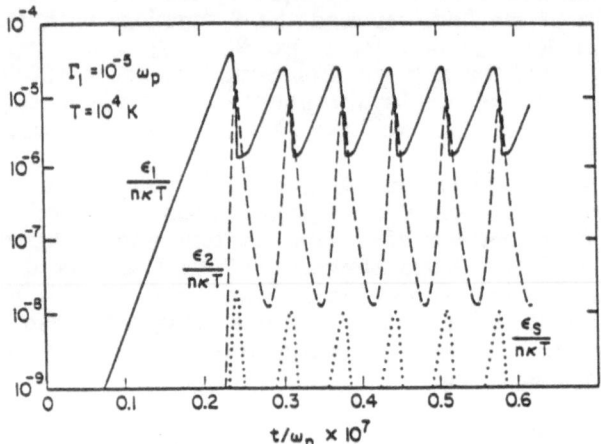

FIG. 1. -Solutions for the wave energy levels $W_1 = \epsilon_1/n\kappa T$, $W_2 = \epsilon_2/n\kappa T$, and $W_s = \epsilon_s/n\kappa T$ obtained by solving eqs. (21)-(23) are shown as a function of time (t) for two-stream growth rate $\Gamma_1 = 10^{-5}w_p$ and $T = 10^4$ K (Rose et al. 1984)

For the solution shown in Figure 1, the amplitude of the resonant two-stream wave grows to $W_{1(max)} = \epsilon_1/n\kappa T \simeq 0.42 \times 10^{-4}$ and then decreases, because after the threshold for the OTS instability is reached, wave energy is rapidly fed into nonresonant waves (W_2) and to a lesser extent into ion acoustic waves (W_s) more rapidly than it can be gained as a result of the two-stream instability. An observer in the reference frame of the background plasma observes quasi-periodic changes in the amplitudes of the waves (relaxation oscillations). As wave energy is transferred from resonant to nonresonant waves, electron tails are produced as a consequence of Landau damping. The energy that is transferred into ion-acoustic waves is dissipated as heat. For the particular choice of parameters $\Gamma_1 = 10^{-5}w_p$, $\gamma = 10^3$, $n_p = 10^4$ cm^{-3} and $n_b/n_p = 10^{-4}$, the cycle is repeated in approximately 0.38 seconds and $<a(d\epsilon_1/dt) \simeq 0.17 \times 10^{-12}$ ergs cm^{-3} s^{-1}. It follows that

$$L_p \simeq \frac{\gamma}{2} \frac{<cn_b mc^2>}{<a(d\epsilon_1/dt)>} \simeq 7.2 \times 10^{18} \text{ cm} \tag{27}$$

and

$$n_p L_p \simeq 7.2 \times 10^{22} \text{cm}^{-2}.$$

Figure 2 shows solutions to equations (21)-(23) with similar two-stream instability growth rates but with higher background plasma temperatures

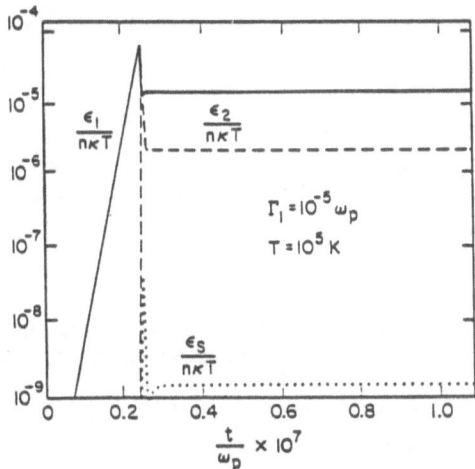

FIG. 2.-$W_1 = \epsilon_1/n\kappa T$, $W_2 = \epsilon_2/n\kappa T$, and $W_s = \epsilon_s/n\kappa T$ are shown as a function of time (t) for $\Gamma_1 = 10^{-5}\omega_p$ and $T = 10^5$K (Rose et al. 1984)

than the solution shown in Figure 1. These solutions demonstrate that a (quasi) steady state solution is attained for sufficiently high T and that resonant and nonresonant wave saturation levels increase as a function of T. A (quasi) steady state solution is also attained for a background plasma whose temperature is $T = 10^5$ K if the growth rate of the two-stream instability is decreased sufficiently as compared to the value assumed in obtaining the solution shown in Figure 1.

4. DISCUSSION

We now compare the propagation length due to plasma processes with that from individual particle interactions.

The propagation length for plasma losses of a beam of relativistic particles with $\gamma = 10^3$ and $\Gamma_1 = 10^{-5}\omega_p$ moving through a plasma with a density $n_p = 10^4$ and a temperature $T = 10^4$, and with $n_b/n_p = 10^{-4}$ is, using equation (27), found to be $L_p = 7 \times 10^{18}$ cm.

Since radiative losses dominate over collisional losses, We need only calculate the radiative losses. Evaluating the various numerical constants, we find

$$dE_{rad}/dx = -4.2 \times 10^{-28} \, N \, E \text{ ergs/cm}, \tag{28}$$

which yields a propagation length of order $L_p(\text{rad}) = 2 \times 10^{23}$ cm.

We may calculate the inverse Compton losses by making appropriate assumptions about the nature of the radiation field in the cloud. If the dominant radiation field in the cloud is due to optically thin thermal bremsstrahlung, the energy density of the radiation can be easily estimated.

The rate of production of photons due to thermal bremsstrahlung is

$$\dot{n}_{ph} = 1.4 \times 10^{-27} \; T^{1/2} \; n_e^2 \; / \; h\nu \; \text{cm}^{-3}\text{s}^{-1}, \qquad (29)$$

where T is the temperature, n_e is the electron density in the ambient medium, and hν is the average energy of the photons. Therefore the number density of the photons whose frequency is sufficiently high that they are not reabsorbed is

$$n_{ph} = \dot{n}_{ph} \; (R/c), \qquad (30)$$

where R is the radius of the cloud of ambient material in the vicinity of the beam, and c is the speed of light. For a cloud with T=10⁴K, R = 3 x 10¹⁹cm, and $n_p = 10^4$, and $n_{ph} = 1$.

The propagation length due to inverse Compton scattering losses for an electron-positron beam will be

$$L_{pIC} = mc^2/(\sigma_T \gamma \; n_{ph} \; h\nu), \qquad (31)$$

where σ_T is the Thomson cross section, mc^2 is the rest energy of the particle, and hν is the energy of the photon before it is inverse Compton scattered. For the parameters cited above, $L_{pIC} = 8 \times 10^{19}$ cm.

We note that collective (plasma) losses in the beam dominate over inverse Compton and radiative losses. The inverse Compton X-ray emission represents a rate of energy loss that is < 10 % of the rate at which the beam deposits energy into the ambient medium.

To obtain an order of magnitude estimate of the relative fraction of the X-ray and γ-ray photons produced, we note that for an electron-proton beam, the propagation length is $10^3 \; L_p$. The electron-proton beam thus traverses a distance of 7 x 10²¹ cm of material in loosing an appreciable fraction of its energy. Typical cross sections for p-p reactions to produce pions are of the order of 1 - 2 x 10⁻²⁶ cm². Therefore, the mean-free-path (MFP) for pion production is of the same order as the propagation length due to inverse Compton emission. Thus the two processes produce X-ray and γ-ray photons at roughly the same rate and this rate is < 10 % of the rate at which the beam looses energy because of collisionless effects.

If the cloud of material is optically thick, inverse Compton losses can be the dominant energy loss mechanism for the beam (Rose et al. 1987).

The X-ray and γ-ray spectra produced by this model have been calculated by Beall et al. (1987) and Bednarek et al. (1989), and work on refining the extant calculations is proceeding.

It is an obvious corollary to the foregoing discussion that collective processes can transfer significant amounts of momentum to the ambient medium through which the beam propagates. This may be relevant to the question of the genesis of the giant radio lobes in radio galaxies and to the evolution of the ejecta of some supernovae.

5.   CONCLUSIONS

We have calculated the relative rates at which individual particle and collective (plasma) energy loss mechanisms extract energy from a beam of relativistic particles propagating through an ambient medium.   In most parameter ranges plasma processes are the dominant energy loss mechanism for the beam, although if the beam propagates through an optically thick cloud, inverse Compton losses can dominate.   Typically, the radiative losses are < 10% of the plasma losses.

The foregoing discussion suggests a natural explanation for the relative quiescence of the relativistic portion of the beam-jet structure in radio galaxies.   It is also likely that the material accelerated by the beam plays a role in the evolution of the giant radio lobes.

REFERENCES

Beall, J.H. 1989, Proc. of the Vulcano Workshop: Frontier
Objects in Astrophysics and Particle Physics
(Italian Physical Society: Bologna).

Beall, J.H., Bednarek, W., Karakula, S., and Tkaczyk, W. 1987,
Proc. of the 20th Intnl. Cosmic Ray Conference,
Vol. 1 (NAUKA: Moscow)

Bednarek, W., Karakula, S., Tkaczyk, W., and Giovannelli, F.
1989 in press

Bludman, S.A., Watson, K.M., and Rosenbluth, M.N. 1960, Phys.
Fluids, 3, 747.

Boyd, T.J.M., and Sanderson, J.J. 1969, Plasma Dynamics
(New York: Barnes & Noble).

Breizman, B., and Ryutov, D. 1971, Soviet Phys. - JETP. 33, 220.

Brown, R.L., Johnston, K.J., and Lo, K.Y. 1981, Ap. J., 250, 155.

Chevalier, R.A., and Soker, N. 1989, Ap.J., 341, 867.

Dawson,J., and Oberman, C. 1962, Phys. Fluids, 5, 517.
1963, Phys. Fluids, 6, 394.

Feigelson, E.D. et al. 1981, Ap.J., 251, 31.

Freund, H.P., Haber I., Palmadesso, P., and Papadopoulos, K. 1980,
Phys. Fluids, 23, 518.

Geldzahler, B.J., and Shaffer, D.B. 1982, Ap. J. Lett. 260, L69.

Godfrey, B.B., Shanahan, W.R., and Thode, L.E., 1976, Phys.
Fluids, 18, 46.

Jackson, J.D. 1960, Classical Electrodynamics (New York: Wiley).

Kato, T. 1976, Ap. J. Suppl., 30, 397.

Kazanas, D. and Ellison, D.C. 1986, Nature 319, 380.

Lovelace, R.V.E., MacAuslan, J., and Burns, M. 1979,
AIP Conf. Proc.,56, 399.

Papadopoulos, K. 1975, Phys. Fluids, 18, 1769.

Rose, W.K., Guillory, J., Beall, J.H., and Kainer, S. 1984,
Ap. J. 280, 550.

Rose, W.K., Beall, J.H., Guillory, J., and Kainer, S. 1987,
Ap. J., 314, 95.

Rowland, H.L. 1977, Ph.D. thesis, University of Maryland.

Rukhadze, A.A., and Rukhlin, V.G. 1972, Soviet Phys. - JETP, 34, 93.

Schamel, H., Lee, Y.C., and Morales, G.J. 1975,
Phys. Fluids, 19, 849.

Scott, J.S., Holman, G.D., Ionson, J.A., and Papadopoulos, K.
1980, Ap. J., 239, 769.

Shapiro, S.L., Lightman, R.P., and Eardley, D.M. 1976,
Ap. J. 204, 187.

Smith, B., Goldstein, M., and Papadopoulos, K. 1976,
Solar Phys., 46, 515.

Spitzer, L. 1956, Physics of Fully Ionized Gases (New York: Wiley).

Stocke, J.T. 1978, A.J., 83, 348.

Thode, L.E., and Sudan, R.N. 1973, Phys. Rev. Letters, 30, 732.

Ulvestad, J.S., Wilson, A.S., and Sramek, R.A. 1981,
Ap.J., 247, 419.

Vestrand, T.W. 1983, Ap. J. 271, 304.

Walker, R.C. 1984, Physics of Energy Transport in Extragalactic
    Radio Sources: Proc. of NRAO Workshop No. 9, A.H. Bridle and
    J. A. Eilek, eds.
Wilson, A.S., and Willis, A.G. 1980, Ap.J., 240, 429.
Zakharov, V.E. 1972, Soviet Phys. - JETP, 35, 908.

AN INTRODUCTION TO RELATIVISTIC PLASMAS IN ASTROPHYSICS

ROLAND SVENSSON
NORDITA
Blegdamsvej 17
DK-2100 Copenhagen Ø, Denmark
(bitnet: svensson@dknbi51)

ABSTRACT. The theory of physical processes in relativistic plasmas is reviewed on an introductory level. The basic properties of the physical processes, especially electron-positron pair production in photon-photon interactions, are briefly presented. The effects pair production introduces in thermal and nonthermal radiation models as well as in more realistic accretion scenarios are discussed using simple theoretical arguments. Finally, the suggested existence of pair dominated relativistic plasmas in specific classes of astrophysical objects is reviewed.

1. Introduction

In this workshop, both astrophysical and laboratory plasmas under a wide variety of conditions were discussed. Figure 1, covering most of the possible conditions in the Universe, shows the regions in the density-temperature plane discussed by some of the authors in these proceedings that put their emphasis on the physical processes occurring in hot plasmas, rather than on the cosmic objects themselves. These regions are in general nonrelativistic with electron temperatures less than 10^9 K.

In this review, the emphasis will be on tenuous relativistic plasmas, i.e. plasmas where the electrons have mildly relativistic or relativistic energies and the electron temperature is $> 10^9$ K. Such plasmas occur near black holes, where stellar or interstellar matter accrete either onto galactic $(1 - 10 M_\odot)$ or supermassive $(10^7 - 10^9 M_\odot)$ black holes.

The gravity of black holes drives gaseous bulk motions, the kinetic energy of which gets dissipated through e.g. collisionless shocks or magnetic reconnection. The dissipated energy is channeled either to heat the bulk of the (thermal) matter or to heat only a small fraction of the particles to relativistic (nonthermal) energies. The heated matter cools by emitting radiation. The emphasis will be on the physical processes responsible for emitting the radiation and for turning a fraction of this radiation into electron-positron pairs. The effects these pairs have on the accretion process are also discussed. Excellent general reviews of accretion scenarios onto black holes in which relativistic pair plasmas may form are those of Rees *et al.* (1982), Begelman, Blandford, and Rees (1984), and Rees (1984). Some important aspects of such scenarios are the following:

W. Brinkmann et al. (eds.), Physical Processes in Hot Cosmic Plasmas, 357–381.
© 1990 *Kluwer Academic Publishers.*

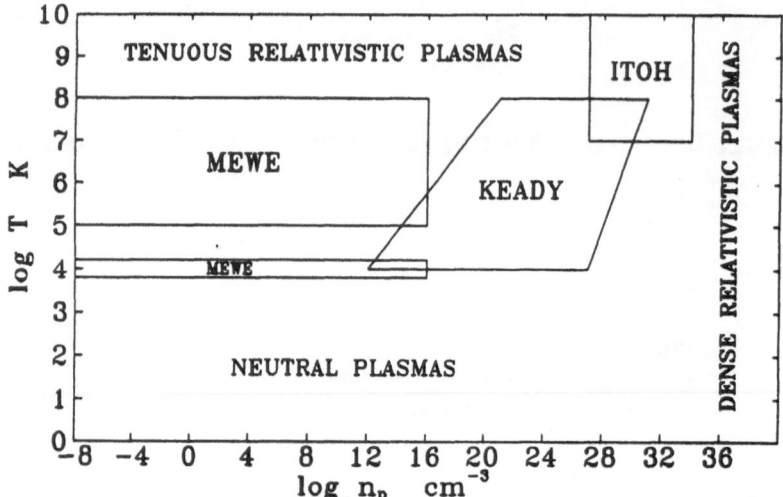

Figure 1. The density-temperature plane for cosmic plasmas ranging from the mean density of the universe, $\approx 10^{-6}\,\mathrm{cm}^{-3}$, up to nuclear densities. Various regions discussed in these proceedings emphasizing physical processes are outlined.

(i) The size, R, of the emitting gas region should be larger than about $3R_S = 6GM/c^2 \approx 10^7$ and 10^{14} cm for black hole masses of 10 and $10^8 M_\odot$, respectively. Here, R_S, is the Schwarzschild radius.

(ii) The luminosity, L, is limited by the Eddington luminosity, $L_{\mathrm{Edd}} = 2\pi m_p c^3 R_S/\sigma_T$ at which the gravitational force on the protons in an ionized hydrogen gas is balanced by the radiation pressure on the free electrons. The Eddington luminosity is $\approx 10^{39}$ and 10^{46} erg s^{-1} for black hole masses of 10 and $10^8 M_\odot$, respectively. In a pure electron-positron gas the Eddington limit is a factor $m_p/m_e \approx 2 \times 10^3$ smaller.

(iii) The ratio of these two quantities, the *compactness* L/R, is less than $L_{\mathrm{Edd}}/3R_S \approx 10^{32}$ erg s^{-1} cm^{-1} for gravitationally bound pair free plasmas. The compactness is a parameter of crucial importance in hot compact gases and it is often written in a dimensionless form,

$$\ell \equiv \frac{L}{R}\frac{\sigma_T}{m_e c^3} = \frac{2\pi}{3}\frac{m_p}{m_e}\left(\frac{L}{L_{\mathrm{Edd}}}\right)\left(\frac{3R_S}{R}\right). \tag{1.1}$$

Here, σ_T is the Thomson scattering cross section. The maximum Eddington limited value of ℓ is $(2\pi/3)(m_p/m_e) \approx 3600$. In physical units

$$\frac{L}{R} = 4 \times 10^{28}\ell\,\mathrm{erg\,s}^{-1}\,\mathrm{cm}^{-1}. \tag{1.2}$$

(iv) The temperature of virialized matter at a distance r from the black hole is $T \approx 0.1(m_p c^2/k)(R_S/r) \approx 10^{12}(R_S/r)$ K. Thermal electrons (and positrons) at the virial temperature are relativistic ($kT > m_e c^2$) for $r < 10^3 R_S$.

The physical location of pair producing relativistic plasmas within the standard accretion scenarios are (with increasing accretion rate):

(i) The geometrically thick hot two-temperature disk (Shapiro, Lightman and Eardley 1976) or ion torus (Rees et al. 1982) containing ions at approximately the virial temperature and mildly relativistic electrons.

(ii) The corona of the standard geometrically thin cool accretion disk (Shakura and Sunyaev 1972).

(iii) The corona of the geometrically thick cool radiation pressure dominated torus.

Energy may also be released nonthermally throughout these regions as well as in any (anisotropic, jet-like) outflow.

2.   Brief History of Relativistic Pair Plasmas

The most important advance in the theory of relativistic plasmas during the 1980's may have been the final realization that electron positron pairs are produced in great abundance in certain circumstances affecting both the properties of the plasma and the spectral shape of the emerging radiation. The importance of photon-photon pair production in compact plasmas was pointed out by Jelley (1966), Herterich (1974), and Lightman, Giacconi and Tananbaum (1978). The crucial parameter determining whether pairs are produced or not is the compactness, i.e., the luminosity to size ratio, introduced in a different context by Cavaliere and Morrison in 1980. Early pioneering studies in the 1970s were either limited in scope (Bisnovatyi-Kogan, Zel'dovich and Sunyaev, 1971, on thermal pair production) or in error (Bonometto and Rees, 1971 on nonthermal pair production). The field lay essentially dormant until the comprehensive treatments of thermal pair production in relativistic plasmas by Lightman (1982) and Svensson (1982) stimulated a series of papers of increasing complexity on thermal pairs. The interest in nonthermal pair production was renewed by Guilbert, Fabian and Rees (1983), Fabian (1984), and Kazanas (1984), but it was the work by Zdziarski and Lightman (1985) that formulated the problem in such way that the seemingly complicated nonthermal pair cascades and their resulting photon spectra could be understood in simple terms (see e.g. Svensson 1986, 1987). Here we review the results obtained so far regarding thermal and nonthermal pair production in compact plasmas. More detailed reviews of work prior to 1986 can be found in Svensson (1986) and Zdziarski and Lightman (1987).

3.   Physical Processes in Relativistic Plasmas

Only three radiation processes are considered to be important in compact hot plasmas. They are (i) bremsstrahlung, (ii) cyclo/synchrotron radiation, and (iii) Compton scattering. Two further processes become relevant when the plasma is *compact* enough in a sense to be defined below. Then electron-positron pairs are created in photon-photon pair production and destroyed in pair annihilations. Some relevant properties of these basic processes are summarized here.

3.1.   RADIATION PROCESSES

3.1.1. *Bremsstrahlung.* Electrons or positrons decelerating in the field of charged particles emit radiation with a spectral intensity index $\alpha \sim$ a few tenths extending up to the energy of the particle. Bremsstrahlung is a very inefficient cooling process. The cooling rate of mildly relativistic electrons of typical energy $E \sim m_e c^2$ is approximately

$$\frac{dE}{dt} \sim 4\alpha_f \sigma_T c n_p m_e c^2, \tag{3.1}$$

where $\alpha_f \approx 1/137$ is the fine structure constant, and n_p is the proton density. The luminosity from a uniform spherical cloud with electron density n_- and proton density n_p at a mildly relativistic temperature is simply

$$L \approx \frac{4\pi R^3}{3} n_- n_p \alpha_f \sigma_T m_e c^3, \tag{3.2}$$

which can be rewritten as

$$\ell \equiv \frac{L}{R}\frac{\sigma_T}{m_e c^3} \approx 10\alpha_f \tau_p^2, \tag{3.3}$$

where the 'proton' optical depth, $\tau_p = n_p \sigma_T R$, is the Thomson scattering optical depth of the ionization electrons. Compactnesses larger than 0.1 in a mildly relativistic plasma cloud require a Thomson scattering optical depth larger than unity making Comptonization of soft bremsstrahlung photons to be the dominant cooling process. When Comptonization is important, other sources (e.g., thermal cyclotron photons, nonthermal synchrotron photons or external black body photons) often dominate bremsstrahlung as soft photon producers.

3.1.2. *Cyclotron/Synchrotron Radiation.* A nonrelativistic electron gyrating in a magnetic field of strength, B, radiates a line at the gyrofrequency,

$$\nu_B = \frac{eB}{2\pi m_e c}. \tag{3.4}$$

Let ϵ_B be the corresponding photon energy. At sufficiently large particle energy higher harmonics start becoming important and at relativistic particle energies most of the power is radiated at very large harmonic numbers, i.e., at photon energies $\epsilon \gg \epsilon_B$. Three relativistic effects (the Doppler effect, a factor γ^2; beaming of the emitted radiation, a factor γ; and the relativistic increase of the mass, a factor $1/\gamma$; see, e.g., Rybicki and Lightman 1979) cause the typical observed photon energy, ϵ, radiated in the synchrotron process by an electron with the Lorentz factor γ to be a factor γ^2 larger then ϵ_B,

$$\epsilon \sim \gamma^2 \epsilon_B. \tag{3.5}$$

The emissivity of a single electron peaks at the energy given by eq. (3.5). It is then clear that the photons emitted by a power law distribution of electrons

$$N(\gamma) \propto \gamma^{-p}, \tag{3.6}$$

will also be distributed as a power law. For p not too small the emission at a given photon energy is dominated by electrons with a Lorentz factor $\gamma \sim (\epsilon/\epsilon_B)^{1/2}$. The number particles in an interval near γ is approximately $\gamma N(\gamma)$. As the spectral emissivity $j(\epsilon)$ only depends on γ through the factor ϵ/γ^2 the radiated intensity simply becomes

$$I(\epsilon) \propto \gamma N(\gamma)|_{\gamma \propto \epsilon^{1/2}} \propto \epsilon^{-(p-1)/2} , \qquad (3.7)$$

i.e., the intensity index $\alpha \equiv -d\ln I(\epsilon)/\ln \epsilon$ becomes

$$\alpha = \frac{p-1}{2} , \qquad (3.8)$$

a very important result (see, e.g., Rybicki and Lightman 1979 for further details).

3.1.3. *Compton Scattering.* Compton scatterings occur in a variety of forms, thermal and nonthermal, upscattering and downscattering.

Low energy (soft) photons that Compton-scatter on more energetic thermal electrons will on average receive a fractional energy increase per scattering,

$$\frac{\Delta \epsilon}{\epsilon} = \frac{4kT}{m_e c^2} . \qquad (3.9)$$

The diffusive escape of the photons takes $\sim \tau^2$ scatterings when $\tau > 1$ making a total energy gain of

$$\left(\frac{\Delta \epsilon}{\epsilon}\right)_{tot} \sim \left(\frac{4kT}{m_e c^2}\right) \tau^2 \equiv y , \qquad (3.10)$$

where the Compton y-parameter has been defined. When $y \gtrsim 1$, this *upscattering* of soft photons produces power law spectra extending up to kT at which energy there is an exponential turnover. The analytical theory at nonrelativistic temperatures ($kT \lesssim m_e c^2$ or $T \lesssim 5 \times 10^9$ K) and for $\tau > 1$ gives the spectral index of the power law as

$$\alpha \sim \left(\frac{9}{2} + \frac{1}{y}\right)^{1/2} - \frac{3}{2} \qquad (3.11)$$

(e.g., Rybicki and Lightman 1979; Sunyaev and Titarchuk 1980). Detailed emergent spectra at mildly relativistic temperatures require numerical calculations, in general Monte Carlo simulations (see, e.g., Pozdnyakov *et al.* 1983; Górecki and Wilczewski 1984). When $y \gg 1$ almost all soft photons upscatter to $\sim kT$ and thermalize with the electrons into a Wien distribution before escaping.

If, on the other hand, higher energy photons scatter on colder electrons the photons will experience a fractional energy loss per scattering of the magnitude, $\Delta \epsilon/\epsilon \sim -\epsilon/m_e c^2$ and the total energy loss after the typical number of scatterings, τ^2, it takes for a photon to escape becomes

$$\left(\frac{\Delta \epsilon}{\epsilon}\right)_{tot} \sim -\left(\frac{\epsilon}{m_e c^2}\right) \tau^2 , \qquad (3.12)$$

This is larger than unity for $\epsilon > m_e c^2/\tau^2$ and thus the relatively few photons in the X-ray spectrum above $m_e c^2/\tau^2$ downscatter to $\epsilon \sim m_e c^2/\tau^2$ while the spectrum at lower energies is is unaffected by *downscattering*. The spectrum emerging from a cold homogeneous cloud, where a power law photon spectrum is uniformly injected, steepens at $\sim m_e c^2/\tau^2$ from the injected spectral index α to $\alpha + 1/2$ (Sunyaev and Titarchuk 1980). The more concentrated the photon injection is towards the center of the cloud, the steeper the spectrum above $m_e c^2/\tau^2$ becomes. Downscattering is fully effective for particles cold enough $(kT \ll m_e c^2/\tau^2)$. Compton upscattering is then negligible as the Compton parameter y is much less than unity.

When the electron is relativistic, i.e., when the Lorentz factor, γ, of the electron is much larger than unity, the low energy photons experience a very large fractional energy gain of the order of γ^2. The final energy of a photon with initial energy ϵ_i is given by eq. (3.5) if ϵ_B is replaced with ϵ_i. In fact, the synchrotron process can be viewed as the electron scattering the virtual photons of the magnetic field each having the energy ϵ_B.

The Compton cooling rate of a relativistic electron scattering on soft photons is

$$\frac{dE}{dt} = m_e c^2 \dot{\gamma}_C \sim \text{(reaction rate)(density of target photons)(energy gain)}$$

$$\sim (\sigma_T c)(n_{\text{target}})(\epsilon_i \gamma^2) \tag{3.13}$$

$$\sim \sigma_T c U_{\text{rad}} \gamma^2,$$

where U_{rad} is the radiation energy density and where equation (3.5) has been used for the energy gain. For synchrotron cooling, U_{rad} is replaced with the magnetic energy density $U_B = B^2/8\pi$. As the radiation energy density can be related to the luminosity, L, of a source of size, R, through $U_{\text{rad}} \sim L/4\pi R^2 c$ one can write the Compton cooling time scale as

$$t_{\text{cool}} = \frac{E}{dE/dt} \sim \left(\frac{R}{c}\right) \frac{10}{\gamma \ell}. \tag{3.14}$$

This means that for compactnesses greater than about 10 all electrons cool before they can escape $(t_{\text{esc}} \gtrsim R/c)$. To avoid a pileup of cool particles reacceleration (or annihilation if the particles are pairs) is necessary.

3.2. PAIR PRODUCTION AND ANNIHILATION

Two additional processes are of uttermost importance in compact high energy plasmas: (i) photon-photon pair production, $\gamma\gamma \to e^+ e^-$, in which two photons interact to produce an electron-positron pair, and (ii) the reverse process, pair annihilation $e^+ e^- \to \gamma\gamma$, in which an electron-positron pair self-destructs to produce two photons.

The pair production process is a threshold process that requires that the product, $\epsilon_1 \epsilon_2$, of the energies of the two interacting photons to exceed $(m_e c^2)^2$ in order to produce an $e^+ e^-$ pair. The cross section rises sharply to a maximum of order σ_T and then has the same Klein-Nishina decline at $\epsilon_1 \epsilon_2 \gg 1$ as the Compton scattering process.

It was early realized (Jelley 1966; Herterich 1974) that a source could be optically thick to photon-photon pair production at the γ-ray energy, ϵ, if the photon density, n_γ, above the energy, $(m_e c^2/\epsilon)m_e c^2$, corresponding to the threshold condition is large enough. The optical depth of a source of luminosity L and radius R with most photons at $\epsilon \sim m_e c^2$ is simply

$$\tau_{\gamma\gamma} \sim n_\gamma(\epsilon \sim m_e c^2)\sigma_T R \sim \left(\frac{L}{4\pi R^2 c m_e c^2}\right)\sigma_T R, \tag{3.15}$$

which is greater than unity for

$$\ell = \frac{L}{R}\frac{\sigma_T}{m_e c^3} > 4\pi \simeq 10 \quad \text{or} \quad \frac{L}{R} > 4 \times 10^{29} \text{erg s}^{-1} \text{cm}^{-1}. \tag{3.16}$$

More detailed considerations give an energy and spectral shape dependent expression, but equation (3.15) shows the correct order of magnitude. The compactness can be rewritten in terms of the Eddington luminosity, L_{Edd}, and the Schwarzschild radius, R_S, of a black hole (see eq. [1.1]).

Two conditions must then be satisfied in order for ample electron-positron pair production to occur in a source. First, photons above the threshold energy, $m_e c^2$, must be emitted. This is naturally the case in nonthermal mechanisms and occurs in thermal plasmas when $kT \simeq m_e c^2$. Second, the photons must produce pairs, demanding $\ell \gtrsim 10$, which is satisfied for $L > 10^{-3} L_{Edd}$ if the radiation source is at a few Schwarzschild radii (Lightman, Giacconi, and Tananbaum 1978).

The pair annihilation process is of direct observational interest as it produces photons. The spectrum is most simply expressed as a single integral (Svensson 1983; Dermer 1984). At nonrelativistic temperatures the line is narrow with the relative width,

$$\frac{\Delta \epsilon}{\epsilon} \sim \left(\frac{kT}{m_e c^2}\right)^{1/2}, \qquad T \ll 5 \times 10^9 \text{K}, \tag{3.17}$$

being due to the Doppler effect. It is centered on the rest mass energy, $m_e c^2 = 511$ keV. At relativistic temperatures the line mimics the annihilating particle distribution and is broad,

$$\frac{\Delta \epsilon}{\epsilon} \sim 1, \qquad T \gg 5 \times 10^9 \text{K}, \tag{3.18}$$

and is centered at the thermal energy kT. The pair annihilation rate is constant at nonrelativistic temperatures being approximately $0.4 n_+ n_- c \sigma_T$ where n_+ is the density of pairs. Similar to the Compton scattering and the photon-photon pair production processes the annihilation cross section also has a Klein-Nishina decline at relativistic energies making the annihilation rate to decrease as T^{-2} at relativistic temperatures. This causes the bremsstrahlung emission from annihilating pairs to dominate the annihilation emission as soon as kT is greater than a few $m_e c^2$ (corresponding to $T > 10^{10}$ K) making the line invisible.

4. Thermal Pair Plasmas

When the electron temperature of a plasma reaches approximately $m_e c^2/k \approx 10^9$–10^{10}K, i.e., when the kinetic energies become of the same order as the electron rest mass energy then pair production may become important. The basic simplifying assumption in the study of thermal plasmas is that the particle distribution is thermal, i.e., the particles have a Maxwell-Boltzmann velocity distribution characterized by one single parameter, the temperature. Created pairs are assumed to instantaneously join the thermal distribution. This may not be the case in situations where, e.g., the energy exchange time scale with a nonthermal radiation field or the cooling time scale is much faster than the particle relaxation time scales (e.g., Baring 1987; Dermer and Liang 1989).

Thermal pair equilibrium in static atmospheres has by far received most attention (e.g., Lightman 1982; Svensson 1982, 1984; Zdziarski 1984, 1985, 1986a; Kusunose and Takahara 1983, 1985; Sikora and Zbyszewska 1986), while only a few studies have been made of time-dependent properties (Guilbert and Stepney 1985; Kusunose 1987; Takahara 1988; White and Lightman 1990). Here only the most important results of those studies will be discussed.

4.1. THE MAXIMUM TEMPERATURE

The steady pair density in pair equilibrium plasmas is determined by solving the pair balance equation,

$$(\text{pair production rate}) = (\text{pair annihilation rate}). \tag{4.1}$$

If photons densities are not large enough, the dominant pair production process will be particle-particle pair production. It was noted by Bisnovatyi-Kogan, Zel'dovich and Sunyaev (1971) that above a critical temperature the pair annihilation rate could not keep up with the pair production. The pair production rate at relativistic temperatures when pairs dominate the plasma is

$$n_+^2 c\sigma_T \alpha_f^2 \frac{6}{\pi^2} \left(\ln \frac{kT}{m_e c^2} \right)^3 , \tag{4.2}$$

while the annihilation rate is

$$n_+^2 c\sigma_T \frac{3}{16} \left(\frac{kT}{m_e c^2} \right)^{-2} \ln \frac{kT}{m_e c^2} . \tag{4.3}$$

The pair production rate dominates above $kT \approx 25 m_e c^2$. This maximum temperature is seen as a vertical boundary in Figure 2. The region to the right of this boundary is forbidden for pair equilibrium plasmas unless the pairs have some means to escape.

4.2. WIEN PAIR EQUILIBRIUM

In a plasma cloud where the y parameter is large enough for most photons to scatter into a Wien distribution there is a unique maximum compactness being a function of temperature only (Svensson 1984). The pair balance at $kT < m_e c^2$ when photon-photon pair production dominates is simply

$$n_\gamma^2 c\sigma_T f(T) = 0.4 n_+ n_- c\sigma_T , \tag{4.4}$$

where n_γ is the photon density in the Wien peak and $f(T)$ is the pair production cross section averaged over the normalized Wien spectra (eq. [B6a] in Svensson 1984). The ratio of the photon density to the pair density in a pair dominated plasma then becomes

$$\frac{n_\gamma}{n_+} \sim f^{-1/2}(T) \sim 2 \left(\frac{kT}{m_e c^2} \right)^{3/2} e^{m_e c^2/kT} . \tag{4.5}$$

The luminosity from the cloud can be written as

$$L = (\text{volume})(\text{mean photon energy})(\text{photon production rate})$$

$$\sim \left(\frac{4\pi R^3}{3} \right) (3kT) \left(\frac{dn_\gamma}{dt} \right) , \tag{4.6}$$

where the photon production rate can be related to the photon to pair ratio by solving a simple radiative transfer equation where photon production is balanced by photon escape,

$$\frac{dn_\gamma}{dt} \simeq \frac{n_\gamma}{t_{esc}} \simeq \frac{n_\gamma}{\tau_T R/c} \simeq \left(\frac{n_\gamma}{n_+}\right)\frac{c}{R^2\sigma_T}. \tag{4.7}$$

Using n_γ/n_+ from equation (4.5) and eliminating the photon production rate in equation (4.6) gives a simple expression for the maximum compactness in a pair dominated cloud,

$$\ell = \frac{L}{R}\frac{\sigma_T}{m_e c^3} \simeq 10\left(\frac{kT}{m_e c^2}\right)^{5/2} e^{m_e c^2/kT}. \tag{4.8}$$

This unique compactness is seen in Figure 2 as the low temperature asymptotic curve towards which all curves converge. The region above the maximum compactness curve is forbidden for clouds in pair equilibrium.

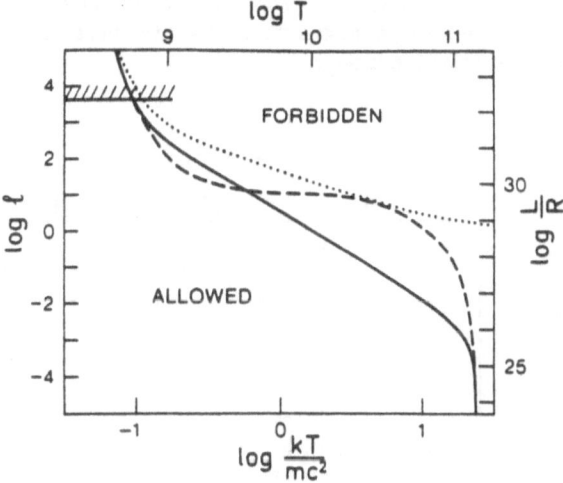

Figure 2.   The maximum compactness, ℓ, as a function of the temperature, T (from Svensson 1986). The regions above the curves are forbidden for spherical clouds in thermal pair equilibrium. The dominant radiation process is Comptonized bremsstrahlung for the dashed curve (Svensson 1984), and Comptonized soft photons with slope $\alpha = 0.7$ for the other two curves (Zdziarski 1985). The dominant pair destruction mechanism is pair escape for the dotted curve and pair annihilation for the other two. The pair free Eddington limit (eq. [3.17]) is violated in the shaded region.

4.3. THE PAIR EQUILIBRIUM CURVE

Besides the pair dominated equilibrium cloud there also exist the possibility of having a cloud in pair equilibrium with the pair density much smaller than the density of the ionization electrons. The small amount of pairs present does not affect the cloud luminosity. In the case of bremsstrahlung at relativistic temperatures, $kT > m_e c^2$, the compactness from equation (3.3) (which was derived for $kT \sim m_e c^2$) is modified into

$$\ell \approx 0.1 \tau_p^2 \frac{kT}{m_e c^2} \ln \frac{kT}{m_e c^2} \qquad kT > mc^2 . \qquad (4.9)$$

The curve corresponding to this low compactness would appear in the 'allowed' region in Figure 2. At some temperature it would join the maximum compactness curve determined by pair dominated pair equilibrium. As the compactness of an initially pair free source increases, the temperature first increases (eq. [4.9]) until the plasma becomes pair dominated. Increasing ℓ further causes the temperature to decrease (eq. [4.8]). In a pair free source the temperature increases in order for the electrons to be able to radiate a larger luminosity (i.e. larger compactness if the radius is constant). Once the source is pair dominated each particle, of course, radiates at a lower rate at lower temperatures, but the large number of pairs radiate a larger total luminosity.

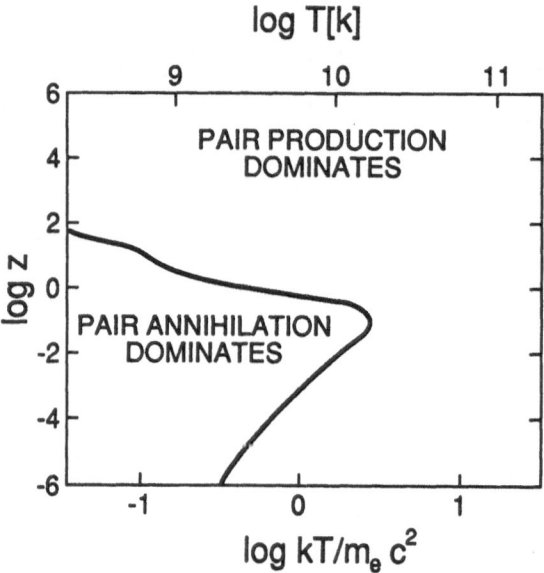

Figure 3. The dimensionless pair density $z = n_+/n_p$ in a thermal plasma in pair equilibrium (from Svensson 1984). The proton optical depth is $\tau_p = n_p \sigma_T R = 1$. Pair production dominates to the right of the equilibrium curve, while pair annihilation dominates to the left.

Figure 3 shows a typical pair equilibrium curve with the relative pair fraction, $z = n_+/n_p$, plotted as a function of temperature. At high enough temperatures there is no equilibrium solution as the pair annihilation due to the Klein-Nishina decline of the cross section cannot keep up with the pair production. The upper pair dominated branch corresponds to the maximum compactness discussed in the previous section, while the lower branch corresponds to a 'pair free' state with the compactness being given by equation (4.9) when no prolific source of soft photons is present.

4.4. THE PAIR ANNIHILATION LINE

A pair annihilation line is in general *not* seen from a uniform cloud in pair balance. At relativistic temperatures other emission processes dominate (see §3.1.1), and at nonrelativistic temperatures in pair dominated plasmas the optical depths to Compton scattering and photon-photon pair production are large enough to modify the annihilation line into a Wien tail (see Svensson 1984; Zdziarski 1984, 1985, 1986a).

5. Nonthermal Pair Plasmas

Part or most of the power released in a source may be used to accelerate a minor fraction of the particles in the source to relativistic energies much larger than the typical (nonrelativistic or mildly relativistic) particle energy. The acceleration mechanisms in astrophysical sources are not well understood. It is therefore common in treatments emphasizing radiation processes not to worry about the details of the particle acceleration but rather to make simple assumptions regarding the injection rate of particles of a given energy per unit volume.

The injected relativistic particles cool by emitting synchrotron radiation or by inverse Compton scattering on low energy photons. In doing so the particle energy distribution changes. A steady particle injection results in a steady energy distribution of particles. The emitted radiation has a spectral shape that is easily related to the particle energy distribution.

If secondary particles, e.g., electrons-positron pairs are produced they must be included when calculating the steady particle distribution and the resulting radiation spectrum. The radiation from the pairs may produce further generation of pairs resulting in a pair-photon cascade.

In this section we discuss what radiation spectra result from nonthermal particle injection and show how pair cascades modify these spectra.

5.1. THE STEADY ELECTRON AND PHOTON DISTRIBUTION

The relaxation of the particle distribution, $N(\gamma)\,\mathrm{cm}^{-3}\,\mathrm{s}^{-1}$, is governed by the kinetic equation

$$\frac{\partial N(\gamma)}{\partial t} + \frac{\partial}{\partial \gamma}\left[\dot{\gamma}N(\gamma)\right] = Q(\gamma), \qquad (5.1)$$

where $Q(\gamma)\,\mathrm{cm}^{-3}\,\mathrm{s}^{-1}$ is the injection rate of relativistic particles with Lorentz factor γ, and $\dot{\gamma}$ is the cooling rate given by equation (3.13), and where it has been assumed that other loss terms such as particle escape are negligible. Equation (5.1) can be viewed as a particle conservation equation in energy space with $\dot{\gamma}$ being the velocity along the energy axis.

In steady state $\partial N(\gamma)/\partial t = 0$ and the solution to equation (5.1) becomes

$$N(\gamma) = \frac{1}{\dot{\gamma}} \int_{\gamma}^{\infty} Q(\gamma) d\gamma. \qquad (5.2)$$

For monoenergetic injection at γ_{max} we have

$$N(\gamma) \propto \frac{1}{\dot{\gamma}} \propto \gamma^{-2}, \qquad \gamma < \gamma_{max}, \qquad (5.3)$$

and for power law injection,

$$Q(\gamma) \propto \gamma^{-\Gamma}, \qquad (5.4)$$

the distribution becomes

$$N(\gamma) \propto \gamma^{-\Gamma-1}, \qquad \Gamma > 1, \qquad (5.5)$$

for 'steep' injection. 'Flatter' injection ($\Gamma < 1$) is equivalent to the monoenergetic case.

The radiation spectrum produced by the steady particle distribution through either synchrotron or inverse Compton cooling has an intensity index (eq. [3.8]),

$$\alpha = \frac{p-1}{2} = \begin{cases} 1/2, & \text{monoenergetic or } \Gamma < 1; \\ \Gamma/2 > 1/2, & \Gamma > 1. \end{cases} \qquad (5.6)$$

In the simplest nonthermal models the spectral slopes are therefore never flatter than 0.5. The unique slope $\alpha = 0.5$ is expected for for all injection slopes $\Gamma < 1$. Several complications may, however, change this picture. They are i) secondary particle injection resulting in pair cascades, ii) Comptonization by thermal particles, and iii) higher order Compton scatterings.

5.2. PAIR CASCADES

If the radiation spectrum extends beyond $m_e c^2$ then electron-positron pairs may be produced in photon-photon interactions where γ-rays with energies, $\epsilon > m_e c^2$, mainly interact with X-rays slightly above the pair production threshold, $(m_e c^2)^2/\epsilon$, producing electrons and positrons with a mean energy $\epsilon/2$.

Figure 4 shows the effect of pair cascades for increasing injection compactness, ℓ_i, in the case of monoenergetic injection of primary particles cooling on a black body luminosity, L_{bb}, in the UV range (see, e.g., Zdziarski and Lightman 1985; Fabian et al. 1986; Svensson 1987; Lightman and Zdziarski 1987; Done and Fabian 1989; Done, Ghisellini and Fabian 1990; Coppi 1990). The black body photons could emerge from a geometrically thin, optically thick accretion disk around a supermassive black hole. Here, we consider the case when the blackbody photons dominate the radiation energy density, while the case with the X-rays dominating will be discussed in §5.4.

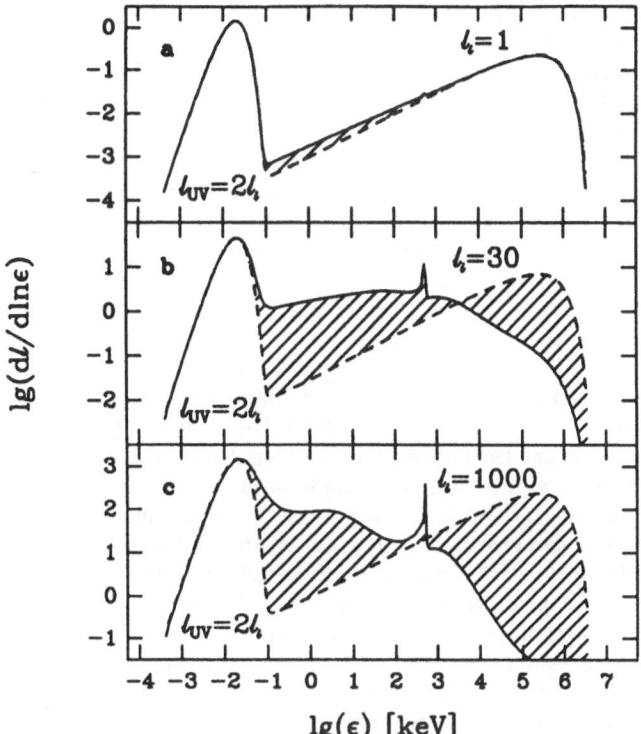

Figure 4. The effects of e^+e^- pair cascades on the spectra from nonthermal plasmas for increasing compactness (from Svensson and Zdziarski 1989). The monoenergetic pairs are injected at $\gamma_i = 7.5 \times 10^3$ with the compactness ℓ_i. Blackbody UV photons are injected at $kT = 5.1$ eV with the compactness ℓ_{UV}. The solid curves give the pair-reprocessed spectra. The dashed curves in all panels correspond to the spectra without pair reprocessing, as would be the case for $\ell \ll 1$. The X-ray spectral index is then $\alpha_X = 0.5$ and the spectrum breaks in the high energy γ-ray regime above $(4/3)\gamma^2 kT$. The solid curve in panel a gives the pair-reprocessed spectrum when pairs begin to be important. The optical depth to pair production $\tau_{\gamma\gamma} \lesssim 1$ for the entire spectrum. The pairs produced in the optically thin regime soften the X-ray spectrum causing $\alpha_X = 0.57$ (2–10 keV). Panel b shows the case with strong nonthermal pair-reprocessing. The pair absorption optical depth is unity near the intersection of the solid and dashed curves. The spectral index is $\alpha_X = 0.85$. The optical depth of thermal pairs is still relatively small, $\tau_{pair} = 1.6$. Panel c shows the case in which both nonthermal reprocessing and thermal downscattering are important. Now $\tau_{\gamma\gamma} = 1$ at $\epsilon \simeq 500$ keV, $\tau_{pair} = 13$, $\alpha_X = 1.3$. The X-ray spectrum is steepened above a few keV due to downscattering (see §5.3). The spectra were calculated using the model of Lightman and Zdziarski (1987).

For injection compactnesses, ℓ_i, less than unity the optical depth the X-rays provide for the γ-rays is much less than unity and there is no significant pair production. The steady particle spectrum has slope $p = 2$ and consequently the X-ray intensity index of the upscattered black body photons is $\alpha_X = 0.5$.

For compactnesses of order unity the optical depth at the highest γ-ray energies becomes of order unity and there is a slight absorption at those energies. The small amount of absorbed γ-ray power used to make pairs is reradiated in the relatively low luminosity X-ray range where the additional radiation easily dominates. The X-ray luminosity changes appreciably as well as the slope steepens to $\alpha_X \sim 0.6$ (Fig. 4a).

For even larger injection compactness, ℓ_i, more and more of the γ-ray luminosity is redistributed into the X-ray range (Fig. 4b, c) and the spectrum pivots about $\epsilon \sim m_e c^2$. The produced pairs cool to nonrelativistic energies where they thermalize and eventually annihilate at the equilibrium Compton temperature which they obtain by exchanging energy with the radiation field in Compton scatterings. The Thomson depth of these cool pairs becomes larger than unity for injection compactnesses large enough ($\gtrsim 20$). Then the interactions between the pairs and the photons will have a noticeable effect on the radiation field, i.e., the UV and X-ray radiation will get Comptonized. The UV black body photons will get upscattered producing a steep soft X-ray power law spectrum. The hard X-rays at energies $\epsilon > m_e c^2 / \tau_T^2$ will lose energy by scattering on the cool pairs causing a spectral break in which the spectrum steepens towards higher energies (see Fig. 4c).

The limiting spectral slope of the X-rays neglecting Comptonization by cool pairs is $\alpha \simeq 1$. This occurs when the compactness is large enough for essentially all γ-ray photons to get absorbed. Such cascades are called *saturated* (Svensson 1986).

It is easy to show why $\alpha \approx 1$ for saturated pair cascades (Svensson 1986, 1987). The slope of the steady distribution of primary particles is $p = 2$ and the radiated photons from these particles have a spectral intensity index $\alpha_1 = 0.5$, where the subscript indicates that this is the first generation of photons. Now, if every photon makes a pair where the electron and the positron approximately share the photon energy, then the slope, Γ_1, of the injection rate of the first generation of pairs, $Q(\gamma)$ cm^{-3} s^{-1}, is the same as the production rate of the pair producing photons, $\dot{n}(\epsilon) \propto I(\epsilon)/\epsilon$, i.e., $\Gamma_1 = \alpha_1 + 1 = 1.5$. From equation (5.5) we find that the steady distribution of the first pair generation have slope $p_1 = \Gamma_1 + 1 = 2.5$ and the second photon generation they produce has an intensity index $\alpha_2 = (p_1 - 1)/2 = 0.75$. If the cascade continues the following generations have $p_3 = 2.75$, $p_4 = 2.875$, $p_5 = 2.9375$, and $\alpha_3 = 0.875$, $\alpha_4 = 0.9375$ and so on, approaching $p = 3$ and $\alpha = 1$, which are the limiting slopes for a saturated pair cascade. The limiting slopes correspond to equal power per logarithmic energy interval being injected as pairs and being radiated as photons.

5.3. THERMAL COMPTONIZATION

The Thomson depth of cooled nonrelativistic pairs,

$$\tau_{\text{pair}} = (n_+ + n_-)\sigma_T R \approx 2n_+ \sigma_T R, \qquad (5.7)$$

can be obtained by balancing the pair production and the annihilation rates. The annihilation rate per unit volume is simply $(3/8)n_+ n_- c\sigma_T \approx (3/8)n_+^2 c\sigma_T$ while the pair production rate per unit volume is

$$\frac{Y}{2m_e c^2}\left(\frac{L_i}{4\pi R^3/3}\right). \qquad (5.8)$$

The injected power L_i per unit volume divided by $2m_ec^2$ gives the maximum number of pairs that possibly can be created. The actual production rate is obtained by multiplying with the pair yield, Y, i.e., the fraction of injected power that is converted into pair rest mass. Balancing the rates gives the important relation (Guilbert, Fabian, and Rees 1983),

$$\tau_{\text{pair}} \sim (Y\ell_i)^{1/2} . \tag{5.9}$$

The pair yield is an increasing function of ℓ_i but reaches a constant value of about 10 % for saturated pair cascades (Svensson 1986). The rest of the injected power emerges as continuum X-ray radiation. Also the pair rest mass eventually emerges as radiation (in the form of annihilation photons). For saturated cascades, $\ell_i \gtrsim 40$, $\tau_{\text{pair}} \sim 2(\ell_i/40)^{1/2}$, which is greater than unity.

Compton downscattering on the optically thick thermal pairs causes a dip in the spectrum between $\sim m_ec^2/\tau$ and m_ec^2 (see Fig. 4c and §§3.1.3, 5.2). Compton upscattering by the thermal pairs is more important the smaller the ratio L_{bb}/L_i (see the section below). The blackbody and soft X-ray photons get upscattered into a thermally Comptonized power law for $\epsilon < kT$ and into a Wien hump at $\epsilon \sim$ a few kT (§3.1.3).

5.4. HIGHER ORDER COMPTON SCATTERINGS

A more complicated situation occurs when the radiation energy density is dominated by the nonthermal X-rays and not by the black body photons, i.e., $L_i > L_{bb}$ (see, e.g., Zdziarski and Lamb 1986). Then the particles will mainly cool on the X-rays. As a relativistic electron with Lorentz factor γ only loses energy efficiently when scattering photons in the Thomson limit, i.e., photons with energy $\epsilon < m_ec^2/\gamma$, the total radiation energy density, U_{rad}, in the cooling rate should be replaced with the radiation energy density below the energy m_ec^2/γ. Then there is an additional γ-dependence besides γ^2 in the cooling rate and the steady particle and photon distributions obtained above are no longer valid. Furthermore, when $L_X > L_{bb}$ higher order Compton scatterings are important. Although complicated in detail one can easily obtain an estimate of the X-ray spectral index. Let the Thomson optical depth of nonthermal particles near $\gamma \sim 1$ be τ. Then the nonthermal luminosity of photons that have scattered on electrons near $\gamma \sim 1$ is of the order τL_{bb}. These scattered photons hardly gain any energy and emerge near the energy of the blackbody photons, ϵ_{bb}. The injected particle power, L_i, emerges as radiation. Both due to pair production and due to the decline of the Compton scattering cross section at $\epsilon \sim m_ec^2$ for nonrelativistic and mildly relativistic particles one expects most of the luminosity, L_i, to be at $\epsilon \sim m_ec^2$. Then the nonthermal intensity index is simply estimated to be

$$\alpha_X \sim 1 - \frac{\ln L_i/\tau L_{bb}}{\ln m_ec^2/\epsilon_{bb}}, \tag{5.10}$$

where the term of unity corrects for the fact that the ratio $L_i/\tau L_{bb}$ corresponds to a ratio of integrated intensities and not to an intensity ratio. For L_i/L_{bb} sufficiently large α_X becomes smaller than 0.5. This was studied for a wide range of parameters by Done and Fabian (1989, see their Fig. 2).

When the compactness is large and $L_i \gg L_{bb}$ the optically thick *thermal* pairs will get heated to a temperature, T, at which the Compton parameter $y > 1$. Then upscattering by thermal pairs becomes important and a Wien hump appears in the X-ray spectrum at ϵ of a few kT (Zdziarski, Coppi and Lamb 1990).

5.5. SYNCHROTRON PAIR CASCADES

For large enough magnetic fields and low enough photon densities, the synchrotron cooling of nonthermal electrons dominates. If the photon spectrum extends beyond $m_e c^2$, $e^+ e^-$ pairs may be produced. Pair production attenuates the spectrum above $m_e c^2$ redistributing the attenuated power into the X-rays causing a softening of the spectrum below $m_e c^2$ (Stern 1985; Ghisellini 1989). For large enough ℓ, the cooled thermal pairs become optically thick and downscatter part of the synchrotron spectrum. A pair annihilation line would also appear.

Electrons below a certain energy, γ_{abs}, typically produce radiation that is self-absorbed by other low energy electrons. This reabsorption acts as a thermalizing mechanism. As a result a quasi-Maxwellian electron distribution forms below the critical energy, $\gamma_{abs} \sim (\epsilon_{abs}/\epsilon_B)^{1/2}$, of the particles radiating at the self-absorption energy, ϵ_{abs} (Ghisellini, Guilbert and Svensson 1988; de Kool, Begelman and Sikora 1989). The spectrum below the self-absorption frequency has approximately a Rayleigh-Jeans behavior with $\alpha = -2$.

The brightness temperature at the self-absorption frequency is expected to be $T_b \sim \gamma_{abs} m_e c^2 / k \sim 10^{11}$ K, as $\gamma_{abs} \sim 30$ in a large range of parameters typical for compact sources.

5.6. SUMMARY

For small compactness, $\ell_i < 1$, the emerging X-ray radiation has $\alpha_X = 0.5$. When pair cascades are important, i.e., for $\ell_i > 1$, the X-ray spectra steepens from $\alpha_X = 0.5$ to 1.

Spectral breaks or turnovers are introduced at,

i) $\epsilon \sim \gamma_{max}$, as photons can scatter no further;

ii) $\epsilon \sim \gamma_{max}^2 \epsilon_{bb}$, the maximum energy of once scattered black body photons with initial energy ϵ_{bb};

iii) $\epsilon \sim m_e c^2$ due to absorption in photon-photon pair production. If the compactness is not large enough this break occurs at the energy $\epsilon > m_e c^2$ where the absorption optical depth is unity;

iv) $\epsilon \sim m_e c^2 / \tau_{pair}^2$ due to down scattering on cool pairs if $\tau_{pair} > 1$.

It is important to note that a *thermal* plasma component of cooled pairs occurs in purely *nonthermal* models.

6. Stability of Pair Plasmas

It is important to know if the pair equilibria discussed in Section 4 are stable or not. If they are unstable for some parameter range this might explain some of the X-ray variability in compact X-ray sources.

The pair equilibrium studies considered two equations, the pair balance equation,

$$\frac{dn_+}{dt} = (\text{pair production}) - (\text{pair annihilation}) , \qquad (6.1)$$

and the photon balance equation,

$$\frac{dn(\epsilon)}{dt} = (\text{photon production}) - (\text{photon absorption}) - (\text{photon escape}). \qquad (6.2)$$

It was shown in these studies that one could solve for the relative fraction of pairs, n_+/n_p, and of photons, $n(\epsilon)/n_p$, in terms of two parameters, the 'proton' optical depth, $\tau_p = \sigma_T n_p R$ (i.e., the Thomson scattering optical depth due to the ionization electrons), and the electron and pair temperature, T_e.

When the timescale for the photon density to achieve steady state is shorter than the pair equilibrium time scale, then the stability properties during *isothermal* perturbations can be found by an inspection of the pair equilibrium curve in Figure 2. To the right of the pair equilibrium curve pairs are produced faster than they can annihilate and to the left of the curve annihilation dominates. Increasing the pair density slightly on the upper pair equilibrium branch causes the plasma to move into a state where further pairs are produced and a pair runaway occurs. Similarly, decreasing the pair density slightly leads to a situation with continued annihilation. The opposite conditions holds on the low pair density branch. The same conclusions can be reached in a more formal way. The pair production rate is independent of n_+ for $n_+ \ll n_p$ (the lower branch) and is $\propto n_\gamma^2 \propto (n_+^2)^2 \propto n_+^4$ for $n_+ \gg n_p$ (the upper branch). The pair annihilation rate is $\propto n_+$ for $n_+ \ll n_p$ and $\propto n_+ n_- \propto n_+^2$ for $n_+ \gg n_p$. The pair production rate increases much faster than the annihilation rate with increasing pair density when $n_+ \gg n_p$. The upper pair dominated branch is therefore unstable to *isothermal* pertubations.

However, it is unrealistic that the temperature, T_e, remains constant during perturbations. To know how T_e evolves requires the heating mechanism to be specified.

Hot ($T_e \gtrsim 10^9$ K) thermal plasmas occur in two-temperature accretion disks where the protons are heated by the dissipation mechanism driving the inflow. The protons then transfer their energy to the electrons in Coulomb collisions. By some efficient cooling mechanism the electrons lose this energy at the same rate as they are heated. If the cooling mechanism is efficient enough the electrons will be kept at a temperature much smaller than the proton temperature, T_p. The crucial assumption in the two-temperature models is that no plasma instability exists that is more efficient than Coulomb scattering in transferring energy from the protons to the electrons (see, however, Begelman and Chiueh 1988).

With this accretion scenario in mind, the stability of pair equilibria has been studied adding two energy equations, one for the electron temperature,

$$\frac{dT_e}{dt} = (\text{electron heating}) - (\text{electron cooling}), \qquad (6.3)$$

and one for the proton temperature,

$$\frac{dT_p}{dt} = (\text{proton heating}) - (\text{proton cooling}). \qquad (6.4)$$

Besides the 'proton' optical depth parameter, τ_p, there enters a second parameter related to the proton heating. Generally, one assumes a constant heating rate per unit volume (independent of the number of pairs present). The convenient second parameter turns out to be the compactness, ℓ, of the source.

The most detailed study so far is the linear stability analysis by Björnsson and Svensson (1989). They find the range of unstable parameters to be small ($10 \lesssim \ell \lesssim 20$ and $\tau_p \lesssim 2$). This study assumed the dominant radiation process to be Comptonization of bremsstrahlung photons and that other sources of low energy photons are lacking. Their treatment of the radiative transfer (a one-zone model) and of the Comptonization was, however, very simple.

Kusunose (1987) treated the radiation transfer in great detail. He made nonlinear perturbations (the protons were heated instantaneously) and the time-dependent evolution of the system was then followed numerically. In all cases where the system could be reliably followed it returned to equilibrium. Kusunose's choices of parameters ℓ and τ_p fall outside the unstable region of Björnsson and Svensson and the two studies are thus consistent.

These instability studies suffer from not being placed in a realistic accretion disk scenario where both the size of the region (i.e., the height, h, of the accretion disk) and the 'proton' optical depth (i.e., the surface density, $\Sigma = m_p n_p h$, of the disk) may evolve during a possible pair runaway.

7. Pair Production in Accretion Flows

7.1. THE MAXIMUM NUMBER OF PAIRS PER ACCRETED PROTON

Early considerations by Liang (1979) indicated that pairs may be important in accretion flows. A simple estimate (Svensson 1987) of the maximum number of nonthermal pairs per accreted proton onto a black hole is obtained as follows: With an accretion efficiency, η, of 10% there is about $0.1 m_p c^2 \sim 100$ MeV per proton available to make radiation and ultimately at most 100 pairs. A fraction say 50%, may end up as nonthermal radiation and another fraction $Y < 10\%$ of this becomes pairs, i.e., 5 pairs or 10 particles per proton. Formally, we can write

$$\left(\frac{2n_+}{n_p}\right)_{\text{max}} = \eta \frac{m_p}{m_e} \frac{L_{\text{nonthermal}}}{L_{\text{total}}} Y . \qquad (7.1)$$

For thermal plasmas the luminosity ratio is replaced by the corresponding thermal one, but the pair yield is harder to estimate.

7.2. PAIRS IN TWO-TEMPERATURE ACCRETION FLOWS

Kusunose and Takahara (1988) discovered that including the effects of pairs makes two-temperature accretion disks non-steady above some critical accretion rate $\dot{M}_{\text{cr}}$. Defining $\dot{m}$ as the dimensionless accretion rate $\dot{m} \equiv \dot{M} c^2 / L_{\text{Edd}}$ it was found in this and subsequent papers on the subject (Kusunose and Takahara 1989; White and Lightman 1989, 1990) that the *pair critical accretion rate*, $\dot{m}_{\text{cr}}$, increases from 0.3 to 1 as the viscosity parameter, α_{vis}, changes from 0.01 to 1.

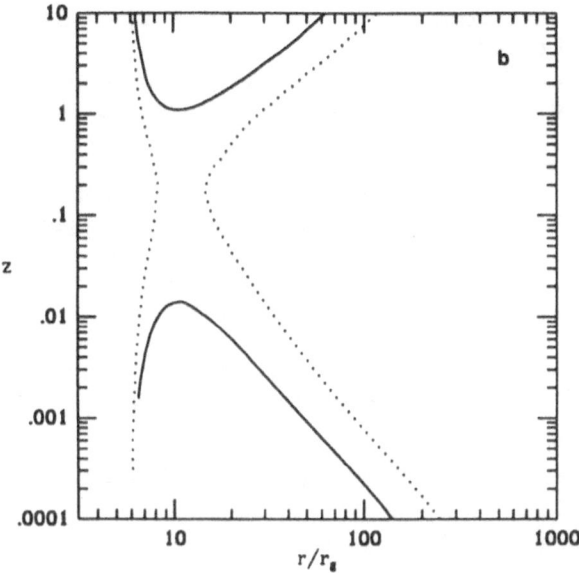

Figure 5. The pair ratio, $z = n_+/n_p$, as a function of the distance distance from the black hole for $\dot{m} = 0.3$ (sub pair-critical; *solid curve*) and $\dot{m} = 0.65$ (super pair-critical; *dotted curve*). From White and Lightman (1989).

Figure 5 shows the behavior of $z = n_+/n_p$ as a function of distance from the black hole for one sub pair-critical accretion rate ($\dot{m} = 0.3$, solid curve) and one super pair-critical accretion rate ($\dot{m} = 0.65$, dotted curve). The absence of steady solutions first occur at $r \approx 10GM/c^2$ for $\dot{m} > \dot{m}_{cr}$ as seen in the figure. The low pair density solution for $\dot{m} < \dot{m}_{cr}$ is just the standard two-temperature accretion with a small amount of pairs present. The high pair density solution appears unphysical as the proton temperature is found to exceed the virial temperature over a large range of radii.

The thermal radiation mechanism dominating near $\dot{m} \sim 1$ is Comptonization of soft photons. As noted by White and Lightman (1989) the studies made so far find a pair critical accretion rate when the soft photons are produced *internally*, i.e., by the same particles that do the Comptonization. These models could be called *soft self-Compton* models and include Comptonization of bremsstrahlung photons (Tritz and Tsuruta 1989; White and Lightman 1989) and Comptonization of thermal cyclotron photons (Kusunose and Takahara 1989). A pair critical accretion rate is not found when the soft photons are produced *externally* (Tritz and Tsuruta 1989; White and Lightman 1989). White and Lightman conclude that it is the nonlinearity introduced by the *internal* photon production that is the cause for the absence of steady solutions above the pair critical accretion rate.

Even when the solutions appear steady at all accretion rates, there is a critical accretion rate above which the protons will cool on an inflow time scale and the assumption of a hot steady two-temperature disk breaks down (Begelman, Sikora and Rees 1987, Tritz and Tsuruta 1989). Inside some *cooling radius* (that increases with $\dot{m}$) the protons will cool and the disk will collapse to a geometrically thin accretion disk. The presence of pairs makes the protons loose their energy at a faster rate and the cooling and the collapse occurs at larger radii and lower accretion rates.

Park and Ostriker (1989) also found two possible self-consistent solutions in spherical accretion flows at a given accretion rate for $\dot{m} > 3$ where radiation pressure and preheating due to Compton interactions must be accounted for. One solution is the standard low pair density, low luminosity accretion solution. In the other solution, a reasonable amount of pairs are produced ($n_+ \sim n_p$) that can drain the protons of their energy more efficiently and consequently radiate a larger luminosity. The larger density of photons accounts for the enhanced pair production.

7.3. PAIR-INDUCED COLLAPSE OF ACCRETION DISKS

Besides equations (6.1)–(6.4), White and Lightman (1990) included two additional time-evolving equations, the equation of motion for the disk thickness, h,

$$\frac{d^2 h}{dt^2} = (\text{pressure forces}) - (\text{gravitational forces}), \tag{7.2}$$

and the conservation of mass for the surface density, Σ,

$$\frac{d\Sigma}{dt} = (\text{mass inflow}) - (\text{mass outflow}), \tag{7.3}$$

and followed the evolution of the two-temperature accretion disk starting with the steady solutions discussed above. They found that at an accretion rate slightly less than $\dot{m}_{\rm cr}$ the low pair density solution was stable while the high pair density solution had a pair runaway that drained the protons of their energy resulting in a collapse of the disk. They found a similar behavior for the low pair density solution at an accretion rate marginally above $\dot{m}_{\rm cr}$.

A different time-dependent treatment considering phenomenological spherical accretion was made by Moskalik and Sikora (1986). They included equations (6.1)-(6.3) and a mass conservation equation similar to equation (7.3). The proton temperature was kept constant and the mass inflow into the plasma region was assumed to be constant while the outflow (interpreted as inflow into a black hole) took place on the proton cooling time scale. A flaring limit cycle behavior of the luminosity was found for $\ell \approx 10$.

8. Pair Plasmas in Compact Objects

8.1. ACTIVE GALACTIC NUCLEI (AGNs)

Great efforts (e.g., Fabian et al. 1986; Zdziarski 1986b; Lightman and Zdziarski 1987; Sikora et al 1987; Dermer 1988; Done and Fabian 1989; Done, Ghisellini and Fabian 1990) have been made to explain some or all of the following three properties of X-ray spectra from AGNs:

(i) the spectral index $\alpha_{2-10{\rm keV}}$ ranges from about 0.6 to 0.8 (see compilation in Done and Fabian 1989),

(ii) the compactness $\ell_{2-10\text{keV}} \sim 0.1 - 10$ (from observed luminosity and time variability, see compilation in Done and Fabian 1989),

(iii) the MeV-break required by observations of the cosmic X-ray background.

Lightman and Zdziarski (1987) and Done and Fabian (1989) conclude that Compton-cooled pair cascades initiated by monoenergetic injection cannot account for these properties.

A good fit (Svensson and Zdziarski, unpublished) is obtained by having most electron power injected at Lorentz factors of 200, allowing EUV photons from the UV-bump to be scattered no further then to about 1 MeV with little resulting pair production. The power law injection below Lorentz factors of 200 must have $\Gamma = 2\alpha_{2-10\text{keV}}$ (see eq. [5.6]). Thus strong constraints are placed on the particle injection and the ultimate answer to why the X-ray spectra behave as they do may lie in the acceleration mechanism, not the radiation mechanism.

Dermer (1988) suggests that a two-temperature thermal model where the hot protons besides heating the electrons also produce pions that decay into nonthermal electrons and γ-rays. The nonthermal electrons provide the soft photons that the thermal electrons Comptonize into the X-rays. The calculated X-ray compactnesses and spectral indices agree with the the observed ones.

8.2. THE COSMIC X-RAY BACKGROUND

Leiter and Boldt (1982) found that subtracting the contributions by AGNs with typical X-ray spectral index $\alpha_X \approx 0.7$ gives a flat residual background spectrum with $\alpha_X \approx 0$ and an exponential roll-over at 25 keV. This spectrum, containing most of the energy density of the background, has a spectral shape different from known possible sources. Leiter and Boldt (1982) and Boldt and Leiter (1987) proposed the existence of a class of high-redshift ($z \approx 5$) unresolved precursor AGNs, whose spectra have a thermal character with $kT \approx 25(1 + z)$ keV ≈ 150 keV. These objects would evolve from large to small compactnesses undergoing a change to the present-day nonthermal ($\alpha_X \approx 0.7$) spectrum when the compactness dropped below unity. Zdziarski (1988) investigated spectra from thermal Comptonization of cyclotron photons in spherical accretion flows at large compactnesses and large redshifts finding spectral shapes and temperatures in agreement with the residual spectrum.

Quite a different nonthermal scenario was suggested by Fabian, Done, and Ghisellini (1988). As more and more pairs are produced and share the available released power in a source, the individual particle energy after acceleration must decrease. This *pair loading* stops when the typical radiated (e.g., Compton scattered) photon has an energy of about 500 keV then being unable to pair produce. The objects (also precursor AGNs) must be at $z \approx 10$–30 in order for the MeV break to redshift to 25 keV.

8.3. GAMMA RAY BURSTS

Risetimes of a few ms, observed fluxes, and distances greater than 20 pc imply compactnesses greater than unity and pairs are expected to be produced (see, e.g., Zdziarski 1987). The flat X-ray spectra ($\alpha_X \approx 0$) and the steeper γ-ray spectra ($\alpha_X \approx 0.5 - 2$) above the MeV break cannot be explained by thermal models. Zdziarski and Lamb (1986) considered repeated Compton scatterings in a photon starved source (i.e., a source with few available soft photons) and got reasonable spectral fits.

9.   Summary

The main points in this review are:
  (i) Plasmas are heated to relativistic temperatures in compact astrophysical objects.
  (ii) Pairs are expected when the X-ray compactness ℓ is greater than unity.
  (iii) Inclusion of pair effects allows a *large* variety of spectral shapes with little predictive power in the steady models considered so far.
  (iv) Even so, it is hard to explain X-ray and γ-ray spectra from compact objects in detail with the *steady models*.
  (v) Most effort has been spent on studying radiation (two-body) processes in relativistic plasmas. Little is known about the importance of plasma processes and collective effects.

10.   References

Baring, M. (1987) 'The effect of cooling processes on the tail of a relativistic thermal pair distribution', *M. N. R. A. S.*, **228**, 695–712.

Begelman, M. C., Blandford, R. D., and Rees, M. J. (1984) 'Theory of extragalactic radio sources ', *Rev. Mod. Phys.*, **56**, 255–351.

Begelman, M. C., Chiueh, T. (1988) 'Thermal coupling of ions and electrons by collective effects in two-temperature accretion flows', *Ap. J.*, **332**, 872–890.

Begelman, M. C., Sikora, M., and Rees, M. J. (1987) 'Thermal and dynamical effects of pair production on two-temperature accretion flows', *Ap. J.*, **313**, 689–698.

Bisnovatyi-Kogan, G. S., Zel'dovich, Ya. B., and Sunyaev, R. A. (1971) 'Physical processes in a low-density relativistic plasma', *Sov. Astr.*, **15**, 17–22.

Björnsson, G. and Svensson, R. (1989) 'The stability of electron-positron pair equilibria and the time variability in active galactic nuclei', preprint.

Boldt, E., and Leiter, D. (1987) 'Constraints on possible precursor AGN sources of the cosmic X-ray background', *Ap. J. (Letters)*, **322**, L1–L4.

Bonometto, S., and Rees, M. J. (1971) 'On possible observable effects of electron pair production in QSOs', *M. N. R. A. S.*, **152**, 21–35.

Cavaliere, A., and Morrison, P. (1980) 'Extreme nonthermal radiation from active galactic nuclei', *Ap. J. (Letters)*, **238**, L63–L66.

Coppi, P. S. (1990) 'Time-dependent models of magnetized pair plasmas', preprint.

de Kool, M., Begelman, M. C., and Sikora, M. (1989) 'Self-absorbed synchrotron sources in active galactic nuclei', *Ap. J.*, **337**, 66–77.

Dermer, C. D. (1984) 'The production spectrum of a relativistic Maxwell-Boltzmann gas', *Ap. J.*, **280**, 328–333.

Dermer, C. D. (1988) 'Hot ion model for the X-ray spectra of active galactic nuclei', *Ap. J. (Letters)*, **335**, L5–L8.

Dermer, C. D., and Liang, E. P. (1989) 'Electron thermalization and heating in relativistic plasmas', *Ap. J.*, **339**, 512–528.

Done C., and Fabian, A. C. (1989) 'The behavior of compact nonthermal sources with pair production', *M. N. R. A. S.*, **240**, 81–102.

Done C., Ghisellini, G., and Fabian, A. C. (1990) 'Pair loading in compact sources', *M. N. R. A. S.*, submitted.

Fabian, A. C. (1984) 'The origin of the X-ray background', in *X-Ray and UV-emission from active galactic nuclei*, eds. W. Brinkmann and J. Trumper, MPE Report 184, Garching, pp. 232–242.

Fabian, A. C., Blandford, R. D., Guilbert, P. W., Phinney, E. S., and Cuellar, L. (1986) 'Pair-induced spectral changes and variability in compact X-ray sources', *M. N. R. A. S.*, **221**, 931–945.

Fabian, A. C., Done, C., and Ghisellini, G. (1988) 'Photon-starved, pair-loaded compact sources and the X-ray background', *M. N. R. A. S.*, **232**, 21p–26p.

Ghisellini, G. (1989) 'Pair production in steady synchrotron self Compton models', *M. N. R. A. S.*, **238**, 449–479.

Ghisellini, G., Guilbert, P. W., and Svensson, R. (1988) 'The synchrotron boiler', *Ap. J. (Letters)*, **334**, L5–L8.

Górecki, A., and Wilczewski, W. (1984) 'A study of Comptonization of radiation in an electron plasma using the Monts Carlo method', *Acta Astr.*, **34**, 141–160.

Guilbert, P. W., Fabian, A. C., and Rees, M. (1983) 'Spectral and variability constraints on compact sources', *M. N. R. A. S.*, **205**, 593–603.

Guilbert, P. W., and Stepney, S. (1985) 'Pair production, Comptonization and dynamics in astrophysical plasmas', *M. N. R. A. S.*, **212**, 523–544.

Herterich, K. (1974) 'Absorption of gamma rays in intense X-ray sources', *Nature*, **250**, 311–313.

Jelley, J. V. (1966) 'Absorption of high-energy gamma-rays within quasars and other radio sources', *Nature*, **211**, 472–475.

Kazanas, D. (1984) 'Photon-photon absorption and the uniqueness of the spectra of active galactic nuclei', *Ap. J.*, **287**, 112–115.

Kusunose, M. (1987) 'Relativistic thermal plasmas: time development of electron-positron pair concentration', *Ap. J.*, **321**, 186–198.

Kusunose, M., and Takahara, F. (1983) 'Pair equilibrium in a relativistic plasma with magnetic field', *Progr. Theor. Phys.*, **69**, 1443–1456.

Kusunose, M., and Takahara, F. (1985) 'Electron-positron pair equilibrium in a mildly relativistic plasma', *Progr. Theor. Phys.*, **73**, 41–53.

Kusunose, M., and Takahara, F. (1988) 'Two-temperature accretion disks with electron positron pair production', *Publ. Astr. Soc. Japan*, **40**, 435–448.

Kusunose, M., and Takahara, F. (1989) 'Two-temperature accretion disks in pair equilibrium: effects of unsaturated Comptonization of soft photons', *Publ. Astr. Soc. Japan*, in press.

Leiter, D., and Boldt, E. (1982) 'Spectral evolution of active galactic nuclei: a unified description of the X-ray and gamma-ray backgrounds', *Ap. J.*, **260**, 1–19.

Liang, E.P. (1979) 'Electron-positron pair production in hot unsaturated Compton accretion models around black holes', *Ap. J.*, **234**, 1105–1112.

Lightman, A. P. (1982) 'Relativistic thermal plasmas: pair processes and equilibria', *Ap. J.*, **253**, 842–858.

Lightman, A. P., Giacconi, R., and Tananbaum, H. (1978) 'X-ray flares in NGC 4151:

a thermal model and constraints on a black hole', *Ap. J.*, **224**, 375–380.

Lightman, A. P., and Zdziarski, A. A. (1987) 'Pair production and Compton scattering in compact sources and comparison to observations of active galactic nuclei', *Ap. J.*, **319**, 643–661.

Moskalik, P., and Sikora, M. (1986) 'Pair production instabilities as a source of X-ray flares from accreting black holes', *Nature*, **319**, 649–652.

Park, M.-G., and Ostriker, J. P. (1989) 'Spherical accretion onto black holes: a new higher efficiency type of solution with significant pair production', *Ap. J.*, **347**, in press.

Pozdnyakov, L. A., Sobol', I. M., and Sunyaev, R. A. (1983) 'Comptonization and the shaping of X-ray source spectra: Monte Carlo calculations', in *Soviet Scientific Review*, Section E2, Harwood, London, pp. 189–331.

Rees, M. J. (1984) 'Black hole models for active galactic nuclei', *Ann. Rev. Astr. Ap.*, **22**, 471–506.

Rees, M. J., Begelman, M. C., Blandford, R. D., and Phinney, E. S. (1982) 'Ion-supported tori and the origin of radio jets', *Nature*, **295**, 17–21.

Rybicki, G. R., and Lightman, A. P. (1979) *Radiative Processes in Astrophysics*, McGraw-Hill, New York.

Shakura, N. I., and Sunyaev, R. A. (1973) *Astr. Ap.*, **24**, 337.

Shapiro, S. L., Lightman, A. P., and Eardley, D. M. (1976) 'A two-temperature accretion disk model for Cygnus X-1: structure and spectrum', *Ap. J.*, **204**, 187–199.

Sikora, M., Kirk, J. G., Begelman, M. C., and Schneider, P. (1987) 'Electron injection by relativistic protons in active galactic nuclei', *Ap. J. (Letters)*, **320**, L81–L85.

Sikora, M., and Zbyszewska, M. (1986) 'A stability analysis of electron-positron pair equilibria of a two-temperature plasma cloud', *Acta Astr.*, **36**, 255–273.

Stern, B. E. (1985) 'On the possibility of efficient production of electron-positron pairs near pulsars and accreting black holes', *Sov. Astr.*, **29**, 306–313.

Sunyaev, R. A., and Titarchuk, L. G. (1980) 'Comptonization of X-rays in plasma clouds. Typical radiation spectra.', *Astr. Ap.*, **86**, 121–138.

Svensson, R. (1982) 'Electron-positron pair equilibria in relativistic plasmas', *Ap. J.*, **258**, 335–348.

Svensson, R. (1983) 'The thermal pair annihilation spectrum: a detailed balance approach', *Ap. J.*, **270**, 300–304.

Svensson, R. (1984) 'Steady mildly relativistic thermal plasmas: processes and properties', *M. N. R. A. S.*, **209**, 175–208.

Svensson, R. (1986) 'Physical processes in active galactic nuclei", in *Radiation Hydrodynamics in Stars and Compact Objects*, IAU Coll. 89, eds. D. Mihalas and K.-H. Winkler, Springer, New York, pp. 325–345.

Svensson, R. (1987) 'Nonthermal pair production in compact X-ray sources: first order Compton cascades in soft radiation fields', *M. N. R. A. S.*, **227**, 403–451.

Svensson, R., and Zdziarski, A. A. (1989) 'An introduction to pair plasmas in astrophysics', in *STScI-GSFC Workshop on Ultrahot Plasmas and Electron-Positron Pairs in Astrophysics*, eds. A. A. Zdziarski and D. Kazanas, STScI Publ., Baltimore, pp. 1-22.

Takahara, F. (1988) 'Time development of relativistic thermal plasmas', *Publ. Astr.*

Soc. Japan, **40**, 499–510.

Tritz, B., and Tsuruta, S. (1989) 'Effects of electron-positron pairs on accretion flows', *Ap. J.*, **340**, 203–215.

White, T., and Lightman, A. P. (1989) 'Hot accretion disks with electron-positron pairs', *Ap. J.*, **340**, 1024–1037.

White, T., and Lightman, A. P. (1990) 'Instabilities and time evolution of hot accretion disks with electron-positron pairs', *Ap. J.*, in press.

Zdziarski, A. A. (1984) 'Spectra from pair-equilibrium plasmas', *Ap. J.*, **283**, 842–847.

Zdziarski, A. A. (1985) 'Power-law X-ray and gamma-ray emission from relativistic thermal plasmas', *Ap. J.*, **289**, 514–525.

Zdziarski, A. A. (1986a) 'Monte Carlo spectra from pair-equilibrium, weakly magnetized thermal plasmas', *Ap. J.*, **303**, 94–100.

Zdziarski, A. A. (1986b) 'On the origin of the infrared and X-ray continua of active galactic nuclei', *Ap. J.*, **305**, 45–56.

Zdziarski, A. A. (1987) 'Physics of continuum spectra of gamma-ray bursts', in *Thirteenth Texas Symposium on Relativistic Astrophysics*, ed. M. Ulmer, World Scientific, Singapore, pp. 553–562.

Zdziarski, A. A. (1988) 'Steady state thermal Comptonization in compact sources and the cosmic X-ray background', *M. N. R. A. S.*, **233**, 739–758.

Zdziarski, A. A., and Lamb, D. Q. (1986) 'Gamma-ray burst spectra from photon-deficient Compton scattering by nonthermal electrons', *Ap. J. (Letters)*, **309**, L79–L82.

Zdziarski, A. A., and Lightman, A. P. (1985) 'Nonthermal electron-positron pair production and the "universal" X-ray spectrum of active galactic nuclei', *Ap. J. (Letters)*, **294**, L79–L83.

Zdziarski, A. A., and Lightman, A. P. (1987) 'Relativistic plasmas in active galactic nuclei', in *Variability in Galactic and Extragalactic X-Ray Sources*, ed. A. Treves, Associazione per l'avanzamento dell'astronomia, Milano, pp. 121–135.

Zdziarski, A. A., Coppi, P. S., and Lamb, D. Q. (1990) 'Physical processes in photon-starved nonthermal pair plasmas', preprint.

X-RAYS AND GAMMA-RAYS AT COSMOLOGICAL DISTANCES

ANDRZEJ A. ZDZIARSKI
Space Telescope Science Institute
3700 San Martin Dr.
Baltimore, MD 21218, USA
(BITNET: *zdziarski@stsci*)

ROLAND SVENSSON
NORDITA
Blegdamsvej 17
DK-2100 Copenhagen Ø, Denmark
(BITNET: *svensson@dknbi51*)

ABSTRACT. Our recent work on absorption and reprocessing of γ-rays at cosmological redshifts as well as the current status of this field of research are discussed. We consider Compton scattering and pair production by γ-rays on cosmic baryonic matter, and photon-photon scattering and photon-photon pair production by γ-rays and Compton scattering of relativistic pairs on the cosmic blackbody background. We point out the cosmological importance of photon-photon scattering. The region where the universe is transparent to γ-rays and the regions of dominance of the elementary processes in the photon–energy-redshift plane are determined. The problem of cosmological γ-ray reprocessing is relevant, e.g., for observational γ-ray astronomy, for which the signatures of the cosmological origin of discrete and background sources need to be determined, and for studies of the effects that pair cascades, caused by the decay of unstable particles from a hot Big Bang, have on the primordial nucleosynthesis and on the shape of the cosmic microwave background spectrum.

1. Introduction

This paper discusses the current status of the research on physical processes occurring during propagation of γ-rays at cosmological redshifts. These physical processes are interactions with the cosmic thermal background and baryonic matter, which change the spectra and directions of the originally emitted γ-rays.

The universe is transparent up to a redshift of $z \sim 10^3$ in the photon energy range from ~ 100 keV to ~ 100 MeV, and becomes opaque at $z \ll 1$ only above ~ 100 TeV (e.g., Arons and McCray 1969; Stecker 1971, 1975). Thus, it is possible that some γ-rays originating at very high redshifts reach the earth. Studying the physical processes that reprocess γ-rays at cosmological redshifts may lead to finding a way of distinguishing cosmological γ-rays from local ones. This reprocessing may lead to the formation of universal spectra and time profiles of the observable γ-rays. Studies of such effects appear particularly important with the advent of the Gamma Ray Observatory, scheduled for launch in 1990. Possible discrete

W. Brinkmann et al. (eds.), Physical Processes in Hot Cosmic Plasmas, 383–394.
© *1990 Kluwer Academic Publishers.*

sources of cosmological γ-rays include superconducting cosmic strings (Babul, Paczyński and Spergel 1987; Paczyński 1988; Vilenkin 1988).

The cosmic γ-ray background may also be due to, e.g., matter-antimatter annihilation at cosmological distances (e.g., Stecker 1975; Montmerle 1975; Desért and Schatzman 1986), or superconducting strings (Thompson and Madau 1989). The possible distortion of the cosmic blackbody background (Matsumoto et al. 1988) may be also of cosmological origin and related to cosmological X-rays and γ-rays (Stone and Field 1988; Fukugita, Kawasaki, and Yanagida 1989; Field and Walker 1989; Thompson and Madau 1989).

Reprocessing of γ-rays may also be important at redshifts at which the reprocessed radiation is not directly observable. Long-lived unstable particles (e.g., gravitinos, heavy neutrinos) have been predicted in some theories to be produced in the initial stages of the hot Big Bang. The particles decay at some redshift giving rise to a large number of γ-rays (Ellis, Nanopoulos and Sarkar 1985; Juszkiewicz, Silk and Stebbins 1985; Audouze, Lindley and Silk 1985; Lindley 1985; Domínguez-Tenreiro 1987; Kawasaki and Sato 1987; Dimopoulos et al. 1988), e.g., from the decay of neutral pions, or from Compton scattering by relativistic electrons from charged muon decay. The γ-rays interact mainly with the cosmic blackbody background producing e^+e^- pairs, giving rise to electromagnetic cascades with alternating generations of γ-rays and e^+e^- pairs. If this process occurs at $z \sim 10^6$–10^7, the reprocessed photons have energies in the MeV range and are most effective at photo-dissociating light elements (e.g., deuterium). Thus, this process affects the primordial abundances. Also, the energy release associated with the decay affects the rate of universal expansion.

When studying the reprocessing of γ-rays, only photons and baryonic matter are of interest. The photon temperature is proportional to $(1 + z)$,

$$\Theta = \Theta_0(1 + z) \simeq 4.55 \times 10^{-10} T_{2.7}(1 + z), \tag{1}$$

where $\Theta \equiv kT/m_e c^2$ is the dimensionless temperature, m_e is the electron mass, and $T_{2.7} = T_0/2.7\,\text{K}$. Hereafter, the subscript 0 refers to the quantities at $z = 0$. The photon distribution is given by the Planck law, with the photon density $\propto (1 + z)^3$. The temperature of matter equals that of equation (1) at $z \gtrsim 10^3$, i.e., before the recombination epoch. Up to $z \sim 10^8$, the matter is nonrelativistic, and its temperature does not affect the reprocessing of γ-rays. The average electron density is

$$n_e = n_{e,0}(1 + z)^3 = 2.46 \times 10^{-7} \Omega_{0.1} h_{50}^2 \ \text{cm}^{-3}(1 + z)^3, \tag{2}$$

where $\Omega_{0.1} = (\Omega_b/0.1)$, and Ω_b is the fraction of the closure density of the universe in baryons. The densities of hydrogen and helium are given by $n_{\text{H}} = (6/7)n_e$, and $n_{\text{He}} = (1/14)n_e$, assuming the mass fraction of He of 25%.

The photon energy evolves with the redshift due to the universal expansion,

$$\epsilon = \epsilon_0(1 + z), \tag{3}$$

where $\epsilon \equiv E/m_e c^2$ is the dimensionless photon energy.

High energy photons in the presence of blackbody photons and cold matter interact predominantly in the ways listed in Table 1. The table also gives the regions of dominance of the processes in the photon-energy–redshift plane, see § 2 and Figure 2 below.

The optical depth of the Universe to an interaction process from a matter-dominated redshift z to $z = 0$ is given by an integral (e.g., Weinberg 1972) over the interaction probability per unit length, $d\tau/dl$,

$$\tau(\epsilon_0, z) = \frac{c}{H_0} \int_0^z \frac{d\tau/dl}{(1+z')^2(1+\Omega z')^{1/2}} dz',$$ (4)

where Ω is the ratio of the total density to the density required to close the universe, c/H_0 is the present Hubble length,

$$\frac{c}{H_0} = 1.85 \times 10^{28} h_{50}^{-1} \text{ cm},$$ (5)

$h_{50} \equiv H_0/(50 \text{ km s}^{-1}\text{Mpc}^{-1})$, and H_0 is the Hubble parameter. If the major contribution to integral (4) comes from the high-redshift limit, Zdziarski and Svensson (1989, hereafter abbreviated as ZS89) showed that

$$\tau(\Omega, \Omega z \gg 1) \simeq \Omega^{-1/2} \tau(\Omega = 1),$$ (6)

implying that in the high-redshift limit it suffices to calculate τ in the $\Omega = 1$ case only.

TABLE 1. INTERACTION PROCESSES FOR γ-RAYS

Number	Name of the process	Reaction	Region
1	Photoionization of atoms	$\gamma A \rightarrow A^+ e^-$	II, IIa
2	Compton scattering on electrons	$\gamma e \rightarrow \gamma e$	III, IIa, Ia
3	Production of pairs on atoms	$\gamma A \rightarrow A e^+ e^-$	IV
4	Production of pairs on free electrons	$\gamma e \rightarrow e e^+ e^-$	V
5	Production of pairs on free nuclei	$\gamma Z \rightarrow Z e^+ e^-$	V
6	Scattering on blackbody photons	$\gamma\gamma \rightarrow \gamma\gamma$	VI
7	Production of single pairs on photons	$\gamma\gamma \rightarrow e^+ e^-$	VIIa, b
8	Production of double pairs on photons	$\gamma\gamma \rightarrow e^+ e^- e^+ e^-$	VIII

2. Regions of Dominance of the Elementary Processes

ZS89 have calculated the optical depths of the universe to the processes listed in Table 1. The various dotted and dashed lines in Figure 1 (from ZS89) show contours of unit optical depths to the processes of Table 1 as functions of the observed photon energy, ϵ_0. The lower solid curve gives z corresponding to the total optical depth for absorption and scattering equal to unity. The upper solid curve corresponds to the total absorption and energy loss optical depth equal to unity. Below the lower solid curve (region I), unreprocessed radiation

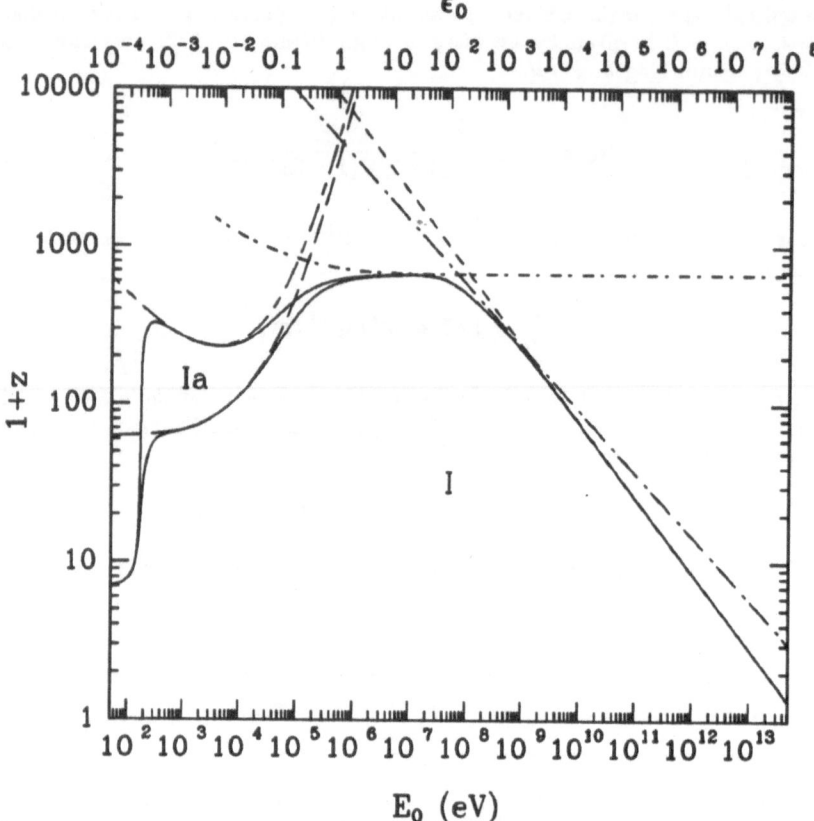

Figure 1. The contours of unit optical depth at matter dominated redshifts for photon-photon pair production (short dashes), photon-photon scattering (dots and long dashes), photon-matter pair production (dots and short dashes), Compton scattering (long dashes), and Compton energy loss (short and long dashes). The lower solid curve corresponds to a combined scattering and absorption optical depth of unity, while the upper solid curve corresponds to a combined energy loss and absorption optical depth of unity. The contour of unit optical depth for photoionization is not shown, but its opacity is included in the (solid) curves corresponding to the total optical depth of unity. Those two solid curves are important for discrete and diffuse sources, respectively.

from discrete sources can be observed. Below the upper solid curve (the union of regions Ia and I), radiation from sources contributing to the isotropic background can reach the earth. The assumed cosmological parameters are $\Omega = \Omega_{0.1} = h_{50} = 1$.

Below an energy of ~ 200 eV, the universe is optically thick to photoionization of helium beyond some reionization redshift, assumed equal to 6 in the calculations of Figure 1 (ZS89). In a relatively narrow range of energy of 200 eV $\lesssim E \lesssim 500$ eV, the unit optical depth to scattering and absorption is due to Thomson scattering and the former is reached at $z = 62$ for our choice of cosmological parameters (see eq. [11] in § 3.4 below). Above this energy, the curve of $\tau = 1$ turns upward, reflecting the Klein-Nishina reduction of the Compton cross section. Around ~ 100 keV, pair production on atoms (see § 3.3) becomes the dominant source of opacity. From ~ 100 keV to ~ 300 MeV, there is a γ-ray window of reduced opacity, as pointed out by Arons and McCray (1969) and Stecker (1975). Photon-photon scattering dominates the opacity from ~ 50 MeV to ~ 1 GeV (see § 3.2; ZS89; Svensson and Zdziarski 1990). Above ~ 1 GeV, the main source of opacity is photon-photon pair production (see § 3.1). Above ~ 100 TeV, this process causes the universe to be opaque locally.

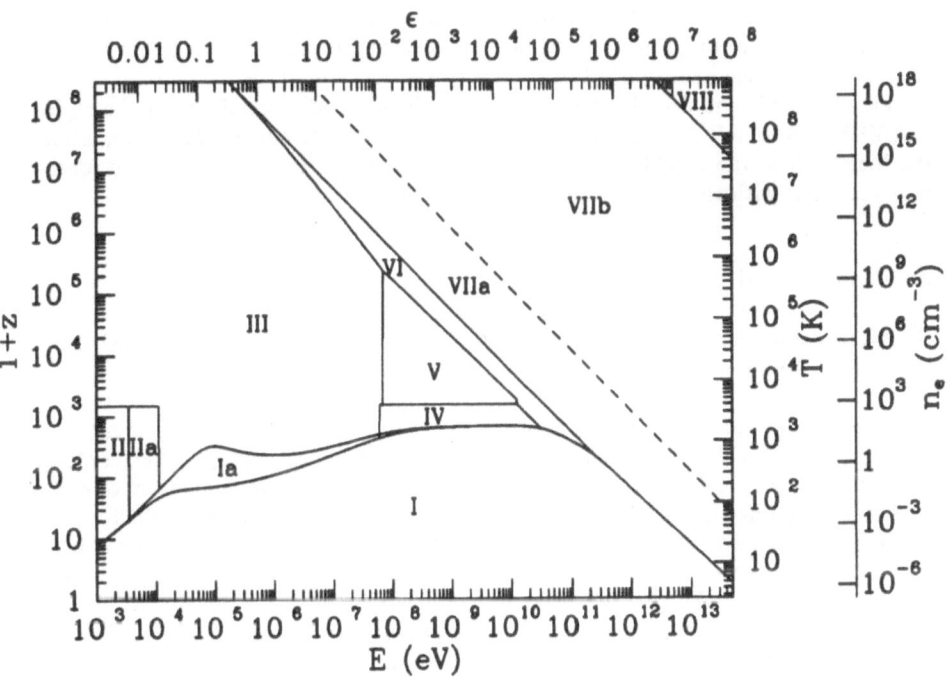

Figure 2. The plane ϵ–z divided into regions of dominance of the various absorption and scattering processes. The regions labeled from I to VIII correspond to: I, τ(scattering and absorption) < 1; I+Ia, τ(energy loss and absorption) < 1; II, photoionization dominant over Compton scattering; II+IIa, photoionization dominant over Compton energy loss; III, dominant Compton scattering; IV, pair production on atoms; V, pair production on ions and free electrons; VI, photon-photon scattering; VIIa, b, single photon-photon pair production; VIII, double pair production. The right-hand axes give the radiation temperature and electron density, of the universe.

Figure 2 (from ZS89) shows the regions in the ϵ–z plane in which the various radiative processes discussed in § 2 dominate. Here ϵ is the energy measured at z. The cosmological parameters are the same as in Figure 1. The bottom solid and dashed curves give the contours of unit optical depth to combined scattering and absorption, and energy loss and absorption, respectively. From the regions below those curves (region I and the union of I and Ia, respectively) the radiation emitted by discrete and diffuse sources, respectively, can be received at the earth. Regions II and the union of II and IIa correspond to the dominance of He photoionization, for discrete and diffuse sources, respectively. Compton downscattering dominates in region III. Radiation emitted in regions IV and V will be absorbed in pair-producing interactions with baryonic matter in neutral and ionized state, respectively. The boundary at $z = 1500$ corresponds to the recombination epoch.

Photon-photon scattering dominates in region VI. For $\Omega_b = 0.1$ this region extends as a diagonal strip for $300 \lesssim z \lesssim 10^8$, with the widths of the strip in ϵ and z being as large as a factor of about 4. For $\Omega_b = 0.01$, the maximum widths of the diagonal region would increase to a factor of 10 in both ϵ and z. Note that this process can be important down to relatively low energies of $\epsilon \sim 1$.

Finally, single and double photon-photon pair production dominate in the regions VII and VIII, respectively. The region VII is further divided into one in which interactions with the Wien tail of the blackbody distribution dominate (VIIa) and one where most of the interactions is with the peak of the blackbody distribution (VIIb).

3. Cosmological γ-Ray Cascades

3.1. PAIR-PHOTON CASCADES FROM PHOTON-PHOTON PAIR PRODUCTION AND COMPTON UPSCATTERING

Gamma-rays can interact with photons of the cosmic blackbody background producing positron-electron pairs. In the center-of-momentum reference frame, the threshold for pair production corresponds to $\epsilon_{CM} = 1$. The cross section reaches its maximum of $\sim 0.3\sigma_T$ at $\epsilon_{CM} \approx 1.4$, and decreases like $\sim 1/\epsilon$ far above the threshold. Here $\sigma_T \simeq 6.65 \times 10^{-25}$ cm^2 is the Thomson cross section. Because of the very large density of the cosmic blackbody photons, absorption of γ-rays occurs even at such low energies that pairs are produced only in interactions with the photons deep in exponential tail of the blackbody distribution. Therefore, the condition for the optical depth of unity corresponds to $\epsilon \times (10^1-10^2)\Theta = 1$, with only a weak dependence on the redshift and cosmological parameters. An approximate expression for the optical depth was derived by Fazio and Stecker (1970). More exact expressions were derived by ZS89. The optical depth of unity corresponds approximately to (ZS89; Fazio and Stecker 1970)

$$1 + z \simeq 8.8 \times 10^3 \epsilon^{0.485} , \qquad (7)$$

at $1.1 \leq z \leq 300$ and $h_{50} = \Omega = 1$. The proximity of the exponent to $1/2$ reflects the effect of pair absorption by the exponential blackbody tail, discussed above.

At $z \ll 1$ the universe becomes opaque at energies $\gtrsim 100$ TeV. At $E \simeq 1$ PeV, which corresponds to the peak of the cross section, the mean free path is as short as $\sim 8 T_{2.7}^{-3}$ kpc. Thus, this process is important both locally and cosmologically.

The cross section for pair production decreases with energy as $1/\epsilon$. Thus, one would expect that at some very high γ-ray energy the universe would again become transparent to photon-photon pair production. However, at a lower energy the process of double pair production becomes dominant, causing the universe to be opaque at all above-threshold energies (Brown, Mikaelian, and Gould 1973). The cross section for double pair production, $\sim \alpha_f{}^2 \sigma_T$, is independent of energy. Here α_f is the fine structure constant. Absorption of γ-rays by double pair production dominates over single pair production at $\epsilon \gtrsim 6.7 \times 10^5/\Theta$ (see Fig. 2), and the optical depth of unity at those energies corresponds to (ZS89, Brown, Mikaelian and Gould 1973),

$$z \simeq 0.021 h_{50} T_{2.7}{}^{-3} . \tag{8}$$

Photon-photon pair production on soft (blackbody) photons produces electrons and positrons with energies comparable to the γ-ray energy. The electrons and positrons immediately scatter blackbody photons, as the cross sections for pair production and Compton scattering are similar. Close to the threshold for pair production, there is an approximate energy equipartition among the members of the pair, and at energies far above the threshold one member of the pair receives most of the energy. Scattering of the electron (positron) that received most of the γ-ray energy is now in the Klein-Nishina limit, in which the up-scattered photon receives most of the electron energy. Thus, consecutive pair production and Compton scattering produce a γ-ray with energy only slightly less than the energy of the original γ-ray. This means that a γ-ray with energy far above the threshold will initiate an electromagnetic cascade consisting of many generations of e^+e^- pairs and γ-rays. The cascade continues until all γ-rays have energies below the threshold for pair production. When this happens, the electrons still continue losing energy through Compton scatterings, producing photons with energies decreasing with decreasing electron energy.

The process of pair cascades has been simulated by means of Monte Carlo methods (Aharonian, Kirillov-Ugryumov and Vardanian 1985; Protheroe 1986; Domínguez-Tenreiro 1987). Recently, Zdziarski (1988) has reconsidered the problem and derived analytical expressions for the redistribution functions for pair production and Compton scattering on blackbody photons. The problem then becomes much more tractable. The reprocessed spectra can be obtained by iteratively solving the kinetic equation describing the cascade. As the mean free path for either pair production or Compton scattering is many orders of magnitude shorter than the local Hubble length, the cascade process can be considered without including the effects of the cosmological expansion on the energies and occupation numbers.

Figure 3 shows examples of reprocessing of isotropic monoenergetic γ-rays by pair cascades on blackbody photons (Zdziarski 1988). The primary γ-rays are injected at energies 10 and 100 times larger than the absorption threshold energy, ϵ_{th}, which was assumed to correspond to $1/(30\Theta)$. One sees that the resulting spectra do not depend on the injection energy. Also, they are very similar for electrons or e^+e^- pairs rather than photons being the primary injection. The obtained photon power laws are approximately $\dot{n}(\epsilon) \propto \epsilon^{-1.8}$ and $\dot{n}(\epsilon) \propto \epsilon^{-1.5}$, above and below $\sim \epsilon_{th}/30$, respectively. The latter dependence is characteristic of injection of monoenergetic relativistic electrons in the Thomson limit. It appears here because the pair cascade converts a large fraction of the original γ-ray (or high-energy electron) into a number of electrons with energies just at the boundary of the Thomson limit (Zdziarski 1988; Svensson 1987). We find that when the absorption threshold energy corresponds to $\epsilon_{th}\xi\Theta = 1$, $\xi \gg 1$, then the break in the power law occurs at $\sim \epsilon_{th}/\xi$, again independent of the injection energy of primary γ-rays or electrons. Typically, $\xi \sim 20$ for the lower boundary of the region of dominance of photon-photon pair production in the

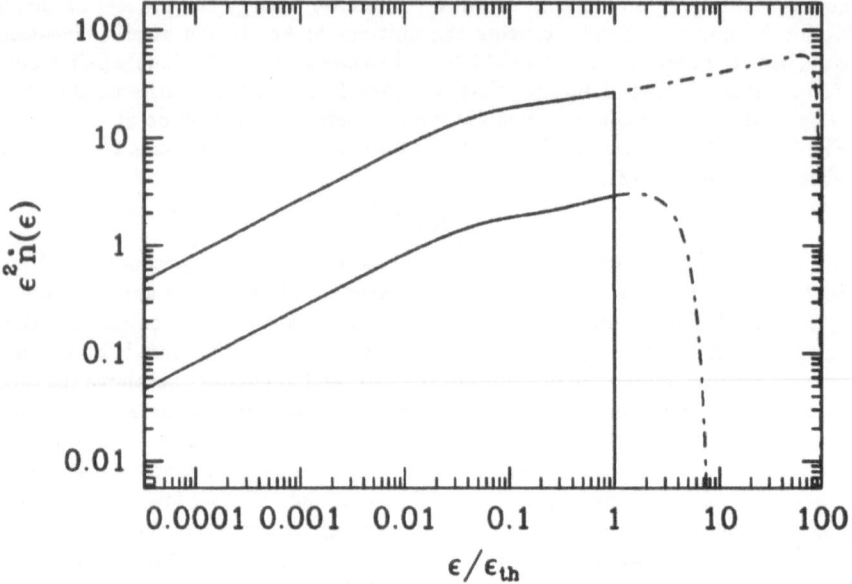

Figure 3. Photon spectra from pair cascades on blackbody photons (solid curves). The lower and upper curves correspond to monoenergetic γ-ray injection at energies 10 and 100 times above the threshold for pair absorption, ϵ_{th}, respectively. The dot-dashed curves show the photon production rates in the optically thick region, where all produced photons are absorbed.

energy-redshift plane (see Fig. 2).

Cascades from discrete sources emitting beams of pairs and γ-rays in the direction of the earth are studied by Zdziarski, Paczyński and Svensson (1990). They calculate time, redshift, and angle-dependent spectra from an instantaneous electron injection and derive general constraints on the observed fluxes of time-varying γ-rays from large redshifts. Those constraints turn out to be in good agreement with the observed fluxes from cosmic γ-ray bursts.

3.2. PHOTON-PHOTON SCATTERING

The cosmological importance of this process was first pointed out by ZS89 and Svensson and Zdziarski (1990). They calculated the scattering probability per unit length, and the optical depth of the universe to the process. Both quantities are proportional to the cube of the photon energy up to the energy corresponding to the threshold for photon-photon pair production. The maximum cross section is of the order $0.01\alpha_f^2\sigma_T$. As photon-photon pair production has a cross section $\sim \alpha_f^{-2}$ times larger, it dominates the scattering completely above the threshold. The redshift corresponding to unit optical depth is given by,

$$1 + z \simeq 4.80 \times 10^3 \, T_{2.7}^{-4/5} h_{50}^{2/15} \Omega^{1/15} \epsilon_0^{-2/5} . \qquad (9)$$

The regions where the universe is transparent to this process and where photon-photon scattering is the dominant process are shown in Figures 1 and 2, respectively. For blackbody photons, photon-photon scattering dominates over photon-photon pair production at (ZS89) $\epsilon < 1/(22\Theta)$. Below this energy photon-photon scattering can dominate over photon-matter pair production for a decade or less, depending on the cosmological parameters (see Figs. 1 and 2).

The effect of γ-ray reprocessing by this process was studied in detail by Svensson and Zdziarski (1990). They found the exact redistribution function for scattering on isotropic blackbody photons and solved analytically the kinetic equation describing photon transfer due to repeated photon-photon scattering. A single scattering on blackbody photons converts a γ-ray into two γ-rays, with an approximate energy equipartition between the two new photons. Repeated scatterings of a power law γ-ray photon spectrum lead to the attenuation of the photons with energies in the optically thick region. The absorbed energy reappears as up-scattered blackbody photons forming a hump in the spectrum at the boundary of the optically thick region. As a specific example, Figure 4 (Svensson and Zdziarski 1990) shows the effect of reprocessing by photon-photon scattering of the pair-photon cascade spectrum in Figure 3.

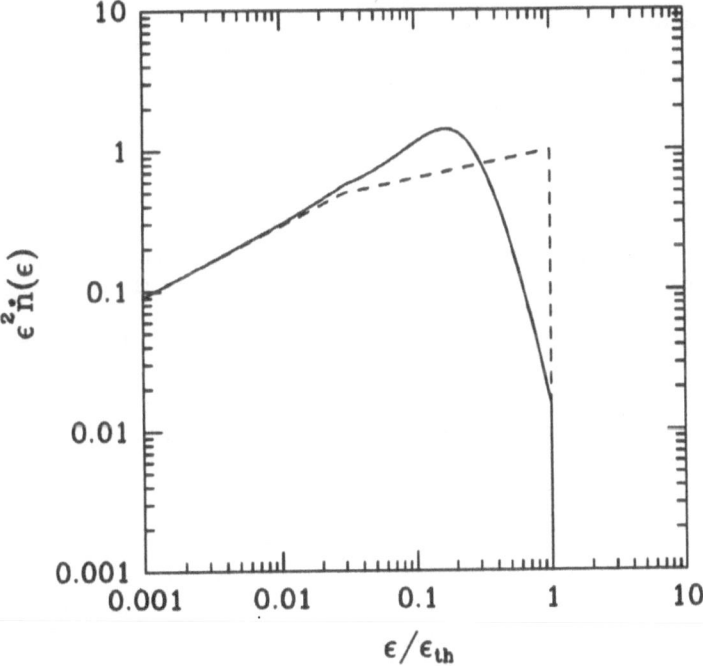

Figure 4. Effects of reprocessing by photon-photon scattering on black body photons. The pair-photon cascade spectrum from Fig. 3 (dashed curve here) is reprocessed into the spectrum shown by the solid curve. The absorbed energy appears as a hump at the photon energies where the optical depth to photon-photon scattering is of order unity.

3.3. PHOTON-MATTER PAIR PRODUCTION

Pair production takes place on free electrons and ions before the recombination epoch and on atoms after it. The cross sections for those processes differ and the exact calculations of the optical depth in the two cases need to be carried out separately. Such calculations are performed in ZS89. The pair production cross section is $\sim \alpha_f \sigma_T$. Pair production on neutral matter with cosmological abundances dominates over Compton scattering at $E \gtrsim 60$ MeV.

As the cross section for photon-matter pair production varies slowly in the high-energy limit, $\epsilon \gg 1$, one can approximate it as constant (Stecker 1971, 1975). In fact, the cross section for pair production on atoms approaches a constant in the high-energy limit. One then obtains an expression for $z(\tau = 1)$ (see ZS89),

$$1 + z \simeq 670 \Omega^{1/3} (\Omega_{0.1} h_{50})^{-2/3}, \quad \epsilon_0 z \gg 825. \tag{10}$$

The regions where the universe is transparent to this process and where pair production on matter dominates are shown in Figures 1 and 2.

The process of γ-ray absorption by photon-matter pair production was included in an approximate manner along with Compton scattering by Arons (1971) in his treatment of cosmological radiative transfer of isotropic γ-rays.

We note that pairs produced by high-energy γ-rays are relativistic, and they can up-scatter photons of the cosmic blackbody background giving rise to new X-rays and γ-rays. This effect is needed to be included in future studies of cosmological reprocessing of γ-rays.

3.4. COMPTON SCATTERING

The physical circumstances of Compton scattering here differ from those of Compton scattering discussed in § 3.1. There a relativistic electron scattered on the soft photons of the blackbody background, whereas now a hard γ-ray scatters on the cold electrons of the baryonic matter.

The total cross section for Compton scattering is given by the Klein-Nishina formula (Jauch and Rohrlich 1980). In the nonrelativistic limit, $\epsilon \ll 1$, this process is called Thomson scattering. The Compton cross section is constant in the Thomson limit ($= \sigma_T$) and decreases with energy as $\sim 1/\epsilon$ at $\epsilon \gtrsim 1$. The optical depth of the universe to Thomson scattering is then constant as well and was given by Gunn and Peterson (1965). The Thomson depth of unity is reached for $\epsilon_0 \ll 10^{-2}$ at

$$1 + z \simeq 63 \Omega^{1/3} (\Omega_{0.1} h_{50})^{-2/3}. \tag{11}$$

The exact Compton scattering optical depth in the case $\Omega = 1$, and the optical depth at any Ω in the high-energy (Klein-Nishina) limit, $\epsilon \gg 1$, have been calculated by ZS89. The former result can be used for any Ω at $\Omega z \gg 1$ by using equation (6). As $\tau = 1$ is reached only at $z \gg 1$ (see eq. [11]), this range of z is most important. Figure 1 shows the contour of the unit optical depth in the energy-redshift plane, and Figure 2 shows the region in which Compton scattering is the dominant process.

Steady state spectra from a continuous injection of γ-rays in an infinite scattering medium of nonrelativistic electrons are calculated in Zbyszewska and Zdziarski (1990).

Figure 5 shows the total injection rate of the primary and scattered photons (solid curve) from a primary injection photon injection (dashed curve) with a broken power law spectrum due to a pair cascade (cf. Fig. 3) with the maximum energy of ~ 25 MeV ($\epsilon = 50$). One sees that the number of photons above the threshold for deuterium photodestruction, 2.225 MeV (see discussion in § 1), is increased approximately by $\sim 50\%$. These calculations assume that the blackbody temperature is low enough so that the electrons energized by Compton scatterings radiate photons with energies below the γ-ray regime.

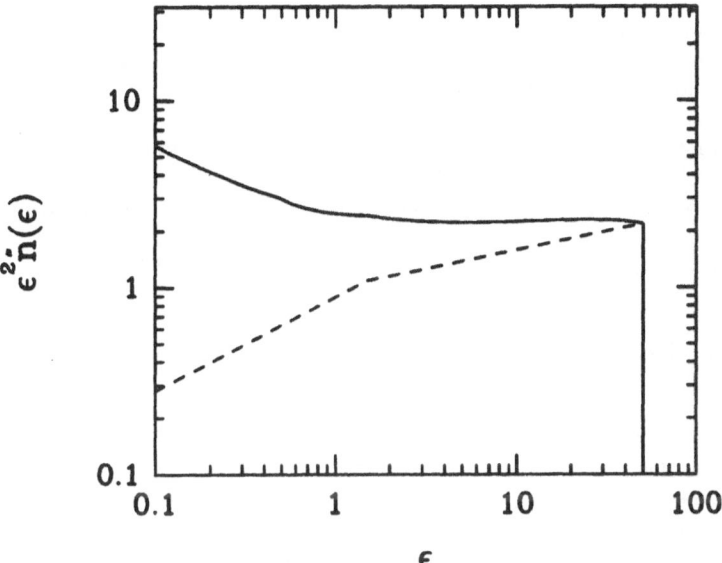

Figure 5. Effect of reprocessing by Compton scattering in an infinite medium of nonrelativistic electrons. The pair-photon cascade photon production rate from Fig. 3 (dashed curve here) is now enhanced due to the effect of repeated Compton scatterring (solid curve). The dimensionless photon energy ϵ is in units of 0.511 MeV (§ 1).

For radiation from a discrete source, a single Compton scattering will deflect the photon from the line of sight. The optical depth for scattering then provides a measure of observability of cosmological γ-rays from discrete sources. On the other hand, isotropic background X-rays may originate at redshifts larger than those given by $z(\tau = 1)$ in the Thomson regime, as pointed out by Arons and McCray (1969) and Rees (1969). This is because a photon loses only a small fraction of its energy per scattering at $\epsilon \ll 1$. ZS89 calculated the optical depth of the universe for the relative energy loss in Compton scattering.

394

4. References

Aharonian, F. A., Kirillov-Ugryumov, V. G., and Vardanian, V. V. 1985, *Astr. Sp. Sci.*, 115, 201.
Arons, J. 1971, *Ap. J.*, 164, 437; *Ap. J.*, 164, 457.
Arons, J., and McCray, R. 1969, *Ap. J. (Letters)*, 158, L91.
Audouze, J., Lindley, D., and Silk, J. 1985, *Ap. J. (Letters)*, 293, L53.
Babul, A., Paczyński, B., and Spergel, D., 1987, *Ap. J. (Letters)*, 316, L49.
Brown, R. W., Mikaelian, K. O., and Gould, R. J. 1973, *Astr. Lett.*, 14, 203.
Désert, F. X., and Schatzman, E. 1986, *Astr. Ap.*, 158, 135.
Dimopoulos, S., Esmailzadeh, R., Hall, J, and Starkman, G. 1988, *Ap. J.*, 330, 545.
Domínguez-Tenreiro, R. 1987, *Ap. J.*, 313, 523.
Ellis, J., Nanopoulos, D., and Sarkar, S. 1985, *Nucl. Phys. B*, 259, 175.
Fazio, G. G., and Stecker, F. W. 1970, *Nature*, 226, 136.
Field, G. B., and Walker, T. P. 1989, preprint.
Fukugita, M., Kawasaki, M., and Yanagida, T. 1989, *Ap. J. (Letters)*, 342, L1.
Gunn, J. E., and Peterson, B. A. 1965, *Ap. J.*, 142, 1633.
Jauch, J. M., and Rohrlich, F. 1980, *The Theory of Photons and Electrons*, (Berlin: Springer).
Juszkiewicz, R., Silk, J., and Stebbins, A. 1985, *Phys. Letters*, 158B, 463.
Kawasaki, M., and Sato, K. 1987, *Phys. Letters*, 189B, 23.
Lindley, D. 1985, *Ap. J.*, 294, 1.
Matsumoto, T., Hayakawa, S., Matsuo, H., Murakami, H., Sato, S., Lange, A. E., and Richards, P. L. 1988, *Ap. J.*, 329, 567.
Montmerle, T. 1975, *Ap. J.*, 197, 285.
Paczyński B. 1988, *Ap. J.*, 335, 525.
Protheroe, R. J. 1986, *M. N. R. A. S.*, 221, 769.
Rees, M. J. 1969, *Ap. Letters*, 4, 113.
Stecker, F. W. 1969, *Ap. J.*, 157, 507.
Stecker, F. W. 1971, *Cosmic Gamma Rays* (Washington: NASA, Scientific and Technical Publication Office).
Stecker, F. W. 1975, in *Origin of Cosmic Rays*, ed. J. L. Osborne and A. W. Wolfendale (Dordrecht: Reidel), p. 267.
Stone, A. D., and Field, G. B. 1988, *B. A. A. S.*, 20, 973.
Svensson, R. 1987, *M. N. R. A. S.*, 227, 403.
Svensson, R., and Zdziarski, A. A. 1990, *Ap. J.*, 349, in press.
Thompson, C., and Madau, P. 1989, in *STScI-GSCF Workshop on Ultra-Hot Plasmas and Electron-Positron Pairs in Astrophysics*, ed. A. A. Zdziarski and D. Kazanas (Baltimore: STScI Publ.), p. 48.
Vilenkin, A. 1988, *Nature*, 332, 610.
Weinberg, S. 1972, *Gravitation and Cosmology*, (New York: Wiley).
Zbyszewska, M., and Zdziarski, A. A. 1990, in preparation.
Zdziarski, A. A. 1988, *Ap. J.*, 335, 786; 342, 1212.
Zdziarski, A. A., Paczyński, B., and Svensson R. 1990, in preparation.
Zdziarski, A. A., and Svensson R. 1989, *Ap. J.*, 344, 551 (ZS89).

Synchrotron self-absorption as a thermalizing mechanism

G. GHISELLINI[1] & R. SVENSSON[2]

1: *Institute of Astronomy, Madingley Road, Cambridge CB3 0HA, U.K.*
2: *NORDITA, Blegdamsvej 17, Copenhagen, Denmark*

ABSTRACT. Synchrotron self-absorption is a powerful heating mechanism for low energy particles emitting and absorbing synchrotron photons. They are driven towards a thermal distribution in a few cooling times. The inhibition of synchrotron cooling has important consequences for the overall spectrum of the emitted radiation.

1. Introduction

Synchrotron emission is well understood (*e.g.* Ginzburg & Syrovatskii 1965, Rybicki & Lightman 1979) and thought to be responsible of a significant part of the radiation we receive from a variety of cosmic objects, such as supernova remnants, radio jets, compact radio sources and active galactic nuclei. However, the inverse process of synchrotron absorption has not received the same attention in the past, and here we present new results concerning the interplay between synchrotron photons and the relativistic particles responsible for both the emission and the absorption.

Low energy, but still relativistic, particles absorb as well as emit synchrotron radiation. Therefore they gain as well as lose energy by exchanging photons with the self absorbed radiation field. This process can be thought of as an efficient way for electrons to exchange energy between themselves. Given enough time, they can therefore thermalize.

In the following we derive the absorption rate of an individual particle, showing that mildly relativistic particles always absorb more energy than they emit.

The kinetic equation for the particle distribution is discussed, considering continuos injection of new particles. As expected, low energy particles attain a quasi-Maxwellian distribution, in a timescale approximately equal to a few cooling times.

This thermalizing process is important if the cooling time is shorter than the escape time of the particles. This is the case in sources of high luminosity to size ratio, such as active galactic nuclei. Depending on the details of the injection mechanism, which provides the necessary supply of power, synchrotron self-absorption can have important effects on the way the bulk of the radiation is emitted, constraining the

W. Brinkmann et al. (eds.), Physical Processes in Hot Cosmic Plasmas, 395–400.
© 1990 *Kluwer Academic Publishers.*

shape of the overall spectrum.

2. The energy balance

Suppose that a population of relativistic particles of energy γmc^2 and density $N(\gamma)$ is homogeneously distributed in a spherical source of radius R and magnetic field B. By emitting synchrotron radiation, particles lose energy at a rate $\dot{\gamma}_e(\gamma) \propto B^2 p^2$, where $p \equiv \sqrt{\gamma^2 - 1}$ is the momentum in units of mc. The simplest way to derive the absorption rate $\dot{\gamma}_a(\gamma)$ of the single particle is by means of the absorption coefficient, given by

$$\kappa_\nu = \frac{1}{8\pi m\nu^2} \int_{\gamma_{min}}^{\gamma_{max}} \frac{N(\gamma)}{\gamma p} \frac{d}{d\gamma} [\gamma p\, j(\nu, \gamma)] d\gamma \tag{1}$$

where $j(\nu)$ is the emissivity of a single particle (see e.g. Rybicki & Lightman 1979). The absorption optical depth $[\equiv R\kappa(\nu)]$ is greater than unity below the self-absorption frequency ν_t. The electrons emitting mainly at ν_t have an energy $\gamma_t \sim (\nu_t/\nu_B)^{1/2}$, where ν_B is the cyclotron frequency.

If $I(\nu)$ is the synchrotron intensity, the total power absorbed by all electrons is $\int \kappa(\nu)I(\nu)d\nu$. Therefore the power absorbed by one particle is

$$\dot{\gamma}_a(\gamma) = \frac{1}{\gamma p} \frac{d}{d\gamma} \left[\gamma p \int_0^\infty \frac{I(\nu)}{2m\nu^2} j(\nu, \gamma) d\nu \right] \tag{2}$$

Eq. (2) can be evaluated analytically in special cases. This was first done by Rees (1967), who assumed that $N(\gamma)$ is a power law of slope n [i.e. $N(\gamma) \propto \gamma^{-n}$]. In this case, and at energies $\gamma < \gamma_t$, both the absorption and the cooling rate have the same energy dependence, and they exactly balance for the particular slope $n = 3$ of the electron distribution. For flatter $(n < 3)$ slopes, particles absorb more energy than they emit, while the opposite is true for $n > 3$.

This, however, seems to lead to a paradox. The fact that radiation field is self absorbed implies, by definition, that all the emitted power is also absorbed, and so the integrated (over energy) losses and gains of the ensemble of electrons with $\gamma < \gamma_t$ should balance for any slope of their energy distribution. We then require

$$\int_1^{\gamma_t} N(\gamma) \left[\dot{\gamma}_a(\gamma) - \dot{\gamma}_e(\gamma) \right] d\gamma = 0 \tag{3}$$

It is obvious that $[\dot{\gamma}_a - \dot{\gamma}_e]$ must change sign at some $\gamma_c < \gamma_t$. For flat distributions $(n < 3)$ the losses of a relatively larger number of particles close to γ_t can balance the gains of the rest of the particles at lower energies. For steep distributions $(n > 3)$ the gains of the relatively larger number of particles close to $\gamma \sim 1$ can balance the losses of all the higher energy particles. Therefore, the value of γ_c must be close to γ_t for flat distributions $(n < 3)$, while it must be close to unity if $n > 3$.

For $n < 3$ the requirement of eq. (3) can be satisfied once the transition from the optical thick and optical thin part of $I(\nu)$ [to be used in eq. (2)] is properly accounted for. On the other hand, for $n > 3$, the solution of the paradox requires a careful examination of the absorption rate for mildly relativistic particles. In the limit $p \to 0$ the emission process becomes cyclotron, whose spectrum can be approximated with a δ-function at the cycloton frequency: $j(\nu, \gamma) \propto p^2 \delta(\nu - \nu_B)$. If we substitute this into eq. (2), we obtain a *constant* absorption rate. Since the cooling rate $\dot{\gamma}_e \propto p^2$ is an increasing function of p, we have the remarkable result that low energy electrons always gain rather than lose energy. In this case losses balance gains at $\gamma_c \sim 1$, as required by eq. (3) (see Fig. 1).

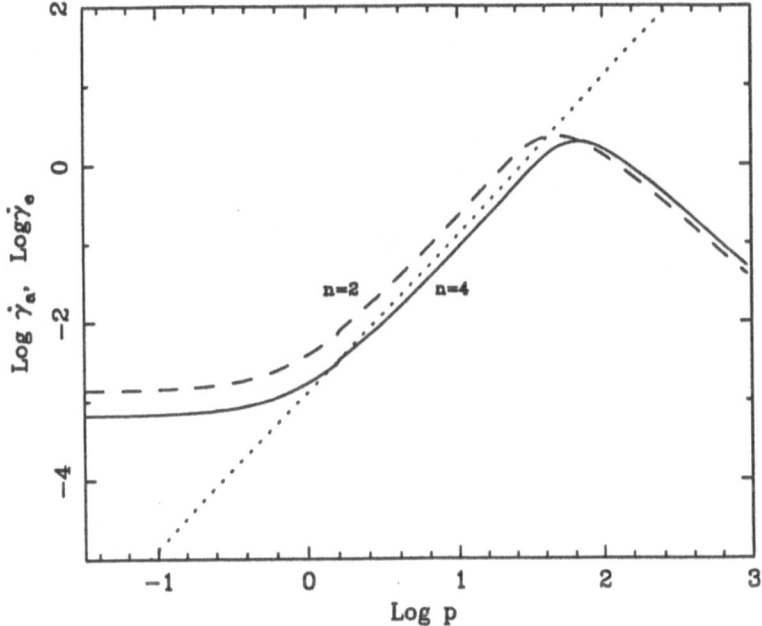

Figure. 1. *The synchrotron cooling rate (dotted line) is compared with the absorption rate for two different power law particle distributions with slopes $n = 2$ and $n = 4$. Note that the absorption rate becomes constant at low energies, is proportional to p^2 (just as the cooling rate) at intermediate energies, and declines at high energies, where absorption is not important.*

3. The kinetic equation

The existence of a particular energy γ_c, where gains and losses balance, forces the particles to accumulate at this energy. A peak is therefore expected to develop, and particle will diffuse out of the peak by stochastic energy changes. The kinetic equation describing the time evolution of the particle distribution takes the form of

398

a Fokker Planck equation. This was first discussed by McCray (1969), who found three possible equilibrium solutions for $N(\gamma)$, corresponding to power laws of index $n = 2$ and $n = 3$ and a Maxwellian distribution.

The same problem was rediscussed by Ghisellini, Guilbert & Svensson (1988), taking into account transrelativistic effects, neglected by McCray. They found that the only possible steady solution is a quasi-Maxwellian at low energies, with a power law tail at energies where absorption is unimportant. Power law solutions for low energy electrons are transient, and are an approximate description of the $N(\gamma)$ distribution only at very early stages of its evolution (at times smaller than the cooling time for electrons with $\gamma = \gamma_c$).

The effects of other cooling processes, such as the inverse Compton process, are

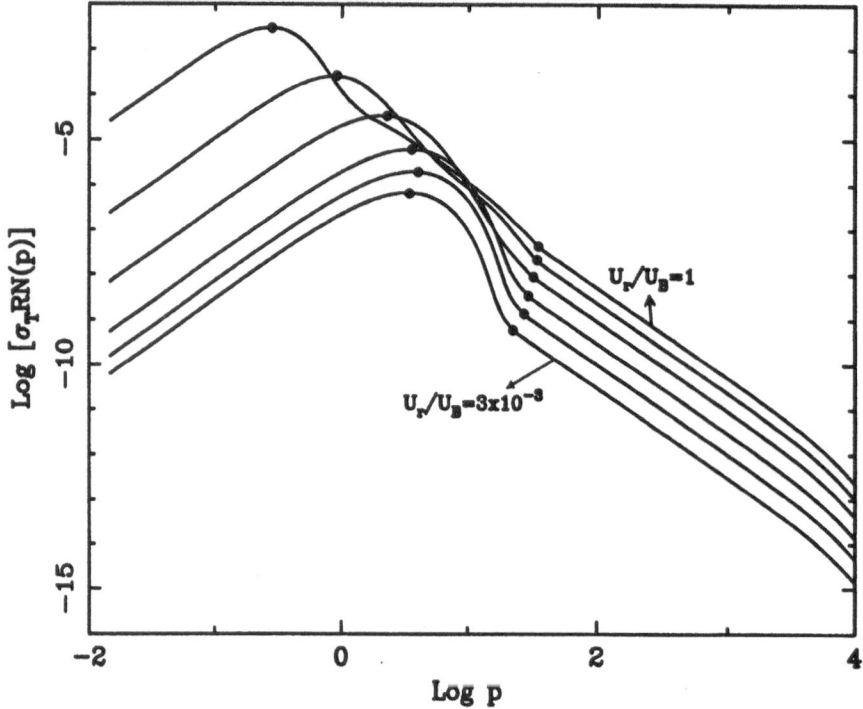

Figure 2. *The steady particle distribution, $N(p)$, corresponding to $U_r/U_B = 3 \times 10^{-3}$, 0.01, 0.03, 0.1, 0.3, and 1 as a function of momentum p. Particles are continuously injected at high energies, and reach a steady state distribution as a result of radiative cooling, synchrotron heating, and escape. Dots indicate the values of $p_t \sim \gamma_t$ and the energy of the peak. The distributions respond to changes in the amount of inverse Compton cooling: when this effect is not important the energy of the peak is closely related to γ_t, while it shifts to lower energies for increasing dominance of inverse Compton cooling.*

discussed by de Kool, Begelman & Sikora (1989), and Svensson & Ghisellini (1989). When important, the additional losses caused by inverse Compton cause the peak of the quasi-Maxwellian distribution to shift towards lower energies. Fig. 2 shows examples of steady particle distributions for increasing importance of inverse Compton losses. In this figure we assume that particles are continuously injected at high energies. As can be seen, the energy of the peak of the quasi-Maxwellian part is associated with the value of γ_t until inverse Compton losses become important, causing the energy of the peak to shift at lower energies. A detailed discussion of the importance of inverse Compton losses will be presented elsewhere together with analytical estimates (Svensson & Ghisellini 1989).

4. Discussion

The thermalizing process of synchrotron self-absorption operates through the exchange of photons. It can be very efficient in relativistic, magnetized plasmas, where low energy particles are driven towards a relativistic or transrelativistic Maxwellian distribution on a very short timescale. Contrary to Coulomb collisions, synchrotron reabsorption does not depend on the particle density, and becomes more efficient at high energies (the absorption timescale is equal to the cooling timescale, and therefore inversely proportional to the energy of the particle).

Even if the optically thin synchrotron spectrum is a power law, it is likely that the underlying electron distribution is a power law only at energies greater than γ_t. The optically thick part of the synchrotron spectrum is predicted to be Rayleigh-Jeans like ($I(\nu) \propto \nu^2$), and not to follow the $\nu^{5/2}$ law. However, any inhomogeneity of the magnetic field or the particle density will make the self absorbed spectrum even flatter than ν^2. The very steeply rising radio spectra seen in radio quiet quasars are probably due to other processes, and not to synchrotron self-absorption (Sanders et al. 1989).

Synchrotron reabsorption can have dramatic consequences on the overall spectrum, produced not only by the synchrotron process, but also by the self Compton mechanism. Usually, the relative importance of the two processes is measured by the ratio of the magnetic to radiation energy density U_B/U_r. However, the relative importance of Compton cooling can be greatly enhanced if the synchrotron cooling is inhibited by self-absorption. This is particularly true if the low energy particles are the ones responsible for the bulk of the emitted luminosity.

If new electrons are continuously supplied to the source by an injection mechanism, and their cooling time is shorter than the escape time, we can balance the injected luminosity with the radiated luminosity. This luminosity balance condition qualitatively determines the shape of the spectrum at all energies. Suppose monoenergetic particles are injected at some $\gamma_{max} > \gamma_t$. High energy particles ($\gamma > \gamma_t$) cool by emitting optically thin synchrotron and inverse Compton radiation, while low energy particles ($\gamma > \gamma_t$) cool by the small amount of synchrotron radiation they emit at optically thin frequencies above ν_t or by the inverse Compton process.

The fraction of the injected luminosity that the particles with $\gamma < \gamma_t$ must radiate is approximately γ_t/γ_{max}. For large values of U_B/U_r this fraction of the luminosity is radiated as an 'excess' above the optically thin synchrotron power law close to ν_t. On the other hand, for small values of U_B/U_r, the inverse Compton mechanism is dominant, and particles with $\gamma < \gamma_t$ produce an inverse Compton 'excess' just above the maximum frequency of the synchrotron spectrum (Svensson & Ghisellini 1989).

If the mean energy of the injected particle is smaller than γ_t, the bulk of the luminosity is emitted by the inverse Compton process through multiple scatterings. In this case the (strongly self absorbed) synchrotron radiation only provides seeds soft photons to be scattered at high energies. Again, balancing injected and radiated luminosity suffices to find the overall spectral index, which cannot be steeper than unity (Ghisellini 1989).

The lack of Faraday depolarization and the upper limits of Faraday rotation in variable radio sources set strong constraints on the relative number of low energy particles (Wardle 1977, Jones & O'Dell 1977). Wardle (1977) suggested that a *relativistic* Maxwellian particle distribution could naturally account for the required lack of low energy particles. Here we have shown that such a thermalized distribution is a natural consequence of synchrotron self-absorption.

References

De Kool, M. Begelman, M.C., & Sikora, M., (1989) *Astrophys. J.* **337**, 66.

Ghisellini, G., Guilbert, P.W., & Svensson. R., 1988, *Astrophys. J.*, **334**, L5

Ghisellini, G., 1989, *Mon. Not. R. astr. Soc.*, **236**, 341.

Ginzburg, V.L., & Syrovatskii, S.I., 1965, *Ann. Rev. Astr. Astrophys.*, **3**, 297

Jones, T.W, & O'Dell, S.L., 1977, *Astron. Astroph.*, **61**, 291

McCray, R., 1969, *Astrophys. J.*, **156**, 329

Rees, M.J., 1967, *Mon. Not. R. astr. Soc.*, **136**, 279.

Rybicki, G.B., & Lightman, A.P., 1979, *Radiative processes in astrophysics*, ed. Wiley.

Sanders, D.B., Phinney, E.S., Neugebauer, G., Soifer, B.T., Matthews, K. & Green, R.F., 1989, preprint.

Svensson, R., & Ghisellini, G., 1989, in preparation

Wardle, J.F.C., 1977, *Nature*, **269**, 563

Reacceleration and Pair Loading in Compact Sources

C. DONE, G. GHISELLINI & A.C. FABIAN
Institute of Astronomy
Madingley Road
Cambridge CB3 0HA
England

ABSTRACT. High energy observations of Active Galactic Nuclei provide constraints on possible acceleration mechanisms. The lack of an observed Compton downscattering break in the X–ray spectrum implies a small optical depth. This is in conflict with the large optical depth predicted by *all* models in which a compact source is powered by continuous injection of highly relativistc electrons. Reacceleration is *not* a solution in compact sources as the optical depth will increase from its initial value through pair production.

1. Introduction

In Active Galactic Nuclei (AGN) the power output from the central source dominates the starlight from the surrounding galaxy. Fig. 1 shows a schematic view of this non–stellar continuum. The spectrum peaks in the UV and near IR in what are thought to be thermal features from an accretion disc and hot dust respectively. The rest of the spectrum can be described by two power–laws, which we will assume to be non–thermal in origin.

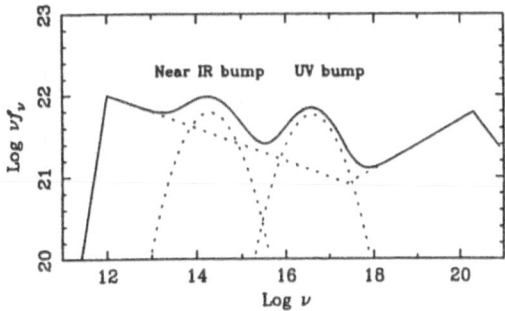

Fig. 1. Schematic view of the non–stellar continuum of a radio–quiet (Seyfert 1) AGN.

W. Brinkmann et al. (eds.), Physical Processes in Hot Cosmic Plasmas, 401–406.
© 1990 *Kluwer Academic Publishers.*

It is immediately apparent that AGN emit roughly the same power in each frequency range over nearly 10 decades, so the X-rays form more than 10% of the total luminosity of the source.

The high energy flux is often observed to vary rapidly. Changes in the X-ray flux on timescales of hours are common, and in a few objects the flux has doubled in as short as five minutes. As the source cannot vary faster than the light travel time across the emitting region, the large luminosities seen must be produced in a very small region. This means high photon densities and thus there is a high probability that photons collide and produce electron–positron pairs. The probability of this happening, P, is proportional to the photon density of the source and the path length $\sim R$, so

$$P \propto \frac{L}{4\pi R^2 c} \times R \propto \frac{L}{R}.$$

The luminosity to size ratio is thus an important parameter of an AGN. It is usual to redefine it as a dimensionless compactness parameter, ℓ, (Guilbert, Fabian & Rees 1983)

$$\ell = \frac{L}{R} \frac{\sigma_T}{m_e c^3}$$

where σ_T is the Thomson cross–section, m_e is the electron rest mass and c is the speed of light. The source is completely optically thick to γ-rays crossing the source when $\ell \geq 60$.

Fig. 2 is a histogram of observed compactness parameters for a sample of AGN (Done & Fabian 1989). There are indeed some objects with $\ell > 60$. The observed compactness is likely to be an underestimate of the intrinsic ℓ as the source is unlikely to vary on its absolute minimum timescale. Consequently it is reasonable to assume that pair production processes are important in a large fraction of X-ray variable AGN.

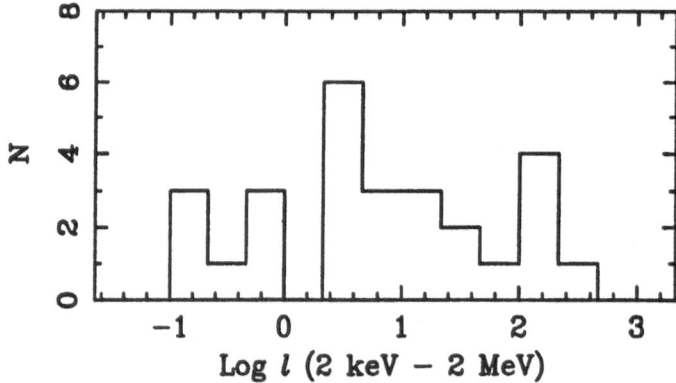

Fig. 2. Histogram of the compactness parameter, ℓ, from values given in Done & Fabian (1989).

2. Constraints from the High Energy Spectrum

The 2–10 keV X-ray spectra of Seyfert galaxies are well described by a power–law of energy index 0.7 ± 0.2 (Turner & Pounds 1989). Observations at higher energies, up to 100–200 keV (Rothschild *et al.* 1983), are consistent with an extrapolation of this power law, though a turnover at 50 keV cannot be ruled out at the 90% confidence level. γ-ray data for AGN are sparse, but those that do exist indicate that the spectrum must turn over between 200 keV and 3 MeV (Bignami *et al.* 1979, Bassani & Dean 1984), as sketched in Fig. 1. This must be a general feature, for if the spectra continued unbroken past 3 MeV, the flux from AGN overproduces the observed diffuse γ-ray background (Rothschild *et al.* 1983). This is the 'standard' description of the spectrum, but another less well known result is that it shows very little evidence for cool particles. If the source is optically thick to Thomson scattering *i.e.* $\tau_T > 1$, then thermal Comptonization changes the spectrum. High energy photons are downscattered, leading to a break in the spectrum at $\sim 511/\tau_T^2$ keV. Since there is no evidence for a break in the well measured 2–10 keV spectrum, $\tau_T < 7$. The observations extending out to at least 50 keV require $\tau_T < 3$

The constraints on the optical depth are important because of the 'dead electron problem'. In highly compact sources, $\ell \gg 1$, there is a large photon density. Relativistic particles of energy $\gamma m_e c^2$ cool by inverse Compton scattering on a timescale $t_{cool} \sim R/c(\ell\gamma)^{-1} \ll R/c$. This is very short, of the order of seconds in many cases. To see a steady spectrum there must be a continuous energy supply. Particles injected into the source cool before they escape from the region and build–up a cool electron population. (This is a different situation from that in extended radio sources, where the photon density is low and the cooling time is long.)

To make a highly luminous source, we can either continually inject a few particles with large energies, or many particles with smaller energies. For a given source power, a lower energy per particle means a higher injection rate. Even if particles escape (or are accreted) at close to light speed, this is not fast enough to stop a build–up of the optical depth in a compact source. In this situation, $\tau_T \sim \ell/\gamma$, so compact ($\ell > 1$) optically thin ($\tau_T < 1$) sources cannot be powered by low energy particle injection. On the other hand, increasing the energy per particle is *not* a solution in compact sources. The number of particles increases through pair production if the cooling electrons produce γ-rays, so the optical depth rises. Balancing injection and pair production against annihilation and escape requires that $\ell < 200$ for $\tau < 3$ (Done, Ghisellini & Fabian 1989). This is a generous overestimate of ℓ, as it assumes the particles escape radially at the speed of light, whereas the path will be a random walk between scatterings, and will increase considerably if there is a tangled magnetic field. The value of 200 for ℓ is the maximum for any continuous high energy particle ($\gamma > 150$) injection mechanism which is consistent with the spectrum.

Fig. 2 shows that there are several AGN with an observed compactness parameter of a few hundred. The largest has $\ell = 230$, which is just over the limit derived above.

Observations can rule out continuous injection of highly relativistic electrons as a means to power the extremely compact AGN. This includes the 'standard' explanation, in which particles are continually injected as $Q(\gamma) \propto \gamma^{-1.4}$ giving $\alpha = 0.7$, and also proton–proton collisions. If the relativistic electrons have less energy, $\gamma \sim 40$, then the limit can be extended to $\ell < 450$ as fewer pairs are produced because of the reduced γ-ray flux. However, the spectrum could then not extend to 2 MeV, but turns over below $m_e c^2$. Future observations will determine the extent of the high energy spectrum and so decide whether *any* continuous particle injection mechanism is consistant with the low optical depth.

3. Reacceleration and Loading

Cavaliere and Morrison (1980) pointed out the problem of cool electrons building up within a source, and suggested that cool particles were reaccelerated to relativistic energies. Rather than inject particles, there is some mechanism that accelerates the cool particles present in the source. Again this is consistent at low compactness, but pair production causes problems at high ℓ. Cool particles are reaccelerated to relativistic energies, and cool down again by inverse Compton scattering to produce γ-rays. If the source is compact, then the γ-rays produce pairs which join with the initial distribution. Each reacceleration step produces more pairs, so the optical depth begins to rise.

If the cooled pairs are reaccelerated to the same energy as the initial particles, then the total energy output of the source increases. The source has an even larger compactness, and even more pairs are produced which can be reaccelerated! This process must saturate as the source cannot have infinite energy available. It is analogous to the battery circuit in Fig. 3.

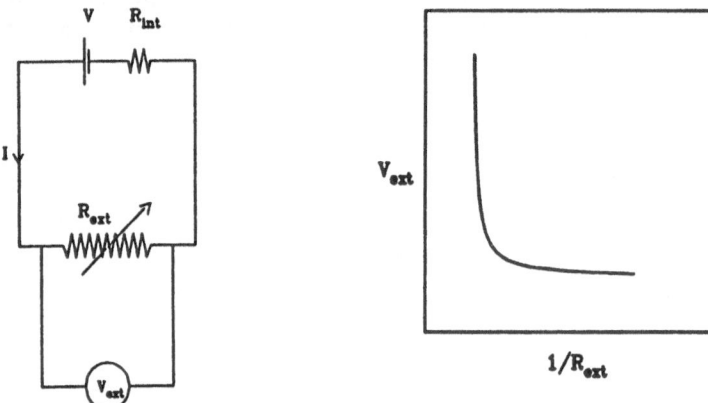

Figs. 3 and 4. The voltage, V_{ext}, across the external resistor, R_{ext}, decreases as the resistance is lowered. A larger fraction of the total voltage is spent across the internal battery resistance.

If the resistance is decreased then the current, I, and so the power, VI, spent over the resistor increases. This continues until the maximum power output is reached, where the internal battery resistance is equal to the resistance of the external circuit. Then a further increase in the current causes the voltage across the resistor to drop, as shown in Fig. 4. There is an increased load on the battery, with most of the battery power used internally to maintain the current.

We expect the reacceleration mechanism to be loaded by any new particles that join the cool population. The energy to which the particles can be accelerated decreases as their number increases, to keep the total power output constant. Suppose there is an initial optical depth, τ_i, and that these particles are reaccelerated to energies γ_i. In a highly compact source, pairs are produced which increase τ so the maximum reacceleration energy decreases, as shown in Fig. 5. They still produce γ-rays, though at slightly lower energies, so more pairs are produced, increasing further the optical depth. This feedback loop stops when the pairs can be annihilated as fast as they are produced, leading to a final energy, γ_f, and optical depth, τ_f.

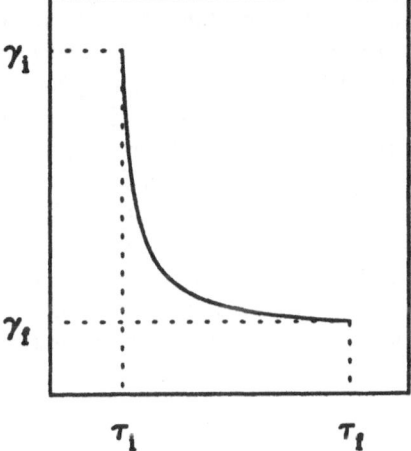

Fig. 5. The few initial particles, τ_i, can be given a large energy γ_i. As the number of particles increases to τ_f through pair production so their mean energy, γ_f, must decrease for the source to work at a constant power.

Loading is a process that we expect if highly compact sources are powered by reaccelerated cool particles. It is a variant on the reacceleration mechanism and so is subject to the same problems with a build-up of the optical depth due to pair production. Neither reacceleration nor loaded reacceleration can solve the 'dead electron problem'. Loading does, however, produce spectra with a γ-ray break at the same energy as that observed in AGN, as the low energy particles produce few γ-rays.

4. Clumping

The constraints on the optical depth come from the assumption of a homogeneous source. The problems can be circumvented if the cool electrons are clumped into many small regions of high optical depth, embedded in the optically thin source. If the covering fraction of these dense regions is small, then they intercept only a little of the flux. The Comptonization break is diluted by the unscattered emission.

A problem with this approach is that fragmentation must take place on very short timescales. Annihilation is a very rapid process, so these clumps must form before the pairs annihilate if they are to make a difference to the total optical depth. Radiation pressure may be able to do this, or some more exotic form of plasma instability.

5 Summary

If AGN produce γ-rays, and are not beamed, then the high compactness observed mean that pairs will be created. If the emission region is homogeneous, then the highly compact sources powered by continual injection of relativistic particles will have a large optical depth through pair production. Reacceleration of cool particles is not a solution to this problem, as it is the pair production, not the injection mechanism that determines the optical depth. If reacceleration does take place, then the energy given to each electron decreases as pairs increase the number of particles. The source must be loaded to avoid infinite power.

Future high energy observations *e.g.* from GRO, will place more stringent constraints on possible acceleration mechanisms. If the AGN spectra do not break below 511 keV, this will limit the optical depth to $\tau_T < 1$.

The 'dead electron problem' can be avoided if the cool material in the emission region is clumped. Possible mechanisms for this fragmentation are certainly worth further investigation.

References

Bassani, L. & Dean, A.J., 1983. *Space Sci. Rev.*, **35**, 367.

Bignami, G.F., Fichtel, C.E., Hartman, R.C. & Thompson, D.J., 1979. *Astrophys. J.*, **232**, 649.

Cavaliere, A. & Morrison, P., 1980.. *Astrophys. J.*, **238**, L63.

Done, C. & Fabian, A.C., 1989. *Mon. Not. R. astr. Soc.*, in press.

Done, C., Ghisellini, G. & Fabian, A.C., 1989. In preparation.

Guilbert, P.W., Fabian, A.C. & Rees, M.J., 1983. *Mon. Not. R. astr. Soc.*, **205**, 593.

Rothschild, R.E., Mushotzky, R.F., Baity, W.A., Gruber, D.E., Matteson, J.L. & Peterson, L.E., 1983. *Astrophys. J.*, **269**, 423.

Turner, T.J. & Pounds, K.A., 1989. *Mon. Not. R. astr. Soc.*, in press.

Quantum Effects in Magnetized Electron-Positron Plasmas: Synchrotron Pair Cascades and Gamma-ray Burst Spectra

MATTHEW G. BARING

Max Planck Institut für Astrophysik
Karl Schwarzschild Strasse 1
D-8046 Garching bei München, West Germany

ABSTRACT: Modelling of gamma-ray burst spectra with optically thin synchrotron emission in strong magnetic fields is examined. The properties of burst sources that suggest the existence of electron-positron pairs and strong magnetic fields in the emission region are summarized. In such a field, interesting quantum processes occur, such as the creation of an electron-positron pair by a free photon and the splitting of a photon in two. These have a profound influence on the synchrotron emission modelling of the continuum of burst sources. The basic properties of these processes of these sources are reviewed. The spectrum is attenuated at gamma-ray energies, which may or may not be observed. Pairs that are created by photons can reradiate synchrotron photons at lower energies that may create further pairs. The process can continue forming a pair cascade. The equilibrium spectra from such cascades are calculated to see what modifications the pair creation makes to the hard X-ray synchrotron spectrum: it is found that the spectral index does not change from that expected without any pair production. The distribution of angles of radiation and particles with respect to the direction of the magnetic field is also considered, and it is found that accurate fitting of hard gamma-ray burst spectra above 1 MeV is possible.

1. Introduction

Quantum processes in strong magnetic fields are an interesting and esoteric field of study in physics. They are presently well beyond any possibility of laboratory detection, and so represent a purely theoretical prediction of quantum electrodynamics that awaits verification. However, the physical environments needed to establish these processes are readily supplied by astrophysical laboratories such as neutron stars. These stars are expected to have fields of the order of 10^{12} Gauss or more (Pacini, 1968), which are close to the quantum critical field strength of $B_c = m_e^2 c^3 / e\hbar = 4.413 \times 10^{13}$ Gauss, for which the excitation energy of an electron in the field between consecutive transverse energy states is comparable to its rest mass energy. The quantum features in such a field will then be prevalent if particles in the emission region typically have energies that are at least mildly relativistic. The compact nature of neutron stars and their large associated gravitational field guarantee this relativistic requirement. Neutron stars therefore provide a good environment to study quantum processes in strong magnetic fields, and test the predictions of QED. Conversely, such is the current confidence in the theory of quantum electrodynamics that often its unvalidated predictions are used with axiomatic certainty in modelling the emission from

W. Brinkmann et al. (eds.), Physical Processes in Hot Cosmic Plasmas, 407–424.
© 1990 *Kluwer Academic Publishers.*

cosmic neutron stars. Examples of these exotic quantum processes are the creation of an electron-positron pair by a photon and the splitting of a photon into two photons. These processes are prevalent when the photons have gamma-ray energies: therefore gamma-ray burst sources are the best class of cosmic objects for the study of these processes.

From the astrophysical viewpoint, gamma-ray burst sources, generally thought to be special members of the family of neutron star like celestial objects, are just as curious and exotic as the quantum processes that are likely to occur in their strong magnetic fields. First detected in 1969 (announced in Klebesadel, Strong and Olson, 1973), they have remained an enigma in many ways to the present day. This is primarily caused by the difficulty of their detection and the collection of sufficiently useful observational data from such rapid and short-lived transients. Nevertheless much progress has been made in the understanding of their nature. For them to be associated with neutron stars, they almost certainly have to be sources within our galaxy. However, conclusive evidence to determine the distances to these objects is yet to be obtained. Why they should be transient and possibly not recurrent sources provokes unusual and imaginative models from theorists for their energy supply, with no particular model being uniquely preferred by the astrophysical community. The existence of strong magnetic fields in the emission region of burst sources is itself a subject of controversy, though the recent observations of cyclotron absorption features in bursts detected by the GINGA satellite (Murakami, et al. 1988; Fenimore, et al. 1988) confirm the earlier predictions of fields in excess of 10^{12} Gauss in these sources (Mazets, et al. 1981). Another contentious issue is the role of electron-positron pairs in burst sources. They are expected in most models of the emission because of the relativistic energetics in a neutron star environment, but observational evidence for their existence is limited and conflicting. Resolution of all of these issues will be substantially aided by the coming generation of X-ray and gamma-ray satellite detectors to be put into operation in the next decade.

Much of the most interesting and useful information from burst detections is the data of their energy spectra. It casts light on the nature of the emission mechanism of radiation emission in these sources, which has been the focus of many theoretical investigations including the work presented in this paper. Here the synchrotron emission model in a strong magnetic field is considered in an attempt to fit spectra at gamma-ray energies. This popular model is modified in this paper to include and investigate the effects of pair production on the continuum component of the source energy spectra. It will be shown that pair production does not necessarily attenuate the gamma-ray spectrum as sharply as was previously predicted, and that good spectral fits to observational data can be obtained. A review of the properties of gamma-ray burst sources (GRBs) is first presented, followed by a summary of the main features of interesting and relevant quantum processes in strong magnetic fields. The details of the synchrotron pair cascade model of burst spectra are then enunciated, including the effects of angular dispersion of particles and radiation. This model assumes a non-thermal supply of relativistic electrons that provide the basic sources of synchrotron radiation. The gamma-rays are then absorbed in the magnetic field producing electrons and positrons (pairs) which can then reradiate via the synchrotron process. This sequence continues, forming a *pair cascade*. Other processes such as photon splitting complicate the picture. The pair cascade model is assessed to be very suitable for some sources, indicating that the exotic quantum processes in strongly magnetized electron-positron pair plasmas may be very important in the understanding of the emission mechanism of gamma-ray bursts.

2. Observational and Theoretical Properties of Gamma-ray Bursts

Gamma-ray burst sources are so named because their emission is detected only in the hard X-ray and soft gamma-ray range (typically between about 20 KeV and 10 MeV) for durations of the order of less than a second to a minute or so. Their transient nature, with emission only at such high energies makes their detection difficult, and only limited observational data has been available to date. The quality and usefulness of data is poor compared with that for other sources in other energy bands so that theoretical understanding of these high energy transients is more difficult. Gamma-ray detectors have large fields of view (often as much as 180 degrees) so that location of the position of the sources in the sky is difficult. Usually large 'error boxes' for gamma-ray burst source locations are obtained (e.g. see Mazets, 1988) and identification of bursts with others sources detected in different energy bands is nearly impossible. Therefore these sources remain very inaccessable to close scrutiny, and remain very enigmatic. Optical 'flashes' have been found in archival plates within some gamma-ray burst error boxes (Schaefer, 1984): these may be optical counterparts to the gamma-ray sources occuring at different times. Burst sources, although transient, are sometimes recurrent, and the recurrence time of a typical burst is expected to be as long as 1000 years (e.g. Mazets, 1988).

Probably the most important question concerning gamma-ray bursts is their distance from us. Are they galactic or extragalactic? Because they cannot be positioned accurately in the sky or positively identified with other sources, distance measurement is impossible. The rapid time variability of burst emission (e.g. see Mazets et al. 1979), suggests compact emission regions as small as 10^6 cm, which would be consistent with a neutron star site for gamma-ray burst emission. With such a site, the luminosity of the sources would then be below the Eddington luminosity of about 10^{38} erg/sec only if the bursts were at about one kiloparsec distant from us. The sources would then be galactic. The uniformity of the directional distribution of sources in the sky is not inconsistent with such a 'local' distance. Neither is a cosmological location, as was proposed by Paczyński (1986). The location of the supernova remnant N49 in the Large Magellanic Cloud within the erro box of the event of the 5th March, 1979 is suggestive of an extragalactic source (Lamb, 1984), which would then have an incredibly high luminosity (for a neutron star) of 10^{44} erg/sec. This source is also special in another way since it has periodic emission. Nearly all sources have largely random time profiles.

If the popular opinion that the sources are galactic neutron stars is accepted, then the high luminosity approaching the Eddington limit implies a high gamma-ray radiation compactness (Guilbert, Fabian and Rees, 1983), which would lead to copious production of electron-positron pairs. Therefore a relativistic pair plasma would be expected in the emission region of a source. The association of magnetic fields in excess of 10^{12} Gauss with neutron stars sites for pulsars (e.g. see Pacini, 1968) indicates that such fields might be present in burst emission regions. These intense fields would probably be needed to effect some confinement of the relativistic plasma that produces the emission for the duration of a burst event (Lamb, 1984). These pieces of circumstantial evidence for strong magnetic fields in gamma-ray bursts were corroborated by the observation of low energy absorption features between 30 KeV and 70 keV in up to thirty percent of the Konus sources (Mazets, et al. 1981). The interpretation of these features as emanating from cyclotron emission/absorption lines leads to estimates of the magnetic field in the sources of around $10^{12} - 5 \times 10^{12}$ Gauss (see Mazets et al. 1981; Liang, Jernigan and Rodrigues, 1983; and Brainerd and Lamb, 1987).

The existence of these features has been questioned by some authors because of the difficulties of unfolding the spectra from raw instrumental data (Fenimore, *et al.* 1982). A further puzzle was why a second harmonic did not appear in most observations when it should be of the same strength as the first (Bussard and Lamb, 1982). Another problem with spectral observations is the apparent lack of absorption of high energy photons (above $2m_ec^2$) in a strong magnetic field, producing electron-positron pairs (the means of establishing a pair plasma in the emission region). The spectral turnover expected from this process is sharp (*e.g.* Baring, 1988b) and will occur when the product of the photon energy and the magnetic field are sufficiently large (*e.g.* Daugherty and Harding, 1983; Baring, 1988b). It will be shown in this paper that this turnover may not necessarily be very sharp, given certain angular distributions of particles and radiation. The doubts about the strength of magnetic fields in burst sources were significantly reduced with the observation of double absorption features in the low energy spectra of two burst sources detected by the Ginga satellite (Murakami, *et al.* 1988; Fenimore, *et al.* 1988) last year. The features are located at about 20 keV and 40 keV and are of roughly the same strength, thereby strongly suggesting that they are cyclotron in origin, with magnetic fields of about 2×10^{12} Gauss.

Other spectral properties include the presence of emission line features at between 400 and 500 keV in about seven percent of the Konus data (Mazets, *et al.* 1981), and also detected by other instruments (Barat, *et al.* 1984; Heuter, 1984). The natural interpretation of these lines was that they were due to electron-positron annihilation, redshifted below 511 keV by the gravitational field of the neutron star (the redshift is typical of that expected for neutron stars — see Liang, 1986). Most of these lines were narrow, corresponding to temperatures of around 50 keV (Ramaty and Mészáros, 1981), which is much cooler than that of the continuum emission. These lines are absent from the data from the SMM mission (Nolan, *et al.* 1983; Nolan, *et al.* 1984). The continuum normally extends from below the cyclotron frequency to a few hundred keV, or for harder spectra, into the MeV range. No signature of sharp pair absorption turnovers is evident at gamma-ray energies, above the (flux) limits of detection. No emission is found at low X-ray energies. This X-ray paucity places severe constraints on acceptable models (Epstein, 1986): since the neutron star surface is expected to degrade radiation to X-ray energies, the emission region must either be distant from it, or the escaping radiation must be beamed in a special way. The Konus continua were generally well fit by thermal models, using either bremsstrahlung or synchrotron emission. The SMM spectra are power-laws and so require non-thermal models: either inverse Compton scattering or synchrotron emission by relativistic electrons. The synchrotron emission mechanism in a region above the stellar surface is popular because of its efficiency, its prediction of the low energy features and also because it is compatible with a neutron star environment. It is the model under investigation here.

3. Quantum Processes in Strong Magnetic Fields

Before considering a specific synchrotron model of the continuum emission of burst sources (the pair cascade model), it is appropriate to review the basic and interesting properties of some relevant QED processes in strong magnetic fields. The first question that one may ask is when do quantum effects become important, since simpler classical models have often been used in gamma-ray burst modelling and are preferable unless unavoidable. The answer is that quantum mechanics becomes important when the quantization of an electron's motion has an observable influence on the description of the interactions it participates in. The transverse motion of an electron in a magnetic field is quantized, giving electronic energy levels E_n characterized by the quantum number n according to the formula

$$E_n^2 = (m_e c^2)^2 + (p_\| c)^2 + 2ne\hbar cB \quad , \tag{1}$$

where $p_\|$ is the electron momentum parallel to the magnetic field B (and is not quantized). The discreteness of energy states is then observable when the difference between the energies of consecutive states approaches the rest mass energy of the electron: *i.e.* when B approaches the quantum critical field strength of $B_c = m_e^2 c^3/e\hbar = 4.413 \times 10^{13}$ Gauss. The magnetic field then provides additional freedoms: electron momentum perpendicular to the field is not conserved in a QED interaction. This non-conservation of momentum occurs only on a microscopic scale: the residual momentum is taken up by the particles that sustain the field (it is said that the field 'absorbs' momentum), and on a macroscopic scale momentum *is* conserved. This changes the kinematics of classical processes and permits some interactions that are forbidden in field-free regions (such as the production of electron-positron pairs by a single photon) to be possible and even probable in strong magnetic fields. Quantum effects in synchrotron radiation, single-photon pair production and photon splitting will now be discussed. For simplicity, throughout this paper, electron and photon energies will be assumed dimensionless, having been scaled by $m_e c^2$, and the magnetic field strength will be expressed in units of B_c.

3.1. SYNCHROTRON RADIATION

Synchrotron emission is the only source of conversion of the energy of pairs to the photon population in an electron-positron/photon system such as in the pair cascades considered in the next section, and hence is the most important process for determining the burst spectrum. In the case of classical emission, when $B \ll B_c$, the photons produced by an emitting electron have energies of the order of $\gamma^2 B$. For an ultra-relativistic electron population, the criterion that this emission is classical is that the photon energy ε be much less than the initial electron energy γ, which is then equivalent to $\gamma B \ll 1$. Therefore, since classical emission is only possible for $B \ll 1/\gamma$, when $B \sim 1$, the emission is always quantum (*i.e.* $\varepsilon \sim \gamma$) and the produced photons assume a significant fraction of the particle kinetic energy. For gamma-ray burst fields of the order of $B \sim 0.1$, quantum effects in the synchrotron emission become important for relativistic pair distributions, though at low energies the emission is quasi-classical. The emission probability decreases below the classical value when the produced photon assumes a significant fraction of the electron energy because of the recoil of the photon. These features are evident in the form of the emission spectrum for relativistic electrons, given for example in Sokolov and Ternov (1968, see also Baring, 1988a):

$$j(\gamma,\varepsilon) = \frac{1}{\pi\sqrt{3}} \frac{e^2}{\hbar c} \frac{(m_e c^2)^2}{\hbar} \frac{\varepsilon}{\gamma^2} \left\{ \int_y^\infty K_{5/3}(x)\,dx + \frac{\varepsilon^2}{\gamma(\gamma-\varepsilon)} K_{2/3}(y) \right\} \quad , \tag{2a}$$

where

$$y = \frac{2}{3B\sin\alpha} \frac{\varepsilon}{\gamma(\gamma-\varepsilon)} \quad . \tag{2b}$$

Here the electron pitch angle is $\sin\alpha$ and the photon angle is integrated over. Spectra from Eq. (2) are illustrated in Brainerd and Petrosian (1987), and also Baring (1988a), illustrating the reduction of the emissivity due to photon recoil, as described by the argument of the Bessel functions in Eq. (2). The first term in Eq. (2) reduces to the familiar classical emissivity (the second becomes small) when $B \ll 1$.

The ultra-relativistic form in Eq. (2) can be integrated over a power-law distribution of electrons and positrons, $n_e(\gamma) \sim \gamma^{-p}$, to give a broken power-law for the emission spectrum. The limiting forms in the classical ($\varepsilon B \sin \alpha \ll 1$) and ultra-quantum ($\varepsilon B \sin \alpha \gg 1$) regimes are given in Brainerd and Petrosian (1987) and Baring (1988a), and have the dependences

$$j(\varepsilon, p) \sim \begin{cases} [B \sin \alpha]^{(p+1)/2} \varepsilon^{-(p-1)/2}, & \varepsilon B \sin \alpha \ll 1; \\ [B \sin \alpha]^{2/3} \varepsilon^{-(p-2/3)}, & \varepsilon B \sin \alpha \gg 1. \end{cases} \tag{3}$$

These predict a steepening of the spectrum if $p > 1/3$. This is illustrated in Baring (1988b). The fact that the index almost doubles in the quantum limit is a consequence of the photon energy scaling as $\varepsilon \sim \gamma$ compared with the classical limit dependence $\varepsilon \sim \gamma^2$. The decline in the synchrotron emission rate also contributes to the steepening. For an arbitrary distribution $n_e(\gamma)$ of pairs, the total synchrotron emissivity can be found by integrating Eq. (2a) weighted by the distribution. It should also be noted that quantum effects do not modify the cyclotron energy by much when $B = 0.1$ (Baring, 1988b).

3.2. SINGLE-PHOTON PAIR PRODUCTION

In strong magnetic fields, two processes are important for attenuation of the high energy portion of a radiation spectrum. The first of these is the production of electron-positron pairs by single photons "colliding" with the virtual photons of the magnetic field, which is well-known in the literature since it was proposed as the mechanism for the creation of pairs in pulsar models by Sturrock (1971). This process is forbidden in a field-free region because energy and momentum conservation cannot simultaneously be satisfied, but is quite probable at pulsar field strengths. This interaction is a first-order QED process and is related to synchrotron radiation by a crossing symmetry. The second process is the even more exotic interaction of photon splitting, and will be discussed shortly. Both these interactions are purely quantum in nature, with no classical counterparts. Therefore they only become probable when the magnetic field approaches the quantum-critical value of 4.413×10^{13} Gauss, i.e. in a neutron star environment.

The production of a pair by a single photon has a threshold energy of $2m_e c^2$. Below this, no pair production is possible. Above the threshold, the reaction rate is an exponentially rising function of the photon energy ε (e.g. see Erber, 1966), which is very strongly dependent on the size of the magnetic field: a field of greater than 2×10^{12} Gauss permits copious pair production in a neutron star magnetosphere, while a field of less than 10^{12} Gauss gives almost none. The optical depth for this process is only weakly dependent on the size of the region, and for the higher field strengths, it is very much greater than unity. All these features are evident from the well-known form for the optical depth (see Erber, 1966) well above the threshold:

$$\tau_\gamma(\varepsilon, \theta) \sim \frac{e^2}{\hbar c} \frac{m_e c^2}{\hbar} \frac{3\sqrt{3}}{16\sqrt{2}} B_\perp t_{esc} \exp\left\{-\frac{8}{3\varepsilon B_\perp}\right\}, \quad 2 \ll \varepsilon \ll 1/B_\perp. \tag{4}$$

Here $t_{esc} = c/R$ is the photon escape timescale for a region of size R and $B_\perp = B \sin \theta$, where θ is the angle of propagation of the photon with respect to the direction of the magnetic field. When the photon energy is very high ($\varepsilon \gg 1/B_\perp$), the optical depth decays as $\varepsilon^{-1/3}$, which for very strong fields only drops below unity at extremely high energies. This energy range is not of interest to the problem considered here. The formula in Eq. (4) is inaccurate by at least an order of magnitude near threshold (Baring, 1988a), since it is an approximation where the pairs produced are assumed to be ultra-relativistic (and hence displays an apparent threshold of $\varepsilon = 0$). For numerical computations of the optical depth near threshold, the "mildly relativistic" result derived in Baring (1988a) is more suitable: it treats the kinematics of the process exactly. The exponential form in Eq. (4) indicates that the depth is a rapidly rising function of the photon energy: this implies that the gamma-ray burst spectral attenuation due to this process should be very dramatic (Baring, 1988b). Another property of the optical depth that is evident from Eq. (4), and that is central to the anisotropy considerations of Sect. 5, is that the depth is a strong function of the angle θ of propagation of the photon with respect to the field. It is a maximum when $\theta = \pi/2$ and decreases dramatically when the photon moves at a small angle to the field. Near the threshold, the production spectrum of pairs maps the exponential rise of the optical depth in Eq. (4), but at very high energies it has a power-law dependence for power-law photons (Baring, 1989a).

3.3. PHOTON SPLITTING

The importance of photon splitting in a neutron star magnetosphere was outlined by Usov and Shabad (1983), who predicted that the polarization properties of the process would provide a diagnostic for observation of pulsar field strengths. The importance of this process for the modification of gamma-ray burst spectra was discussed by Mitrofanov et al. (1986) and by Baring (1988b). It is a third order QED process and is effectively the Compton scattering of a photon off a virtual photon of the magnetic field. It is strictly forbidden in a field free zone. Simple analytic results for the spectral rate of splitting exist for photon energies $\varepsilon \ll 1$ and $\varepsilon \gg 1$ in the weak field limit, $B \ll 1$. The region of most interest in the study of pair cascades in burst sources is just below the pair creation threshold where the splitting is not swamped by pair production and can therefore degrade the emission spectrum. At low photon energies, Adler (1971) showed that the optical depth (dependent on the photon polarization states) has the form

$$\tau_{sp}(\varepsilon, \theta) \propto \{B \sin \theta\}^6 \varepsilon^5 \quad , \quad \varepsilon \ll 1/(B \sin \theta) \quad . \tag{5}$$

and is of the order of unity for $\varepsilon \approx 1$ in a gamma-ray burst environment (Baring, 1988b). Since it can act below the pair production threshold, it can yield a "precursor" to the pair absorption turnover expected in gamma-ray spectra (Baring, 1989a) and that the spectral break may be of the order of an index of five. This "precursor" turnover, when observed, may be a useful observational diagnostic for the magnetic field in gamma-ray burst sources.

4. Steady-state Synchrotron Pair Cascades

The simplest form of synchrotron emission mechanism model for the continuum spectrum of gamma-ray burst sources involves the solution of the steady-state kinetic equations for the electron and photon distribution functions allowing for the processes of synchrotron radiation and single-photon pair production in the strong magnetic field. The influence of photon splitting, while significant, is only discussed briefly. For simplicity, the emission region is assumed to be optically thin to Compton scattering so that the complications of scattering in strong fields are avoided. An equilibrium assumption is not inappropriate since the emission from burst sources has a variability on timescales comparable to or greater than the light-crossing time for neutron stars (*e.g.* see Lamb, 1984), which is much longer than the timescales of the cascade processes of synchrotron radiation and pair production. The aim of this section is to describe the spectra that are produced self-consistently within a steady-state scenario.

How is a pair cascade formed? Well, here the gamma-ray burst emission region is assumed to be of size 10^6 cm containing uniform magnetic field. Electrons or pairs are injected into the emission region at a fixed rate with a power-law distribution. They cool by synchrotron radiation and do not escape, but form a cold component which annihilates. The photons they create produce pairs if their energy energy exceeds $2m_e c^2$ and these first-generation pairs subsequently reradiate synchrotron photons. Multiple generations occur from electrons injected at high energies. This process of sequential synchrotron emission and pair production is a pair cascade. Each generation of reprocessed radiation and particles occurs at lower energies so that the net effect is to absorb gamma-ray rays and bolster the X-ray spectrum. It is intended here to find the changes that occur in the hard X-ray spectral index. Historically the ideas of a pair cascade have been most developed in the theory of pulsars (involving curvature radiation instead of the synchrotron process) and of the non-thermal X-ray and gamma-ray continua of AGN (with inverse Compton scattering as the process of radiation production). The criterion for a cascade to develop in AGN is that the radiation field be compact (since pairs are then created by the two-photon field-free process). Here, since the single-photon pair production process operates, the magnetic field strength must be large, namely over 10^{12} Gauss. Also the cascade is linear in the photon distribution since pairs are produced by a single photon — this feature simplifies the mathematical analysis of such a cascade considerably. The acceleration mechanism for the relativistic electrons is not considered here. It could be due, for example, to shocks or to electric fields.

4.1. THE CASCADE EQUATIONS

The source of energy in this model, the injection of the relativistic electrons, is assumed to occur at a constant rate, which defines one parameter of the system. The maximum injection energy γ_{max} and the injection index p are two further parameters. In the pair cascade model considered here, γ_{max} is set to be 10^8. This value is quite possible for motion along the field, but is rather high for motion transverse to the field lines even for the most efficient acceleration mechanisms. It is chosen purely for mathematical convenience, for the illustration of high energy spectral results. The nature of the pair cascade reprocessing is barely changed by lowering γ_{max} to 10^2. The steady-state equations defining the pair cascade are derived from the kinetic equations for the evolution of the photon and pair distributions. They assume quite simple forms, but have to be solved by a mixture of analysis and numerical procedure. The equation for the photon number density $n(\varepsilon)$ is

$$0 = \dot{n}_s(\varepsilon) - \left\{ 1 + \tau_\gamma(\varepsilon) \right\} \frac{n(\varepsilon)}{t_{esc}} \quad . \tag{6}$$

Here $\dot{n}_s(\varepsilon)$ is the synchrotron emission rate, $\tau_\gamma(\varepsilon)$ is the optical depth to single-photon pair production. The rate for photon splitting has been omitted. The escape timescale is $t_{esc} = h/c$ where h is the height of the emission/absorption region. If $n_e(\gamma)$ is the number density of electrons and positrons, then their steady-state equation takes the form

$$0 = P(\gamma) + Q(\gamma) + S_s(\gamma) - n_e(\gamma)R_s(\gamma) \quad . \tag{7}$$

$Q(\gamma)$ is the rate of injection of electrons, $P(\gamma)$ is the rate of production of pairs by photons in the magnetic field, $S_s(\gamma)$ is the rate of "scattering" of pairs into a given energy band by synchrotron emission, and $n_e(\gamma)R_s(\gamma)$ is the rate of "scattering" of pairs out of an energy band due to the synchrotron process. This equation is for the non-thermal relativistic component of the pair distribution. The cooling at low energies and formation of a thermal component is difficult to treat analytically, and so the thermal component is considered separately. Its density is maintained at a level that balances thermal two-photon annihilation with the non-thermal single-photon pair production. The description of the solution of Eqs. (6) and (7) is the objective of this section.

4.2. THE PAIR DISTRIBUTION

Solution of Eq. (6) for the photon distribution is straight-forward, given the pair distribution. Solving Eq. (7) for the steady-state pair distribution is, however, analytically impossible except in special limiting cases. Nevertheless, it is possible to proceed via a semi-analytic method by interpolating numerically between these cases. This saves time computationally since an integral equation does not then have to be solved numerically. Equation (7) can be cast into the form

$$0 = P(\gamma) + Q(\gamma) - n_e(\gamma) \int_1^\gamma d\gamma' \, R_s(\gamma, \gamma') + \int_\gamma^\infty d\gamma' \, n_e(\gamma') \, R_s(\gamma', \gamma) \tag{8}$$

where $R_s(\gamma, \gamma') = j(\gamma, \gamma - \gamma')/(\gamma - \gamma')$ is the synchrotron reaction rate, $j(\gamma, \varepsilon)$ is given in Eq. (2) and $\gamma' = \gamma - \varepsilon$ is the electron energy after synchrotron emission. The upper limit of infinity on the second integral (the production spectrum) is obviously replaced by the maximum energy of the injected pairs if this is not infinite. Equation (8) incorporates an assumption that the pairs are isotropic at all times (which simplifies the analysis and which will be relaxed in the next section). Synchrotron cooling anisotropizes the distribution, making it more parallel to the field, but this makes no qualitative change to the cascade process. In the classical limit, the emission gives photons of energies very much less than γ so that the radiation leads to differential changes in the electron's energy. The integral Eq. (8) can then be written as a first-order Fokker-Planck equation. This differential equation is then easily solved to give

$$n_e(\gamma) = -\frac{1}{\dot{\gamma}_s} \int_\gamma^\infty \left[Q(\gamma') + P(\gamma') \right] d\gamma' \quad . \tag{9}$$

This form of solution has been derived in the theory of Thomson pair cascades (see Lightman and Zdziarski, 1987 for example). If the electrons are ultra-relativistic, as is assumed here, the cooling rate is $-\dot{\gamma}_s \propto B^2 \gamma^2$. The form of Eq. (9) is quite easily explained. Since the cooling is differential in the electron energy, every electron injected with energy greater than γ will eventually cool to energy γ and then further. Therefore the population level at γ is simply the ratio of the rate of injection above γ to the rate at which the particles cool through energy γ. Hence the form of Eq. (9). Clearly, for power-law injection (assuming that the pair-production term $P(\gamma)$ is negligible) the cooling steepens the pair distribution by index unity at ultra-relativistic energies. For monoenergetic injection, the pair index is roughly two, which is the same for injection by a power-law with index unity. This just reflects the common feature of these two situations: that most of the energy is injected at the maximum value of γ.

When quantum effects become important ($\gamma B \gg 1$), the cooling is not continuous but discrete, and so such a simple situation does not exist. Equation (8) cannot then be rewritten as a Fokker-Planck equation. Another approach is needed and is detailed in Baring, 1989a). The first step is to note that the production spectrum assumes a simple form in the ultra-quantum limit. The Bessel functions in the emissivity assume their power-law asymptotic form and it is straight-forward to show that, to leading order in $1/(\gamma B)$, Eq. (8) assumes the form $R_s(\gamma, \gamma') = R_\infty(\gamma)\beta(\gamma'/\gamma)/\gamma$, with $R_\infty(\gamma) \sim \gamma^{-1/3}$ (Baring, 1989a). This simple form for the synchrotron reaction kernel (R_∞ is the reaction rate in the ultra-quantum limit), is a consequence of the fact that the average energy of a synchrotron photon asymptotically approaches a fixed fraction of the electron energy in the ultra-quantum limit. Setting $q(\gamma) = P(\gamma) + Q(\gamma)$, and $f(\gamma) = R_\infty(\gamma)n_e(\gamma)$ to define the pair distribution, the steady-state equation for the pair distribution (Eq. (8)) becomes

$$q(\gamma) = f(\gamma) - \int_0^1 \frac{dt}{t} h(t) f(\gamma/t) \ . \tag{10}$$

By observation, for power-law injection $q(\gamma) = N\gamma^{-p}$, a power-law solution for $f(\gamma)$ results, and since $R_\infty(\gamma) \sim \gamma^{-1/3}$, the pair distribution is $n_e(\gamma) \sim \gamma^{-(p-1/3)}$. Therefore, in the quantum limit, the cooling flattens the spectrum by an index of one-third instead of steepening it by unity as in the classical case. This flattening reflects the decrease in the reaction rate (from being constant in the classical case to declining as $\gamma^{-1/3}$ in the quantum limit — see Baring, 1989a), and the quantum nature of the emission i.e. the average energy lost per interaction by an electron varies as γ^2 in the classical limit and as γ in the quantum.

For general particle injections, Eq. (10) can be solved by Laplace or Mellin transorm techniques: the relevant details can be found in Baring (1989a). There it is shown that the solution for the electron-positron distribution in the ultra-quantum limit is

$$n_e(\gamma) = \frac{1}{R_\infty(\gamma)} \left\{ q(\gamma) + \int_\gamma^\infty \frac{d\xi}{\xi} q(\xi) \Phi(\gamma/\xi) \right\} \ , \tag{11a}$$

with

$$\Phi(x) \approx \frac{\rho_1}{x} + \rho_2 \Big[\log(1/x)\Big]^{-1/3} + \rho_3 \Big[\log(1/x)\Big]^{-2/3} \ , \quad 0 < x < 1 \ , \tag{11b}$$

for $\rho_1 = 1.899$, $\rho_2 = 0.334$ and $\rho_3 = 0.354$. This approximation to the Green's function in Eq. (11a) is derived in Baring (1989a) and is accurate to better than three percent over the whole domain $0 < x < 1$. For numerical computations of the pair distribution it is sufficient to use the asymptotic solutions in Eqs. (9) and (11a) in their domains of validity, and to interpolate between them in the transition region (typically $5 < \gamma < 200$ for large $\gamma_{\max}$ and $B \sim 0.1$). The accuracy of this approximation is, at worst, twenty percent (Baring, 1989a). The form of Eq. (11) can be explained as follows. The presence of the injection term $q(\gamma)$ corresponds to particles being injected with energy γ and being scattered to lower energies. The term involving an integral over the injection distribution above γ gives the contribution from particles with higher energies scattered into and then out of the channel with energy γ. Not all electrons so injected pass through this energy channel, since the cooling is discrete. Noting the dominance of the x^{-1} term in Eq. (11b), the integral term in Eq. (10) has the appearance of an integrated injection rate divided by the quantum cooling rate, *i.e.* it is similar to the result in Eq. (9).

4.3. PHOTON SPECTRA

Having derived solutions for the particle distributions, photon spectra can now be obtained. Before including the effects of pair production, it is instructive to obtain the spectra from pure synchrotron cascades. Neglecting pair production, the injection of electrons appearing in the equations for the equilibrium electron distribution can be simply prescribed with no need to determine any reprocessing. The photon spectrum is then a solution of Eq. (6), setting the photon absorption depth $\tau_\gamma(\varepsilon)$ to zero. For power-law injection, the asymptotic limits for the electron distribution obtained from Eqs. (9) and (11) are power-laws, so the emission spectrum results in Eq. (5) are useful. These yield limiting forms in the classical ($\varepsilon B \ll 1$) and ultra-quantum ($\varepsilon B \gg 1$) regimes for the steady-state photon *number* distribution of

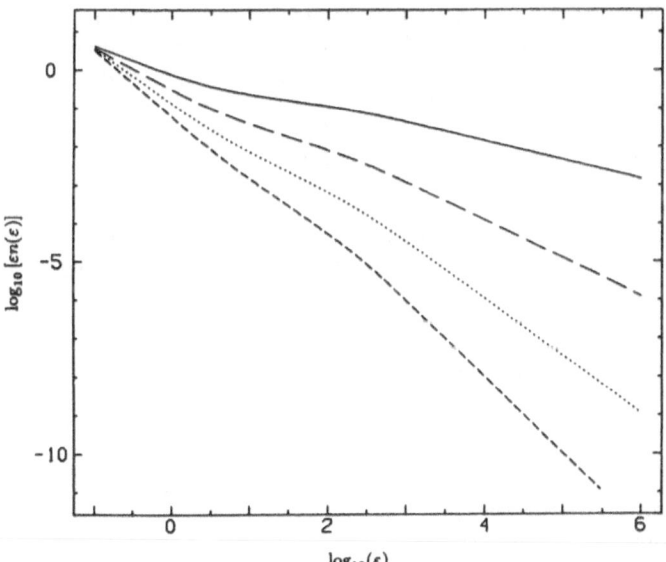

Figure 1 The synchrotron emission spectra for power-law electron injection (see Eq. (12)). The electron injection index is p, giving classical spectral indices (for $\varepsilon < 1$) of $p/2 = 0.75$, 1.00, 1.25 and 1.50 in order from top to bottom. The corresponding quantum indices are $p-1 = 0.5$, 1.0, 1.5 and 2.0 when $\varepsilon \gg 1$. The quantum spectrum is flatter than the classical for $p < 2$ and steeper otherwise.

$$n(\varepsilon) \propto \begin{cases} \varepsilon^{-(p+2)/2}, & \varepsilon B \ll 1; \\ \varepsilon^{-p}, & \varepsilon B \gg 1, \end{cases} \qquad (12)$$

with the coefficients being given in Eq. (27) of Baring (1989a). Therefore, the spectrum will steepen in the quantum regime if the injection index p exceeds two; otherwise it will flatten. Numerical computations confirm this behaviour and are illustrated in Fig. 1. Whether or not the index increases in the quantum regime reflects whether or not the increase in the electron population due to the discrete cooling dominates the reduction in the emissivity (caused by the photon recoil). The curves in Fig. 1 display a 'transition shelf,' which reflects the competition between the decrease in spectral emission and the increase in the equilibrium electron population in the quantum regime. Notice that the photon *number* distribution index is precisely the same as the electron injection index. This is a consequence of the flattening in the quantum regime of the electron distribution being described exactly by the decline in the synchrotron cooling rate, together with the fact that the emitted photons assume significant fractions of the particles' energies. For monoenergetic particle injection, the steady-state electron distribution has an index of two in the classical regime and of 2/3 in the quantum limit. Therefore it can be deduced that the corresponding limiting indices for the photon spectrum are 1/2 and zero, respectively. These also correspond to the case of power-law injection of index unity.

For the environment of a gamma-ray burst source, the magnetic field is sufficiently large that the absorption depth of photons to pair production is much greater than unity. This leads to dramatic attenuation of the gamma-ray spectrum somewhere above the threshold $\varepsilon = 2$. The absorption depth rises exponentially above threshold roughly according to Eq. (4), peaks at $\varepsilon \sim 1/B$, and then declines very slowly as the power-law of index 1/3. The photon spectrum turns over when the depth equals unity: for an emission region of size 10^6 cm and field $B = 0.05$, this occurs at $\varepsilon \approx 2.5$. The radiation from the produced pairs can potentially alter the spectral slope below the turnover: the degree of this alteration is the main concern here. Successive generations of photons and pairs are produced by electrons injected at high energies as the cascade process continues until the threshold is reached. The net effect is that virtually no photons will be emitted above threshold, but the absorbed ones are reprocessed to energies below $\varepsilon = 2$. The emerging photon spectrum is found by numerically solving Eqs. (6) and (7). Simple analysis can reveal the nature of the resulting spectra. First consider energies well above the pair production threshold. Although not much of the photon luminosity is emitted at these energies, for flat electron injection spectra, most of the power is injected at these energies, and so the nature of the radiation spectrum below threshold will be strongly influenced by the reprocessing at high energies. Assume a power-law injection of electrons or pairs with index p. What is the expected production spectrum of first order pairs? In this ultra-quantum regime, the equilibrium electron index is $p - 1/3$ (see Sect. 4.2), giving an unattenuated photon distribution index of p (it can be said that the photon distribution "traces" the injection of electrons as their indices are the same), as in Eq. (12). The decaying optical depth to pair production at these energies (of index $-1/3$) gives an equilibrium photon spectral index of $p - 4/3$, using Eq. (6), which then results in the creation of a first order pair generation with index p (see Baring, 1989a). Therefore it follows that each successive generation of pairs in the ultra-quantum regime has the same index, and this equals the electron injection index. Therefore the successive pair generations "trace" the injection. The pair production rate traces the photon distribution for exactly the same reason that the photon production rate traces the injection: the attenuation of photons modifies the photon distribution (to give its equilibrium form) in exactly the same fashion that the pair production rate declines. This "tracing" is an important feature of this linear cascade, since it implies that for power-law injection, the photon distribution at energies well above the absorption turnover has the same index as the electron injection. Generally a decrease in the injection index gives an increase in the contribution of pairs, and this contribution can dominate the injection (Baring, 1989a).

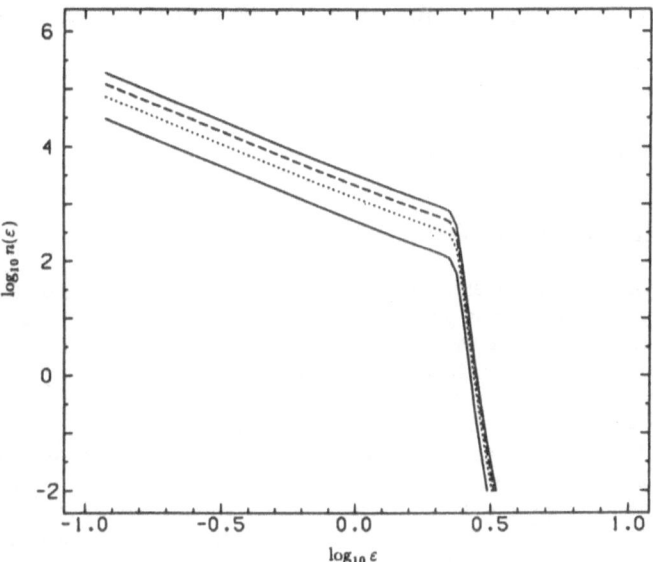

Figure 2 The low-energy portion of the photon number spectrum for an injection index of $p = 1.75$ and magnetic field of $B = 0.05$. The various orders of pair reprocessing are shown: from bottom to top, the first, third and fifth, with the equilibrium spectrum illustrated as the solid curve at the top. Each order is represented as the sum of all lower orders. The turnover is effectively a truncation. Other parameters for this cascade are $\gamma_{max} = 10^8$ and a height of $h = 10^6$ cm for the photon absorption/emission region.

The reprocessing at high energies is a major factor in determining the spectrum at much lower energies. The high energy spectrum is of little consequence on its own because its small magnitude makes it effectively unobservable in sources. The cascading at high energies merely provides a source of mildly relativistic pairs as the end product. However, the photons created at low gamma-ray energies $\varepsilon < 2$ are produced by pairs with energies $\gamma \sim 10 - 20$. These are in turn created by photons with energies $\varepsilon \sim 20 - 40$ almost in the power-law portion of the spectrum. Therefore 'moderate' energy pairs that form the lower portion of the quantum tail of the cascade (and hence have an index of p) produce the reprocessed low-energy gamma-ray photons and hard X-rays below the pair absorption threshold, via classical synchrotron emission. It follows that the spectral index for the higher order contributions (all of them) of reprocessed radiation is $p/2$ just below the turnover, and is precisely the spectral index of radiation in the classical regime when the contribution from reprocessing is small. This important result, that *the reprocessing of the radiation spectrum due to the pair cascade does not change the hard X-ray spectral index,* is a consequence of the higher order pair generations "tracing" the injection, and the fact that the quantum production of pairs influences the non-quantum regime of synchrotron photon production. The reprocessing contribution for different orders is plotted in Fig. 2, illustrating the unchanging spectral index. Note that this unchanging behaviour contrasts with classical regime of Compton pair cascades, where the pair reprocessing can change the X-ray spectral index significantly (see for example Lightman and Zdziarski, 1987). Also note that a 'pile-up' of particles will occur at mildly relativistic energies that will affect the spectrum at X-ray energies lower than those considered here. In Fig. 2, $\gamma_{max} = 10^8$ was chosen. Lowering this to values as low as 10^2 does not change the nature of the cascade reprocessing very much: the spectral index below the turnover remains unaltered and only the overall contribution of the reprocessed radiation is reduced. Since photon splitting acts below the pair absorption turnover, it can create a precursor turnover at slightly lower energies that may be an observable diagnostic for the magnetic field in gamma-ray burst sources. This is discussed in Baring (1989a).

5. The Influence of Distribution of Photon Angles

Until now, the synchrotron model has made the simplifying assumptions of isotropy of particles and propagation of photons perpendicular to the field. The latter has a particularly strong influence on the shape of the spectrum: having a single photon angle guarantees an extremely sharp pair absorption turnover. This section aims to investigate how relaxing these assumptions can modify the gamma-ray cascade spectrum. Since the processes of synchrotron emission and pair production have the greatest rates when the species move perpendicularly to the field, and this situation has been suitably included in the model, the main conclusion of the last section still holds: that the inclusion of the cascade reprocessing does not change the X-ray spectral index. Anisotropy of electrons can easily arise in realistic models of the emission region. Acceleration mechanisms caused by shocks and electric fields can give anisotropy of particles (discussed in more detail in Baring, 1989b), and the synchrotron cooling process is itself an agent of anisotropization. These factors all tend to leave more particles moving more parallel to the magnetic field lines. In modelling anisotropic electron distributions it is desirable to avoid the specifics of any particular model. So here, a parametrization of the anisotropy that is independent of the acceleration model is needed. A convenient, though not unique, choice is that the angular distribution of electrons take the form $\phi(\alpha) \propto (\sin \alpha)^{-\lambda}$ for $\lambda \geq 0$. This choice has the advantages that it renders angular integrations of the spectrum analytically tractable and that it gives anisotropies favouring the directions nearly parallel to the field. The distribution may be restricted by a maximum pitch angle α_{max}: $0 \leq \alpha \leq \alpha_{max} \leq \pi/2$. Other parameterizations of the electron anisotropy are conceivable, but less convenient, and may not yield the same insight into spectral modifications by electron and photon anisotropy.

In determining the effect of the distribution of photon angles on the emergent spectrum, the details of the cascade reprocessing are superfluous. The spread of photon angles mutes the severity of the spectral turnover. It is clear from the form of Eq. (4) that for a given angle θ of photon propagation, the turnover energy (where the optical depth equals unity) is a strongly varying function of θ: in fact $\varepsilon_t \sim 1/\sin\theta$. Because this turnover occurs at energies much less than $1/B_\perp$, the unattenuated portion of the spectrum corresponds to synchrotron radiation in the classical regime (see Baring, 1989a). In a realistic observational situation, a single angle θ is not detected by an observer, but rather a spread of angles since the field lines curve and the emission is expected from a region that is a significant portion of the magnetosphere. Therefore it is quite reasonable to assume that the observed spectrum would consist of a sum over possible photon angles θ of the classical synchrotron spectrum in Eq. (3), with each angle having a different turnover:

$$J(\varepsilon,p) \propto \varepsilon^{-(p-1)/2} \int_0^{\alpha_{max}} d\theta \, \frac{\{\sin\theta\}^{(p-2\lambda+3)/2}}{1 + \tau_\gamma(\varepsilon,\theta)} \quad . \tag{13}$$

Here it has been assumed that the angle of emission of synchrotron photon θ is the same as the pitch angle α of the emitting electron, which is true for E.R. electrons. Therefore the maximum angle of photon propagation is α_{max}. Anisotropy of the electrons is assumed when either $\alpha_{max} < \pi/2$ or $\lambda > 0$. The pair production optical depth is given by Eq. (4).

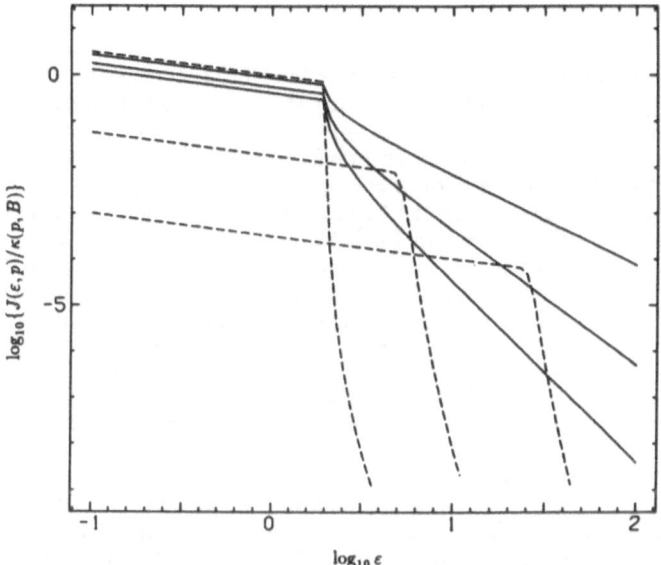

Figure 3   The synchrotron spectrum $J(\varepsilon,p)$ (suitably scaled) from a power-law electron distribution with index $p = 2$ in a magnetic field of $B = 0.1$, attenuated by single-photon pair production according to Eq. (13) in a region of size 10^6 cm. The dashed curves show the "truncated" spectra from emission of photons at different values of the angle θ of the photon direction with respect to the field. For these curves, $\sin\theta = 1.0, 0.2$, and 0.04, with the values decreasing as the cutoff occurs at higher energies. The solid lines are the spectra, integrated over all angles, in Eq. (13) for three different values of the electron anisotropy parameter: $\lambda = 0, 1$ and 2 (increasing as the high energy portions of the spectra flatten). Here $\alpha_{max} = 1$.

The behaviour of this spectrum is easily deduced. Since the attenuation for any given angle θ of photon propagation is effectively a truncation, then the denominator in the integrand in Eq. (13) can be well approximated by a step function. For each θ, as $\sin\theta$ decreases, the cutoff ε_t increases approximately as $1/\sin\theta$. Therefore, the sum of contributions from different θ is a sum of truncated continua that extend to higher energies as θ decreases. The net result is an envelope of all of these contributions (see Fig. 3), and gives a turnover to the spectrum that is significant, but much less marked than the truncation observed for each value of θ. This is the main result of this section: that *the distribution of angles of propagation of photons in the field gives a much more gradual turnover than is expected for mono-directional photons.* The magnitude of the unattenuated portion of each contribution decreases as θ decreases. Analytically, it is possible to derive the spectral index of the hard portion of the spectrum. For a given energy $\varepsilon > 2$ of emission, there is clearly a maximum value of θ, for which radiation of that energy is not attenuated by pair production. This value is given by $\theta_{max} = \zeta(\varepsilon) \sim \arcsin\{8/3\varepsilon B\}$ for $\varepsilon \gg 2$, and is the value which gives the greatest contribution to the radiation at energy ε. When ε is small enough so that $\tau_\gamma < 1$, then $\zeta(\varepsilon) = \alpha_{max}$. For large photon energies, Eq. (13) is then easily integrable giving

$$J(\varepsilon, p) \sim \varepsilon^{-(p-\lambda+2)} \quad , \quad \varepsilon \gg 8/(3B \sin \alpha_{max}) \quad . \tag{14}$$

This steepening above the turnover is illustrated in Fig. 3: the spectral break is independent of the magnetic field strength. Increasing the anisotropy parameter λ flattens the turnover, as is evident from Fig. 3 and Eq. (14), and is caused by a pile-up of photons at small angles θ. For $\lambda = 0$, the spectrum breaks by an index of $(p+5)/2$. Therefore, an anisotropic electron distribution with $\lambda = (p+5)/2$ gives no steepening of the spectrum at all, just a dip that occurs where the turnover in the spectrum would normally be: at $\varepsilon \sim 8/(3B \sin \alpha_{max})$. The spectra in Fig. 3 are for a field of $B = 0.1$. The spectral slopes at low and high energies do not change when the field is changed, however their magnitudes do. Lowering the magnetic field reduces the attenuation and so increases the hard component of the spectrum, as can be deduced from the form of Eq. (13).

While the parameter λ adjusts the sharpness of the spectral turnover, α_{max} determines the energy, at which it occurs. These parameters can encompass a broad range of models and will now be used for the spectral fitting of a gamma-ray burst spectrum. Since the observations of gamma-ray burst spectra in the hard gamma-ray region are generally not at all precise, with large error bars, the choice of the parameters α_{max} and λ is somewhat subjective. The index p is easier to determine because the hard X-ray observations have smaller uncertainties. The parameters quoted here should therefore be regarded as estimates awaiting better spectral data. The value of α_{max} is the hardest to determine (*i*) because spectral data do not always have a clearly-defined position for the turnover, and (*ii*) because the magnetic field is an unknown quantity if no low-energy absorption features are observed: the field strength is needed to determine α_{max} from the energy of the turnover. A fit to the spectrum of the 25 January, 1982 event, an SMM observation, is given in Fig. 4. The data are taken from Nolan *et al.* (1984), where the spectrum was modelled with a thermal synchrotron continuum. There is clearly a turnover at around $\varepsilon_t \sim 10$, which corresponds to $\alpha_{max} \sim 0.2$ if the magnetic field is about $B = 0.05$. The soft and hard (energy flux) spectral indices are $q_s = 0.2$ and $q_h = 2.6$ respectively, which yield $p = 1.4$ and $\lambda = 0.8$. The value of λ is not that large, and could be chosen smaller and still fit the data (remembering that the highest energy data point is an upper bound only). The value of α_{max} suggests a significant, though not unreasonable, beaming of the injected electrons: various injection mechanisms may give such an anisotropy. It is also important to note that a value of $\lambda \sim 2$ is expected from synchrotron cooling of an isotropic injection distribution (see Appendix C of Brainerd and Lamb, 1987), so significant values of λ are possible in realistic emission models. Other sources with marginal high energy spectral breaks can have their spectra fit by this model (Baring, 1989b).

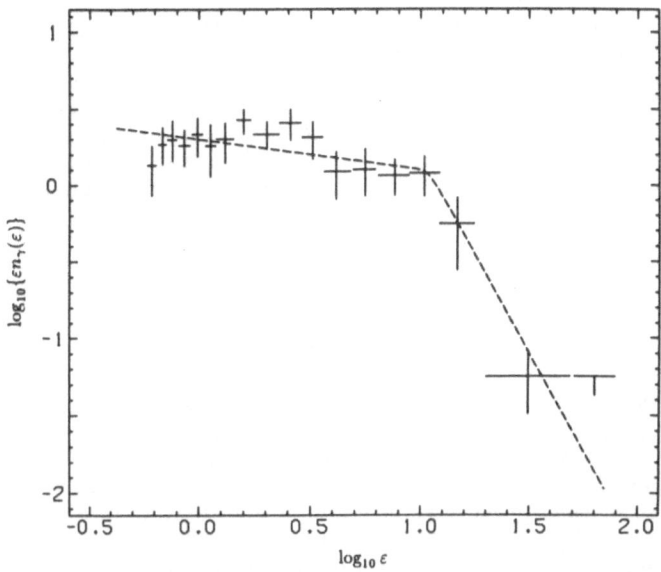

Figure 4 A fit to the spectrum of the 25 January, 1982 event. The observational data are taken from Nolan *et al.* (1984), and were multiplied by the photon energy to give an energy flux spectrum. The fit from Eq. (13) was scaled accordingly and the electron anisotropy parameters are $\alpha_{max} = 0.2$ and $\lambda = 0.8$. The electron power-law index was $p = 1.4$. The data point of highest energy, being purely an upper bound, was not considered in the fit. The magnetic field chosen for the spectral fit was $B = 0.05$.

6. Summary

In this paper quantum effects that occur in the strong magnetic fields that can arise in gamma-ray burst emission regions and their influence on source spectra were studied. The properties of gamma-ray burst sources were reviewed. A synchrotron pair cascade model for burst emission was developed, in which the steady-state kinetic equations for the electron-positron and photon distributions were solved self-consistently in the strong field. In such a field, single-photon pair production and photon splitting become very probable. The features of these processes were discussed in some detail in Sect. 3. In Sect. 5, the spectra from synchrotron-cooled equilibrium distributions were obtained. Including pair production leads to the important result that for equilibrium, all pair generations in the cascade have the same spectral index at high energies as the electron injection. This "tracing" of the injection then implies that the inclusion of pair reprocessing *does not alter the hard X-ray spectral index*, which therefore retains its classical value of $p/2$ for an electron injection index of p. The energy and sharpness of the pair absorption turnover at $\varepsilon \sim 2$ is not modified by the reprocessing. Photon splitting produces a spectral break just below the pair absorption turnover, which if observed, provides another possible way to obtain an estimate for the magnetic field. It was found that allowing for a distribution of photon propagation angles in a strong magnetic field, the turnover in synchrotron spectra due to photon absorption by pair production was broadened from the truncation expected for a single photon direction, to a power-law tail. Two parameters α_{max} and λ (specifying injection anisotropy: see Sect. 5), defined a wide range of broken power-laws that can be used to fit spectral observations. The spectrum of the event of 25 January, 1982 and 5 August, 1984 was fit quite easily, indicating that attenuation of the gamma-ray tail by pair production may be compatible with the extended hard spectra such as in the SMM observations. The model can be tested with future, more-refined gamma-ray observations.

REFERENCES:

Adler, S. L.: 1971 *Ann. Phys.* **67** , 599

Barat,C., *et al.* .: 1984 *Astrophys. J. (Letters)* **286** , L11

Baring, M. G.: 1988a *Mon. Not. Roy. astr. Soc.* **235** , 51

Baring, M. G.: 1988b *Mon. Not. Roy. astr. Soc.* **235** , 79

Baring, M. G.: 1989a *Astron. Astrophys.* in press

Baring, M. G.: 1989b *Mon. Not. Roy. astr. Soc.* submitted

Brainerd, J. J., and Lamb, D. Q.: 1987 *Astrophys. J.* **313**, 231

Brainerd, J. J., and Petrosian, V.: 1987 *Astrophys. J.* **320**, 703

Bussard, R. W., and Lamb, F. K.: 1982 in *Gamma Ray Transients and Related Astrophysical Phenomena* eds. Lingenfelter, R. E., *et al.* , (AIP, New York) p 189

Daugherty J. K., and Harding, A. K.: 1983 *Astrophys. J.* **273** , 761

Epstein, R., 1986 in *Radiative Hydronamics in Stars and Compact Objects* eds. Winkler, W., H., and Mihalas, D., (Springer, Heidelberg)

Erber, T.: 1966 *Rev. Mod. Phys.* **38**, 626

Fenimore, E. E., *et al.* .: 1988 *Astrophys. J. (Letters)* **335** , L71

Guilbert, P. W., Fabian, A. C., and Rees, M. J.: 1983 *Mon. Not. Roy. astr. Soc.* **205** , 593

Heuter, G. J.: 1984 in *High Energy Transients in Astrophysics* ed. Woosley, S. E., (AIP, New York) p 373

Klebesadel, R. W., Strong, I. B., and Olson, R. A.: 1973 *Astrophys. J. (Letters)* **182** , L85

Lamb, D. Q.: 1984 *Ann. New York Acad. Science* **422**, 237

Liang, E. P.: 1986 *Astrophys. J. (Letters)* **308** , L17

Liang, E. P., Jernigan, T. E., and Rodrigues, R.: 1983 *Astrophys. J.* **271**, 766

Lightman, A. P., and Zdziarski, A. A.: 1987 *Astrophys. J.* **319** , 643

Mazets, E. P.: 1988 *Adv. Space Res.* **8** , 669

Mazets, E. P., *et al.* .: 1979 *Nature* **282** , 587

Mazets, E. P., *et al.* .: 1981 *Nature* **290** , 378

Mitrofanov, I. G., *et al.* .: (1986) *Sov. Astron.* **30** , 659

Murakami, T., *et al.* .: 1988 *Nature* **335** , 234

Nolan, P. L., *et al.* .: 1983 in *Positron-Electron Pairs in Astrophysics,* eds: Burns, M. L., *et al.* , (AIP, New York) p 59

Nolan, P. L., *et al.* .: 1984 in *High Energy Transients in Astrophysics* ed. Woosley, S. E., (AIP, New York) p 399

Pacini, F.: 1968 Nature **219**, 145

Paczyński, B.: 1986 *Astrophys. J. (letters)* **308** ,L43

Ramaty, R., and Mészáros, P.: 1981 *Astrophys. J.* **250** , 384

Schaefer, B. E.: 1984 *Nature* **294** , 722,

Sokolov, A. A., and Ternov, I. M.: 1968 *Synchrotron Radiation* (Pergamon Press, Oxford)

Sturrock, P. A.: 1971 *Astrophys. J.* **164** , 529

Usov, V. V., and Shabad, A. E.: 1983 *Sov. Astron. Letters* **9** , 212

GAMMA-RAYS FROM ACCRETION PROCESSES AND RELATIVISTIC BEAMS

F. Giovannelli ^, W. Bednarek", S. Karakula"^, W. Tkaczyk"

^ Istituto di Astrofisica Spaziale, CNR,
C.P. 67, I 00044 Frascati, Italy.
" Institute of Physics, University of Lodz,
Ul. Nowotki 149/153, PL 90-236 Lodz, Poland.

ABSTRACT. In this paper we review our recent works in studying the production of γ- rays from collapsed objects. The main discussed process is always the π^0-decay. We simply changed the geometry of the sources, starting from the spherical symmetry to a strong anisotropy like in the beamed sources. Firstly we starded the calculations for the processes producing γ-rays via π^0-decay in the spherical accretion of matter onto collapsed objects and then we moved in studying the interactions of a monoenergetic, one dimensional beam, with a Lorentz factor γ and velocity of the beam β, with the surrounding matter and radiation.
   In the spherical accretion onto massive objects, the matter may be heated up to temperature as high as 10**12 K. In such a relativistic plasma, inelastic collision of protons may produce π^0 which are the γ-ray source. We determined the γ- ray energy production spectra in the comoving plasma system for different temperature and expected γ- ray energy spectra for the case of spherically symmetric accretion of matter onto a black hole. The predictions of the model gauged to the quasar 3C 273 are presented. The photon-photon interaction has been taken into account. From the best fit, a dimension of the X-ray source of few times 10**17 cm is derived.
   The photon energy spectra from the electron-proton (e-p) beam interactions with the matter and radiation were calculated. We obtained the photon energy spectra from Inverse Compton Scattering (ICS) of an arbitrary background radiation by relativistic electron-beams and photon energy spectra from p + p interactions via π^0-decay for relativistic proton-beams. The results for selected beam Lorentz factors and selected angles between the emitted photon direction and the beam axis are presented. The calculated theoretical spectra from ICS of the background photons by relativistic electrons and from the interaction of beamed relativistic particles with the surrounding matter have been succesfully used to fit the spectrum of Cyg X-1, in the energy range greater than 1 MeV, and that of Geminga in the COS B energy range, respectively.

1. Introduction

   The SAS II and COS B measurements of γ-ray fluxes (E > 100 MeV)

W. Brinkmann et al. (eds.), Physical Processes in Hot Cosmic Plasmas, 425–442.
© 1990 *Kluwer Academic Publishers.*

indicate that their main part has a galactic origin, since an excess of the flux from the galactic plane is observed. The possibility that this radiation originates from interactions of cosmic ray protons and electrons with diffuse matter and electromagnetic radiation (diffuse component) has been discussed in many papers, but the discovery of the γ-ray sources indicates that the diffuse component is probably not dominant.

Up-to-date, 25 γ-ray sources have been discovered (Swanenburg et al., 1981). Only few of them were identified with known astrophysical objects. Information on the properties of the galactic γ-ray sources and proposed models to explain the production of the gamma quanta in such objects were deeply discussed by Bignami and Hermsen (1983). In the past, several detailed studies on spherically sysmmetric accretion of matter onto massive objects have been done by Michel (1972), Kolykhalov and Sunyaev (1979), Giovannelli, Karakula and Tkaczyk (1982a; 1982b).

In particular, under the hypothesis that in the spherical accretion onto massive objects, the matter may be heated up to temperature as high as $10^{**}12$ K, inelastic collision of protons may produce π^os which generate, via decay, γ- rays. Then, we determined the γ-ray energy production spectra in the comoving plasma system for different temperatures and expected γ-ray energy spectra in the case of spherically symmetric accretion of matter onto a black hole.

Recently, because of the many observational evidences of highly anisotropic emission and non-spherical structure of several astronomical objects (Bridle and Perley, 1984; Rees, 1985) and because of the fact that in the central region of twelve extragalactic radio-sources, superluminal motion has been discovered (Cohen and Urwin, 1984) meaning that the particles are moving with Lorentz's factor much greater than 1 at small angles with respect to the line of sight, we started a study of the interaction of relativistic particles emitted by the central engine of such sources with the surrounding radiation and matter. We noted that in the case of Active Galactic Nuclei (AGNs), isotropic background radiation can be produced by ionized gas distributed in clouds (Frank, King and Raine, 1985). So, we have computed the photon energy production spectra for monoenergetic, one-dimensional beam with a selected Lorentz's factor (γ) and relative velocity of the beam (β), as follows: i) the ICS of the electron-beam with the background photons; ii) the interaction of the proton-beam with the matter.

In this paper, we present the results of such calculations and the successful tentatives of fitting the COS B data from 3C 273, the hard tail of the spectrum of Cyg X-1 in the MeV region, under the hypothesis that its origin is non-thermal (Porcas, 1985), and the spectrum of Geminga in the COS B energy range.

2. Gamma-ray energy production spectra in a relativistic plasma

2.1. Gamma-rays from accreting matter onto collapsed objects

In the spherical accretion onto massive objects, the matter may be heated up to temperature as high as $10^{**}12$ K (Kolykhalov and Sunyaev,

1979). In such a hot plasma, the thermal bremsstrahlung (e-e and e-p) and π^0-decay from inelastic collisions of protons are the main γ-ray sources.

One can determine the γ-ray energy production spectra from the π^0-decay and from bremsstrahlung for different temperatures (Giovannelli et al., 1982b). Then, in order to fit the experimental data, it is necessary to evaluate the expected γ-ray energy spectra.

To determine the γ-ray energy production spectra in an isothermal plasma at temperature T, the following assumptions were made (Giovannelli, Karakula and Tkaczyk, 1982b):
a) the plasma is fully ionized;
b) the electron and proton momentum distribution is maxwellian;
c) the electron and proton angular distribution is isotropic;
d) the characteristics of the interactions p + p $\longrightarrow$ π^0+ anything were derived from approximations of experimental data (Barashenkov and Tonev, 1972).

The numerical calculations were made attending the following steps in determining the γ-ray energy production spectra for thermal and non thermal models, respectively, as follows:

2.1.1. Thermal model
 i) the π^0-momentum production spectrum as function of the proton temperature;
 ii) the γ-ray energy production spectra as secondary products of the π^0 - decay;
iii) the γ-ray energy production spectra from bremsstrahlung (e-e and e-p), using Gould's corrected formula (Gould, 1980, 1981; Gorecki and Kluzniak, 1981).

The considered range of the plasma temperature was 10**11 - 10**13 K. Fig. 1 shows the energy production spectra in the comoving plasma system from π^0-decay and from bremsstrahlung (e-e and e-p) for different selected temperatures and unit concentration of the plasma. The spectrum from π^0-decay has the maximum at 67.5 MeV. The production rate is highly dependent on the plasma temperature and increases with it. The width of the spectrum also increases with the temperature. The γ-ray spectra from bremsstrahlung are practically always over those from π^0-decay (if the temperature of the electrons and protons is the same).

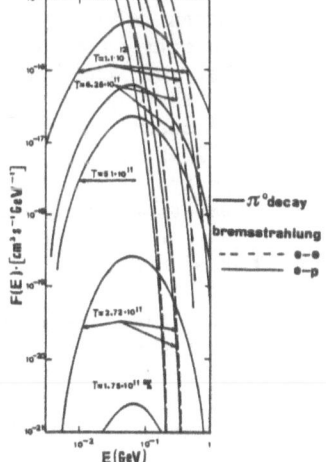

Figure 1. The energy production spectra in the comoving plasma system from π^0-decay and from bremsstrahlung (e-e and e-p) for different selected temperatures and unit concentration of the plasma.

2.1.2. Non thermal model

i) the γ-ray energy production spectrum resulting from π^0-decay in the case of a power law proton spectrum (Tkaczyk, 1978; Stecker, 1971). The energy spectrum of the protons is like that of cosmic rays (A*E**(-6)) as observable in the solar environment;

ii) the γ-ray energy production spectrum from bremsstrahlung (e-p). The electron energy spectrum is a power-law-type which has a break at about 2 GeV (n(E)$\propto$E**(-1.6) for E < 2 GeV and n(E)$\propto$E**(-2.6) for E $\geqslant$ 2 GeV (Tkaczyk, 1978)).

We assumed the ratio between protons and electrons equal to about 100, the same as it was observed in the cosmic ray spectrum near the solar system. For the γ-ray sources correlated with quasars the large ratio L_γ /Lx = 100 - 1000 was observed (Ulrich, 1981; Moffat et al., 1983). If in that type of sources the X-rays are coming from electron bremsstrahlung and γ-rays from protons (π^0-decay), the density of electrons should be lower than that of the protons. This indicate that γ-ray sources can originate cosmic rays.

Fig. 2 shows the γ-ray production spectra from π^0-decay and from bremsstrahlung (solid and dashed lines, respectively), for the non thermal model.

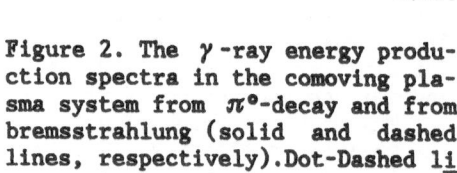

Figure 2. The γ-ray energy production spectra in the comoving plasma system from π^0-decay and from bremsstrahlung (solid and dashed lines, respectively).Dot-Dashed line is the sum of the two spectra.

From the kinematics it is clear that the γ-ray spectrum from π^0-decay has a maximum at the energy of 67.5 MeV, like in the thermal model.

2.2. The expected gamma-ray spectra from collapsed objects

To determine the expected γ-ray energy spectra from collapsed objects it is necessary to evaluate the temperature, concentration and velocity of the plasma near a black hole. In order to do this, the system of equations (Bernoulli, continuity and adiabatic equations) describing the motion of the plasma must be solved (Michel, 1972):

$$\left(\frac{\varkappa}{\varkappa-1}\theta+1\right)^2\left(1-\frac{1}{r}+u^2\right)=\text{const}\ ;\ \ nur^2=\text{const}\ ;\ \ \frac{\theta}{n^{\varkappa-1}}=\text{const}$$

where:
$\theta = KT/m_p c^2$

n – concentration of the plasma

$r = R/R_g$ – distance from collapsed object (black hole) in gravitational radius units

u – R-component of four velocity

$\varkappa = 5/3$

To determine the temperature as function of the distance from a black hole, different values of u_0^2 were taken for a given r_0. Calculations were done for $u_0^2 = 2.6*10**(-5)$ and $6*10**(-5)$ corresponding to Mach's numbers roughly 1 and 2, respectively by Giovannelli, Karakula and Tkaczyk (1982b). In the case of spherical symmetric accretion of matter onto a Schwarzschild black hole, they calculated the observable γ-ray energy spectrum at infinity, transforming the comoving spectrum and taking into account all the relativistic effects such as Doppler and gravitational redshifts. Fig. 3 shows the energy spectrum $Q(E_\gamma)$ observed at infinity, multiplied by $Rg/\dot{M}^2$ for Mach's numbers 1 and 2 (curves a and b, respectively).

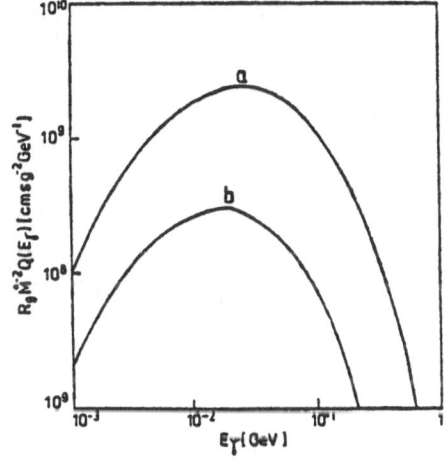

Figure 3. The energy spectrum $Q(E_\gamma)$ observed at infinity, multiplied by $Rg/\dot{M}^2$. Curves a and b refer to Mach's numbers about 1 and 2, respectively.

But a question is still not completely settled: under which conditions can the protons reach a sufficiently high energy to produce neutral and charged pions via proton-proton reaction taking into account the radiative losses and the general relativistic effects ?

The answer is far to be definitive. Therefore the problem of the conditions of the plasma around a black hole deserves further discussions. From our point of view, we have considered two possibilities:

i) a thermal process in which the protons may fall adiabatically, while the radiative losses could limit the electron temperature to lower values. This means that electrons and protons are thermally decoupled near a black hole;

ii) a non-thermal process in which a shock around a black hole is producing a power law spectrum of high energy protons. considering that the electrons may loss a large quantity of energy, they cannot be directly accelerated. The secondary electrons produced via π^0-decay have an equilibrium spectrum. It can be determined by means the magnetic field, energy density of the electromagnetic radiation and dimensions of the source (Protheroe and Kazanas, 1983).

We have not solved the equilibrium equation; we simply assumed that the electrons have a cosmic ray-type spectrum.

2.2.1. The energy spectrum of 3C 273

We have attempted the fit of the experimental data of the quasar (QSO) 3C 273 by using our simple adiabatic accretion model in the case of a thermal proton momentum spectrum. The X-ray data are coming from the experiments of Worral et al. (1979), Primini et al. (1979), Pietsch et al. (1981), Bezler et al. (1984). The γ-ray data are coming from COS B satellite (Bignami et al., 1981; Pollock et al., 1981). The adopted values for the parameters of 3C 273 are: mass $M = 2.5*10**9$ Me and distance d = 860 Mpc (Ulrich, 1981). A two temperature plasma model has been used.

We have considered two possiblities:

i) the X-ray and γ-ray sources are separated;
ii) the X-ray and γ-ray originate in the same hot region surrounding the collapsed object.

i) In this case the photon-photon interaction does not exist. The best fit of the experimental data with our thermal model gives an electron maximum temperature of $1.35*10**11$ K with the electron temperature profile described by $Te = Tp$ for $R > 10Rg$ (R - the distance from the collapsed object; Rg - its gravitational radius), while for $R \leqslant Rg$, $Te = Tp(R = 10Rg) = $ constant. The derived mass accretion rate is $\dot{M} = $ 280 Me/yr. The critical distance at which $Te < Tp$ is about 10Rg, corresponding to $7.5*10**15$ cm (Karakula, Tkaczyk and Giovannelli, 1984);

ii) in this case the photon-photon interaction must be considered. So, we have computed the γ-ray spectra expected from π^0-decay for different selected radii (Rs) of the X-ray source from $10**18$ to $10**16$ cm, taking into account that the dimensions of the central compact object in 3C 273 have a lower limit of $> 1.5*10**15$ cm (Elvis et al., 1980; Bignami et al., 1981) and an upper limit of $\leqslant 4*10**17$ cm (White and Ricketts, 1979).

The expected bremsstrahlung energy spectra for electron temperature profiles having different selected values of the maximum temperature (Tmax) have been computed too.

Fig. 4 shows the X and γ-ray energy spectra together with the experimental data already mentioned. The observed X-ray power-law index (1.5) is obtained with the theoretical spectra for Tmax = $5*10**9$ K. From this normalization we derive a mass accretion rate $\dot{M}$ = 730 Me/yr.

It is evident that the COS B data are now below the energy spectrum predicted from π^0-decay, without the photon-photon interaction (dot-dashed line), but above the curve predicted by the energy γ-ray spectrum from the central hot region of 3C 273 surrounded by the

X-ray source with radius of 10**17 cm. This radius is comparable with the distance of 316Rg (= 2.4*10**17 cm) obtained for Tmax = 5*10**9 K.

So, from our model best fitting the experimental data, we can conclude that the radius of the X-ray emitting region must be few times 10**17 cm, which is in excellent agreement with that derived by White and Ricketts (1979) from X-ray variability of 3C 273 measured with Ariel V Sky Survey Instrument (⩽ 4*10**17 cm). The non-thermal model does not fit the experimental data.

3. Gamma-ray energy production spectra from beam interactions with matter and radiation

There are many observational evidences of highly anisotropic emission and non-spherical structure of some astronomical objects. Jet-like features were detected in more than 100 extragalactic sources (Bridle and Perley, 1984) and in objects of our galaxy: SS 433, Sco X-1, Cyg X-3 (Rees, 1985). In the central region of 12 extragalactic radio-sources, superluminal motion has been discovered (Cohen and Urwin, 1984). This means that particles are moving with Lorentz factor $\gamma \gg 1$ at small angle to the line of sight. Overmore, X and γ-ray emissions have been detected from sources with jets, such as 3C 273, NGC 1265, M87, etc. Relativistic particles emitted by the central engines of such sources can interact with the surrounding radiation and matter.

In the case of active galactic nuclei (AGNs), isotropic background radiation can be produced by ionized gas distributed in clouds (Frank, King and Raine, 1985).

In the case of galactic sources, relativistic particles, emitted by pulsars or black holes, can interact with the radiation and matter from supernova remnants (SNRs), or in close binary systems with the thermal radiation and matter from the companion star.

So, we have calculated the photon energy production spectra for a monoenergetic-one-dimension beam with a Lorentz factor γ and the relative velocity of the beam β as follows:
 i) the inverse compton scattering (ICS) of an electron-beam with the background photons;
ii) the interaction of the proton-beam with the matter.

3.1. The ICS of an electron-beam with the background photons

The comptonization of a photon gas by relativistic electrons has been studied in an astrophysical context by several authors (Blumental and Gould, 1970; Sunyaev and Titarchuk, 1980; Reynolds, 1982; Canfield, Howard and Liang, 1987; Begelman and Sikora, 1987). These authors have usually made some simplifications, such as Thomson limit, highly relativistic electrons, low optical depth, etc., because in the general case the solution is very complicate.

We have calculated the photon energy spectra (Q) at a selected angle α,

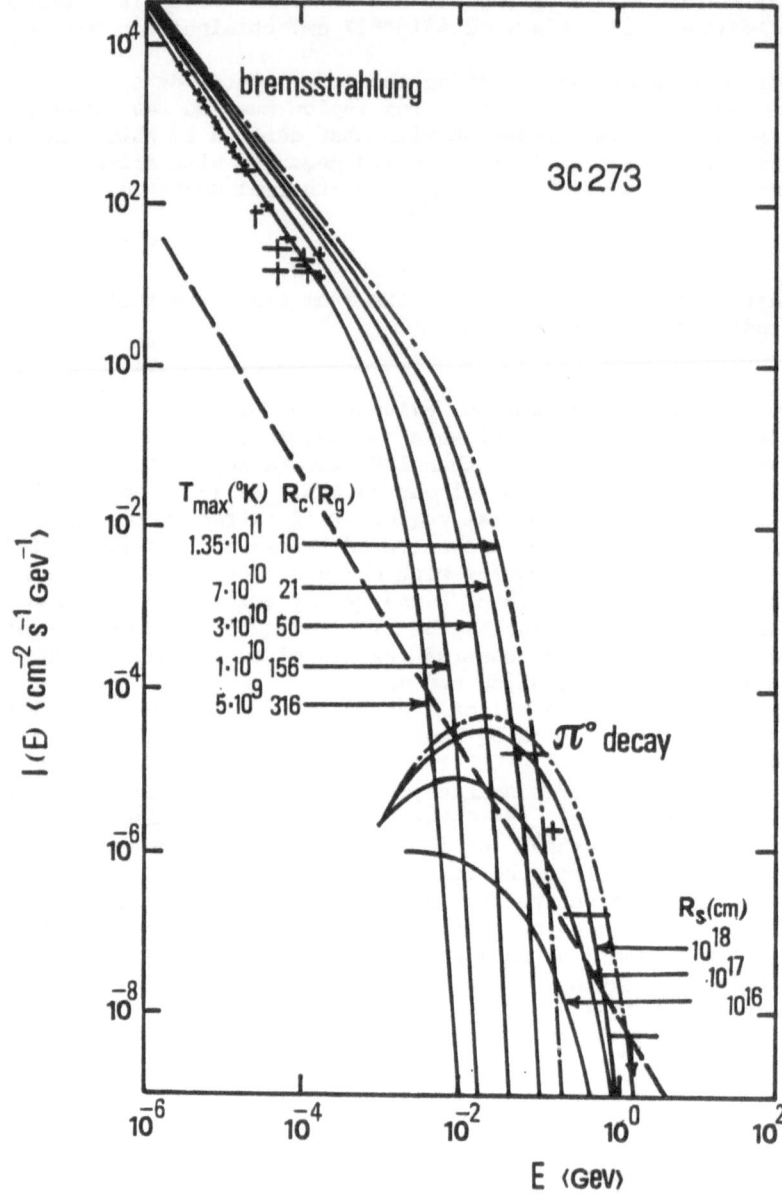

Figure 4. The X and γ-ray expected spectra from 3C 273 for different selected maximum electron temperatures and dimensions (Rs) of the X-ray source. Dot-dashed and thin lines refer to the thermal model in which X and γ-rays originate in different places or in the same place, respectively. Dashed line refers to the non-thermal model. The experimental data are overlapped.

between the direction of the emitted photons and the beam axis from the ICS of a relativistic electron-beam with an arbitrary background photon spectrum. The numerical calculations were made for low optical depth using the following formula (for an extensive discussion of this problem see Bednarek et al. 1990a):

$$\frac{dN}{d\varepsilon_1\, d\Omega\, dt\, dV} = \frac{n_e}{\gamma^2(1-\beta\cdot\cos\alpha_1)} \iiint c\cdot n'(\varepsilon',\alpha')\; \frac{d^2\sigma}{d\Omega_1'\cdot d\varepsilon_1'}\cdot\frac{d\Omega_1'}{d\Omega^+}\cdot\frac{d\Omega'}{d\Omega_\mu}\cdot d\Omega_\mu\cdot d\varepsilon',$$

where: $\dfrac{d\Omega_1'}{d\Omega^+}=1$, $\dfrac{d\Omega'}{d\Omega_\mu}=1$ Jacobians,

$\dfrac{1}{\gamma^2(1-\beta\cdot\cos\alpha_1)}$ Jacobian of the transformation from the electron rest frame to the observer frame,

$n'(\varepsilon',\alpha')$ - distribution of the background photons in the electron rest frame,

$\dfrac{d^2\sigma}{d\Omega_1'\, d\varepsilon_1'}$ - Klein-Nishina cross-section,

n_e - electron concentration of the beam.

Figs. 5 and 6 show the comptonized black body (10**4 K and 10**7 K, respectively) photon energy spectra normalized to a value of the background photon energy density equal to 1 MeV/cm**3, for beam Lorentz factor values $\gamma = 100$, 1000, and for two values the angle $\alpha_1 = 0$ and 10 degrees, and for unit electron-beam concentration ($n_e = 1$). Fig. 7 shows the comptonized black body (10**7) spectra for $\gamma = 1000$ and selected angles α_1.

In general, the intensities of these comptonized black body spectra, for small angles α_1 (with respect to $1/\gamma$), are inversely proportional to γ (Figs. 5 and 6). For large angles α_1 (with respect to $1/\gamma$), the intensities are determined by the angle α_1 itself (Fig. 7), and for fixed values of this angle, they are inversely proportional to γ^3 (Figs. 5 and 6). The maximum intensity in the peak, for a fixed angle, is roughly independent on γ as shown in Fig. 6.

Fig. 8 shows the comptonized thermal bremsstrahlung (10**6 K) for $\gamma = 100$, 1000 and two angles $\alpha_1 = 0$ and 10 degrees. In general, these comptonized bremsstrahlung spectra have a shape similar to that of the background spectrum because the main contribution to the spectra comes from a very small area around the beam axis. The position of the cut-offs in the comptonized bremsstrahlung spectra are determined like in the case of comptonized black body spectra, because these features are caused by the kinematics of ICS. The intensities of the bremsstrahlung comptonized spectra for small angles are proportional to γ , while for large angles these intensities are proportional to γ^3 (see Fig. 8).

434

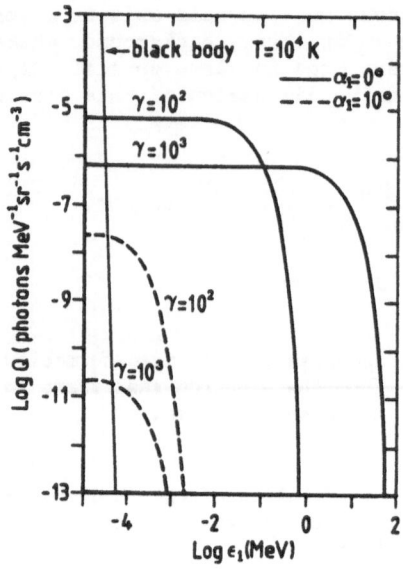

Figure 5. The comptonized black body (10**4 K) spectra.

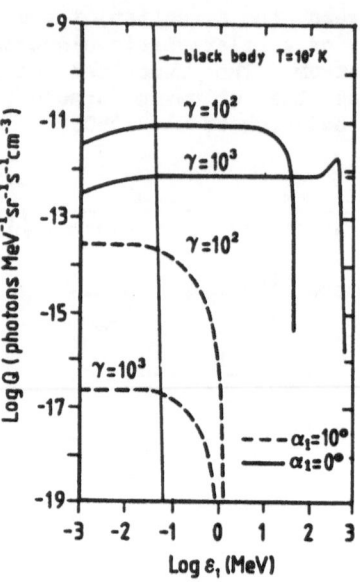

Figure 6. The comptonozed black body (10**7 K) spectra.

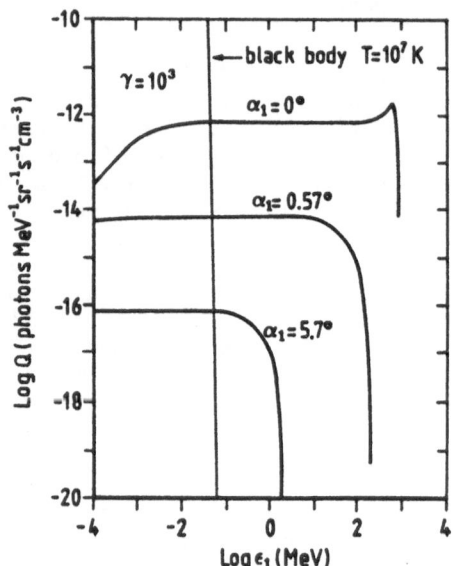

Figure 7. The comptonized black body (10**7 K) spectra for electron-beam Lorentz factor $\gamma = 1000$ and selected angles $\alpha_1 = 0, 0.57, 5.7$ degrees.

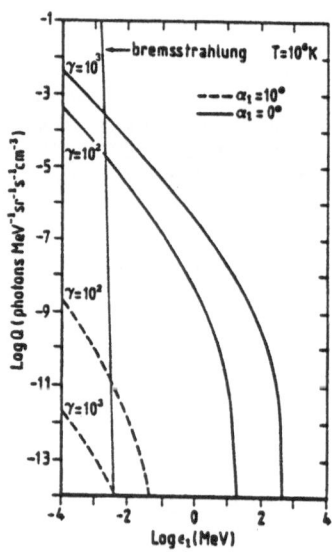

Figure 8. The comptonized thermal bremsstrahlung (10**6 K) spectra for electron-beam Lorentz factor $\gamma = 100, 1000$ and angles $\alpha_1 = 0, 10$ degrees.

3.2. The interaction of a proton-beam with the matter

The relativistic proton beam, passing trough a hydrogen cloud, loses its energy by proton-proton inelastic interactions. The high energy photons can originate by π^0-decay produced in the proton-proton reaction.

We have analyzed this process and obtained the photon spectra (Q) from a monoenergetic beam and hydrogen cloud interaction for a given angle α_1 between the direction of the emitted photons and the beam axis. These spectra can be descibed by the following formula:

$$\frac{dN}{dE_\gamma \cdot d\Omega \cdot dt \cdot dV} = \beta \cdot c \cdot n_b \cdot n_H \iiint \frac{d^3\sigma(p_{\pi^0}, \cos\Theta_{\pi^0}, \varphi_{\pi^0})}{dp_{\pi^0} \cdot d(\cos\Theta_{\pi^0}) \cdot d\varphi_{\pi^0}} \cdot P(E_\gamma, \cos\Theta, \varphi) \cdot dp_{\pi^0} \cdot d(\cos\Theta_{\pi^0}) d\varphi_{\pi^0}$$

where: n_H - hydrogen concentration in the cloud,
  n_b - proton concentration in the beam,
  c - light velocity,
  p_{π^0} - π^0 momentum in the observer frame,
  $\Theta_{\pi^0}, \varphi_{\pi^0}$ angles between the π^0emission direction and beam axis in the observer frame,

$$\frac{d^3\sigma(p_{\pi^0}, \cos\Theta_{\pi^0}; \varphi_{\pi^0})}{dp_{\pi^0} \cdot d(\cos\Theta_{\pi^0}) \cdot d\varphi_{\pi^0}} = \frac{p_{\pi^0}^2}{E_{\pi^0}} \left(E_{\pi^0} \cdot \frac{d^3\sigma}{dp^3}\right) \quad \text{- cross section in polar coordinates,}$$

$E_{\pi^0} \cdot \dfrac{d^3\sigma}{dp^3}$ - invariant cross section (Ref. 13),

$P(E_\gamma, \cos\Theta, \varphi)$- energy and angular distributions of photons in the observer frame,

$E_\gamma, \cos\Theta, \varphi$ - energy and angles of the emitted photons in the observer frame.

For the details of this analysis, see the paper by Bednarek et al. (1990a). The differential photon spectra from π^0-decay have been calculated numerically according to the former formula in the case of beam Lorentz factor $\gamma \gg 1$ and for unit concentrations of the proton beam and hydrogen cloud.

Fig. 9 shows the differential photon spectra from π^0-decay for proton beam Lorentz factor $\gamma = 10$, 100, 1000 and two values of the angle $\alpha_1 = 0$ and 10 degrees. Fig. 10 shows the differential photon spectra from π^0-decay for proton-beam Lorentz factor $\gamma = 10{,}000$ and selected angles $\alpha_1 = 0$, 0.57, 5.7 degrees.

In general the shapes and intensities of these spectra are very sensitive to γ and/or to α_1. From Fig. 9, we can note that for small angles α_1 (with respect to $1/\gamma$), the intensities, the widths and positions of the maxima of the photon spectra grow proportionally to γ , since in this case the momenta of π^0s, which generate these photons , are proportional

436

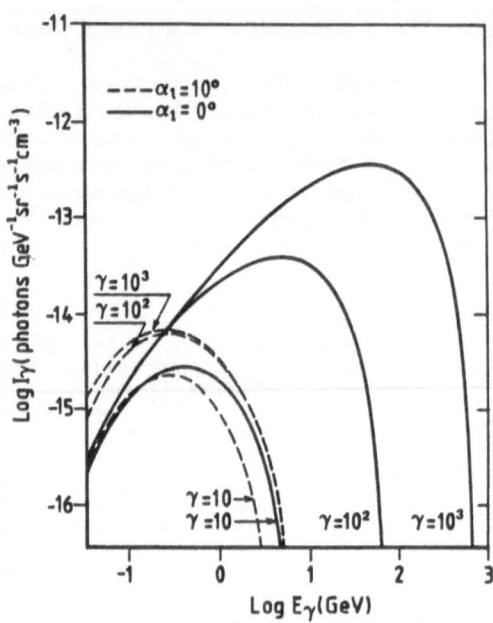

Figure 9. The photon spectra from π^0-decay for proton-beam Lorentz factor $\gamma = 10$, 100, 1000 and two values of the angle between the direction of the emitted photons and the beam axis: $\alpha_1 = 0$ and 10 degrees, for unit concentrations of the proton-beam and hydrogen cloud.

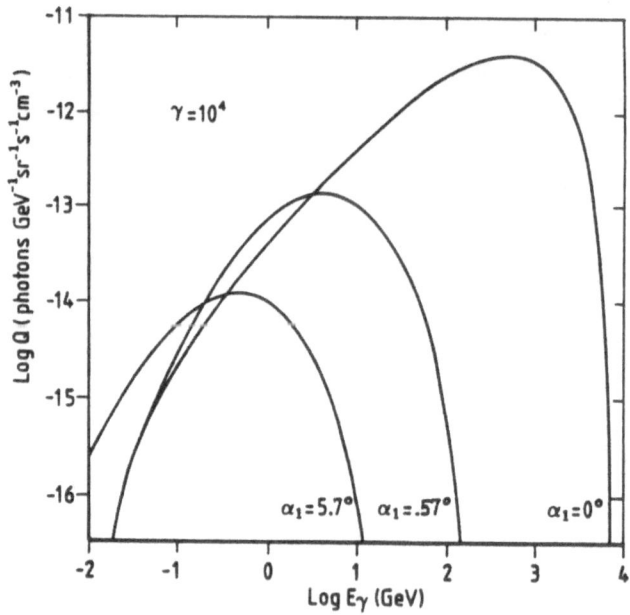

Figure 10. The photon spectra from π^0-decay for the proton-beam Lorentz factor $= 10,000$ and selected angles α_1, for unit concentrations of the proton-beam and hydrogen cloud.

to γ. For large angles α_1 (with respect to $1/\gamma$), the shapes and intensities of the photon spectra are practically independent on γ, since the momenta of π^0s are roughly the same (Bednarek et al., 1990a: Appendix B). In a certain range of the photon energies, the intensities of the spectra for γ = constant are comparable in the rising part for different values of α_1. We note that for large angles, the position of the maximum intensities of the photon spectra are determined by the values of the angle α_1 and are independent on γ. All this means that sources with different intensities could give roughly the same experimental results in the low energy region of Fig. 9, simply because the axis of their beams form a different angle with the line of sight.

4. Origin of the high energy γ-rays from Cyg X-1 and Geminga

We want to show how the calculated photon energy spectra can be useful used in fitting the experimental data from two of the most interesting high energy γ-ray sources: Cyg X-1 and 2CG 195+04 (Geminga).

4.1. Cyg X-1

Cyg X-1 is the most promising galactic black hole candidate because of its general behaviour and morphology, difficult to explain in terms of neutron star models. Cyg X-1 lies in a binary system, in which the primary star is the blue supergiant HDE 226868. The characteristics of this system as well as its multi-frequency behaviour can be found in the reviews papers by Oda (1977) and Liang and Nolan (1984).

Up-to-date several suggestions attempting to explain the complex shape and behaviour of the spectrum of Cyg X-1 have been published. Many thermal mechanisms have been discussed in the literature (see the references in the paper by Bednarek et al., 1990b). On the contrary, we suggest that the photon spectrum of Cyg X-1 in the MeV region is non-thermal. In fact, from radio measurements performed by Woodsworth et al. (1980), a non-thermal nature of the spectrum was unambigously derived. They proposed a model in which the radio emission is coming from relativistic electrons in a stellar wind. So, following the expression of the photon

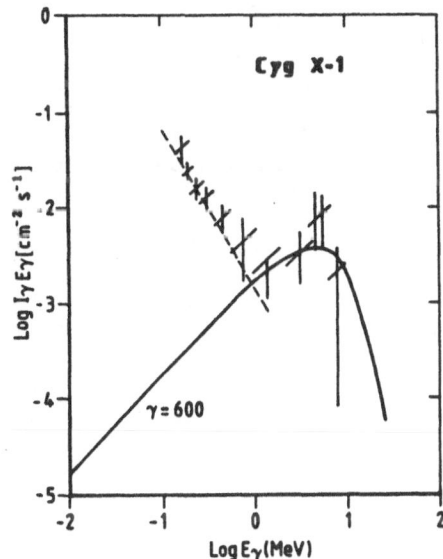

Figure 11. The best fit of the experimental data of Cyg X-1.

energy production spectra derived in § 3.1, we have developed a model in which the MeV photon spectrum originates from ICS of the background photons by relativistic electrons.

The experimental high energy spectrum of Cyg X-1 above 1 MeV (McConnell et al., 1987) can be fitted very well with a comptonized black body spectrum (Tbb = 3*10**4 K), coming from the supergiant companion HDE 226868, by relativistic electrons with Lorentz factor γ = 600, emitted to our direction. Fig. 11 shows the best fit of the experimental data with our calculated spectrum at energies above 1 MeV and with the derived spectrum from the single-temperature inverse Compton model (Sunyaev and Titarchuk, 1980) at energies below 1 MeV. The orbital parameters of Cyg X-1/HDE 226868 system have been taken by the papers by Ninkov, Walker and Yang (1987a, 1987b).

From the observed intensity of the photons above 1 MeV, the lower limit on the emissivity of the relativistic electrons can be estimated.It is equal to 1.5*10**38 electrons/s, corresponding to an electron luminosity of 7.0*10**34 erg/s.

4.2. Geminga (2CG 195+04)

Geminga is one of the COS B sources for which many works have been done in order to find the optical counterpart as well as to interpret the nature of its spectrum at energies greater than 1 MeV. Until now, several suggestions on the probable optical counterpart have been done. They essentially are works supporting the galactic origin, but one supporting the extragalactic origin of Geminga (for the references see the paper by Bednarek et al. (1990b). With the fit of our model to the experimental data of Geminga, we will go indirectly to support its galactic origin.

In our model, we propose that the high energy γ-rays coming from Geminga originate in the interaction of beamed relativistic particles with the surrounding matter. Such a mechanism of high energy photon production can be realized in close binary systems, where beamed relativistic particles emitted by a collapsed object (e.g. neutron star) interact with the matter from the stellar companion.

In § 3.2 we have obtained the photon spectra from the interaction of a monoenergetic proton beam with

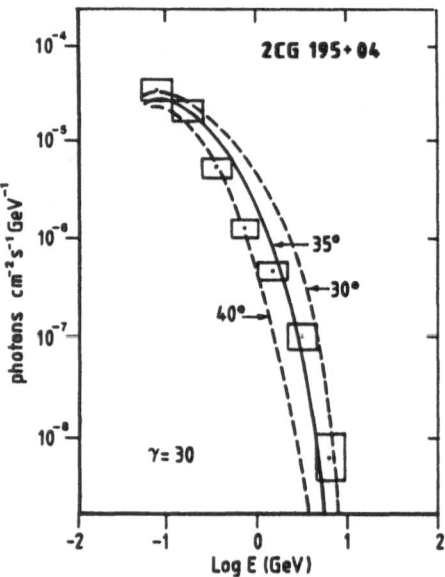

Figure 12. The best fit of the experimental data of Geminga.

hydrogen cloud for a given angle α_1 between the direction of the emitted photons and the beam axis. We compare in Fig. 12 the theoretical photon production spectra with the spectral measurements of Geminga from COS B satellite (Masnou et al. 1981). One can see that the photon spectrum, calculated for an angle α_1 equal to 35 degrees and the proton-beam Lorentz factor $\gamma \geqslant 30$, fits very well the COS B data. In order to show how the fit is sensible to the angle α_1, we have also reported the theoretical curves for 30 and 40 degrees, with the same value of the Lorentz factor. From the best fit, we obtain the emissivity of the proton beam, necessary to produce such a spectrum, equal to:

$$Np = 7.5*10**(-4)*d**2/Xp \text{ (particles/s)}$$

where Xp is the column density (in g/cm**2) of the particles crossed by the proton-beam. This corresponds to a proton luminosity Lp = mp* γ *Np, where mp is the proton rest energy.

So, from the energetical point of view, our model excludes the identification of Geminga with the QSO 0630+180 placed at z = 1.2 (Moffat et al., 1983), and favours that with a galactic source (1E 0630+ 178) as suggested by Bignami, Caraveo and Lamb (1983) and recently by Halpern and Tytler (1988).

Our model is definitively valid in the case that Geminga is a binary system, where a density Xp of several grams/cm**2 is expected.

5. Discussion and conclusions

We have shown that the π^o-decay may be the source of high energy γ-rays, if the accreting plasma onto collapsed objects can be heated up to temperatures as high as 10**12 K. We calculated the γ-ray energy production spectra in such a relativistic plasma under the simplest hypothesis, namely: i) the plasma is fully ionized; ii) the electron and proton momentum distribution is maxwellian; iii) the electron and proton angular distribution is isotropic. Then, solving the equations describing the plasma motion around a black hole, we evaluated the temperature, concentration and velocity of the plasma near the black hole.

Using a two temperature plasma model, we have fitted the COS B data from 3C 273 considering two possibilities: a) the X-ray and γ-ray are separated; b) the X-ray and γ-rays originate in the same place.

In the first case the photon-photon interaction does not exist. The best fit of the experimental data with our thermal model gives an electron maximum temperature of 1.35*10**11 K with the electron temperature profile described by Te = Tp for R > 10Rg, while for R $\leqslant$ Rg, Te = Tp(R = 10Rg) = constant. The derived mass accretion rate is Ṁ = 280 Me/yr. The critical distance at which Te < Tp is about 10Rg, corresponding to 7.5*10**15 cm.

In the second case we considered the photon-photon interaction and we have computed the γ-ray spectra expected from π^o-decay for different selected radii of the X-ray source from 10**18 to 10**16 cm, taking into account that the dimensions of the central compact object in 3C 273 have a lower limit of > 1.5*10**15 cm, and an upper limit of ≤ 4*10**17 cm. We calculated also the expected bremsstrahlung energy spectra for electron temperature profiles having different selected values of the maximum temperature.

The best fit to the experimental data gives a maximum temperature of the electrons equal to 5*10**9 K and a radius of the X-ray region surrounding the collapsed object of few times 10**17 cm. This value is in very good agreement with the evaluation made by White and Ricketts (1979), which gives a dimension of ≤ 4*10**17 cm derived from variations in the X-ray KeV region. Our estimated value of the dimensions of the source is also in agreement with the lower limit of > 1.5*10**15 cm, derived from short time scale variability in the soft X-ray region (Bignami et al., 1981). From our fit we derived also a mass accretion rate of the order of 730 Me/yr which is not far to be reasonable, taking into account the large mass of the central black hole of 3C 273.

Since many experimental evidences of strong anisotropies from many cosmic sources were detected, we moved our interest in searching to solve the problems of the interactions between relativistic electron and positron beams with the surrounding matter and radiation.

The expression for the inverse Compton photon spectrum has been derived for an arbitrary distribution of photons, various Lorentz's factors of the beam and selected angles between the photon emission direction and the beam axis, by using Klein-Nishina cross section.
Numerical calculations have been performed and presented for optically thin thermal bremsstrahlung and black body radiation in the case of isotropic distribution. Such ambient photon distributions probably surround AGNs and close binary systems.

In general, for an isotropic distribution of ambient photons, the shape of comptonized spectra is determined by the shape of the background spectrum, but the mean energy of the comptonized spectra strongly depends on the electron-beam Lorentz's factor and on the angle between the direction of the observation and the beam axis.
The photon energy spectrum from π°-decay produced in the interaction of the proton beam with the matter could be important in the case of AGNs with directly stable jets placed in small angles to the observer (e.g. 3C 273). High energy photons can be produced near the central core where the matter is denser. The characteristic shape of the spectra (with maxima strongly dependent on the beam Lorentz factor for small angles and roughly stable for greater angles) can be changed by absorption of high energy photons ($\gamma + \gamma \longrightarrow e^+ e^-$) and contributions from other processes (bremsstrahlung, ICS, in the case of electron-proton beam).

The results presented in this paper concern only monoenergetic beams. We are working, now, on a more realistic problem from astrophysical point of view, taking into account non-monoenergetic beams and their propagation. Neverthless, we have shown - even if in the simple case of monoenergetic beam - how the fit of experimental data with the computed spectra can be good for Cyg X-1, which is one of the best black hole candidates.

The expression for the photon spectrum from proton-proton interaction via π°-decay for a relativistic proton beam has been calculated too. The results were given for selected beam Lorentz factors and selected angles between the direction of the emitted photons and the beam axis.

We have applied the results of our calculations to Geminga, in order to fit the experimental data. The best fit of the COS B data shows that in this source the very high energy photons originate in the interaction of beamed relativistic particles with Lorentz factor equal to about 30 and the angle between the beam axis and the direction to the observer equal to 35 degrees. We derived the emissivity of the proton beam $Np = 7.5*10**(-4)*d**2/Xp$ particles/s, corresponding to a proton luminosity $Lp = mp*\gamma*Np$. So, if Geminga should be an extragalactic source, such as the QSO 0630+180 placed at $z = 1.2$, these values would drastically grow up proportionally to $d**2$. Then from the energetical point of view, our model excludes the identification of Geminga with the QSO 0630+180 and, on the contrary, favours its galactic origin. Overmore, our model is definitively valid in the case that the source is in a binary system, where a density of several grams/$cm**2$ is expected.

References

Barashenkov, W.S. and Tonev, W.D. (1972), Wzaimodejstvija wysokoenergeti-cheskih chastic i atomnyh jader s jadrami, Moscow.

Bednarek, W., Giovannelli, F., Karakula, S., and Tkaczyk, W. (1990a), Astron. Astrophys. (in press).

Bednarek, W., Giovannelli, F., Karakula, S., and Tkaczyk, W. (1990b), Astron. Astrophys. (in press).

Begelman, M.C. and Sikora, M. (1987), Astrophys. J. <u>322</u>, 650.

Bezler, M., Kendziorra, E., Staubert, R., Hasinger, G., Pietsch, W., Reppin, C., Trumper, J., and Voges, W. (1984), Astron. Astrophys. <u>136</u>, 351.

Bignami, G.F., Bennet, K., Buccheri, R., Caraveo, P.A., Hermsen, W., Kanbach, G., Lichti, G.G., Masnou, J.L., Mayer-hasselwander, H.A., Paul, J.A., Sacco, B., Scarsi, L., Swanenburg, B.N., and Wills, R.D. (1981), Astron. Astrophys. <u>93</u>, 71.

Bignami, G.F., Caraveo, P.A., and Lamb, R.C. (1983), Astrophys. J. Letters <u>272</u>, L9.

Bignami, G.F. and Hermsen, W. (1983), Ann. Rev. Astron. Astrophys. <u>21</u>, 67.

Blumental, R.G. and Gould, R.J. (1970), Rev. Mod. Phys. <u>42</u>, 237.

Bridle, A.H. and Perley, R.A. (1984) Ann. Rev. Astron. Astrophys. <u>22</u>, 319.

Canfield, E., Howard, W.M., and Liang, E.P. (1987), Astrophys. J. <u>323</u>, 565.

Cohen, M.H. and Urwin, S.C. (1984), in VLBI and Compact Radio Sources, (eds. R. Fanti, K. Kellerman, and G. Setti), Dordrecht: Reidel, IAU Symp. No. 110, p. 95.

Elvis, M., Feigelson, E., Griffiths, R.E., Henry, J.P., and Tananbaum, H. (1980), in High-lights of Astronomy, (ed. H. van der Laan).

Frank, J., King, A.R., and Raine, D.J. (1985), in Accreting Power in Astrophysics, Cambridge University Press, U.K.

Giovannelli, F., Karakula, S., and Tkaczyk, W. (1982a), Astron. Astrophys. 107, 376.

Giovannelli, F., Karakula, S., and Tkaczyk, W. (1982b), Acta Astron. 32, 121.

Gorecki, A. and Kluzniak, W. (1981), Acta Astron. 31, 457.

Gould, R.J. (1980), Astrophys. J. 238, 1026.

Gould, R.J. (1981), Astrophys. J. 243, 677.

Halpern, J.P. and Tytler, D. (1988), Astrophys. J. 330, 201.

Karakula, S., Tkaczyk, W., and Giovannelli, F. (1984), Adv. Space Res. Vol. 3, No. 10-12, p. 335.

Kolykhalov, P.I. and Sunyaev, R.A. (1979), Soviet Astron. 23, 183.

Liang, E.P. and Nolan, P.L. (1984), Space Sci. Rev. 38, 353.

Masnou, J.L., Bennett, K., Bignami, G.F., Bloemen, J.B.G.M., Buccheri, R., Caraveo, P.A., Hermsen, W., Kanbach, G., Mayer-Hasselwander, H.A., Paul, J.A., and Wills, R.D. (1981), Proc. 17th Int. Cosmic Ray Conf. (Paris) 1, 177.

McConnell, M.L., Owens, A., Chupp, E.L., Dunphy, P.P., Forrest, D.J., and Vestrand, W.T. (1987), Proc. 20th Int. Cosmic Ray Conf. (Moscow) 1, 58.

Michel, F.C. (1972), Astrophys. Space Sci. 15, 153.

Moffat, A.F.J., Schlickeiser, R., Shara, M.M., Sieber, W., Tuffs, R., and Kuhr, M. (1983), Astrophys. J. Letters 271, L45.

Ninkov, Z., Walker, G.A.H., and Yang, S. (1987a), Astrophys. J. 321, 425.

Ninkov, Z., Walker, G.A.H., and Yang, S. (1987b), Astrophys. J. 321, 438.

Oda, M. (1977), Space Sci. Rev. 20, 757.

Pietsch, W., Reppin, C., Trumper, J., Voges, W., Lewin, W., Kendziorra, E., and Staubert, R. (1981), Astron. Astrophys. 94, 234.

Pollock, A.M.T., Bignami, G.F., Hermsen, W., Kanbach, G., Lichti, G.G., Masnou, J.L., Swanenburg, B.N., and Wills, R.D. (1981), Astron. Astrophys. 94, 116.

Porcas, R.W. (1985), in Active Galactic Nuclei, (ed. J.E. Dyson), Manchester University Press, U.K., p. 20.

Primini, F.A., Cooke, B.A., Dobson, C.A., Howe, S.K., Scheepmaker, A., Wheaton, W.A., and Lewin, W.H.G. (1979), Nature 278, 234.

Protheroe, R.J. and Kazanas, D. (1983), Astrophys. J. 265, 620.

Rees, M. (1985), Proc. 19th Int. Cosmic Ray Conf. (La Jolla), 9, 1.

Reynolds, S.P. (1982), Astrophys. J. 256, 38.

Stecker, F.W. (1971), in Cosmic Gamma Rays, Baltimore: Mono Book Corp., NASA SP-249.

Sunyaev, R.A. and Titarchuk, L.G. (1980), Astron. Astrophys. 86, 121.

Swanenburg, B.N., Bennet, K., Bignami, G.F., Buccheri, R., Caraveo, P., Hermsen, W., Kanbach, G., Lichti, G.G., Masnou, J.L., Mayer-Hasselwander, H.A., Paul, J.A., Sacco, B., Scarsi, L., and Wills, R.D. (1981), Astrophys. J. Letters 243, L69.

Tkaczyk, W. (1978), Postepy Astronomii Tom XXVI, Zeszyt 2.

Ulrich, M.H. (1981), Space Sci. Rev. 28, 89.

White, G.L. and Ricketts, M.J. (1979), Mon. Not. R. astr. Soc. 187, 757.

Woodsworth, A.W., Higgs, L.A., and Gregory, P.C. (1980), Astron. Astrophys. 84, 379.

Worral, D.M., Mushotzky, R.F., Boldt, E.A., Holt, S.S., and Serlemitsos, P.J. (1979), Astrophys. J. 232, 683.

Subject Index